Studies in Fuzziness and Soft Computing

Volume 332

Series editor
Janusz Kacprzyk, Polish Academy of Sciences, Warsaw, Poland
e-mail: kacprzyk@ibspan.waw.pl

About this Series

The series "Studies in Fuzziness and Soft Computing" contains publications on various topics in the area of soft computing, which include fuzzy sets, rough sets, neural networks, evolutionary computation, probabilistic and evidential reasoning, multi-valued logic, and related fields. The publications within "Studies in Fuzziness and Soft Computing" are primarily monographs and edited volumes. They cover significant recent developments in the field, both of a foundational and applicable character. An important feature of the series is its short publication time and world-wide distribution. This permits a rapid and broad dissemination of research results.

More information about this series at http://www.springer.com/series/2941

Plamen Angelov · Sotir Sotirov
Editors

Imprecision and Uncertainty in Information Representation and Processing

New Tools Based on Intuitionistic Fuzzy Sets and Generalized Nets

Springer

Editors
Plamen Angelov
School of Computing and Communications
Lancaster University Bailrigg
Lancaster
UK

Sotir Sotirov
Intelligent Systems Laboratory
"Prof. Dr. Asen Zlatarov" University
Bourgas
Bulgaria

ISSN 1434-9922 ISSN 1860-0808 (electronic)
Studies in Fuzziness and Soft Computing
ISBN 978-3-319-79925-4 ISBN 978-3-319-26302-1 (eBook)
DOI 10.1007/978-3-319-26302-1

Springer Cham Heidelberg New York Dordrecht London

Softcover reprint of the hardcover 1st edition 2016

Printed on acid-free paper

Springer International Publishing AG Switzerland is part of Springer Science+Business Media (www.springer.com)

To Professor Krassimir T. Atanassov,
a great researcher, scholar,
friend and human being

Preface

Professor Krassimir Todorov Atanassov is a Bulgarian mathematician with outstanding contributions in the areas of fuzzy logic and fuzzy mathematics, uncertainty analysis, mathematical modelling, and decision making, as well as some areas of number theory, notably the arithmetical functions and Fibonacci objects.

In 1982, Professor Atanassov proposed a novel mathematical formalism for the description, simulation and control of parallel processes, named by him the Generalized Nets which represents a generalization of the well-known concept of the Petri Nets and all their hitherto existing extensions and modifications. During the course of years, Professor Atanassov has developed the main theoretical foundations and analytic tools for the Generalized Nets, and—in collaboration with many specialists from various fields of science and practice—has stimulated and developed applications of this mathematical apparatus in various areas of science, business and technology, notably artificial intelligence, medicine, telecommunication, transportation, chemical and petrochemical industries, and many more. He is one of the very few people in Bulgaria who holds two different Doctor of Sciences degrees, both from the Bulgarian Academy of Sciences; his first D.Sc. in Technical Sciences was granted to him in 1997 for his research in the area of Generalized Nets.

Another significant field of Professor Atanassov's research interests is the theory of *fuzzy sets* proposed in 1965 by Professor Lotfi A. Zadeh, who later originated the idea of soft computing. In 1983, Professor Atanassov proposed an essential and far reaching extension of the concept of a fuzzy sets, called an *intuitionistic fuzzy set*, in which to the degree of membership (belongingness) of an element to a (fuzzy) set, which is from the unit interval, there is assigned an additional degree that of non-membership (non-belongingness) of an element to a (intuitionistic fuzzy) set, which is also from the unit interval. These two degrees, of membership and non-membership, sum up to a number from the unit interval, not necessarily to 1. The complement of the sum of the degrees of membership and non-membership to 1 constitutes a third degree, that of uncertainty. This opportunity of rendering account of the uncertainty makes the concept that Atanassov pioneered a particularly powerful and flexible instrument in the area of uncertainty analysis and decision making.

It is now a globally recognized scientific field on its own which relates to other fields such as the theory of fuzzy sets, fuzzy logic, mathematical logic, notably multi-valued logic, etc. His second D.Sc. in Mathematical Sciences was awarded in 2000 for his research on intuitionistic fuzzy sets.

For his contributions in the field, in 2013, the International Fuzzy Sets Association (IFSA) elected Professor Atanassov as the IFSA Fellow; and he is the first Bulgarian working in Bulgaria, and the second Bulgarian altogether, who has received this recognition. In 2013, Professor Atanassov was awarded the 'Pythagoras' Prize of the Bulgarian Ministry of Education and Science for established researcher in the field of engineering sciences. In the same year, he was also elected the Corresponding Member of the Bulgarian Academy of Sciences.

Professor Atanassov has authored and co-authored 30 monographs, more than 1,000 publications in international journals and conferences, and has served as a supervisor of more than 20 Ph.D. students. His research is now being followed and developed in multiple countries around the world by various research groups including his own numerous Ph.D. students.

This volume is a small token of appreciation for Professor Atanassov on his 60th anniversary for his great scientific achievement, multifaceted support of research activities and researchers from all over the world, and his constant enthusiasm and readiness to undertake new scientific challenges. We also greatly appreciate his great human qualities and friendship.

We wish to thank all the contributors to this volume for their excellent scientific works which involve many novel research results, insightful and inspiring analyses, as well as relevant applications. We wish to thank Dr. Thomas Ditzinger, Dr. Leontina Di Cecco and Mr. Holger Schaepe from Soringer for their help and support to prepare this volume.

June 2015

Plamen Angelov
Sotir Sotirov

Contents

Part I
Perspectives in Intuitionistic Fuzzy Set

Fuzzy, Intuitionistic Fuzzy, What Next?

Vladik Kreinovich and Bui Cong Cuong

Abstract In the 1980s, Krassimir Atanassov proposed an important generalization of fuzzy sets, fuzzy logic, and fuzzy techniques—intuitionistic fuzzy approach, which provides a more accurate description of expert knowledge. In this paper, we describe a natural way how the main ideas behind the intuitionistic fuzzy approach can be expanded even further, towards an even more accurate description of experts' knowledge.

1 Fuzzy Logic: A Brief Reminder

The main objective of this paper is to describe the main ideas behind intuitionistic fuzzy logic and to describe how these ideas can be expanded. To do that, we need to recall the main motivations and the main ideas behind the original fuzzy logic; for details, see, e.g., [6, 10, 11].

It is important to describe and process expert knowledge. In many practical situations, from medicine to driving to military planning to decisions on whether to accept a paper for publication, we rely on expert opinions.

In every field, there are a few top experts. For example, in every medical area, there are top specialists in this area. In the ideal world, every patient in need of a surgery would be operated by the world's top surgeon, and every person would get an advice from the world's top financial advisor on how to invest his or her savings. Since it is not possible for a few top surgeons to perform all the operations and for top financial advisors to advice everyone, it is desirable to design computer-based system which would incorporate the advice of the top experts—and thus help other

V. Kreinovich (✉)
University of Texas at El Paso, El Paso 79968, USA
e-mail: vladik@utep.edu

B.C. Cuong
Institute of Mathematics, Vietnam Academy of Science and Technology,
18 Hoang Quoc Viet, Cau Giay, Hanoi, Vietnam
e-mail: bccuong@gmail.com

P. Angelov and S. Sotirov (eds.), *Imprecision and Uncertainty in Information Representation and Processing*, Studies in Fuzziness and Soft Computing 332,
DOI 10.1007/978-3-319-26302-1_1

experts provide a better quality advice. Such computer-based systems are often called *expert systems*.

Experts can describe their knowledge in terms of statements $S_1, S_2, \ldots$ (e.g., "if the P/E ratio of a stock goes above a certain threshold t_0, it is recommended to sell it"). In some situations, when we have a query Q—e.g., whether to sell a given stock—we can use one of the expert rules. In many other cases, however, none of the expert rules can lead directly to the desired answer, but a proper combination of the rules can help. For example, in medical expert systems, we rarely have a rule directly linking patient's symptoms with the appropriate treatment, but we have rules which link symptoms with diseases, and we have rules which link diseases with treatments. By combining the corresponding rules, we can get an answer to the query. The part of an expert system which, given a query, tries to deduce the corresponding statement or its negation from the expert rules, is known as an *inference engine*.

Uncertainty of expert knowledge. In using expert knowledge, we need to take into account that experts are usually not 100 % confident that their statements are universally valid. For example, if a patient sneezes and coughs, a medical doctor will conclude that it is most probably cold, flu, or allergy, but the doctor also understands that there is a possibility of some rarer situations with similar symptoms.

A natural way to gauge the experts' uncertainty is to ask the experts to mark their uncertainty on a scale from 0 to some integer n (e.g., on a scale from 0 to 5), so that 0 corresponds to no certainty at all, and n correspond to the absolute certainty. If an expert marks m on a scale from 0 to n, then we claim that the expert's degree of certainty in his/her statement is the ratio m/n.

How to process experts' uncertainty: towards a precise formulation of the problem. Since the experts are not 100 % sure in their statements, we are therefore not sure about the expert system's conclusion either. It is therefore important to make sure that the expert system not only provides a "yes" or "no" (or more complex) answer to a given query, but that the user will also get a degree with which we are confident in this answer.

For example, if a medical expert system recommends a surgery, and the resulting confidence is 99 %, then it is probably a good idea to undergo this surgery. However, if the resulting degree of confidence that this answer is correct is about 50 %, maybe it is better to perform some additional tests so that we may become clearer on the diagnosis.

It is thus important, once we have derived a statement Q from the expert knowledge base $\{S_1, S_2, \ldots\}$, to provide the user with the degree $d(Q)$ that the resulting statement Q is correct. In some cases, there is only one chain of reasoning leading to the conclusion Q, and this chain involves statements $S_{i_1}, \ldots, S_{i_k}$. In this case, all these statements need to be true for Q to be true: if one of the statements in the chain is false, then the whole chain of reasoning collapses. In these cases, Q is true if the statement $S_{i_1} \& \ldots \& S_{i_k}$ is true. Thus, to gauge our degree of belief in Q, we must be able to estimate the degree of belief in a statement $S_{i_1} \& \ldots \& S_{i_k}$.

In general, we may have several derivations of Q—e.g., we may have several different observations supporting the same diagnosis. In this case, Q is deduced if at least one of the corresponding derivation chains is valid, i.e., if a propositional formula of the following type holds:

$$\left(S_{i_1} \,\&\, \dots \,\&\, S_{i_k}\right) \vee \left(S_{i'_1} \,\&\, \dots \,\&\, S_{i'_{k'}}\right) \vee \dots$$

Approximate estimation is needed. In other words, we would like to estimate the degree of belief in different propositional combinations of the original statements S_i. Of course, if we only know the expert's degrees of belief $d(S_1)$ and $d(S_2)$ of different statements S_1 and S_2, we cannot uniquely determine the expert's degree of certainty $d(S_1 \,\&\, S_2)$. For example, if S_1 means that a fair coin falls heads, and $S_2 = S_1$, then it is reasonable to take $d(S_1) = d(S_2) = 0.5$ and, thus, $d(S_1 \,\&\, S_2) = d(S_1) = 0.5$. On the other hand, if we take $S_2 = \neg S_1$, then still $d(S_1) = d(S_2) = 0.5$ but now $d(S_1 \,\&\, S_2) = 0$.

Since we cannot uniquely determine the degrees of certainty in all possible propositional combinations based only on the degrees $d(S_i)$, ideally, we should also find the degrees of certainty in all these propositional combinations. The problem is that for N original statements, we need $>2^N$ different degrees to describe, e.g., the degrees of certainty in different combinations $S_{i_1} \,\&\, \dots\, S_{i_n}$ ($>2^N$ because we have $2^N - 1$ possible non-empty subsets $\{i_1, \dots, i_n\} \subseteq \{1, \dots, N\}$).

Even for middle-size $N \approx 100$, the value 2^N is astronomically high. It is not possible to elicit all these degrees of certainty from the expert. Thus, no matter how much information we elicit, we will always have propositional combinations for which we do not know the corresponding degrees, combinations for which these degrees must be estimated.

How to estimate the corresponding degrees: fuzzy-motivated idea of negation-, "and"- and "or"-operations. A general propositional combination is obtained from the original statement by using the logical connectives $\neg$ ("not"), & ("and"), $\vee$ ("or"). Since we do not know the degrees of all composite statements, we inevitable face the following problem:

- for some statements A and B, we know the expert's degrees of certainty $d(A)$ and $d(B)$ in these statements;
- we need to estimate the expert's degree of certainty in the statements $\neg A$, $A \,\&\, B$ and/or $A \vee B$.

Negation operations. In this situation, to come up with the desired estimate $d(\neg A)$, the only information that we can use consists of a single number $d(A)$. Let us denote the estimate for $d(\neg A)$ corresponding to the given value $d(A)$ by $f_\neg(d(A))$. The corresponding function is usually known as an *negation operation*.

How can we choose this negation operation? Let us first describe some reasonable properties that this function should satisfy. First, we can take into account that $\neg(\neg A)$ usually means the same as A. By applying the negation operation $f_\neg$ to the estimated degree of certainty $d(\neg A) \approx f_\neg(d(A))$, we can estimate the expert's degree of certainty in $\neg(\neg A))$ as $f_\neg(f_\neg(d(A))$. It is reasonable to require that this estimate coincide with the original value $d(A)$: $f_{\neg}(f_{\neg}(d(A)) = d(A)$. This equality must hold for all possible values $a = d(A) \in [0, 1]$, so we must have $f_{\neg}(f_{\neg}(a)) = a$ for all a. In mathematical terms, this means that the function $f_{\neg}(a)$ is an *involution*.

When A is absolutely false, and $d(A) = 0$, then $\neg A$ should be absolutely true, i.e., we should have $f_{\&}(0) = 1$. Similarly, if A is absolute true and $d(A) = 1$, then $\neg A$ should be absolutely false, i.e., we should have $f_{\&}(1) = 0$. In general, the more we believe in A, the less we should believe in $\neg A$, so the function $f_{\&}(a)$ must be decreasing.

The most widely used negation operation is $f_{\&}(a) = 1 - a$, it satisfies all these properties; there are also other negation operations which are sometimes used in fuzzy systems.

"And"-operations. To come up with the desired estimate $d(A \,\&\, B)$, the only information that we can use consists of two numbers $d(A)$ and $d(B)$. Let us denote the estimate for $d(A \,\&\, B)$ corresponding to the given values $d(A)$ and $d(B)$ by $f_{\&}(d(A), d(B))$. The corresponding function is usually known as an *"and"-operation*, or *t-norm*.

How can we choose the "and"-operation? Let us first describe some reasonable properties that the corresponding function $f_{\&}(a, b)$ should satisfy. First, since $A \,\&\, B$ means the same as $B \,\&\, A$, it is reasonable to require that the two estimates $f_{\&}(d(A), d(B))$ and $f_{\&}(d(B), d(A))$ corresponding to different orders of A and B should be the same. This must be true for all possible values of $a = d(A)$ and $b = d(B)$; this means that we must have $f_{\&}(a, b) = f_{\&}(b, a)$ for all real values $a, b \in [0, 1]$. In other words, an "and"-operation must be *commutative*.

Similarly, $A \,\&\, (B \,\&\, C)$ means the same as $(A \,\&\, B) \,\&\, C$. If we follow the first expression, then, to estimate the corresponding degree of certainty, we first estimate $d(A \,\&\, B)$ as $f_{\&}(d(A), d(B))$ and then use the "and"-operation to combine this estimate and the degree of certainty $d(C)$ into an estimate $f_{\&}(f_{\&}(d(A), d(B)), d(C))$. Alternatively, if we follow the second expression, we end up with the estimate $f_{\&}(d(A), f_{\&}(d(B), d(C))$. It is reasonable to require that, since $A \,\&\, (B \,\&\, C) \equiv (A \,\&\, B) \,\&\, C$, these two estimates should coincide, i.e., that the "and"-operation be associative.

The expert's degree of confidence $d(A \,\&\, B)$ that both A and B are true should not exceed the degree of confidence that A is true. Thus, we should have $d(A \,\&\, B) \le d(A)$. It is therefore reasonable to require that $f_{\&}(a, b) \le a$—and thus, that $f_{\&}(0, a) = 0$ for all a.

It is also reasonable to require that when $d(A) = 1$ (i.e., when we are 100 % certain in A), then we should have $A \,\&\, B$ equivalent to B, so $f_{\&}(1, b) = b$ for all b. If we increase our degree of confidence in A and/or B, this should not lead to a decrease in our confidence in $A \,\&\, B$; this means that the "and"-operation should be monotonic: $a \le a'$ and $b \le b'$ implies $f_{\&}(a, b) \le f_{\&}(a', b')$. Finally, small changes in $d(A)$ and $d(B)$ should not lead to a drastic change in $d(A \,\&\, B)$, so the "and"-operation must be continuous.

"Or"-operations. Similarly, if we denote by $f_{\vee}(d(A), d(B))$ the estimate for $d(A \vee B)$, then the corresponding "or"-operation (also known, for historical reasons, as *t-conorm*) must be commutative, associative, monotonic, continuous, and satisfy the properties $f_\vee(0,a) = a$ and $f_\vee(1,a) = 1$ for all a.

Selecting different propositional operations: an empirical task. There are many different negation, "and"-, and "or"-operations which satisfy these properties; for each application area, we select the operations which best describe the reasoning of experts in this area, i.e., for which the resulting estimates for the expert's degrees of confidence in composite statement are the closest to the estimates for $d(\neg A)$, $d(A \& B)$, and $d(A \vee B)$ produced by the experts.

This idea was first implemented for the world's first expert system MYCIN—Stanford's expert system for diagnosing rare blood diseases; see, e.g., [3]. The authors of MYCIN tried different possible "and" and "or-operations and found the one which was the best fit for the actual reasoning of medical experts. It is worth mentioning that when they tried to apply their expert system to a different application area—geophysics—it turned out that the medical-generated "and"- and "or"-operations did not lead to good results, different operations had to be used.

Common misunderstanding. The reason why in fuzzy techniques (and in expert systems in general), we estimate the degree of confidence $d(A \& B)$ by applying an "and"-operation to $d(A)$ and $d(B)$ is *not* because we are under an illusion that the expert's degree of confidence in $A \& B$ is uniquely determined by his/her degrees of confidence in A and B. Everyone understands that there is no uniqueness here, the above example of a coin falling heads or tails is clear. What the "and"-operation produces is an *approximation* to the actual expert's degree of belief in $A \& B$.

We do not use this approximation because we are under some erroneous belief that "and"- and "or"-operations are truth-functional, but simply because we cannot realistically elicit all the degrees of confidence in all the propositional combinations from all the experts, and we therefore need to estimate the unknown degrees of certainty based on the known ones.

2 From Fuzzy to Intuitionistic Fuzzy

How can we improve the traditional fuzzy approach? One of the main ideas behind the traditional fuzzy approach is that, since we cannot elicit the expert's degrees of confidence in all possible propositional combinations of their original statements $S_1, \ldots, S_n$, we:

- extract the degrees of confidence $d(S_i)$ in these statements, and then
- use negation, "and"-, and "or"-operations to estimate the expert's degrees of belief in different propositional combinations.

To make these estimates more accurate, a natural idea is to extract, from the expert, not just his/her degrees of confidence in the original statements, but also degrees of confidence in some propositional combinations of these statements—at least the simplest ones.

This idea naturally leads to intuitionistic fuzzy logic. Which propositional combinations are the simplest? The more original statements are involved in a combination, the more propositional connectives are used, the more complex the statements. From this viewpoint, the simplest propositional combinations are the ones which has the smallest number of the original statements—one—combined by the smallest possible number of possible connectives: one. There are three possible connectives: negation, "and", and "or". "And" and "or" requires at least two original statements to combine (since $S_i \,\&\, S_i$ and $S_i \vee S_i$ mean the same as S_i). So, the only way to have a single original statement is by using negation. Thus, the simplest possible propositional combinations are negations $\neg S_i$.

Thus, to come up with a more adequate description of expert's degree of certainty, a natural next step is not only to elicit the expert's degrees of confidence $d(S_i)$ in their original statements, but also their degrees of confidence $d(\neg S_i)$ in their negations. In other words, to describe the expert's certainty about his/her statement S_i, instead of a single number $d(S_i)$, we now use a *pair* of numbers $d(S_i)$ and $d(\neg S_i)$. This is, in a nutshell, the main idea behind Atanassov's intuitionistic fuzzy logic; see, e.g., [1, 2] (see also [4]).

This idea makes perfect sense. Intuitively, the above idea makes perfect sense. In contrast to the traditional fuzzy logic, this idea enables us to distinguish between two different situations:

- a situation when we know nothing about a statement S, and
- a situation in which we have some arguments in favor of S and equally strong arguments in favor of the opposite statement $\neg S$.

In both situations, we have equally strong arguments in favor of S and in favor of $\neg S$, so it is reasonable to conclude that $d(S) = d(\neg S)$. In the traditional fuzzy logic, when we assume that $d(\neg S) = 1 - d(S)$, this implies that in both situations, we have $d(S) = d(\neg S) = 0.5$. In the intuitionistic fuzzy logic, we describe the situation in which we have no arguments in favor by S by taking $d(S) = 0$, and similarly $d(\neg S) = 0$. Thus, this situation is described differently from the second one when $d(S) = d(\neg S) > 0$.

3 Beyond Intuitionistic Fuzzy

Beyond intuitionistic fuzzy logic: a natural next step. To get an even more adequate description of expert's knowledge, we need to also elicit the expert's degree of confidence in some more complex composite statements.

As we have mentioned, the fewer statements are used in a propositional combination, and the fewer propositional connectives are used, the simpler the combination. If we use one statement S, then the only possible propositional combination is $\neg S$ – which is handled in the intuitionistic fuzzy approach. Thus, if we want to go beyond intuitionistic fuzzy, we need to consider propositional combinations of two original statements S and S'. Among such combinations, the simplest case if when we use a single propositional connective. Thus, the simplest such combinations are combinations of the type $S \,\&\, S'$ and $S \vee S'$.

So, we arrive at the following natural description of the next step: in addition to eliciting, from the experts, their degrees of belief in the original statements S_i, we also elicit their degrees of belief in composite statements $S_i \,\&\, S_j$ and $S_i \vee S_j$. Since we have already included negation, it thus makes sense to also consider the expert's degrees of belief combinations of the type $\neg S_i \,\&\, S_j$, $\neg S_i \,\&\, \neg S_j$, $\neg S_i \vee S_j$, and $\neg S_i \vee \neg S_j$.

The idea in more detail. To describe an imprecise ("fuzzy") property P (e.g., "small"), in the traditional fuzzy logic, to each possible value x of the corresponding quantity, we assign the degree $\mu_P(x) \stackrel{\text{def}}{=} d(P(x))$ to which this quantity satisfies the property P. The corresponding function $\mu_P(x)$ from real values to the interval $[0, 1]$ is known as the *membership function*, or, alternatively, as the *fuzzy set*.

In the intuitionistic fuzzy logic, to describe a property P, we need to assign, for each x, two degrees:

- the degree $d(P(x)) \in [0, 1]$ that the quantity x satisfies the property P, and
- the degree $d(\neg P(x)) \in [0, 1]$ that the quantity x does not satisfy the property P.

This pair of functions forms an *intuitionistic fuzzy set*.

In the new approach, to describe an imprecise property P, we need to also assign, to every pair of values x and x':

- the degree $d(P(x) \,\&\, P(x')) \in [0, 1]$ that both quantities x and x' satisfy the property P;
- the degree $d(P(x) \vee P(x')) \in [0, 1]$ that either the quantity x or the quantity x' satisfies the property P;
- the degree $d(\neg P(x) \,\&\, P(x')) \in [0, 1]$ that the quantity x does not satisfy the property P while the quantity x' satisfies P;
- the degree $d(\neg P(x) \,\&\, \neg P(x')) \in [0, 1]$ that neither x nor x' satisfy the property P;
- the degree $d(\neg P(x) \vee P(x')) \in [0, 1]$ that either x does not satisfy P or x' satisfies P; and
- the degree $d(\neg P(x) \vee \neg P(x')) \in [0, 1]$ that either x or x' does not satisfy the property P.

The resulting collection of functions form the corresponding generalization of the notion of a fuzzy set.

An interesting difference emerges when we want to consider two possible properties P and P'. In both traditional fuzzy approach and intuitionistic fuzzy approach, all we can do is describe these two properties one by one. In the new approach, we also need to describe the relation between the two properties. For example, for each x and x', we can now describe the degree $d(P(x) \,\&\, P(x'))$ to which x satisfies the property P and x' satisfies the property P'.

Comment. The idea of describing such degrees was first formulated—in the probabilistic context—in [5]; see also [9].

This ideas also makes perfect sense. The above idea enables us to describe features of the properties like "small" which are difficult to describe otherwise. For example, while different experts may disagree on which values are small and which are not small, all the experts agree that if x is small and x' is smaller than x, then x' is small as well. In other words, if $x' < x$, then it is not reasonable to believe that x is small but the smaller value x' is not small. In other words, for $P =$ "small" and $x' < x$, the corresponding degree of belief $d(P(x) \,\&\, \neg P(x'))$ should be equal to 0.

This possibility is in contrast to the traditional fuzzy logic, where from $d(P(x)) > 0$ and $d(\neg P(x')) = 1 - P(x') > 0$, we would conclude that $d(P(x) \,\&\, \neg P(x')) \approx f_{\&}(d(P(x)), d(\neg P(x')))$. For most frequently used t-norms such as $f_{\&}(a, b) = \min(a, b)$ and $f_{\&}(a, b) = a \cdot b$, from $d(P(x)) > 0$ and $d(\neg P(x')) > 0$, we deduce that the resulting estimate for $d(P(x) \,\&\, \neg P(x'))$ is also positive—and not equal to 0 as common sense tells us it should.

We can go further. To get an even more adequate representation of expert knowledge, we can also elicit expert;s degrees of belief in composite statements which combine three or more original statements S_i.

4 From Type-1 to Type-2 Fuzzy

Need for type-2: brief reminder. We are interested in situations in which an expert is not 100 % certain about, e.g., the value of the corresponding quantity. In this case, we use, e.g., estimation on a scale to gauge the expert's degree of belief in different statements. The traditional fuzzy approach assumes that an expert can describe his/her degree of belief by a single number.

In reality, of course, the expert is uncertain about his/her degree of certainty—just like the same expert is uncertain about the actual quantity. In this case, the expert's degree of certainty $d(P(x))$ is no longer a single number—it is, in general, a fuzzy set. This construction, in which, to each x, we assign a fuzzy number $d(P(x))$ is known as a *type-2 fuzzy set*; see, e.g., [7, 8].

Need to combine intuitionistic and type-2 fuzzy sets. It is known that, in many practical situations, the use of type-2 fuzzy sets leads to a more adequate description of expert knowledge. Therefore, to achieve even more adequacy, it is desirable to combine the advantages of type-2 and intuitionistic fuzzy set.

At first glance, such a combination is straightforward. At first glance, it looks like the above combination is straightforward: all the above arguments did not depend on the degree $d(S_i)$ being numbers; the exact same ideas— including the possibility to go beyond the intuitionistic fuzzy sets—can be repeated for the case when the values $d(S_i)$ are themselves fuzzy numbers—or, alternatively, intuitionistic fuzzy numbers.

However, as we will see, the relation between intuitionistic and type-2 fuzzy number is more complicated.

Observation: some intuitionistic fuzzy numbers can be naturally viewed as a particular case of type-2 fuzzy numbers. To explain this unexpected relation, let us start with the simplest possible extension of the classical two-valued logic, in which each statement is either true or false. The more possible truth values we add to the original two, the more complex the resulting logic. Thus, the simplest possible non-classical logic is obtained if we add, to the two classical truth values "true" and "false", the smallest possible number of additional truth values—one. A natural interpretation of this new truth value is "uncertain". For simplicity, let us denote the corresponding truth values by T ("true"), F ("false"), and U ("uncertain").

To fully describe the resulting 3-valued logic, we need to supplement the known truth tables for logical operations involving T and F with operations including the "uncertain" degree U.

For negation, this means adding $\neg U$. For each truth value X, the meaning of $\neg X$ is straightforward: if our degree of belief $d(S)$ in a statement S is equal to X, then our degree of belief in its negation $\neg S$ should be equal to $\neg X$. For "uncertain", the truth value $d(S) = U$ means that we are not sure whether the statement S is true or false. In this case, we are equally uncertain about whether the negation $\neg S$ is true or false; thus, $d(\neg S) = U$. In other words, we have $\neg U = U$.

Similarly, if we are uncertain about S, but we know that S' is false, then the conjunction $S \& S'$ is also false; thus, $U \& F = F$. On the other hand, if we know that S' is true (or if we are uncertain about S'), then, depending on whether S is actually true or false, it is possible that the conjunction $S \& S'$ is true and it is also possible that this conjunction is false. Thus, we have $U \& T = U \& U = U$.

If we are uncertain about S, but we know that S' is true, then the disjunction $S \vee S'$ is also true; thus, $U \vee T = T$. On the other hand, if we know that S' is false (or if we are uncertain about S'), then, depending on whether S is actually true or false, it is possible that the disjunction $S \vee S'$ is true and it is also possible that this disjunction is false. Thus, we have $U \vee F = U \vee U = U$.

In the spirit of type-2 logic, instead of selecting one of the three truth values T, F, or U, we can assign *degrees of certainty* $d(T) \geq 0$, $d(F) \geq 0$, and $d(U) \geq 0$ to these three values. One possible way to assign such degrees is to distribute the same fixed amount of degree (e.g., 1) between these three options; in this case, we always have $d(T) + d(F) + d(U) = 1$. Because of this relation, the triple $(d(T), d(F), d(U))$ can be uniquely described by two values $d(T) \geq 0$ and $d(F) \geq 0$ for which $d(T) + d(F) \leq 1$; one can easily see that this is exactly the definition of an intuitionistic fuzzy degree [1, 2].

Moreover, we will show that even some operations on intuitionostic fuzzy degrees can be thus interpreted. Indeed, if we know the triples $(d(T), d(F), d(U))$ and $(d'(T), d'(F), d'(U))$ describing the expert's degree of belief in statements S and S', then the triple $(d''(T), d''(F), d''(U))$ corresponding to the composite statements $S'' = \neg S$, $S'' = S \& S'$, and $S'' = S \vee S'$ can be obtained by using Zadeh's extension principle. Let us describe this in detail.

In the 3-valued logic, $S'' = \neg S$ is true if and only if S is false, and $S'' = \neg S$ is false if and only if S is true. Thus, $d''(T) = d(F)$ and $d''(F) = d(T)$). This is in line with the usual definition of negation in the intuitionistic fuzzy logic, as $f_\neg((d(T), d(F))) = (d(F), d(T))$.

In the 3-valued logic, $S'' = S \,\&\, S'$ is true if and only if S is true and S' is true:

$$S'' \text{ is } T \Leftrightarrow ((S \text{ is } T) \,\&\, (S' \text{ is } T)).$$

We know the degree $d(T)$ to which S is true, and we know the degree $d'(T)$ to which S' is true. Thus, by applying an appropriate "and"-operation (t-norm), we can conclude estimate the desired degree $d''(T)$ that S'' is true as $f_\&(d(T), d(T'))$. In particular, for a frequently used "and"-operation $f_\&(a, b) = a \cdot b$, we get $d''(T) = d(T) \cdot d'(T)$.

Similarly, $S'' = S \,\&\, S'$ is false if and only if:

- either S is false and S' can take any possible value,
- or S' is false and S can take any possible value.

Thus:

$$S'' \text{ is } F \Leftrightarrow (((S \text{ is } F) \,\&\, (S' \text{ is } T)) \vee ((S \text{ is } F) \,\&\, (S' \text{ is } U)) \vee$$

$$((S \text{ is } F) \,\&\, (S' \text{ is } F)) \vee ((S \text{ is } T) \,\&\, (S' \text{ is } F)) \vee ((S \text{ is } U) \,\&\, (S' \text{ is } F))).$$

By using the same "and"-operation and a frequently used "or"-operation $f_\vee(a, b) = \min(a + b, 1)$, we get the estimate

$$d''(F) = \min(d(F) \cdot d'(T) + d(F) \cdot d'(U) + d(F) \cdot d'(F) + d(T) \cdot d'(F) + d(U) \cdot d'(U), 1).$$

Substituting $d(U) = 1 - d(T) - d(F)$ into this formula, we conclude that $d''(F) = d(F) + d'(F) - d(F) \cdot d'(F)$. This is in line with the usual definition of an "and"-operation in the intuitionistic fuzzy case as

$$f_\&((d(T), d(F)), (d'(T), d'(F))) = (f_\&(d(T), d'(T)), f_\vee(d(F), d'(F))),$$

where $f_\vee(a, b) \stackrel{\text{def}}{=} 1 - f_\&(1 - a, 1 - b))$. For $f_\&(a, b) = a \cdot b$, we thus get $f_\vee(a, b) = a + b - a \cdot b$, and therefore, $d''(T) = f_\&(d(T), d'(T)) = d(T) \cdot d(T'))$ and $d''(F) = d(F) + d'(F) - d(F) \cdot d'(F)$, exactly as in the above type-2 formulas.

For $S'' = S \vee S'$, we similarly get

$$S'' \text{ is } F \Leftrightarrow ((S \text{ is } F) \,\&\, (S' \text{ is } F)),$$

and thus, $d''(F) = d(F) \cdot d'(F)$. Also, we get

$$S'' \text{ is } T \Leftrightarrow (((S \text{ is } T) \,\&\, (S' \text{ is } T)) \vee ((S \text{ is } T) \,\&\, (S' \text{ is } U)) \vee$$

$$((S \text{ is } T) \,\&\, (S' \text{ is } F)) \vee ((S \text{ is } F) \,\&\, (S' \text{ is } T)) \vee ((S \text{ is } U) \,\&\, (S' \text{ is } T))),$$

and hence, the degree $d''(T)$ is equal to

$$\min(d(T) \cdot d'(T) + d(T) \cdot d'(U) + d(T) \cdot d'(F) + d(U) \cdot d'(T) + d(F) \cdot d'(T), 1) =$$

$$d(T) + d'(T) - d(T) \cdot d'(T).$$

This is also in perfect accordance with the intuitionistic fuzzy operation $f_\vee((d(T), d(F)), (d'(T), d'(F))) = (f_\vee(d(T), d'(T)), f_\&(d(F), d'(F)))$.

Acknowledgments This work was supported in part by the National Science Foundation grants HRD-0734825, HRD-124212, and DUE-0926721.

References

1. Atanassov, K.: Intuitionistic fuzzy sets. Fuzzy Sets Syst. **20**, 87–96 (1986)
2. Atanassov, K.: Intuitionistic Fuzzy Sets. Springer-Verlag, Heidelberg (1999)
3. Buchanan, B.G., Shortliffe, E.H.: Rule Based Expert Systems: The MYCIN Experiments of the Stanford Heuristic Programming Project. Addison-Wesley, Reading (1984)
4. Cuong, B.C., Kreinovich, V.: Picture Fuzzy Sets–a new concept for computational intelligence problems. In: Proceedings of the Third World Congress on Information and Communication Technologies WICT'2013, pp. 1-6. Hanoi, Vietnam, 15-18 Dec 2013
5. Goodman, I.R.: Fuzzy sets as equivalence classes of random sets. In: Yager, R., et al. (eds.) Fuzzy Sets and Possibility Theory, pp. 327–432. Pergamon Press, Oxford (1982)
6. Klir, G., Yuan, B.: Fuzzy Sets and Fuzzy Logic. (Prentice Hall, Upper Saddle River 1995)
7. Mendel, J.M.: Uncertain Rule-Based Fuzzy Logic Systems: Introduction and New Directions. (Prentice-Hall, Upper Saddle River 2001)
8. Mendel, J.M., Wu, D.: Perceptual Computing: Aiding People in Making Subjective Judgments. IEEE Press and Wiley, New York (2010)
9. Nguyen, H.T., Kreinovich, V.: How to fully represent expert information about imprecise properties in a computer system–random sets, fuzzy sets, and beyond: an overview. Int. J. Gen. Syst. **43**, 586–609 (2014)
10. Nguyen, H.T., Walker, E.A.: A First Course in Fuzzy Logic. Chapman and Hall/CRC, Boca Raton (2006)
11. Zadeh, L.A.: Fuzzy sets. Inf. Control **8**, 338–353 (1965)

Intuitionistic Fuzzy Logic and Provisional Acceptance of Scientific Theories: A Tribute to Krassimir Atanassov on the Occasion of His Sixtieth Birthday

A.G. Shannon

Abstract This essay attempts to outline some essential features of Atanassov's intuitionistic fuzzy logic within the framework of the philosophy of science. In particular, it aims to highlight the brilliance of Atanassov's conceptual and symbolic originality. It also illustrates the danger of the univocal caricaturing of scientific terminology.

1 Introduction

It is a pleasure to pay tribute to my friend and colleague on the occasion of his 60th birthday. We have been research collaborators for almost thirty years, and he has been a generous host on my twenty visits to his beloved Bulgaria over many years. He was a pioneer in the now burgeoning field of computational intelligence with its use of both traditional fuzzy logic [1] and intuitionistic fuzzy logic [2] in a variety of applications. He has also been an internationally renowned creator of new ideas and an insightful solver of problems for almost forty years, particularly in discrete mathematics [3]. These include his work on index matrices [4] and generalized nets [5]. The latter are a major advance on the first form of neural network [6].

This essay is more discursive and expository, rather than technical, particularly in relation to empirical sciences [7] in order to demonstrate the range and scope of Krassimir's fundamental ideas beyond fuzziness in the development of soft computing alone. "Fuzziness" itself is open to misinterpretation. It goes far beyond Russell's notion of vagueness [8] and his study of symbols (which have themselves been an important part of Atanassov's research as we shall show). This leads to a brief discussion on the equivocal misuse of scientific concepts: analogical caricatures make a weak argument weaker, not stronger!

A.G. Shannon (✉)
Faculty of Engineering and Information Technology, University of Technology, Sydney, NSW 2007, Australia
e-mail: Anthony.Shannon@uts.edu.au; tshannon38@gmail.com

P. Angelov and S. Sotirov (eds.), *Imprecision and Uncertainty in Information Representation and Processing*, Studies in Fuzziness and Soft Computing 332,
DOI 10.1007/978-3-319-26302-1_2

2 Science and Pseudoscience

Media reports of recent empirical challenges to the accepted understanding of the nature and speed of light have demonstrated the inadequate critical understanding of the experimental sciences in the popular mind, at a time when that same popular mind is being asked to make important bioethical and technological decisions vicariously through their representatives in parliamentary democracies. Recent public debates to the contrary science is not democratic!

Most science, even mathematics, is conducted in a mode of 'conventionalism' [9], which involves provisional acceptance of hypotheses—the 'probabilism' hinted at by Aquinas [10]. It is a purpose of this note to examine the foundations of this provisional acceptance within the context of intuitionistic fuzzy logic (IFL) [11].

The simplest explanation which fits the facts tends to be the prevailing confirmation in science. Scientists, being human, can be prone to disregard facts which do not fit this prevailing confirmation if their source is from authority less prestigious than the recognized authorities in their field. Argument from authority in science has historically hampered its progress. Scientific progress is usually marked by 'confirmation' [12] or 'refutation' [13], although in practice the working scientist operates within a framework which contains a collection of hypotheses where there can be disagreement between empirical data and individual hypotheses without destroying the theory as a whole [14].

At this working stage of provisional acceptance, somewhere between refutation and confirmation, the empirical support of the theory prevails over any alleged counter-example. Such was Einstein's attitude when he said that "only after a more diverse body of observations becomes available will it be possible to decide with confidence whether the systematic deviations are due to a not yet recognized source of errors or to the circumstances that the foundations of the theory of relativity do not correspond to the facts" [15].

Public intellectuals like Dawkins and Singer, for instance, at times like to use their eminence and expertise in one field to assert authority in another, even seeking though are seeking the truth. Thus I felt in Dawkins' earlier writings that he was almost drifting towards St Anselm's ontological proof for the existence of God. The god that Dawkins and his disciples are now trying to demolish though is the anthropomorphic god of the fundamentalists. His misdirected zeal ironically appeals to anti-theist fundamentalists. Singer too, while more consistently logical, fails to see in his own thinking the faults he attacks in the thinking of others [16]. Not for them the humility inherent in the title of James Franklin's recent book [17], but rather the arrogance of Atkins [18]: "… science has never encountered a barrier, and the only grounds for supposing that reductionism will fail are pessimism on the part of scientists and fear in the minds of the religious". That Dawkins has tried to use mathematical tools in his argumentation and Singer to dismiss them has motivated these comments.

3 Evidence

It is not surprising that moral relativists disparage any "search for certainty". For them the only absolute is that there is no absolute. "It is as if to seek certainty denoted a lack of character, and were a sign of psychological or intellectual immaturity" [19]. Dawkins makes much of his view that "evidence", as he defines it, is missing from religious belief. There are, it is true, some truths, such as the mystery of the Trinity which are inaccessible to reason in terms of existence and content. This does not make it unreasonable to believe them. It depends on whose authority we believe them. In any case a God we could fully understand would not be God—to extend St Anselm [20]. Yet scientists themselves believe some things on the basis of their nature rather than observation alone. Thus we believe it is in the nature of humans to be mortal. While nearly every textbook of introductory logic has the statement "all humans are mortal", we know that all humans who have died must *ipso facto* be mortal, but we do not know it scientifically that all humans are mortal, because, as far as we know, most humans who have ever lived are alive today. We know that we are mortal from the study of natures, which is something we do in mathematics.

Yet for Dawkins the only evidence is scientific evidence, which itself is a metaphysical opinion, not a scientific statement. Moreover, Dawkins has no evidence that there is no evidence. Even the more persuasive Hitchens reduces his evidence to a series of anecdotes [21]. While some might say that these rebuttals are only playing with words, there are more serious underlying scientific issue relevant to the context of this paper.

These have been articulated in a series of papers by McCaughan who distinguishes extrinsic and intrinsic causes to show that even within science confusion of efficient and formality can lead to the domination of physics by mathematics to control all explanation, despite the fact that mathematics can do no more than predict [22]. Statistics too can disguise the existence of goal directed forces, but "goal directed forces eliminate blind chance. In following David Hume, scientists have removed goals or ends from science. This has not eliminated them from nature but left them unrecognised. Blind faith in blind chance just leads to intellectual blindness" [23]. We can see this in the way some evolutionary and generic algorithms are used analogously [24].

4 Genetic Algorithms

Genetic Algorithms (GAs) are an adaptive heuristic search algorithm based on analogies with the evolutionary ideas of natural selection and genetics [25]. Dawkins' dichotomy is that we can have God or evolution but not both [23: 215] and so his goal is to use these algorithms to prove that we cannot have God. The basic techniques of GAs are designed to simulate processes in natural systems necessary for evolution, especially those that seem to follow the principles first laid down by Charles Darwin of "survival of the fittest". That GAs use evolutionary terms can be a trap for the unwary.

GAs are implemented in a computer simulation in which a population of abstract representations of candidate solutions to an optimization problem evolves toward better solutions [26]. The "evolution" usually starts from a population of randomly generated individuals and happens in generations. In each generation, the *fitness* of every individual in the population is evaluated, multiple individuals are stochastically selected from the current population (based on their fitness), and modified (recombined and possibly randomly mutated) to form a new population. The new population is then used in the next iteration of the algorithm. Commonly, the algorithm terminates when either a maximum number of generations has been produced, or a satisfactory fitness (optimal) level has been reached for the population. Once the genetic representation and the fitness (optimal) function are defined, GAs proceed to initialize a population of solutions *randomly*, then improve it through repetitive application of *selection* operators.

For instance, a Generalized Net model (which is essentially a directed graph with choices at the nodes) [27] when combined with IFL (which provides for non-membership as well as membership choices) [28], simultaneously evaluates several "fitness" functions and then ranks the individuals according to their fitness to choose the best fitness function in relation to what is being optimized. GAs require only information concerning the quality of the solution produced by each parameter set (objective function value information). The selection operator could be, for instance, a roulette wheel [29]!

Thus, a GA is an algorithm which has a beginning and which is goal directed in order to eliminate blind chance, but Dawkins, for example, has a goal as the end of his evolutionary algorithm but also, in effect, wants to have no beginning. Hawkins wants to have a beginning, but like Dawkins uses science to sidestep God [30].

5 Intuition

Like Dawkins, Peter Singer steps across into mathematics when he says: "…can we really know anything through intuition? The defenders of ethical intuitionism argued that there was a parallel in the way we know or could immediately grasp the basic truths of mathematics: that one plus one equals two, for instance. This argument suffered a blow when it was shown that the self evidence of the basic truths of mathematics could be explained in a different and more parsimonious way, by seeing mathematics as a system of tautologies, the basic elements of which are true by virtue of the meanings of the terms used. On this view, now widely, if not universally, accepted, no special intuition is required to establish that one plus one equals two- this is a logical truth, true by virtue of the meanings given to the integers 'one' and 'two', as well as 'plus' and 'equals', So the idea that intuition provides some substantive kind of knowledge of right and wrong lost its only analogue" [31].

The broad and very loose statement denying intuitionism as a valid form of knowledge in mathematics is difficult to understand and very contradictory, even without the existence of intuitionism in mathematics [32]. Bertrand Russell a hundred years ago attempted to reduce mathematics to 'tautologies' (logical truths) but it proved impossible.

Working mathematicians simply do not deny intuition. For example, the standard presentation of the foundations of mathematics includes the "axiom of infinity", which says "There exists an infinite set". You just have to take it (by intuition) or leave it. In no way is it a logical truth and no-one the least bit informed maintains it is [33]. Moreover, mathematicians do research by intuitive insights rather than by "symbol shoving" or even logic, though they justify their conclusions with logic acceptable to their peers [34].

Likewise, mathematical notation is more than a form of words; it is a tool of thought [35]. For instance, the relationship between powers and subscripts within the umbral calculus reveals ideas latent in the original mathematical language [36] Here too Atanassov's symbolism has proved to be a powerful tool of thought even if we were only to judge it by the literature it has spawned. To see this we shall touch on some features of IFL.

6 Intuitionistic Fuzzy Logic

We shall now briefly outline the salient features of Intuitionistic fuzzy logic (IFL) by comparison with classical symbolic logic. IFL in many ways is a generalisation of the mathematical intuitionism of Brouwer [32] and the fuzzy sets of Zadeh [37].

In classical terms, to each proposition *p*, we assign a truth value denoted by *1* (truth) or *0* (falsity). In IFL we assign a truth value, $\mu(p)\ \varepsilon[0,1]$, for the degree of truth, and a falsity value, $v(p)\ \varepsilon[0,1]$ [4]:

$$0 \leq \mu(p) + v(p) \leq 1$$

This assignment is provided by an evaluation function V, which is defined over a set of propositions S,

$$V: S \rightarrow [0,1] \times [0,1]$$

such that

$$V(p) = < \mu(p), v(p) >$$

is an ordered pair. If the values V(p) and V(q) of the propositions p and q are known, then V can be extended:

$$V(\neg p) = <\nu(p), \mu(p)>$$
$$V(p \wedge q) = <\min(\mu(p), \mu(q)), \max(\nu(p), \nu(q))>,$$
$$V(p \vee q) = <\max(\mu(p), \mu(q)), \min(\nu(p), \nu(q))>,$$
$$V(p \supset q) = <\max(\nu(p), \mu(q)), \min(\mu(p), \nu(q))>;$$

and, for the propositions $p, q \varepsilon S$:

$$\neg V(p) = V(\neg p),$$
$$V(p) \cap V(q) = V(p \wedge q),$$
$$V(p) \cup V(q) = V(p \vee q),$$
$$V(p) \rightarrow V(q) = V(p \supset q).$$

A *tautology* and an *intuitionistic fuzzy tautology* (IFT) are then defined respectively by

"A is a tautology" if, and only if, $V(A) = <1, 0>$;
"A is an IFT" if, and only if, $V(A) = <a, b> \rightarrow a \geq b$.

Provisional acceptance of a scientific theory means that an individual counterexample of empirical evidence can be related to an individual hypothesis within a theoretical framework in order to modify some of the individual constituents of the theory and thus accommodate the disagreement. This can be written as

(a) $T_1 \equiv ((A \supset C) \wedge \neg C) \supset \neg A$,
(b) $T_2 \equiv ((A \wedge B \supset C) \wedge \neg C \wedge B) \supset \neg A$,
(c) $T_3 \equiv ((A \wedge B \supset C) \wedge \neg C) \supset (\neg A \vee \neg B)$,

for every three propositional forms A, B and C. This leads us to

Theorem T_1, T_2, T_3 are IFTs.

Proof In the interests of brevity, we shall consider (b) only, as it is typical of all three parts.

$$\begin{aligned} V(T_2) &= [(<\mu_A, \nu_A> \wedge <\mu_B, \nu_B>) \rightarrow <\mu_C, \nu_C>] \wedge <\nu_C, \mu_C> \wedge <\mu_B, \nu_B> \rightarrow <\nu_A, \mu_A> \\ &= [<\max(\nu_A, \nu_B, \nu_C), \min(\mu_A, \mu_B, \mu_C)> \wedge <\min(\nu_C, \mu_B), \max(\mu_C, \nu_B)>] \rightarrow <\nu_A, \mu_A> \\ &= <\min(\nu_C, \mu_B), \max(\nu_A, \nu_B, \mu_C), \max(\mu_A, \mu_B, \nu_C)> \rightarrow <\nu_A, \mu_A> \\ &= <\max[\mu_c, \nu_B, \nu_A, \min(\mu_A, \mu_B, \nu_C)], \min[\nu_C, \mu_B, \mu_A, \max(\nu_A, \nu_B, \mu_C)]>, \end{aligned}$$

□

and

$$\max[\mu_C, \nu_B, \nu_A, \min(\mu_A, \mu_B, \nu_C)] - \min[\nu_C, \mu_B, \mu_A, \max(\nu_A, \nu_B, \mu_C)]$$
$$\geq \min(\mu_A, \mu_B, \nu_C) - \min(\nu_C, \mu_B, \mu_A)$$
$$= 0.$$ Therefore, T is an IFT.

7 Concluding Comments

The existence of an additional working *modus operandi* between refutation and confirmation can clarify the way the empirical sciences work. Moreover, the schematic expression of this provisional acceptance of a theory invites an estimation of the truth values in any particular case so that the following type of analysis can be made. Suppose that for the propositional forms A and B:

$$V(A) \leq V(B) \text{ if, and only if, } (\mu_A \leq \mu_B) \wedge (\nu_A \leq \nu_B),$$
$$V(A) > V(B) \text{ if, and only if, } (\mu_A > \mu_B) \wedge (\nu_A < \nu_B).$$

If we assume that μ_A, ν_A, the intuitionistic fuzzy values of A are fixed, then from the form of T_2 we see that T_2 is more reliable as the intuitionistic fuzzy truth of B increases, that is, the bigger μ_C and the smaller ν_c are.

The truth value of T_2 can also increase if any of

- $$(V(A) > V(B)) \vee (V(A) > V(\neg C)), \text{ for fixed } \mu_A;$$
- $$(V(A) < V(B)) \vee (V(B) < V(\neg C)), \text{ for fixed } \nu_A;$$
- $$(V(A) < V(\neg C)) \vee (V(B) < V(\neg C)), \text{ for fixed } \mu_B.$$

On the other hand, T_2 will not be changed if any of

- $$(V(A) \leq V(B)) \vee (V(A) \leq V(\neg C)), \text{ for fixed } \mu_A;$$
- $$(V(A) \geq V(B)) \vee (V(B) \leq V(\neg C)), \text{ for fixed } \nu_A;$$
- $$(V(A) \geq V(\neg C)) \vee (V(B) \geq V(\neg C)), \text{ for fixed } \mu_B.$$

Nevertheless, science should be no more exempt from moral evaluation than any other human activity, especially as it lacks the intellectual certitude of metaphysics and mathematics [38]. The logical analysis of 'provisional acceptance' will not make scientists more logical, but it is important that both scientists and the general public are aware of the nature and scope, including limitations, of science and especially the role of models within science. This is a realm open to research in psychology and philosophical anthropology, namely to relate the conceptual connection between intuition and perception as the link between the internal and external senses and the intellect. In the terminology of evolution, it is a missing link in our knowledge of heuristics.

References

1. Lam, H.K., Ling, S.H., Nguyen, H.T.: Computational Intelligence and Its Applications. Imperial College Press, London (2012)
2. Shannon, A.G., Nguyen, H.T.: Empirical approaches to the application of mathematical techniques in health technologies. Int. J. Bioautomation **17**(3), 125–150 (2013)
3. Atanassov, K.T., Vassia Atanassova, A.G., Shannon, J.C.T.: New Visual Perspectives on Fibonacci Numbers. World Scientific, New York/Singapore (2002)
4. Atanassov, K.: General index matrices. Comptes rendus de l'Académie Bulgare des Sciences **40**(11), 15–18 (1987)
5. Atanassov, K.: Theory of generalized nets: a topological aspect. Methods Oper. Res. **51**, 217–226 (1984)
6. McCulloch, W., Pitts, W.: A logical calculus of the ideas immanent in nervous activity. Bull. Math. Biophysics **5**(4), 115–133 (1943)
7. Polikarov, A., Atanassov, K.: Refutability of physical theories: a new approach. Notes Intuitionistic Fuzzy Sets **8**(2002), 37–41 (2002)
8. Russell, B.: Vagueness. Aust. J. Philos. **1**(1), 84–92 (1923)
9. Poincaré, H.: Science and Hypothesis. Walter Scottp,xxi, London (1905)
10. Maurer, A.: St. Thomas Aquinas: The Division and Methods of the Sciences, pp. 51–52. (Pontifical Institute of Medieval Sciences, Toronto 1958)
11. Atanassov, K.T., Shannon, A.G.: A note on intuitionistic fuzzy logics. Acta Philosophica **7**, 121–125 (1998)
12. Carnap, R.: Logical Foundations of Probability. University of Chicago Press, Chicago, IL (1962)
13. Popper, K.R.: The Logic of Scientific Discovery. Hutchinson, London (1959)
14. Duhem, P.: The Aim and Structure of Physical Theory. Atheneum, New York (1954)
15. Einstein, A.: The Collected Papers, vol. 2, p. 283. Princeton University Press, Princeton, NJ (1989)
16. Abboud, A.J.: The fundamental moral philosophy of Peter singer and the methodology of utilitarianism. Cuadernos Filosofía **18**, 1–105 (2008)
17. Franklin, J.: What Science Knows and How it Knows It. Encounter Books, New York (2009)
18. Atkins, P.: The Limitless Power of Science. In: Cornwell, J. (ed.) Nature's Imagination—The Frontiers of Scientific Vision, p. 125. Oxford University Press, Oxford (1995)
19. Burke, C.: Authority and Freedom in the Church, p. 76. Four Courts Press, Dublin (1988)
20. Charlesworth, M.J. (ed.): St. Anselm's Proslogion. University of Notre Dame Press, Notre Dame, IN (2003). Translator & editor
21. Hitchens, C.: God Is Not Great: How Religion Poisons Everything. Twelve Books, New York (2007)

22. McCaughan, J.B.T.: Capillarity—a lesson in the epistemology of physics. Phys. Educ. **22**, 100–106 (1987)
23. White, J.J.: A Humean Critique of David Hume's Theory of Knowledge. (Gueguen, J.A. (ed.)), Section VII. (University Press of America, New York 1998)
24. Dawkins, R.: The God Delusion. (Houghton Mifflin, Boston 2006)
25. Setnes, M., Roubos, H.: GA-fuzzy modeling and classification: complexity and performance. IEEE Trans. Fuzzy Syst. **8**(5), 509–522 (2000)
26. Roeva, O., Pencheva, T., Shannon, A., Atanassov, K.: Generalized Nets in Artificial Intelligence, Vol. 7: Generalized Nets and Genetic Algorithms. (Prof. M. Drinov Academic Publishing House, Sofia 2013)
27. Atanassov, K.T.: Generalized Nets. World Scientific, Singapore, New Jersey/London (1991)
28. Atanassov, K.T.: Intuitionistic Fuzzy Sets: Theory and Applications. Physica-Verlag, Heidelberg (1999)
29. Pencheva, T., Atanassov, K.T., Shannon, A.G.: Modelling of a roulette wheel selection operator in genetic algorithms using generalized nets. Int. J. Bioautomation **13**(4), 257–264 (2009)
30. McCaughan, J.B.T.: Remove god, lose reason: the sorry cases of hawking and dawkins. Quadrant. **55**(6): 66–71 (2011)
31. Singer, P.: Ethics. Oxford University Press, Oxford (1994)
32. Dummett, M.: Elements of Intuitionism. Clarendon Press, Oxford (1977)
33. Halmos, P.R.: I Want to be a Mathematician. Mathematical Association of America, Washington, DC (1985)
34. Hardy, G.H.: A Mathematician's Apology. (With a Foreword by C.P. Snow). (Cambridge University Press, Cambridge 1967)
35. Iverson, K.E.: Notation as a tool of thought. Commun. Assoc. Comput. Mach. **23**(8), 444–465 (1980)
36. Bucchianico di, A., Loeb, D.E., Rota, G.C.: Umbral Caluclus in Hilbert Space. In: Sagan, B. E., Stanley, R.P. (ed.) Mathematical Essays in Honor of Gian-Carlo Rota, pp. 213–238. (Birkhäuser, Boston 1998)
37. Zadeh, L.: Fuzzy sets. Inf. Control **11**, 338–353 (1965)
38. Sanguineti, J.J.: Logic, pp.194–200. (Sinag-Tala, Manila 1982)

Part II
Intuitionistic Fuzzy Set

On the Atanassov Concept of Fuzziness and One of Its Modification

Beloslav Riečan

Abstract The family $\mathcal{F}$ of intuitionistic fuzzy sets [1–3] is compared with the family $\mathcal{V}$ of interval valued fuzzy sets. Since the spaces are isomorphic, from the measure theory on $\mathcal{F}$ the measure theory on $\mathcal{V}$ can be deduced. In the paper they are mentioned the state representation [7, 8, 44, 50], the inclusion—exclusion property [6, 22, 23] and the existence of invariant state [45].

1 Introduction

Any subset A of a given space Ω can be identified with its characteristic function

$$\chi_A : \Omega \to \{0, 1\}$$

where

$$\chi_A(\omega) = 1,$$

if $\omega \in A$,

$$\chi_A(\omega) = 0,$$

if $\omega \notin A$. From the mathematical point of view a fuzzy set is a natural generalization of χ_A (see [60, 61]). It is a function

$$\varphi_A : \Omega \to [0, 1].$$

B. Riečan (✉)
Faculty of Natural Sciences, Matej Bel University, Tajovsk ého 40,
Bansk á Bystrica, Slovakia
e-mail: Beloslav.Riecan@umb.sk

B. Riečan
Mathematical Institut, Slovak Academy of Sciences, Štef ánikova 49, Bratislava, Slovakia

P. Angelov and S. Sotirov (eds.), *Imprecision and Uncertainty in Information Representation and Processing*, Studies in Fuzziness and Soft Computing 332,
DOI 10.1007/978-3-319-26302-1_3

Evidently any set (i.e. two-valued function on Ω, $\chi_A \to \{0,1\}$) is a special case of a fuzzy set (multi-valued function), $\varphi_A : \Omega \to [0,1]$. There are many possibilities for characterizations of operations with sets (union $A \cup B$ and intersection $A \cap B$). We shall use so called Lukasiewicz characterization [54]:

$$\chi_{A\cup B} = (\chi_A + \chi_B) \wedge 1,$$

$$\chi_{A\cap B} = (\chi_A + \chi_B - 1) \vee 0.$$

(Here $(f \vee g)(\omega) = \max(f(\omega), g(\omega)), (f \wedge g)(\omega) = \min(f(\omega), g(\omega))$.) Hence if φ_A, $\varphi_B : \Omega \to [0,1]$ are fuzzy sets, then the union (disjunction φ_A or φ_B of corresponding assertions) can be defined by the formula

$$\varphi_A \oplus \varphi_B = (\varphi_A + \varphi_B - 1) \wedge 1,$$

the intersection (conjunction φ_A and φ_B of corresponding assertions) can be defined by the formula

$$\varphi_A \odot \varphi_B = (\varphi_A + \varphi_B - 1) \vee 0.$$

In the paper we shall work with the Atanassov generalization of the notion of fuzzy set so-called IF-set (see [2, 3]), what is a pair

$$A = (\mu_A, \nu_A) : \Omega \to [0,1] \times [0,1]$$

of fuzzy sets $\mu_A, \nu_A : \Omega \to [0,1]$, where

$$\mu_A + \mu_A \leq 1.$$

Evidently a fuzzy set $\varphi_A : \Omega \to [0,1]$ can be considered as an IF-set, where

$$\mu_A = \varphi_A : \Omega \to [0,1], \nu_A = 1 - \varphi_A : \Omega \to [0,1].$$

Here we have

$$\mu_A + \nu_A = 1,$$

while generally it can be $\mu_A(\omega) + \nu_A(\omega) < 1$ for some $\omega \in \Omega$. Geometrically an IF-set can be regarded as a function $A : \Omega \to \Delta$ to the triangle

$$\Delta = \{(u,v) \in R^2 : 0 \leq u, 0 \leq v, u + v \leq 1\}.$$

Fuzzy set can be considered as a mapping $\varphi_A : \Omega \to D$ to the segment

$$D = \{(u,v) \in R^2; u + v = 1, 0 \leq u \leq 1\}$$

and the classical set as a mapping $\psi : \Omega \to D_0$ from Ω to two-point set

$$D_0 = \{(0, 1), (1, 0)\}.$$

In the next definition we again use the Lukasiewicz operations.

Definition 1 By an IF subset of a set Ω a pair $A = (\mu_A, \nu_A)$ of functions

$$\mu_A : \Omega \to [0, 1], \nu_A; \Omega \to [0, 1]$$

is considered such that

$$\mu_A + \nu_A \leq 1.$$

We call μ_A the membership function, ν_A the non membership function and

$$A \leq B \Longleftrightarrow \mu_A \leq \mu_B, \nu_A \geq \nu_B.$$

If $A = (\mu_A, \nu_A), B = (\mu_B, \nu_B)$ are two IF-sets, then we define

$$A \oplus B = ((\mu_A + \mu_B) \wedge 1, (\nu_A + \nu_B - 1) \vee 0),$$

$$A \odot B = ((\mu_A + \mu_B - 1) \vee 0, (\nu_A + \nu_B) \wedge 1),$$

$$\neg A = (1 - \mu_A, 1 - \nu_A).$$

Denote by $\mathcal{F}$ a family of IF sets such that

$$A, B \in \mathcal{F} \Longrightarrow A \oplus B \in \mathcal{F}, A \odot B \in \mathcal{F}, \neg A \in \mathcal{F}.$$

Example 1 Let $\mathcal{F}$ be the set of all fuzzy subsets of a set Ω. If $f : \Omega \to [0, 1]$ then we define

$$A = (f, 1 - f),$$

i.e. $\nu_A = 1 - \mu_A$.

Example 2 Let $(\Omega, \mathcal{S})$ be a measurable space [53], $\mathcal{S}$ a σ-algebra, $\mathcal{F}$ the family of all pairs such that $\mu_A : \Omega \to [0, 1], \nu_A : \Omega \to [0, 1]$ are measurable. Then $\mathcal{F}$ is closed under the operations $\oplus, \odot, \neg$.

Example 3 Let $(\Omega, \mathcal{T})$ be a topological space, $\mathcal{F}$ the family of all pairs such that $\mu_A : \Omega \to [0, 1], \nu_A : \Omega \to [0, 1]$ are continuous. Then $\mathcal{F}$ is closed under the operations $\oplus, \odot, \neg$.

Of course, in any case $A \oplus B, A \odot B, \neg A$ are IF-sets, if A, B are IF-sets. E.g.

$$A \oplus B = ((\mu_A + \mu_B) \wedge 1, (\nu_A + \nu_B - 1) \vee 0),$$

hence

$$(\mu_A + \mu_B) \wedge 1 + (\nu_A + \nu_B - 1) \vee 0$$

$$= ((\mu_A + \mu_B) \wedge 1 + (\nu_A + \nu_B - 1)) \vee ((\mu_A + \mu_B) \wedge 1)$$

$$= ((\mu_A + \mu_B + \nu_A + \nu_B - 1) \wedge (1 + \nu_A + \nu_B - 1)) \vee ((\mu_A + \mu_B) \wedge 1)$$

$$\leq ((1 + 1 - 1) \wedge (\nu_A + \nu_B)) \vee ((\mu_A + \mu_B) \wedge 1)$$

$$= (1 \wedge (\nu_A + \nu_B)) \vee ((\mu_A + \mu_B) \wedge 1)$$

$$\leq 1 \vee 1 = 1.$$

2 If Versus IV

If we consider two IF-sets A, B, then B is better than A, if the membership function μ_A is larger then the membership function function μ_B the non-membership function μ_A is smaller then the non-membership function function μ_B.

It is a philosophical background of IF-theory based on some problems inspired by applications [9, 10, 18, 24, 38, 57].

Of course, in the vector space R^2, the usual ordering is given by

$$(x_1, y_1) \leq (x_2, y_2) \Longleftrightarrow (x_1 \leq x_2, y_1 \leq y_2).$$

It leads to so-called interval valued fuzzy sets.

Definition 2 An interval valued fuzzy subset of Ω is a mapping $\bar{A} = (\bar{\mu}_A, \bar{\nu}_A)$ such that $\bar{\mu}_A : \Omega \to [0, 1], \bar{\nu}_A : \Omega \to [0, 1]$ and

$$0 \leq \bar{\mu}_A \leq \bar{\nu}_A \leq 1.$$

If $\bar{A} = (\bar{\mu}_A, \bar{\nu}_A), \bar{B} = (\bar{\mu}_B, \bar{\nu}_B)$ are two IV-sets, then

$$\bar{A} \leq \bar{B} \Longleftrightarrow (\bar{\mu}_A \leq \bar{\mu}_B, \bar{\nu}_A \leq \bar{\nu}_B).$$

If we denote

$$\Delta = \{(u, v) \in R^2 : 0 \leq u, 0 \leq v, u + v \leq 1\},$$

$$\bar{\Delta} = \{(u, v) \in R^2 : 0 \leq u \leq v \leq 1\},$$

then an IF-set is a mapping

$$A : \Omega \to \Delta,$$

and an IV-set is a mapping

$$\bar{A} : \Omega \to \bar{\Delta}.$$

Evidently Δ and $\bar{\Delta}$ are equivalent, e.g.

$$\varphi : \Delta \to \bar{\Delta}, \varphi(u, v) = (u, 1 - v)$$

realize the equivalence. Denote by $\mathcal{F}$ the family of all IF-subsets of Ω, and by $\mathcal{V}$ the family of all IV-subsets of Ω. If $A \in \mathcal{F}, A = (\mu_A, \nu_A)$, and $\bar{A} = \varphi \circ A = (\mu_A, 1 - \nu_A) = (\bar{\mu}_A, \bar{\nu}_A)$, then $\bar{A} = (\bar{\mu}_A, \bar{\nu}_A) \in \mathcal{V}$. Moreover, if

$$\bar{A} = \bar{\varphi}(A) = \varphi \circ A,$$

then

$$\bar{\varphi} : \mathcal{F} \to \mathcal{V}$$

is an equivalence and

$$A \leq B \Longleftrightarrow \bar{A} \preceq \bar{B}.$$

Indeed, $A \leq B$ means $\mu_A \leq \mu_B, \nu_A \geq \nu_B$, hence $1 - \nu_A \leq 1 - \nu_B$, and

$$\bar{A} = (\mu_A, 1 - \nu_A) \preceq \bar{B} = (\mu_B, 1 - \nu_B)$$

in $\mathcal{V}$. Recall that in $\mathcal{F}$

$$(0, 1) \leq A \leq (1, 0)$$

for any $A \in \mathcal{F}$. On the other hand

$$(0, 0) \preceq \bar{A} \preceq (1, 1)$$

for any $\bar{A} \in \mathcal{V}$.

In the measure theory the monotone convergence is important [4, 24, 36, 41, 56, 58, 59]. Then in $\mathcal{F}$

$$A_n \nearrow A \Longleftrightarrow \mu_{A_n} \nearrow \mu_A, \nu_{A_n} \searrow \nu_A.$$

On the other hand in $\mathcal{V}$

$$\bar{A}_n \nearrow \bar{A} \Longleftrightarrow \mu_{A_n} \nearrow \mu_A, \nu_{A_n} \nearrow \nu_A.$$

We have seen that $\mathcal{F}$ and $\mathcal{V}$ are isomorphic as lattices. It is natural to define Lukasziewicz operations on $\mathcal{V}$ by such a way that $\mathcal{F}$ and $\mathcal{V}$ to be isomorph by the help of the isomorphism $\bar{\varphi}$. So

$$\bar{\varphi}(A \oplus B) = \bar{\varphi}((\mu_A + \mu_B) \wedge 1, (\nu_A + \nu_B - 1) \vee 0)$$

$$= ((\mu_A + \mu_B) \wedge 1, 1 - (\nu_A + \nu_B - 1) \vee 0)$$

$$= ((\mu_A + \mu_B) \wedge 1, (1 - \nu_A + 1 - \nu_B) \wedge 1)$$

$$= ((\bar{\mu}_A + \bar{\mu}_B) \wedge 1, (\bar{\nu}_A + \bar{\nu}_B) \wedge 1),$$

$$\bar{\varphi}(A \odot B) = \bar{\varphi}((\mu_A + \mu_B - 1) \vee 0, (\nu_A + \nu_B) \wedge 1)$$

$$= ((\mu_A + \mu_B - 1) \vee 0, 1 - (\nu_A + \nu_B) \wedge 1)$$

$$= ((\mu_A + \mu_B - 1) \vee 0, (1 - \nu_A + 1 - \nu_B - 1) \vee 0|$$

$$= ((\bar{\mu}_A + \bar{\mu}_B - 1) \vee 0, (\bar{\nu}_A + \bar{\nu}_B - 1) \vee 0),$$

hence we shall define the Lukasziewicz operations on $\mathcal{V}$ by the following way.

Definition 3 Let $\bar{A} = (\bar{\mu}_A, \bar{\nu}_A) \in \mathcal{V}, \bar{B} = (\bar{\mu}_B, \bar{\nu}_B) \in \mathcal{V}$. Then

$$\bar{A}\bar{\oplus}\bar{B} + ((\bar{\mu}_A + \bar{\mu}_B) \wedge 1, (\bar{\nu}_A + \bar{\nu}_B) \wedge 1),$$

$$\bar{A}\bar{\odot}\bar{B} = ((\bar{\mu}_A + \bar{\mu}_B - 1) \vee 0, (\bar{\nu}_A + \bar{\nu}_B - 1) \vee 0).$$

Evidently the following proposition holds.

Proposition 1 *If* $A, B \in \mathcal{F}$, *then*

$$\bar{\varphi}(A \oplus B) = \bar{\varphi}(A)\bar{\oplus}\bar{\varphi}(B),$$

$$\bar{\varphi}(A \odot B) = \bar{\varphi}(A)\bar{\odot}\bar{\varphi}(B).$$

Remark 1 If for real numbers $a, b \in R$ we denote $a \oplus b = (a + b) \wedge 1, a \odot b = (a + b - 1) \vee 0$, then for $A, B \in \mathcal{F}$ we have

$$A \oplus B = (\mu_A \oplus \mu_B, \nu_A \odot \nu_B),$$

$$A \odot B = (\mu_A \odot \mu_B, \nu_A \oplus \nu_B),$$

and for $\bar{A}, \bar{B} \in \mathcal{V}$ we obtain

$$\bar{A}\bar{\oplus}\bar{B} = (\bar{\mu}_A\bar{\oplus}\bar{\mu}_B, \bar{\nu}_A\bar{\oplus}\bar{\nu}_B),$$

$$\bar{A}\bar{\odot}\bar{B} = (\bar{\mu}_A\bar{\odot}\bar{\mu}_B, \bar{\nu}_A\bar{\odot}\bar{\nu}_B).$$

Similarly states on $\mathcal{F}$ and $\mathcal{V}$ resp. can be defined in a convenience [11, 13–15, 21, 31, 34, 40, 42, 43]. Of course we shall consider only functions measurable with respect to a σ-algebra $\mathcal{S}$ of subsets of Ω.

Definition 4 A mapping $m : \mathcal{F} \to [0, 1]$ is called a state, if the following properties are satisfied:
(1.1) $m((0, 1)) = 0, m((1, 0)) = 1,$
(1.2) $A \odot B = (0, 1) \Longrightarrow m(A \oplus B) = m(A) + m(B),$
(1.3) $A_n \nearrow A \Longrightarrow m(A_n) \nearrow m(A).$

Definition 5 A mapping $\bar{m} : \mathcal{V} \to [0, 1]$ is called a state, if the following properties are satisfied:
(2.1) $\bar{m}((0, 0)) = 0, \bar{m}((1, 1)) = 1,$
(2.2) $\bar{A}\bar{\odot}\bar{B} = (0, 1) \Longrightarrow \bar{m}(\bar{A}\bar{\oplus}\bar{B}) = \bar{m}(\bar{A}) + \bar{m}(B),$
(2.3) $\bar{A}_n \nearrow \bar{A} \Longrightarrow \bar{m}(\bar{A}_n) \nearrow \bar{m}(\bar{A}).$

Theorem 1 *Let $\bar{m} : \mathcal{V} \to [0, 1]$ be a state. Define $m : \mathcal{F} \to [0, 1]$ by the formula*

$$m(A) = \bar{m}(\bar{\varphi}(A).$$

Then m is a state.

Proof Prove first (1.1). By (2.1)

$$m((1,0)) = \bar{m}(\bar{\varphi}(1,0)) = \bar{m}((1,1) = 1,$$

$$m((0, 1)) = \bar{m}(\bar{\varphi}((0, 1)) = \bar{m}((0, 0)) = 0.$$

Further let $A, B \in \mathcal{F}, A \odot B = (0, 1)$. Then

$$\bar{\varphi}(A)\bar{\odot}\bar{\varphi}(B) = \bar{\varphi}(A \odot B) = \bar{\varphi}((0, 1)) = (0, 0).$$

Therefore by (2.2)

$$\begin{aligned}\bar{m}(\bar{\varphi}(A)\bar{\oplus}\bar{\varphi}(B)) &= \bar{m}(\bar{\varphi}(A)) + \bar{m}(\bar{\varphi}(B)) \\ &= m(A) + m(B).\end{aligned}$$

Of course,

$$\bar{m}(\bar{\varphi}(A)\bar{\oplus}\bar{\varphi}(B)) = \bar{m}(\bar{\varphi}(A \oplus B)) = m(A \oplus B),$$

hence (1.2) is proved.
Finally, let $A_n \in \mathcal{F}, A \in \mathcal{F}, A_n \nearrow A$. Then

$$\bar{\varphi}(A_n) \in \mathcal{V}, \bar{\varphi}(A) \in \mathcal{V}, \bar{\varphi}(A_n) \nearrow \bar{\varphi}(A),$$

and therefore by (2.2)

$$m(A_n) = \bar{m}(\bar{\varphi}(A_n) \nearrow \bar{m}(\bar{\varphi}(A) = m(A),$$

hence (1.3) holds. □

Theorem 2 *Let* $m : \mathcal{F} \to [0,1]$ *be a state. Define* $\bar{m} : \mathcal{V} \to [0,1]$ *by the formula*

$$\bar{m}(\bar{A}) = m(\bar{\varphi}^{-1}(\bar{A})).$$

Then $\bar{m}$ *is a state.*

Proof By (1.1) and the definition

$$\bar{m}((0,0)) = m(\bar{\varphi}^{-1}(0,0)) = m((0,1)) = 0,$$

$$\bar{m}((1,1)) = m(\bar{\varphi}^{-1}(1,1)) = m((1,0)) = 1,$$

hence (2.1) holds.
Now let $\bar{A}, \bar{B} \in \mathcal{V}, \bar{A}\bar{\odot}\bar{B} = (0,1)$. Then $A \odot B = \bar{\varphi}^{-1}(\bar{A}) \odot \bar{\varphi}^{-1}(\bar{B}) = \bar{\varphi}^{-1}(\bar{A}\bar{\odot}\bar{B}) = \bar{\varphi}^{-1}((0,0)) = (0,1)$. Therefore

$$m(A \oplus B) = m(A) + m(B).$$

But

$$\bar{m}(\bar{A}) = m(\bar{\varphi}^{-1}(\bar{A}))) = m(A), \bar{m}(\varphi^{-1}(\bar{B})) = m(B),$$

$$\bar{m}(\bar{A}\bar{\oplus}\bar{B}) = \bar{m}(\bar{\varphi}^{-1}(\bar{A} \oplus \bar{B}) = m(A \oplus B),$$

hence

$$\bar{m}(\bar{A}\bar{\oplus}\bar{B}) = \bar{m}(\bar{A}) + \bar{m}(\bar{B}),$$

and (2.2) holds.
Finally let $\bar{A}_n \nearrow \bar{A}$. Then $A_n = \bar{\varphi}^{-1}(\bar{A}_n) \nearrow \bar{\varphi}^{-1}(\bar{A}) = A$, hence

$$\bar{m}(\bar{A}_n) = m(A_n) \nearrow m(A) = \bar{m}(\bar{A}).$$

□

3 State Representation

One of he main result of the IF-probability theory is the state representation theorem [7, 8, 25, 38, 44, 46, 47, 50]. Since IF-probability theory and IV-probability theory are isomorphic, also IV-state representation theorem holds.

Recall that a state $\bar{m} : \mathcal{V} \to [0, 1]$ is considered with respect to the family $\mathcal{V} = \{\bar{A} = (\bar{\mu}_A, \bar{\nu}_B); \bar{\mu}_A, \bar{\nu}_A : \Omega \to [0, 1]$ are measurable with respect to a σ-algebra $\mathcal{S}$ of subsets of $\Omega, \mu_A \leq \nu_A\}$.

Theorem 3 *Let $\bar{m} : \mathcal{V} \to [0, 1]$ be a state, $\alpha \in [0, 1]$. Then there exist probability measures $P, Q : \mathcal{S} \to [0, 1]$ such that*

$$\bar{m}(\bar{A}) = \int_\Omega \bar{\mu}_A dP + \alpha \int_\Omega (\bar{\nu}_A - \bar{\mu}_A) dQ,$$

and $\alpha = \bar{m}((0, 1))$.

Proof Construct the state $m : \mathcal{F} \to [0, 1]$ by Theorem 1. Then by [7, 8, 44, 50] there exist probability measures $P, Q : \mathcal{S} \to [0, 1]$ such that for any $A = (\mu_A, \nu_A) \in \mathcal{F}$ there holds

$$m(A) = \int_\Omega \mu_A dP + \alpha(1 - \int_\Omega (\mu_A + \nu_A) dQ.$$

By Theorem 2

$$\bar{m}(\bar{A}) = m(\varphi^{-1}(\bar{A})) = m((\mu_A, 1 - \nu_A))$$

$$= \int_\Omega \mu_A dP + \alpha(\int_\Omega (1 - \nu_A - \mu_A) dQ$$

$$= \int_\Omega \bar{\mu}_A dP + \alpha \int_\Omega (\bar{\nu}_A - \bar{\mu}_A) dQ.$$

Moreover $\alpha = m((0, 0)) = \bar{m}(\bar{\varphi}((0, 0))) = \bar{m}((0, 1))$. □

Remark 2 The representation Theorem 3 has been presented by Skřivánek in [56].

4 Inclusion–Exclusion

Inclusion–exclusion principle holds e.g. for any probability measure $P : \mathcal{S} \to [0, 1]$ defined on a σ-algebra $\mathcal{S}$:

$$P(A \cup B) = P(A) + P(B) - P(A \cap B),$$

$$P(A \cup B \cup C) = P(A) + P(B) + P(C)$$

$$-P(A \cap B) - PA \cap B) - P(B \cap C) + P(A \cap B \cap C),$$

etc. We shall present the validity of IF-principle in IV-theory for the Godel operations.

Definition 6 If $\bar{A}, \bar{B} \in \mathcal{V}$ then we define

$$\bar{A} \cup \bar{B} = (\bar{\mu}_A \cup \bar{\nu}_B, \bar{\nu}_A \cup \bar{\nu}_B)$$

$$\bar{A} \cap \bar{B} = (\bar{\mu}_A \cap \bar{\nu}_B, \bar{\nu}_A \cap \bar{\nu}_B)$$

where $f \cup g = \max(f, g), f \cap g = \min(f, g)$.

Theorem 4 *Let $\bar{m} : \mathcal{V} \to [0, 1]$ be a state, $\bar{A}_1, \bar{A}_2, \dots, \bar{A}_n \in \mathcal{V}$. Then*

$$\bar{m}(\bigcup_{i=1}^{n} \bar{A}_i) = \Sigma_{i=1}^{n} \bar{m}(\bar{A}_i) - \Sigma_{i \neq j} \bar{m}(\bar{A}_i \cap \bar{A}_j) + \cdots + (-1)^{(n+1)} \bar{m}(\bar{A}_1 \cap \bar{A}_2 \cap \cdots \cap \bar{A}_n).$$

Proof Again we use the construction of the state $m : \mathcal{F} \to [0, 1]$ presented in Theorem 1:

$$m(A) = \bar{m}(\bar{\varphi}(A)).$$

Define the operations $\vee, \wedge$ in $\mathcal{F}$ by the following way

$$A \vee B = (\mu_A \vee \mu_B, \nu_A \wedge \nu_B),$$

$$A \wedge B = (\mu_A \wedge \mu_B, \nu_A \vee \nu_B).$$

Then

$$\bar{\varphi}(A \vee B) = (\mu_A \vee \mu_B, 1 - \nu_A \wedge \nu_B)$$

$$= (\mu_A \vee \mu_B, (1 - \nu_A) \vee (1 - \nu_B)$$

$$= (\bar{\mu}_A \vee \bar{\mu}_B, \bar{\nu}_A \vee \nu_B) = \bar{A} \cup \bar{B},$$

$$\bar{\varphi}(A \wedge B) = (\mu_A \wedge \mu_B, 1 - \nu_A \vee \nu_B)$$

$$= (\mu_A \wedge \mu_B, (1 - \nu_A) \wedge (1 - \nu_B)$$

$$= (\bar{\mu}_A \wedge \bar{\mu}_B, \bar{\nu}_A \wedge \nu_B) = \bar{A} \cap \bar{B}.$$

By [6, 22, 23] we have

$$m(\bigvee_{k=1}^{n} A_i) = \Sigma_{k=1}^{n} \Sigma_{i_1 < \cdots < i_k} (-1)^{(k+1)} m(A_{i_1} \wedge \cdots \wedge A_{i_k}).$$

Therefore

$$\bar{m}(\bigcup_{i=1}^{n} \bar{A}_i) = m(\bigvee_{k=1}^{n} A_i) = \Sigma_{k=1}^{n} \Sigma_{i_1<\cdots<i_k} (-1)^{(k+1)} m(A_{i_1} \wedge \cdots \wedge A_{i_k})$$

$$= \Sigma_{k=1}^{n} \Sigma_{i_1<\cdots<i_k} (-1)^{(k+1)} \bar{m}(\bar{A}_{i_1} \cap \cdots \cap \bar{A}_{i_k}).$$

□

5 Invariant States

There is well known story about Haar measure [53], such measure μ in an Abelian group $(G, +)$ that $\mu(A + a) = \mu(A)$ for any $a \in G$ and any measurable A.

Assume that Ω is a compact Abelian group, and S is the σ-algebra generated by the family of all compact subsets of Ω. Further let $\mathcal{V}$ consist of all IV-sets $\bar{A} = (\bar{\mu}_A, \bar{\nu}_A)$ with continuous $\bar{\mu}_A, \bar{\nu}_A : \Omega \to [0, 1]$.
For a given element $a \in \Omega$ define a mapping $\bar{\tau}_a : \mathcal{V} \to \mathcal{V}$ by the formula

$$\bar{\tau}_a(\bar{A}) = (\bar{\mu}_{\bar{A}} \circ T, \bar{\nu}_{\bar{A}} \circ T),$$

where $T : \Omega \to \Omega, T(x) = x + a$.

Theorem 5 *There exists exactly one state* $\bar{m} : \mathcal{V} \to [0, 1]$ *such that*

$$\bar{m}(\bar{\tau}_a(\bar{A})) = \bar{m}(\bar{A})$$

for any $\bar{A} \in \mathcal{V}$*, and any* $a \in \Omega$*.*

Proof Define $\tau_a : \mathcal{F} \to \mathcal{F}$ by the formula $\tau_a(A) = (\mu_a \circ T, \nu_a \circ T)$. By [45] there exists exactly one state $m : \mathcal{F} \to [0, 1]$ such that

$$m(\tau_a(A)) = m(A)$$

for any $A \in \mathcal{F}$, and any $a \in \Omega$. Define $\bar{m} : \mathcal{V} \to [0, 1]$ by the equality $\bar{m}(\bar{A}) = m(\bar{\varphi}^{-1}(A))$. If $\bar{A} \in \mathcal{V}$, then

$$\bar{m}(\bar{\tau}_a(\bar{A})) = m(\bar{\varphi}^{-1}(\bar{\tau}_a(\bar{A})))$$

$$= m(\tau_a(A)) = m(A) = \bar{m}(\bar{\varphi}(A)) = \bar{m}(\bar{A}),$$

hence $\bar{m}$ is invariant. Let $\bar{\kappa} : \mathcal{V} \to [0, 1]$ be another invariant state, i.e.

$$\bar{\kappa}(\bar{\tau}_a(\bar{A})) = \bar{\kappa}(\bar{A}).$$

Define $\kappa : \mathcal{F} \to [0, 1]$ by the equality $\kappa(A) = \bar{\kappa}(\bar{\varphi}(A))$. Then

$$\kappa(\tau_a(A)) = \bar{\kappa}(\bar{\varphi}(\tau_a(A))) = \bar{\kappa}(\bar{\tau}_a(\bar{\varphi}(A)))$$
$$= \bar{\kappa}(\bar{\varphi}(A)) = \kappa(A),$$

for any $A \in \mathcal{F}$, and any $a \in \Omega$. Since there exists exactly one invariant state $m : \mathcal{F} \to [0, 1]$, we obtain that $\kappa = m$. Therefore

$$\bar{\kappa}(\bar{A}) = \kappa(\bar{\varphi}^{-1}(\bar{A})) = m(\bar{\varphi}^{-1}(\bar{A})) = \bar{m}(\bar{A})$$

for any $\bar{A} \in \mathcal{V}$. □

6 Conclusion

We have seen that IF-probability theory and IV-probability theory are isomorphs. Therefore all results of IF-probability theory can be reformulated by the help of notions of interval valued notions. As an illustration we have presented three important results: state representation theorem, inclusion exclusion property and Haar measure theorem. Of course, also all results of IV-measure theory can be translated to the IF-measure theory [12, 17, 20, 29, 30, 51, 52].

Acknowledgments The support of the grant VEGA 1/0120/14 is kindly announced.

References

1. Atanassov K.T., Riečan, B.: On two new types of probability on IF-events. Notes on IFS (2007)
2. Atanassov, K.T.: Intuitionistic Fuzzy Sets: Theory and Applications. Studies in Fuzziness and Soft Computing. PhysicaVerlag, Heidelberg (1999)
3. Atanassov, K.T.: On Intuitionistic Fuzzy Sets. Springer, Berlin (2012)
4. Chovanec, F.: Difference Posets and their Graphical Representation (in Slovak). Liptovsk y Mikuláš (2014)
5. Cignoli, L., D'Ottaviano, M., Mundici, D.: Algebraic Foundations of Many-valued Reasoning. Kluwer, Dordrecht (2000)
6. Ciungu, L., Kelemenová, J., Riečan, B.: A new point of view to the inclusion exclusion principle. In: 6th International Conferences on Intelligent Systems IS'12, pp. 142 - 144. Varna, Bulgaria (2012)
7. Ciungu, L., Riečan, B.: General form of probabilities on IF-sets. In: Fuzzy Logic and Applications. Proceedings of WILF Palermo, pp. 101–107 (2009)
8. Ciungu, L., Riečan, B.: Representation theorem for probabilities on IFS-events. Inf. Sci. **180**, 793–798 (2010)
9. De, S.K., Biswas, R., Roy, A.R.: An application of intuitionistic fuzzy sets in medical diagnosis. Fuzzy Sets Syst. **117**, 209–213 (2001)
10. Dvurečenskij, A., Pulmannová, S.: New Trends in Quantum Structures. Kluwer, Dordrecht (2000)

11. Dvurečenskij, A., Rachunek, J.: Riečan and Bosbach states for bounded non-commutative RI-monoids. Math. Slovaca **56**, 487–500 (2006)
12. Foulis, D., Bennett, M.: Eect algebras and unsharp quantum logics. Found. Phys. **24**, 1325–1346 (1994)
13. Georgescu, G.: Bosbach states on fuzzy structures. Soft Comput. **8**, 217–230 (2004)
14. Gerstenkorn, T., Manko, J.: Probabilities of intuitionistic fuzzy events. In: Hryniewicz, O., et al. (eds.) Issues in Intelligent Systems: Paradigms, pp. 58–68 (2005). Intuitionistic fuzzy probability theory 45
15. Grzegorzewski, P., Mrowka, E.: Probability of intuistionistic fuzzy events. In: Grzegorzewski, P., et al. (eds.) Soft Metods in Probability, Statistics and Data Analysis, pp. 105-115 (2002)
16. Hájek, P.: Metamathematics of Fuzzy Logic. Kluwer, Dordrecht (1998)
17. Hanesová, R.: Conditional probability on the family of IF-events with product. In: Atanassov , K.T., et al. (eds.) New Developments Fuzzy Sets, Intuitionistic Fuzzy Sets, Generalized Nets and Related Topics, vol. I: Foundations, pp. 93–98. Warsaw (2012)
18. Jurečková, M.: The addition to ergodic theorem on probability MV-algebras with product. Soft Comput. **7**, 105–115 (2003)
19. Klement, E., Mesiar, R., Pap, E.: Triangular Norms. Kluwer, Dordrecht (2000)
20. Kopka, F., Chovanec, F.: D-posets. Math. Slovaca **44**, 21–34 (1994)
21. Krachounov, M.: Intuitionistic probability and intuitionistic fuzzy sets. In: El-Darzi et al. (eds.) First International Workshop on IFS, pp. 714–717 (2006)
22. Kuková, M.: The inclusion exclusion principle for L-states on IF-events. Inf. Sci. **224**, 165–169 (2013)
23. Kuková, M., Navara, M.: Principles of inclusion and exclusion for fuzzy sets. Fuzzy Sets Syst. **232**, 98–109 (2013)
24. Lašová, L.: The individual ergodic theorem on IF-events. In: Atanassov, K.T., et al. (eds.) New Developments Fuzzy Sets, Intuitionistic Fuzzy Sets, Generalized Nets and Related Topics, vol. I: Foundations, pp. 131–140. Warsaw (2010)
25. Lendelová, K., Petrovičová, J.: Representation of IF-probability for MV-algebras. Soft Comput. **10** 564–566 (2006)
26. Lendelová, K., Riečan, B.: Strong law of large numbers for IF-events. In: Proceedings of the Eleventh International Conference IPMU 2006, Paris, France, pp. 2363-2366, 2–7 July 2006
27. Lendelová, K., Riečan, B.: Weak law of large numbers for IF-events. In: De Baets, B., et al. (eds.) Current Issues in Data and Knowledge Engineering, pp. 309–314 (2004)
28. Lendelová, K.,Michalíková A.: Probability on a Lattice L1. Proceedings East West Fuzzy Colloquium: 12th Zittau Fuzzy Colloquium, Zittau, Germany, **79–83**, 21–23 September 2005
29. Lendelová, K.: Conditional IF-probability. In: Advances in Soft Computing: Soft Methods for Integrated Uncertainty Modelling, pp. 275-283 (2006)
30. Lendelová, K.: Convergence of IF-observables. In: Issues in the Representation and Processing of Uncertain and Imprecise Information—Fuzzy Sets, Intuitionistic Fuzzy Sets, Generalized nets, and Related Topics, EXIT, Warszawa, pp. 232–240 (2005)
31. Lendelová, K.: IF-probability on MV-algebras. Notes Intuition. Fuzzy Sets **11**, 66–72 (2005)
32. Lendelová, K.: A note on invariant observables. Int. J. Theor. Phys. **45**, 915–923 (2006)
33. Mazureková, P., Riečan, B.: A measure extension theorem. Notes IFS **12**, 3–8 (2006)
34. Michalíková, A.: The probability on square. J. Electr. Eng. 12/S **56**, 21-22 (2005)
35. Michalíková, A.: IFS valued possibility and necessity measures. Notes Intuition. Fuzzy Sets **11**(3), 66–72 (2005)
36. Michalíková, A.: Caratheodory outer measure on IF-sets. Mathematica Slovaca **58**(1), 63–76 (2008)
37. Michalíková, A.: Absolute value and limit of the function defined on IF sets. Notes Intuition. Fuzzy Sets **18**, 8–15 (2012)
38. Montagna, F.: An algebraic approach to propositional fuzzy logic. J. Log. Lang. Inf.**9**, 91–124 (2000). Mundici, D. et al. (eds.), Special issue on Logics of Uncertainty
39. Renčová, M.: A generalization of probability theory on MV-algebras to IF-events. Fuzzy Sets and Syst. **161**, 1726–1739 (2010)

40. Renčová M. and Riečan, B.: Probability on IF-sets: an elementary approach. In: First International Workshop on IFS, Generalized Nets and Knowledge Engineering, pp. 8–17 (2006)
41. Riečan, B., Lašová, L.: On the probability on the Kopka D-posets. In: Atanassov, K. et al. (eds.) Developments in Fuzzy Sets, Intuitionistic Fuzzy Sets, Generalized Nets and related Topics I, Warsaw, pp. 167–176 (2010)
42. Riečan, B.: A descriptive definition of the probability on intuitionistic fuzzy sets. In: Wagenecht, M., Hampet, R. (eds.) EUSFLAT '2003, pp. 263–266 (2003)
43. Riečan, B. : On the probability and random variables on IF events. In: Ruan, D., et al. (eds.) Proceedings of the 7th FLINS Conference Genova Applied Artificial Intelligence, pp. 138–145 (2006)
44. Riečan B.: Analysis of Fuzzy Logic Models. In: Koleshko, V.M. (ed.) Intelligent Systems INTECH 2012, pp. 219–244 (2012)
45. Riečan B.: On invariant measures on fuzzy sets. In: Klement, E.P. et al. (eds.) Twelth International Conference on Fuzzy Set Theory FSTA 2014. Liptovský Ján, p. 99 (2014)
46. Riečan, B.: On the product MV-algebras. Tatra Mt. Math. Publ. **16**, 143–149 (1999)
47. Riečan, B.: Representation of probabilities on IFS events. In: Lopez-Diaz, M.C., et al. (eds.) Soft Methodology and Random Information Systems, pp. 243–248. Springer, Berlin (2004)
48. Riečan, B.: Kolmogorov—Sinaj entropy on MV-algebras. Int. J. Theor. Phys. **44**, 1041–1052 (2005)
49. Riečan, B.: On the probability on IF-sets and MV-algebras. Notes IFS **11**, 21–25 (2005)
50. Riečan, B.: On a problem of Radko Mesiar: general form of IF-probabilities. Fuzzy Sets Syst. **152**, 1485–1490 (2006)
51. Riečan, B.: Probability theory on intuitionistic fuzzy events. In: Aguzzoli, S., et al. (eds.) Algebraic and Proof theoretic Aspects of Non-classical Logic Papers in honour of Daniele Mundici's 60th birthday. Lecture Notes in Computer Science, pp. 290–308. Springer, Berlin (2007)
52. Riečan, B., Mundici, D.: Probability in MV-algebras. In: Pap, E. (ed.) Handbook of Measure Theory II, pp. 869–910. Elsevier, Heidelberg (2002)
53. Riečan, B., Neubrunn, T.: Integral Measure and Ordering. Kluwer, Dordrecht (1997)
54. Riečan, B., Petrovičová, J.: On the Lukasiewicz probability theory on IF-sets. Tatra Mt. Math. Publ. **46**, 125–146 (2010)
55. Samuelčík, K., Hollá, I.: Central limit theorem on IF-events. In: Atanassov, K.T. et al. (eds.) Recent Advances in Fuzzy Sets, Intuitionistic Fuzzy Sets, Generalized Nets and Related Topics, vol. I: Foundations, Warsaw, pp. 187–196 (2011)
56. Skřivánek, V.: States on IF events. In: Klement, E.P. et al. (eds.) Twelth International Conference on Fuzzy Set Theory FSTA 2014, Liptovský Ján, p. 102 (2014)
57. Szmidt, E., Kacprzyk, J.: Intuitionistic fuzzy sets in some medical applications. Notes IFS 7, **4**, 58–64 (2001)
58. Valenčáková, V.: A note on the conditional probability on IF-events. Math. Slovaca **59**, 251–260 (2009)
59. Vrábelová, M.: On the conditional probability in product MV-algebras. Soft Comput. **4**, 58–61 (2000)
60. Zadeh, L.A.: Fuzzy sets. Inf. Control **8**, 338–358 (1965)
61. Zadeh, L.A.: Probability measures on fuzzy sets. J. Math. Abal. Appl. **23**, 421–427 (1968)

Intuitionistic Fuzzy Inclusion Indicator of Intuitionistic Fuzzy Sets

Evgeniy Marinov, Radoslav Tsvetkov and Peter Vassilev

Abstract In this paper we introduce a measure for inclusion of two IFSs into each other according to the two main partial orderings in the family of IFSs. This inclusion measure will be observed on few levels. From a set-theoretical point of view, intuitionistic fuzzy point of view and ordinary fuzzy point of view. We also employ the notion of the two modal quasi-orderings, the *necessity* and *possibility*, known for intuitionistic fuzzy sets. All of these inclusion measures can be applied in real world models where intuitionistic fuzzy sets are employed.

1 Introduction to Intuitionistic Fuzzy Sets and their Orderings

A fuzzy set (FS) in X (cf. Zadeh [10]) is given by

$$A' = \{\langle x, \mu_{A'}(x)\rangle | x \in X\} \tag{1}$$

where $\mu_{A'}(x) \in [0, 1]$ is the *membership function* of the fuzzy set A'. As opposed to the Zadeh's fuzzy set, Atanassov (cf. [1, 2]) extended its definition to an intuitionistic fuzzy set (abbreviated IFS) A, given by

$$A = \{\langle x, \mu_A(x), \nu_A(x)\rangle | x \in X\} \tag{2}$$

E. Marinov (✉) · P. Vassilev
Institute of Biophysics and Biomedical Engineering, Bulgarian Academy of Sciences, Acad. G. Bonchev Str., Bl. 105, 1113 Sofia, Bulgaria
e-mail: evgeniy.marinov@biomed.bas.bg

P. Vassilev
e-mail: peter.vassilev@gmail.com

R. Tsvetkov
Faculty of Applied Mathematics and Informatics, Technical University-Sofia, 8 St.Kliment Ohridski Boulevard, Bl. 2, 1756 Sofia, Bulgaria
e-mail: radotzv8@gmail.com

P. Angelov and S. Sotirov (eds.), *Imprecision and Uncertainty in Information Representation and Processing*, Studies in Fuzziness and Soft Computing 332,
DOI 10.1007/978-3-319-26302-1_4

where: $\mu_A : X \to [0,1]$ and $\nu_A : X \to [0,1]$ such that

$$0 \le \mu_A(x) + \nu_A(x) \le 1 \tag{3}$$

and $\mu_A(x)$, $\nu_A(x) \in [0,1]$ denote a *degree of membership* and a *degree of non-membership* of $x \in A$, respectively. An additional concept for each IFS in X, that is an obvious result of (2) and (3), is called

$$\pi_A(x) = 1 - \mu_A(x) - \nu_A(x) \tag{4}$$

a *degree of uncertainty* or also *hesitancy degree* of $x \in A$. It expresses a lack of knowledge of whether x belongs to A or not (cf. Atanassov [1]). It is obvious that $0 \le \pi_A(x) \le 1$, for each $x \in X$. Uncertainty degree turn out to be relevant for both applications and the development of theory of IFSs. For instance, distances between IFSs are calculated in the literature in two ways, using two parameters only (cf. Atanassov [1]) or all three parameters (cf. Szmidt and Kacprzyk [9]). For a detailed discussion about distances and similarities for IFSs one may consult Szmidt [8].

For more detailed information regarding modal operators the reader may refer to [2], Chap. 4.1. "Necessity" and "possibility" operators (denoted $\square$ and $\Diamond$, respectively) applied on an intuitionistic fuzzy set $A \in \text{IFS}(X)$ have been defined as:

$$\begin{aligned} \square A &= \{\langle x, \mu_A(x), \quad 1 - \mu_A(x)\rangle | x \in X\} \\ \Diamond A &= \{\langle x, 1 - \nu(x), \nu_A(x) \quad \rangle | x \in X\} \end{aligned}$$

From the above definition it is evident that

$$\star : \text{IFS}(X) \longrightarrow \text{FS}(X) \tag{5}$$

where $\star$ is the prefix operator $\star \in \{\square, \Diamond\}$, operating on the class of intuitionistic fuzzy sets and $\text{FS}(X)$ denotes the class of fuzzy sets defined over X.

Talking about partial ordering on IFSs, we will by default mean $(\text{IFS}(X), \le)$ where $\le$ stands for the standard partial ordering in IFS(X). That is, for any two A and $B \in \text{IFS}(X)$: $A \le B$ is satisfied if and only if $\mu_A(x) \le \mu_B(x)$ and $\nu_A(x) \ge \nu_B(x)$ for any $x \in X$. On Fig. 1 one may see the triangular representation of the two chosen A and B in a particular point $x \in X$, where $f_A(x)$ stands for the point on the plane with coordinates $(\mu_A(x), \nu_A(x))$. That is, $A \le B$ means exactly that the point $f_B(x)$ must lie in the trapezoidal area (or on its border) defined by the points $\langle \mu_A(x), 0\rangle$, $\langle 1, 0\rangle$, $f_{\Diamond A}(x)$, $f_A(x)$. On the other hand, $B \le A$ is satisfied exactly when point $f_B(x)$ lies in the trapezoidal figure (or on its border) enclosed by the points $f_A(x)$, $f_{\square A}(x)$, $\langle 0, 1\rangle$, $\langle 0, \nu_A(x)\rangle$.

The reader is referred to Marinov [6], where a new partial ordering over the class of IFSs has been introduced. Namely, the so called *π-ordering*, which is actually only a left lattice but not a right one, i.e. it is not a complete lattice. The notion of π-ordering has been employed for the introduction of an *index of indeterminacy* measuring how far (close) is an IFS from (to) the family of the ordinary FSs on the same universe X. A few examples of index of indeterminacy have been introduced

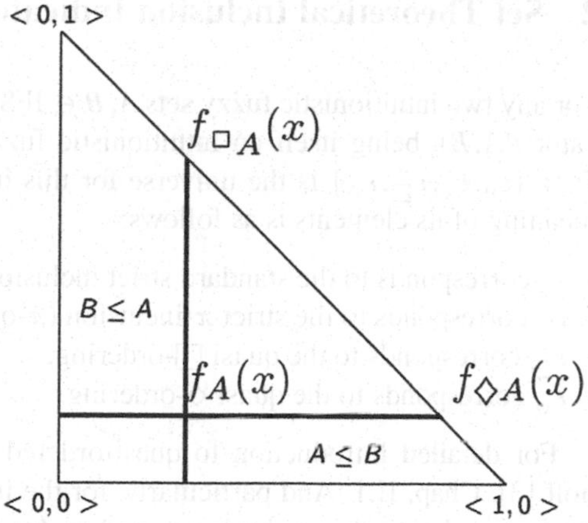

Fig. 1 Triangular representation of the the intuitionistic fuzzy sets A and $B \in$ IFS(X) in a particular point $x \in X$, where $f_A(x)$ stands for the point on the plane with coordinates $(\mu_A(x), \nu_A(x))$. $\Box A$ and $\Diamond A$ stand for the two modal operators "necessity" and "possibility" acting on A

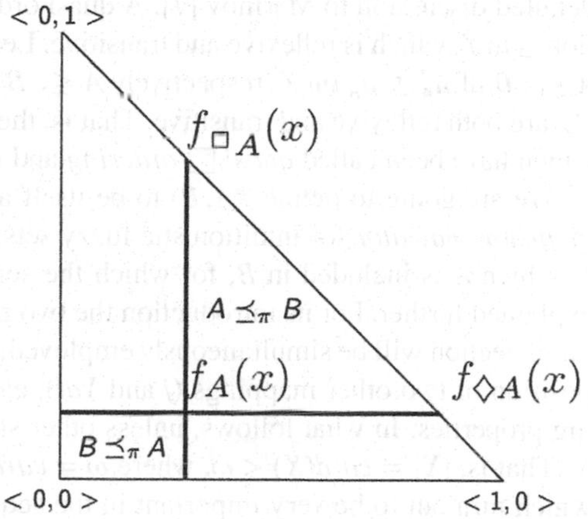

Fig. 2 Triangular representation of the the intuitionistic fuzzy sets A and $B \in$ IFS(X) in a particular point $x \in X$, where $f_A(x)$ stands for the point on the plane with coordinates $(\mu_A(x), \nu_A(x))$. $\Box A$ and $\Diamond A$ stand for the two modal operators "necessity" and "possibility" acting on A

based on the structure and properties of the underlying universe. It has to satisfy three corresponding axioms and should not be confused with the degree of uncertainty called also index of indeterminacy by some authors. In contrast to the standard partial ordering $A \leq B$ between two IFSs A and B, the π ordering $A \preceq_\pi B$ is satisfied iff $\mu_A(x) \leq \mu_B(x)$ and $\nu_A(x) \leq \nu_B(x)$ for all $x \in X$. The triangular representation on Fig. 2 gives us that $A \preceq_\pi B$ iff for all $x \in X$, $f_B(x)$ lies within (or on the border of) the triangular area defined by the points $f_A(x)$,$f_{\Diamond A}(x)$ and $f_{\Box A}(x)$. On the other hand, $B \preceq_\pi A$ is satisfied iff for all $x \in X$, $f_B(x)$ lies within (or on the border of) the rectangular area defined by the points $\langle 0, 0\rangle$, $\langle \mu_A(x), 0\rangle$,$f_A(x)$ and $\langle 0, \nu_A(x)\rangle$. Moreover, the family of maximal elements of (IFS(X), $\preceq_\pi$) consists exactly of the family of ordinary fuzzy sets $FS(X)$ and there is a unique minimal element $0_\pi := \langle 0, 0\rangle$, see Fig. 2.

2 Set Theoretical Inclusion Indicator *j*

For any two intuitionistic fuzzy sets $A, B \in \mathrm{IFS}(X)$ we now define an inclusion indicator $\iota(A, B)$, being itself an intuitionistic fuzzy set, belonging to $\mathrm{IFS}(U_\iota)$, where $U_\iota = \{\varepsilon_0, \varepsilon_\pi, \varepsilon_\Box, \varepsilon_\Diamond\}$ is the universe for this indicator. In the definition of U_ι, the meaning of its elements is as follows:

- ε_0 corresponds to the standard strict inclusion (ordering),
- ε_π corresponds to the strict π-inclusion (π-ordering),
- $\varepsilon_\Box$ corresponds to the quasi $\Box$-ordering,
- $\varepsilon_\Diamond$ corresponds to the quasi $\Diamond$-ordering.

For detailed introduction to quasi-ordered sets the reader could consult Birkhoff [3], Chap. II.1. And particularly, for the introduction of quasi $\Box$-ordering and quasi $\Diamond$-ordering the reader is referred to Atanassov [2], Chap. 4.1 and for a more detailed discussion to Marinov [7]. A quasi ordering is, by definition, a binary relation $\le$ in Y, which is reflexive and transitive. Let us take any $A, B \in \mathrm{IFS}(X)$ and write $A \le_\Box B$ iff $\mu_A \le \mu_B$ on X, respectively $A \le_\Diamond B$ iff $\nu_A \ge \nu_B$ on X. Obviously, $\le_\Box$ and $\le_\Diamond$ are both reflexive and transitive. That is, they are both quasi-orderings in IFS(X) which have been called *quasi* $\Box$*-ordering* and *quasi* $\Diamond$*-ordering*, respectively.

We are going to define $\iota(A, B)$ to be itself an IFS and that is why we call it *IF-inclusion indicator* for intuitionistic fuzzy sets. This indicator expresses the degree to which A is included in B, for which the sense of the notion "included" will be explained further. For its introduction the two partial orderings described in the previous section will be simultaneously employed. We will obtain $\iota = Val \circ j$ as the composition of two other mappings (j and Val), each of which having its own interesting properties. In what follows, unless other stated, we will observe finite universe X. That is, $|X| = card(X) < \omega$, where $\omega = card(\mathbb{N})$. Let us now define two notions, which turn out to be very important in the sequel. Likewise U_ι, we introduce U_j in the following way $U_j = \{\varepsilon_{eq}, \varepsilon_{0,l}, \varepsilon_{\pi,l}, \varepsilon_{\Box,l}, \varepsilon_{\Diamond,l}\}_{l \in \{\mu,\nu\}}$, or more detailed

$$U_j = \{\varepsilon_{eq}, \varepsilon_{0,\mu}, \varepsilon_{0,\nu}, \varepsilon_{\pi,\mu}, \varepsilon_{\pi,\nu}, \varepsilon_{\Box,\mu}, \varepsilon_{\Box,\nu}, \varepsilon_{\Diamond,\mu}, \varepsilon_{\Diamond,\nu}\}.$$

The second important notion deserves to be introduced by a separate definition. For its introduction we will observe functions like $\eta : U_j \longrightarrow \mathscr{P}(X)$ but from a special type.

Definition 1 Let us denote the family of all functions with domain U_j and range $\mathscr{P}(X)$ by $\mathscr{P}(X)^{U_j}$. Then, an important subset of $\mathscr{P}(X)^{U_j}$ can be defined by

$$Split(U_j, X) =$$
$$\{\eta \mid \eta \in \mathscr{P}(X)^{U_j} \;\&\; \cup_{k \in U_j} \eta(k) = X \;\&\; (\forall k, l \in U_j)(k \ne l \Rightarrow \eta(k) \cap \eta(l) = \emptyset)\}.$$

Let us remark, that *Split*(U_j, X) represents exactly the family of all equivalence relations on X, consisting of at most 9 equivalence classes. Each of this equivalence classes turns out to have its own meaning in the terms of the above introduced orderings. Some (but not all) of the equivalence classes corresponding to elements of U_j may prove to be empty subsets of X.

Through the above introduced notions and for any $A, B \in \mathrm{IFS}(X)$, let us define the mapping j, which will be used for the introduction of the most important notion in the current paper, i.e. the IF-inclusion indicator ι. For each mapping $g : X \longrightarrow Y$ and any $Y_1 \subseteq Y$, we denote the preimage of Y_1 by

$$g^{-1}Y_1 = \{x \mid x \in X \,\&\, g(x) \in Y_1\}.$$

Definition 2 In the above introduced notations, where $A, B \in \mathrm{IFS}(X)$, let us define

$$j\colon \mathrm{IFS}(X) \times \mathrm{IFS}(X) \longrightarrow Split(U_j, X), \tag{6}$$

with $j(A, B) \in Split(U_j, X)$ in the following way.

1. For ε_{eq}:
   - $j(A,B)(\varepsilon_{eq}) = (\mu_B - \mu_A)^{-1}\{0\} \cap (\nu_B - \nu_A)^{-1}\{0\}$.
2. For ε_0:
   - $j(A,B)(\varepsilon_{0,\mu}) = (\mu_B - \mu_A)^{-1}(0,1] \cap (\nu_A - \nu_B)^{-1}(0,1]$,
   - $j(A,B)(\varepsilon_{0,\nu}) = (\nu_B - \nu_A)^{-1}(0,1] \cap (\mu_A - \mu_B)^{-1}(0,1]$.
3. For ε_π:
   - $j(A,B)(\varepsilon_{\pi,\mu}) = (\mu_B - \mu_A)^{-1}(0,1] \cap (\nu_B - \nu_A)^{-1}(0,1]$,
   - $j(A,B)(\varepsilon_{\pi,\nu}) = (\mu_A - \mu_B)^{-1}(0,1] \cap (\nu_A - \nu_B)^{-1}(0,1]$.
4. For $\varepsilon_\Box$:
   - $j(A,B)(\varepsilon_{\Box,\mu}) = (\mu_B - \mu_A)^{-1}(0,1] \cap (\nu_B - \nu_A)^{-1}\{0\}$,
   - $j(A,B)(\varepsilon_{\Box,\nu}) = (\mu_A - \mu_B)^{-1}(0,1] \cap (\nu_B - \nu_A)^{-1}\{0\}$.
5. For $\varepsilon_\Diamond$:
   - $j(A,B)(\varepsilon_{\Diamond,\mu}) = (\mu_B - \mu_A)^{-1}\{0\} \cap (\nu_A - \nu_B)^{-1}(0,1]$,
   - $j(A,B)(\varepsilon_{\Diamond,\nu}) = (\mu_B - \mu_A)^{-1}\{0\} \cap (\nu_B - \nu_A)^{-1}(0,1]$.

The reader may easily verify that the above defined $j(A, B)$ is really an element of $Split(U_j, X)$, corresponding to the family of equivalence relations on X with not more than 9 equivalence classes.

Let us now, trough the following remark, give a geometrical interpretation of the above defined subsets of X, $j(A, B)(\varepsilon)$ for each $\varepsilon \in U_j$.

Remark 1 The image of $j(A, B)(\varepsilon)$ for each $\varepsilon \in U_j$ corresponds to a particular area (geometrical figure) from the triangular representation (see Figs. 3 and 4). The position of the point $f_A(x) = (\mu_A(x), \nu_A(x))$ for each x splits the triangular area $\{(t_1, t_2) \mid (t_1, t_2) \in [0, 1] \times [0, 1] \,\&\, t_1 + t_2 \leq 1\}$ in four main figures some of which can be

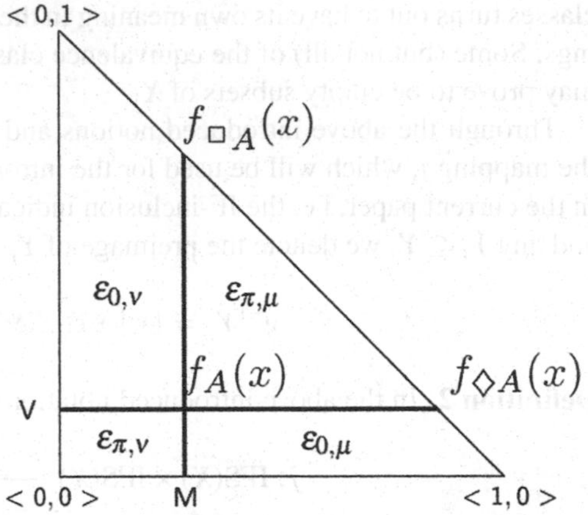

Fig. 3 Illustration of the triangular representation of the the intuitionistic fuzzy set A in a particular point $x \in X$ and the four figures (splitting the triangular area), corresponding to $\varepsilon_{0,\mu}, \varepsilon_{0,\nu}, \varepsilon_{\pi,\mu}, \varepsilon_{\pi,\nu} \in U_j$. The inner contour, i.e. the segments $Vf_{\Diamond A}(x)$ and $Mf_{\Box A}(x)$, is not included in the corresponding figures, although some of the four figures can degenerate into segments or points

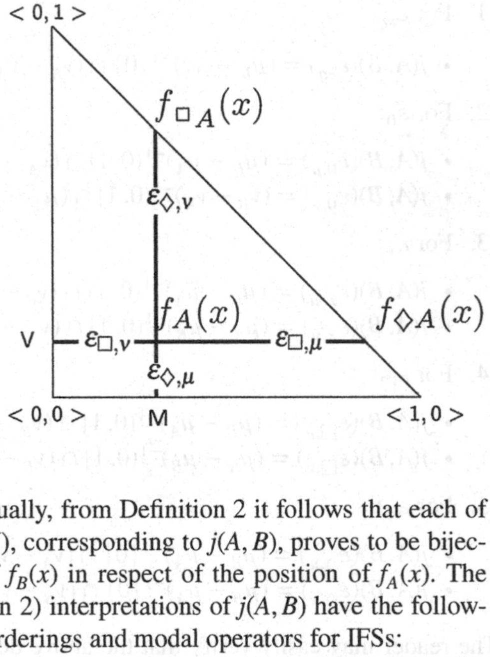

Fig. 4 Illustration of the triangular representation of the the intuitionistic fuzzy set A in a particular point $x \in X$ and the four segments: $f_A(x)f_{\Diamond A}(x)$, $Vf_A(x)$, $Mf_A(x)$, $f_A(x)f_{\Box A}(x)$, corresponding to $\varepsilon_{\Box,\mu}, \varepsilon_{\Box,\nu}, \varepsilon_{\Diamond,\mu}, \varepsilon_{\Diamond,\nu} \in U_j$. Some of the segments can degenerate into points

degenerated into lines or points. Actually, from Definition 2 it follows that each of the equivalence classes of $Split(U_j, X)$, corresponding to $j(A, B)$, proves to be bijectively determined by the position of $f_B(x)$ in respect of the position of $f_A(x)$. The geometrical and analytical (Definition 2) interpretations of $j(A, B)$ have the following correspondence in terms of the orderings and modal operators for IFSs:

1. For ε_{eq}:
   - $j(A, B)(\varepsilon_{eq})$ corresponds to the elements $x \in X$ for which $A = B$, or $f_A(x) = f_B(x)$.
2. For ε_0:
   - $j(A, B)(\varepsilon_{0,\mu})$ corresponds to the elements $x \in X$, for which $f_A(x) < f_B(x)$,
   - $j(A, B)(\varepsilon_{0,\nu})$ corresponds to the elements $x \in X$, for which $f_A(x) > f_B(x)$.

3. For ε_π:
   - $j(A,B)(\varepsilon_{\pi,\mu})$ corresponds to the elements $x \in X$, for which $f_A(x) <_\pi f_B(x)$,
   - $j(A,B)(\varepsilon_{\pi,\nu})$ corresponds to the elements $x \in X$, for which $f_A(x) >_\pi f_B(x)$.
4. For $\varepsilon_\square$:
   - $j(A,B)(\varepsilon_{\square,\mu})$ corresponds to the elements $x \in X$, for which $f_A(x) <_\square f_B(x)$ and $f_A(x) =_\diamond f_B(x)$,
   - $j(A,B)(\varepsilon_{\square,\nu})$ corresponds to the elements $x \in X$, for which $f_A(x) >_\square f_B(x)$ and $A =_\diamond B$.
5. For $\varepsilon_\diamond$:
   - $j(A,B)(\varepsilon_{\diamond,\mu})$ corresponds to the elements $x \in X$, for which $f_A(x) <_\diamond f_B(x)$ and $f_A(x) =_\square f_B(x)$,
   - $j(A,B)(\varepsilon_{\diamond,\nu})$ corresponds to the elements $x \in X$, for which $f_A(x) >_\diamond f_B(x)$ and $f_A(x) =_\square f_B(x)$.

Let us now, in a very intuitive way, introduce in *Split*(U_j, X) a binary relation, denoted by $\preceq_j$.

Definition 3 For any two elements η_1, η_2 of *Split*(U_j, X), let us define the following binary relation $\preceq_j$: $\eta_1 \preceq_j \eta_2$ iff

1. For ε_{eq}:
   - $\eta_1(\varepsilon_{eq}) \subseteq \eta_2(\varepsilon_{eq})$.
2. For ε_0:
   - $\eta_1(\varepsilon_{0,\mu}) \subseteq \eta_2(\varepsilon_{0,\mu})$,
   - $\eta_1(\varepsilon_{0,\nu}) \supseteq \eta_2(\varepsilon_{0,\nu})$.
3. For ε_π:
   - $\eta_1(\varepsilon_{\pi,\mu}) \subseteq \eta_2(\varepsilon_{\pi,\mu})$,
   - $\eta_1(\varepsilon_{\pi,\nu}) \supseteq \eta_2(\varepsilon_{\pi,\nu})$.
4. For $\varepsilon_\square$:
   - $\eta_1(\varepsilon_{\square,\mu}) \subseteq \eta_2(\varepsilon_{\square,\mu})$,
   - $\eta_1(\varepsilon_{\square,\nu}) \supseteq \eta_2(\varepsilon_{\square,\nu})$.
5. For $\varepsilon_\diamond$:
   - $\eta_1(\varepsilon_{\diamond,\mu}) \subseteq \eta_2(\varepsilon_{\diamond,\mu})$,
   - $\eta_1(\varepsilon_{\diamond,\nu}) \supseteq \eta_2(\varepsilon_{\diamond,\nu})$.

Let us remark that the above defined binary relation $\preceq_j$ can be stated in the following more compact way. For $\eta_1, \eta_2 \in$ *Split*(U_j, X) we have that $\eta_1 \preceq_j \eta_2$ iff

1. For ε_{eq}:
   - $\eta_1(\varepsilon_{eq}) \subseteq \eta_2(\varepsilon_{eq})$.
2. For all $k \in \{0, \pi, \Box, \Diamond\}$:
   - $\eta_1(\varepsilon_{k,\mu}) \subseteq \eta_2(\varepsilon_{k,\mu})$,
   - $\eta_1(\varepsilon_{k,\nu}) \supseteq \eta_2(\varepsilon_{k,\nu})$.

The proof of the following theorem is an application of the ordering properties of the set theoretical partial ordering $\subseteq$.

Theorem 1 *The binary relation $\preceq_j$, introduced in Definition 3, is a partial ordering in Split(U_j, X). That is, it satisfies the three axioms for ordering: it is reflexive, transitive and anti-symmetric.*

Proof Let us show that the binary relation $\preceq_j$ is reflexive. Taking any $\eta_1 \in$ *Split*(U_j, X), and because of the reflexivity of the set theoretical partial ordering $\subseteq$, we have that $\eta_1(\varepsilon_{eq}) \subseteq \eta_1(\varepsilon_{eq})$ and $\eta_1(\varepsilon_{k,\mu}) \subseteq \eta_1(\varepsilon_{k,\mu})$, $\eta_1(\varepsilon_{k,\nu}) \supseteq \eta_1(\varepsilon_{k,\nu})$ for all $k \in \{0, \pi, \Box, \Diamond\}$. Therefore, from Definition 3 we obtain that $\eta_1 \preceq_j \eta_1$. The transitivity of $\preceq_j$ is implied directly from the transitivity of $\subseteq$ as well. To show the anti-symmetric property, suppose that for $\eta_1, \eta_2 \in$ *Split*(U_j, X) we have that $\eta_1 \preceq_j \eta_2$ and $\eta_2 \preceq_j \eta_1$ simultaneously. Therefore, from the anti-symmetric property of $\subseteq$ we get that $\eta_1(\varepsilon) = \eta_2(\varepsilon)$ for all $\varepsilon \in U_j$. Hence, $\eta_1 \equiv \eta_2$ on the whole domain of U_j, which yields that $\eta_1 = \eta_2$. The theorem is proved.

As an easy exercise the following propositions about the newly introduced partial ordering $\preceq_j$ is left to the reader.

Remark 2 Let us classify the minimal and maximal elements of the partially ordered set (*Split*(U_j, X), $\preceq_j$). The family of the minimal elements of *Split*(U_j, X) consists of the elements $\eta \in$ *Split*(U_j, X), satisfying

$$\eta(\varepsilon_{k,\mu}) = \emptyset, \text{ for all } k \in \{0, \pi, \Box, \Diamond\}.$$

The family of the maximal elements of *Split*(U_j, X) consists of the elements $\eta \in$ *Split*(U_j, X), satisfying

$$\eta(\varepsilon_{k,\nu}) = \emptyset, \text{ for all } k \in \{0, \pi, \Box, \Diamond\}.$$

Remark 3 Let A, B and C be elements of IFS(X) and $j(A, B) \preceq_j j(B, C)$. Then we have that,

$$j(A, B) \preceq_j j(A, C)$$

More generally, let $A_1, \ldots, A_n \in$ IFS(X) and for all $k = 1, \ldots, n-2$: $j(A_k, A_{k+1}) \preceq_j j(A_{k+1}, A_{k+2})$. Then for any $l \in \{1, \ldots, n-2\}$ we have that,

$$j(A_l, A_{l+1}) \preceq_j j(A_l, A_n)$$

In order to introduce the final notion in this paper, i.e. the IF-inclusion indicator ι, we will need the following mapping,

$$Val : Split(U_j, X) \longrightarrow \text{IFS}(U_\iota),$$

where U_ι has been defined as $U_\iota = \{\varepsilon_0, \varepsilon_\pi, \varepsilon_\square, \varepsilon_\diamond\}$. For each $\eta \in Split(U_j, X)$ we have that $Val(\eta) \in \text{IFS}(U_\iota)$ such that:

1. For ε_0:
   - $\mu_{Val(\eta)}(\varepsilon_0) = \frac{|\eta(\varepsilon_{0,\mu})|}{|X|}$ and $\nu_{Val(\eta)}(\varepsilon_0) = \frac{|\eta(\varepsilon_{0,\nu})|}{|X|}$
2. For ε_π:
   - $\mu_{Val(\eta)}(\varepsilon_\pi) = \frac{|\eta(\varepsilon_{\pi,\mu})|}{|X|}$ and $\nu_{Val(\eta)}(\varepsilon_\pi) = \frac{|\eta(\varepsilon_{\pi,\nu})|}{|X|}$
3. For $\varepsilon_\square$:
   - $\mu_{Val(\eta)}(\varepsilon_\square) = \frac{|\eta(\varepsilon_{\square,\mu})|}{|X|}$ and $\nu_{Val(\eta)}(\varepsilon_\square) = \frac{|\eta(\varepsilon_{\square,\nu})|}{|X|}$
4. For $\varepsilon_\diamond$:
   - $\mu_{Val(\eta)}(\varepsilon_\diamond) = \frac{|\eta(\varepsilon_{\diamond,\mu})|}{|X|}$ and $\nu_{Val(\eta)}(\varepsilon_\diamond) = \frac{|\eta(\varepsilon_{\diamond,\nu})|}{|X|}$

Let us remark that the above definition of the mapping *Val*, that is its image *Val*(η), can be stated in the following compact form. For each $k \in \{0, \pi, \square, \diamond\}$ we can write:

$$\mu_{Val(\eta)}(\varepsilon_k) = \frac{|\eta(\varepsilon_{k,\mu})|}{|X|} \text{ and } \nu_{Val(\eta)}(\varepsilon_k) = \frac{|\eta(\varepsilon_{k,\nu})|}{|X|}.$$

We are going now to show that the so defined *Val*(η) is indeed an element of IFS(U_ι). By the definition of $Split(U_j, X)$, for each of its elements η, we have that $\bigcup_{k \in U_j} \eta(k) = X$ and for all ε_1 and $\varepsilon_2 \in U_j$, $\varepsilon_1 \neq \varepsilon_2$ implies that $\eta(\varepsilon_1) \cap \eta(\varepsilon_2) = \emptyset$. The above statement yields that $\sum_{\varepsilon \in U_j} |\eta(\varepsilon)| = |X|$ and therefore,

$$|\eta(\varepsilon_{eq})| + \sum_{k \in \{0,\pi,\square,\diamond\}} |\eta(\varepsilon_{k,\mu})| + \sum_{k \in \{0,\pi,\square,\diamond\}} |\eta(\varepsilon_{k,\nu})| = |X|.$$

From the above expression, dividing by $|X|$ the two sides of the equation, we get that

$$\frac{|\eta(\varepsilon_{eq})|}{|X|} + \sum_{k \in \{0,\pi,\square,\diamond\}} \frac{|\eta(\varepsilon_{k,\mu})|}{|X|} + \sum_{k \in \{0,\pi,\square,\diamond\}} \frac{|\eta(\varepsilon_{k,\nu})|}{|X|} = 1.$$

The last equation and the definition of the mapping *Val* provides the following interesting property of the elements from $Range(Val) \subset \text{IFS}(U_\iota)$.

Remark 4 For all $\eta \in Split(U_j, X)$, we have that

$$0 \leq \sum_{k \in \{0,\pi,\Box,\Diamond\}} \mu_{Val(\eta)}(\varepsilon_k) + \sum_{k \in \{0,\pi,\Box,\Diamond\}} \nu_{Val(\eta)}(\varepsilon_k) \leq 1. \tag{7}$$

1. The right hand side inequality is equality in the above expression iff $|\eta(\varepsilon_{eq})| = 0$, i.e. $\eta(\varepsilon_{eq}) = \emptyset$.
2. The left hand side inequality is equality in the above expression iff $|\eta(\varepsilon_{eq})| = |X|$, i.e. $\eta(\varepsilon_{eq}) = X$.

The above remark permits us to state the following theorem.

Theorem 2 *The mapping Val* : $Split(U_j, X) \longrightarrow IFS(U_\iota)$ *is well defined. Its range is indeed a subset of* $IFS(U_\iota)$, *i.e. for all* $\eta \in Split(U_j, X)$ *we have that*

$$(\forall \varepsilon \in U_\iota)(\mu_{Val(\eta)}(\varepsilon) + \nu_{Val(\eta)}(\varepsilon) \leq 1).$$

3 IF-inclusion Indicator ι

The reader may compare the notions, defined in this section with Atanassov [1], Chap. 18, Definition 1.8. There an indicator of inclusion of IFSs, $In(A, B)$ for any $A, B \in \text{IFS}(X)$ has been introduced. $In(A, B)$ being itself an element of IFS(X), for which

- $In(A, B) = \langle 1, 0 \rangle$ iff $A \leq B$,
- $In(A, B) = \langle 0, 1 \rangle$ iff $B \leq A$.

Another inclusion indicators and inclusion measures of intuitionistic fuzzy sets can be found in Cornelis [4] and Grzegorzewski [5].

Let us now introduce the most important for this paper notion of IF-inclusion indicator ι through the following definition.

Definition 4 (*IF-inclusion indicator*) The mapping

$$\iota : \text{IFS}(X) \times \text{IFS}(X) \longrightarrow \text{IFS}(U_\iota)$$

will be called *IF-inclusion indicator* and let us define it as the composition of the already defined j and Val, i.e. $\iota = (Val \circ j)$. That is,

$$\iota : \text{IFS}(X) \times \text{IFS}(X) \xrightarrow{j} Split(U_j, X) \xrightarrow{Val} \text{IFS}(U_\iota).$$

Remark 5 Suppose that $\eta = j(A,B) \in Split(U_j, X)$ for some $A, B \in \text{IFS}(X)$ and let us explain what really expresses $Val(\eta) = Val(j(A,B)) = \imath(A,B)$ (see Figs. 2 and 3 for the geometrical interpretation).

1. For ε_0:
   - $\mu_{\imath(A,B)}(\varepsilon_0) = \frac{1}{|X|}|\{x \mid x \in X \,\&\, f_A(x) < f_B(x)\}|$, i.e. the normalized number of elements $x \in X$, such that $f_A(x) < f_B(x)$,
   - $\nu_{\imath(A,B)}(\varepsilon_0) = \frac{1}{|X|}|\{x \mid x \in X \,\&\, f_A(x) > f_B(x)\}|$, i.e. the normalized number of elements $x \in X$, such that $f_A(x) > f_B(x)$.
2. For ε_π:
   - $\mu_{\imath(A,B)}(\varepsilon_\pi) = \frac{1}{|X|}|\{x \mid x \in X \,\&\, f_A(x) <_\pi f_B(x)\}|$, i.e. the normalized number of elements $x \in X$, such that $f_A(x) <_\pi f_B(x)$,
   - $\nu_{\imath(A,B)}(\varepsilon_\pi) = \frac{1}{|X|}|\{x \mid x \in X \,\&\, f_A(x) >_\pi f_B(x)\}|$, i.e. the normalized number of elements $x \in X$, such that $f_A(x) >_\pi f_B(x)$.
3. For $\varepsilon_\square$:
   - $\mu_{\imath(A,B)}(\varepsilon_\square) = \frac{1}{|X|}|\{x \mid x \in X \,\&\, \mu_A(x) < \mu_B(x) \,\&\, \nu_A(x) = \nu_B(x)|$,
   - $\nu_{\imath(A,B)}(\varepsilon_\square) = \frac{1}{|X|}|\{x \mid x \in X \,\&\, \mu_A(x) > \mu_B(x) \,\&\, \nu_A(x) = \nu_B(x)|$.
4. For $\varepsilon_\diamond$:
   - $\mu_{\imath(A,B)}(\varepsilon_\diamond) = \frac{1}{|X|}|\{x \mid x \in X \,\&\, \mu_A(x) = \mu_B(x) \,\&\, \nu_A(x) > \nu_B(x)|$,
   - $\nu_{\imath(A,B)}(\varepsilon_\diamond) = \frac{1}{|X|}|\{x \mid x \in X \,\&\, \mu_A(x) = \mu_B(x) \,\&\, \nu_A(x) < \nu_B(x)|$.

As an easy exercise one can induce from the definition of the IF-inclusion indicator following theorem.

Theorem 3 *In the above notations with $A, B \in IFS(X)$, we have that*

$$\imath(A,B) = \neg\imath(B,A).$$

Remark 6 We note that in (7) for $\eta = j(A,B)$, where $A, B \in \text{IFS}(X)$, the equality

$$\sum_{k\in\{0,\pi,\square,\diamond\}} \mu_{\imath(A,B)}(\varepsilon_k) + \sum_{k\in\{0,\pi,\square,\diamond\}} \nu_{\imath(A,B)}(\varepsilon_k) = 1$$

is satisfied only if we have that for all $x \in X$, $\mu_A(x) \neq \mu_B(x)$ or $\nu_A(x) \neq \nu_B(x)$.

Some combinations of the elements from the universe $U_\imath$ prove to be of interest for practical estimations. Through such combinations we build up estimation expressions not only for the strict $\leq$ and $\leq_\pi$-inclusions, as already done through $\mu_{\imath(A,B)}(\varepsilon_k), \nu_{\imath(A,B)}(\varepsilon_k), k \in \{0, \pi\}$, but for the non-strict ones as well. As already stated,

- ε_0 corresponds to the standard strict inclusion (ordering),
- ε_π corresponds to the strict π-inclusion (π-ordering),
- $\varepsilon_\Box$ corresponds to the quasi $\Box$-ordering,
- $\varepsilon_\Diamond$ corresponds to the quasi $\Diamond$-ordering.

Let us now, employing the above introduced IF-index, give some expressions providing a degree of inclusion (with respect to a partial ordering), for which

- $\{\varepsilon_0, \varepsilon_\Box, \varepsilon_\Diamond\}$ corresponds to the standard inclusion (ordering),
- $\{\varepsilon_\pi, \varepsilon_\Box, \varepsilon_\Diamond\}$ corresponds to the π-inclusion (π-ordering).

And namely, let us take any two $A, B \in \mathrm{IFS}(X)$ and define

1. The degree to which A equals B, $\imath_=(A,B) = \frac{|j(A,B)(\varepsilon_{eq})|}{|X|}$
2. The degree to which A is $\leq$-included into B, $\imath_\leq(A,B) = \mu_{\imath(A,B)}(\varepsilon_0) + \mu_{\imath(A,B)}(\varepsilon_\Box) + \mu_{\imath(A,B)}(\varepsilon_\Diamond) + \imath_=(A,B)$
3. The degree to which A is strictly $\leq$-included into B, $\imath_<(A,B) = \mu_{\imath(A,B)}(\varepsilon_0) + \mu_{\imath(A,B)}(\varepsilon_\Box) + \mu_{\imath(A,B)}(\varepsilon_\Diamond)$
4. The degree to which A is $\preceq_\pi$-included into B, $\imath_{\preceq_\pi}(A,B) = \mu_{\imath(A,B)}(\varepsilon_\pi) + \mu_{\imath(A,B)}(\varepsilon_\Box) + \nu_{\imath(A,B)}(\varepsilon_\Diamond) + \imath_=(A,B)$
5. The degree to which A is strictly $\preceq_\pi$-included into B, $\imath_{\prec_\pi}(A,B) = \mu_{\imath(A,B)}(\varepsilon_\pi) + \mu_{\imath(A,B)}(\varepsilon_\Box) + \nu_{\imath(A,B)}(\varepsilon_\Diamond)$

From (7) it now follows that for each $\rho \in \{\leq, <, \preceq_\pi, \prec_\pi, =\}$,

$$\imath_\rho : \mathrm{IFS}(X) \times \mathrm{IFS}(X) \longrightarrow [0,1]. \tag{8}$$

That is, $\imath_\rho$ for each $\rho \in \{\leq, <, \preceq_\pi, \prec_\pi, =\}$, provides an ordinary fuzzy estimation.

Remark 7 The indicator $\imath_=$ can be expressed as follows

$$\imath_=(A,B) = 1 - \Big(\sum_{k\in\{0,\pi,\Box,\Diamond\}} \mu_{\imath(A,B)}(\varepsilon_k) + \sum_{k\in\{0,\pi,\Box,\Diamond\}} \nu_{\imath(A,B)}(\varepsilon_k)\Big).$$

One may easily verify the following theorem.

Theorem 4 *For any $A, B \in IFS(X)$ the following expressions hold,*

$$\imath_\leq(A,B) = 1 \textit{ iff } A \leq B$$

and

$$\imath_\leq(A,B) = 0 \textit{ iff } \imath_<(B,A) = 1 \textit{ iff } B < A$$

4 Conclusion

In this paper we have introduced a measure for inclusion of two IFSs into each other according to the two main partial orderings in the family of IFSs. This inclusion measure has been observed on two levels. From a set-theoretical point of view, through the introduction of the mapping

$$j\colon \mathrm{IFS}(X)\times\mathrm{IFS}(X)\longrightarrow Split(U_j,X),$$

where $Split(U_j, X)$ corresponds to the family of equivalence relations on X with at least 9 equivalence classes. And the most important and of practical use IF-inclusion indicator,

$$\iota : \mathrm{IFS}(X)\times\mathrm{IFS}(X)\longrightarrow \mathrm{IFS}(U_\iota),$$

introduced in Definition 4. The IF-inclusion indicator ranges over IFSs on a special universe U_ι. It has been employed in the last section for the introduction of expressions for a degree of inclusion of A into B according to any of the two main orderings, where the range now is the interval $[0, 1]$. That is, these expressions give us an ordinary fuzzy estimation. They are all applicable in real world models where intuitionistic fuzzy sets are employed and give a more detailed information about the inclusion of an IFS into another IFS. Many examples can be found in the literature, especially for decision making procedures. This work is going to be extended further for the investigation of the formulas proposed here in a more practical direction.

References

1. Atanassov, K.: Intuitionistic Fuzzy Sets: Theory and Applications. Springer, Heidelberg (1999)
2. Atanassov, K.: On Intuitionistic Fuzzy Sets Theory. Springer, Berlin (2012)
3. Birkhoff, G.: Lattice theory. American Mathematical Society, Providence (1967)
4. Cornelis, C., Kerre, E.: Inclusion Measures in Intuitionistic Fuzzy Set Theory. Symbolic and Quantitative Approaches to Reasoning with Uncertainty. Lecture Notes in Computer Science, vol. 2711, pp. 345–356. Springer, Berlin (2003)
5. Grzegorzewski, P.: On possible and necessary inclusion of intuitionistic fuzzy sets. Inf. Sci. **181**, 342–350 (2011)
6. Marinov E.: π-ordering and index of indeterminacy for intuitionistic fuzzy sets. Modern Approaches in Fuzzy Sets, Intuitionistic Fuzzy Sets, Generalized Nets and Related Topics, Foundations, vol. 1, Systems Research Institute Polish Academy of Sciences, Warsaw (2014)
7. Marinov E.: On modal operators and quasi-orderings for IFSs, Modern Approaches in Fuzzy Sets, Intuitionistic Fuzzy Sets, Generalized Nets and Related Topics, Foundations, vol. 1, Systems Research Institute Polish Academy of Sciences, Warsaw (2014)
8. Szmidt, E.: Distances and Similarities in Intuitionistic Fuzzy Sets. Springer, Switzerland (2014)
9. Szmidt, E., Kacprzyk, J.: Distances between intuitionistic fuzzy sets. Fuzzy Sets Syst. **114**(3), 505–518 (2000)
10. Zadeh, L.A.: Fuzzy sets. Inf. Control **8**, 338–353 (1965)

One, Two and Uni-type Operators on IFSs

Gökhan Çuvalcioğlu

Abstract Intuitionistic Fuzzy Modal Operator was defined by Atanassov in (Intuitionistic Fuzzy Sets. Phiysica-Verlag, Heidelberg, 1999, [2], Int J Uncertain Fuzzyness Knowl Syst 9(1):71–75, 2001, [3]). He introduced the generalization of these modal operators. After this study, Dencheva (Proceedings of the Second International. IEEE Symposium: Intelligent Systems, vol 3, pp 21–22. Varna, 2004, [10]) defined second extension of these operators. The third extension of these was defined by Atanassov in (Adv Stud Contemp Math 15(1):13–20, 2007, [5]). In (Atanassov, NIFS 14(1):27–32 2008, [6]), the author introduced a new operator over Intuitionistic Fuzzy Sets which is generalization of Atanassov's and Dencheva's operators. At the same year, Atanassov defined an operator which is an extension of all the operators. The diagram of One Type Modal Operators on Intuitionistic Fuzzy Sets was introduced first time by Atanassov (Int J Uncertain Fuzzyness Knowl Syst 9(1):71–75, 2001, [3]). The author expanded the diagram of One Type Modal Operators on Intuitionistic Fuzzy Sets with the operator Z (alpha beta gamma). In 2013, the last operators were defined. These operators have properties which are belong to both first and second type modal operators. So, they called uni-type operators. After these operators the diagram of modal operators on intuitionistic fuzzy sets is expanded.

Keywords Uni-type operators · Intuitionistic fuzzy operators · OTMOs

G. Çuvalcioğlu (✉)
Department of Mathematics, University of Mersin, Mersin, Turkey
e-mail: gcuvalcioglu@gmail.com

P. Angelov and S. Sotirov (eds.), *Imprecision and Uncertainty in Information Representation and Processing*, Studies in Fuzziness and Soft Computing 332,
DOI 10.1007/978-3-319-26302-1_5

1 Introduction

The original concept of fuzzy sets in [13], Zadeh was introduced as an extension of crisp sets by enlarging the truth value set to the real unit interval [0, 1]. In fuzzy set theory, if the membership degree of an element x is $\mu(x)$ then the non-membership degree is $1 - \mu(x)$ and thus it is fixed.

Intuitionistic fuzzy sets have been introduced by Atanassov in 1983 [1] and form an extension of fuzzy sets by enlarging the truth value set to the lattice $[0, 1] \times [0, 1]$ is defined as following.

Definition 1 Let $L = [0, 1]$ then $L^* = \{(x_1, x_2) \in [0, 1]^2 : x_1 + x_2 \leq 1\}$ is a lattice with

$$(x_1, x_2) \leq (y_1, y_2) :\Leftrightarrow x_1 \leq y_1, x_2 \geq y_2$$

The units of this lattice are denoted by $0_{L^*} = (0, 1)$ and $1_{L^*} = (1, 0)$.

The lattice $(L^*, \leq)$ is a complete lattice: For each $A \subseteq L^*$,

$$\begin{aligned} \sup A = (&\sup\{x \in [0, 1] : (y \in [0, 1]), ((x, y) \in A)\}, \\ &\inf\{y \in [0, 1] : (x \in [0, 1]), ((x, y) \in A)\}) \end{aligned}$$

and

$$\begin{aligned} \inf A = (&\inf\{x \in [0, 1] : (y \in [0, 1]), ((x, y) \in A)\}, \\ &\sup\{y \in [0, 1] : (x \in [0, 1]), ((x, y) \in A)\}) \end{aligned}$$

As is well known, every lattice L^* has an equivalent definition as an algebraic structure $(L^*, \wedge, \vee, \leq)$ where the meet operator $\wedge$ and the join operator $\vee$ are linked the ordering "$\leq$" by the following equivalence, for $x, y \in L^*$,

$$x \leq y \Leftrightarrow x \vee y = y \Leftrightarrow x \wedge y = x$$

The operators $\wedge$ and $\vee$ (join and meets resp.) on $(L^*, \leq)$ are defined as follows, for $(x_1, y_1), (x_2, y_2) \in L^*$,

$$\begin{aligned} (x_1, y_1) \wedge (x_2, y_2) &= (x_1 \wedge x_2, y_1 \vee y_2) \\ (x_1, y_1) \vee (x_2, y_2) &= (x_1 \vee x_2, y_1 \wedge y_2) \end{aligned}$$

Definition 2 ([1]) An intuitionistic fuzzy set (shortly IFS) on a set X is an object of the form $A = \{<x, \mu_A(x), \nu_A(x)> : x \in X\}$ where $\mu_A(x), (\mu_A : X \to [0, 1])$ is called the degree of membership of x in A, $\nu_A(x), (\nu_A : X \to [0, 1])$ is called the degree of non- membership of x in A, and where μ_A and v_A satisfy the following condition:

$$\mu_A(x) + \nu_A(x) \leq 1, \text{ for all } x \in X.$$

The hesitation degree of x is defined by $\pi_A(x)$ 1 − $\mu_A(x)$ − $\nu_A(x)$

Definition 3 ([1]) An IFS A is said to be contained in an IFS B (notation $A \sqsubseteq B$ if and only if, for all $x \in X$, $\mu_A(x) \leq \mu_B(x)$ and $\nu_A(x) \geq \nu_B(x)$.

It is clear that A = B if and only if $A \sqsubseteq B$ and $B \sqsubseteq A$

Definition 4 ([1]) Let A ∈ IFS and let $A = \{<x, \mu_A(x), \nu_A(x)>: x \in X\}$ then the following set is called the complement of A

$$A^c = \{<x, \nu_A(x), \mu_A(x)>: x \in X\}$$

The notion of Intuitionistic Fuzzy Operators was firstly introduced by Atanassov [1]. The simplest one among them is presented as in the following definition.

Definition 5 ([2]) Let X be a set and $A = \{<x, \mu_A(x), \nu_A(x)>: x \in X\} \in IFS(X)$, $\alpha, \beta \in [0, 1]$ then

(a) $\boxplus A = \{<x, \frac{\mu_A(X)}{2}, \frac{\nu_A(X)+1}{2}>: x \in X\}$
(b) $\boxtimes A = \{<x, \frac{\mu_A(X)+1}{2}, \frac{\nu_A(X)}{2}>: x \in X\}$.

After this definition, in 2001, Atanassov, in [3], defined the extension of these operators as following:

Definition 6 ([3]) Let X be a set and $A = \{<x, \mu_A(x), \nu_A(x)>: x \in X\} \in IFS(X)$, $\alpha, \beta \in [0, 1]$.

(a) $\boxplus_\alpha A = \{<x, \alpha\mu_A(x), \alpha\nu_A(x) + 1 - \alpha>: x \in X\}$
(b) $\boxtimes_\alpha A = \{<x, \alpha\mu_A(x) + 1 - \alpha, \alpha\nu_A(x)>: x \in X\}$.

In these operators $\boxplus_\alpha$ and $\boxtimes_\alpha$; if we choose $\alpha = \frac{1}{2}$, we get the operators $\boxplus$, $\boxtimes$, resp. Therefore, the operators $\boxplus_\alpha$ and $\boxtimes_\alpha$ are the extensions of the operators $\boxplus$, $\boxtimes$, resp. Some relationships between these operators were studied by several authors [10, 12].

In 2004, the second extension of these operators was introduced by Dencheva in [10].

Definition 7 ([10]) Let X be a set and $A = \{<x, \mu_A(x), \nu_A(x)>: x \in X\} \in IFS(X)$, $\alpha, \beta \in [0, 1]$.

(a) $\boxplus_{\alpha,\beta} A = \{<x, \alpha\mu_A(x), \alpha\nu_A(x) + \beta>: x \in X\}$ where $\alpha + \beta \in [0, 1]$.
(b) $\boxtimes_{\alpha,\beta} A = \{<x, \alpha\mu_A(x) + \beta, \alpha\nu_A(x)>: x \in X\}$ where $\alpha + \beta \in [0, 1]$.

The concepts of the modal operators are being introduced and studied by different researchers, [3–6, 10–12], etc.

In 2006, the third extension of the above operators was studied by Atanassov. He defined the following operators in [4].

Definition 8 ([4]) Let X be a set and $A = \{<x, \mu_A(x), \nu_A(x)> : x \in X\} \in IFS(X)$. Then, for $\alpha, \beta, \gamma \in [0,1]$, $\max\{\alpha, \beta\} + \gamma \leq 1$.

(a) $\boxplus_{\alpha,\beta,\gamma}(A) = \{<x, \alpha\mu_A(x), \beta\nu_A(x) + \gamma> : x \in X\}$
(b) $\boxtimes_{\alpha,\beta,\gamma}(A) = \{<x, \alpha\mu_A(x) + \gamma, \beta\nu_A(x)> : x \in X\}$

If we choose $\alpha = \beta$ and $\gamma = \beta$ in above operators then we can see easily that $\boxplus_{\alpha,\alpha,\gamma}, = \boxplus_{\alpha,\beta}$ and $\boxtimes_{\alpha,\alpha,\gamma} = \boxtimes_{\alpha,\beta}$ Therefore, we can say that $\boxplus_{\alpha,\alpha,\gamma}$ and $\boxtimes_{\alpha,\alpha,\gamma}$ are the extensions of the operators $\boxplus_{\alpha,\beta}$ and $\boxtimes_{\alpha,\beta}$, resp. From these extensions, we get the first diagram of One Type Modal Operators (OTMOs) over Intuitionistic Fuzzy Sets (IFSs) as displayed in Fig. 1.

In 2007, after this diagram, the author [7] defined a new operator and studied some of its properties. This operator is named $E_{\alpha,\beta}$ and defined as follows:

Definition 9 ([7]) Let X be a set and $A = \{<x, \mu_A(x), \nu_A(x)> : x \in X\} \in IFS(X)$, $\alpha, \beta \in [0,1]$. We define the following operator:

$$E_{\alpha,\beta}(A) = \{<x, \beta(\alpha\mu_A(x) + 1 - \alpha), \alpha(\beta\nu_A(x) + 1 - \beta)> : x \in X\}$$

If we choose $\alpha = 1$ and write α instead of β we get the operator $\boxplus_\alpha$. Similarly, if $\beta = 1$ is chosen and written instead of β, we get the operator $\boxtimes_\alpha$. In the view of this definition, the diagram of OTMOs on IFSs is figured below (Fig. 2).

These extensions have been investigated by several authors [5, 6, 9]. In particular, the authors have made significant contributions to these operators.

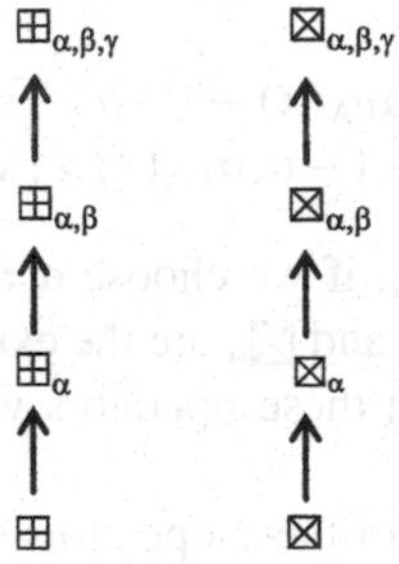

Fig. 1

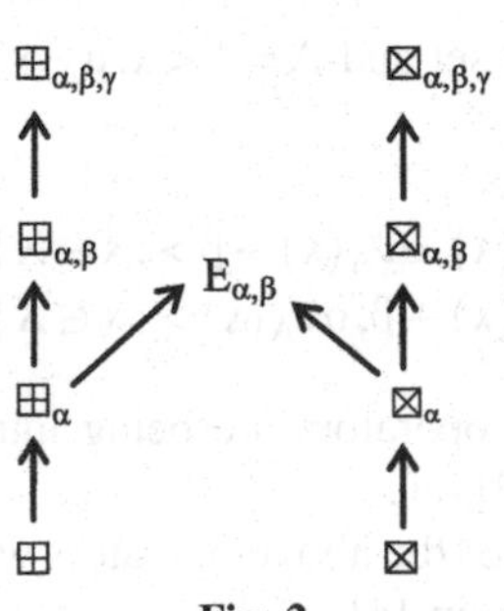

Fig. 2

In 2007, Atanassov introduced the operator $\boxdot_{\alpha,\beta,\gamma,\delta}$ which is a natural extension of all these operators in [5].

Definition 10 ([5]) Let X be a set, A ∈ IFS(X), α, β, γ, δ ∈ [0, 1] such that $\max(\alpha, \beta) + \gamma + \delta \leq 1$ then the operator $\boxdot_{\alpha,\beta,\gamma,\delta}$ defined by

$$\boxdot_{\alpha,\beta,\gamma,\delta}(A) = \{ <x, \alpha\mu_A(x) + \gamma, \beta\nu_A(x) + \delta > : x \in X\}$$

This operator changed the OTMOs' diagram (Fig. 3).

At the end of these studies, Atanassov though that this diagram was completed. However, he realized that it wasn't totally true since there was an operator which was also an extension of two type modal operators.

In 2008, he defined this most general operator $\circledcirc_{\alpha,\beta,\gamma,\delta,\varepsilon,\zeta}$ as following:

Definition 11 ([6]) Let X be a set, A ∈ IFS(X), $\alpha, \beta, \gamma, \delta, \varepsilon, \zeta \in [0, 1]$ such that $\max(\alpha - \zeta, \beta - \varepsilon) + \gamma + \delta \leq 1$ and $\min(\alpha - \zeta, \beta - \varepsilon) + \gamma + \delta \geq 0$ then the operator $\circledcirc_{\alpha,\beta,\gamma,\delta,\varepsilon,\zeta}$ defined by

$$\circledcirc_{\alpha,\beta,\gamma,\delta,\varepsilon,\zeta}(A) = \{ <x, \alpha\mu_A(x) - \varepsilon\nu_A(x) + \gamma, \beta\nu_A(x) - \zeta\mu_A(x) + \delta > : x \in X\}$$

After this definition, the OTMOs' diagram is became as in Fig. 4.

In 2010, the author [8] defined a new operator which is a generalization of $E_{\alpha,\beta}$

Definition 12 ([8]) Let X be a set and $A = \{ <x, \mu_A(x), \nu_A(x) > : x \in X\} \in IFS(X)$, $\alpha, \beta, \omega \in [0, 1]$. We define the following operator:

$$Z^{\omega}_{\alpha\beta}(A) = \{ <x, \beta(\alpha\mu_A(x) + \omega - \omega.\alpha), \alpha(\beta\nu_A(x) + \omega - \omega.\beta) > : x \in X\}$$

The diagram of OTMOs over IFSs is displayed in Fig. 5.

We have defined a new OTMO on IFS, that is generalization of the some OTMOs. The new operator defined as follows:

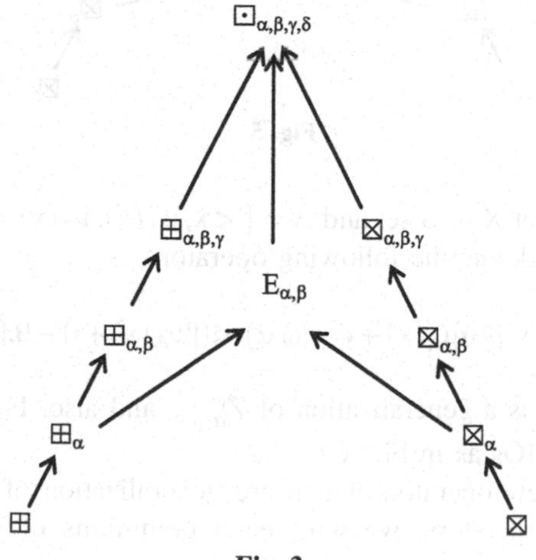

Fig. 3

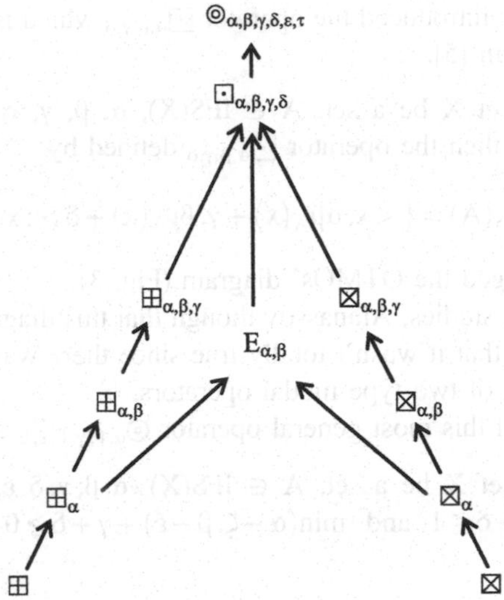

Fig. 4

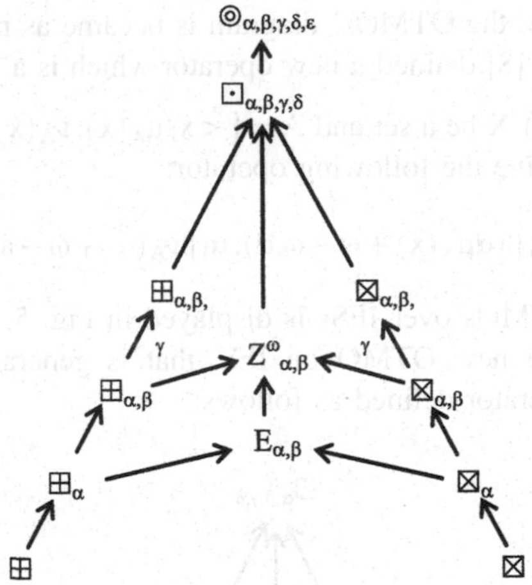

Fig. 5

Definition 13 ([8]) Let X be a set and $A = \{<x, \mu_A(x), \nu_A(x)>: x \in X\} \in IFS(X)$, $\alpha, \beta, \omega, \theta \in [0, 1]$. We define the following operator,

$$Z^{\omega,\theta}_{\alpha,\beta}(A) = \{<x, \beta(\alpha\mu_A(x) + \omega - \omega.\alpha), \alpha(\beta\nu_A(x) + \theta - \theta.\beta)>: x \in X\}$$

The operator $Z^{\omega,\theta}_{\alpha,\beta}$ is a generalization of $Z^{\omega}_{\alpha,\beta}$, and also, $E_{\alpha,\beta}$, $⊞_{\alpha,\beta}$, $⊠_{\alpha,\beta}$. The new diagram of OTMOs as in Fig. 6.

Before defining new operators which are generalization of both one type and second type modal operators, we will recall definitions of second type modal operators.

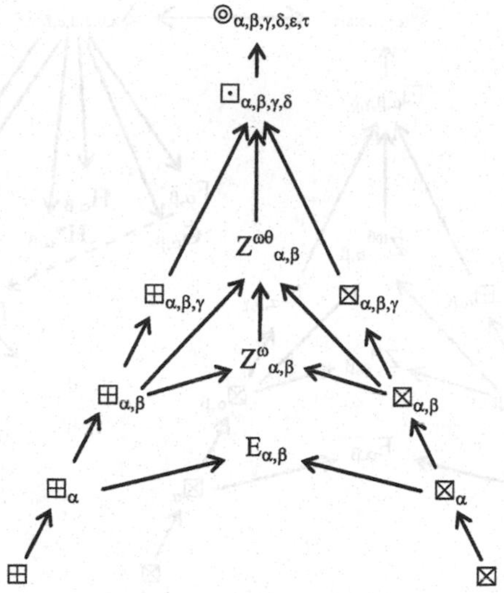

Fig. 6

Definition 14 ([1]) Let X be universal and $A \in IFS(X)$, $\alpha \in [0,1]$ then

$$D_\alpha(A) = \{ <x, \mu_A(x) + \alpha\pi_A(x), \nu_A(x) + (1-\alpha)\pi_A(x)> : x \in X\}$$

Definition 15 ([1]) Let X be universal and $A \in IFS(X)$, $\alpha, \beta \in [0, 1]$ and $\alpha + \beta \le 1$ then

$$F_{\alpha,\beta}(A) = \{ <x, \mu_A(x) + \alpha\pi_A(x), \nu_A(x) + \beta\pi_A(x)> : x \in X\}$$

Definition 16 ([1]) Let X be universal and $A \in IFS(X)$, $\alpha, \beta \in [0, 1]$ then

$$G_{\alpha,\beta}(A) = \{ <x, \alpha\mu_A(x), \beta\nu_A(x)> : x \in X\}$$

Definition 17 ([1]) Let X be universal and $A \in IFS(X)$, $\alpha, \beta \in [0, 1]$ then

(a) $H_{\alpha,\beta}(A) = \{ <x, \alpha\mu_A(x), \nu_A(x) + \beta\pi_A(x)> : x \in X\}$
(b) $H^*_{\alpha,\beta}(A) = \{ <x, \alpha\mu_A(x), \nu_A(x) + \beta(1 - \alpha\mu_A(x) - \nu_A(x))> : x \in X\}$

Definition 18 ([1]) Let X be universal and $A \in IFS(X)$, $\alpha, \beta \in [0, 1]$ then

(a) $J_{\alpha,\beta}(A) = \{ <x, \mu_A(x) + \alpha\pi_A(x), \beta\nu_A(x)> : x \in X\}$
(b) $J^*_{\alpha,\beta}(A) = \{ <x, \mu_A(x) + \alpha(1 - \mu_A(x) - \beta\nu_A(x)), \beta\nu_A(x)> : x \in X\}$

Definition 19 ([1]) Let X be universal and $A \in IFS(X)$ then (Fig. 7)

(a) $\square(A) = \{ <x, \mu_A(x), 1 - \mu_A(x)> : x \in X\}$
(b) $\diamond(A) = \{ <x, 1 - \nu_A(x), \nu_A(x)> : x \in X\}$

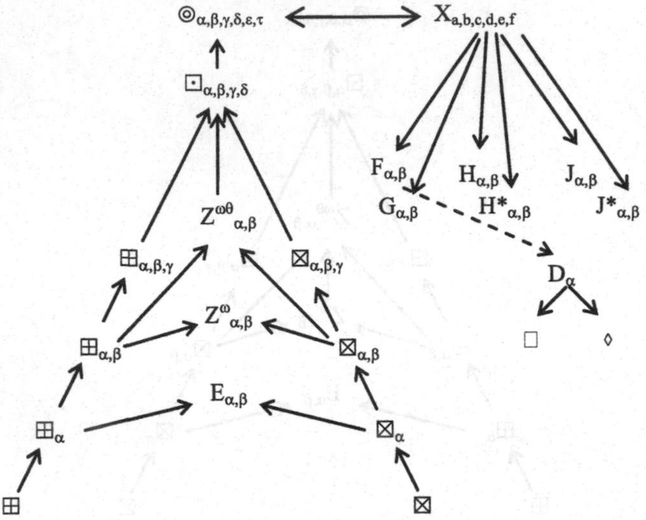

Fig. 7

2 The Uni-type Operators $E^{\omega,\theta}_{\alpha,\beta}$, $B_{\alpha,\beta}$ and $\boxminus_{\alpha,\beta}$

In this section, we give some operators. Some of them satisfy characteristic properties of one and two type operators. Because of these properties, they will be called uni-type operators.

Definition 20 ([9]) Let X be a universal, $A \in IFS(X)$ and $\alpha, \beta, \omega \in [0, 1]$. We define the following operator:

(a) $\boxplus^{\omega}_{\alpha,\beta}(A)=\{\langle x, \beta(\mu_A(x)+(1-\alpha)\nu_A(x)), \alpha(\beta\nu_A(x)+\omega-\omega\beta)\rangle : x\in X\}$

(b) $\boxtimes^{\omega}_{\alpha,\beta}(A)=\{\langle x, \beta(\alpha\mu_A(x)+\omega-\omega\alpha), \alpha((1-\beta)\mu_A(x)+\nu_A(x))\rangle : x\in X\}$

It is clear that;

$$\begin{aligned}&\beta(\mu_A(x)+(1-\alpha)\nu_A(x))+\alpha(\beta\nu_A(x)+\omega-\omega\beta)\\&=\beta(\mu_A(x)+\nu_A(x))+\alpha\omega(1-\beta)\\&\leq\beta+\alpha\omega-\alpha\beta\omega=\beta(1-\alpha\omega)+\alpha\omega\leq 1\end{aligned}$$

It is clear that $\boxplus^{\omega}_{\alpha,\beta}(A)\in IFS(X)$. We can say $\boxtimes^{\omega}_{\alpha,\beta}(A)\in IFS(X)$, too.

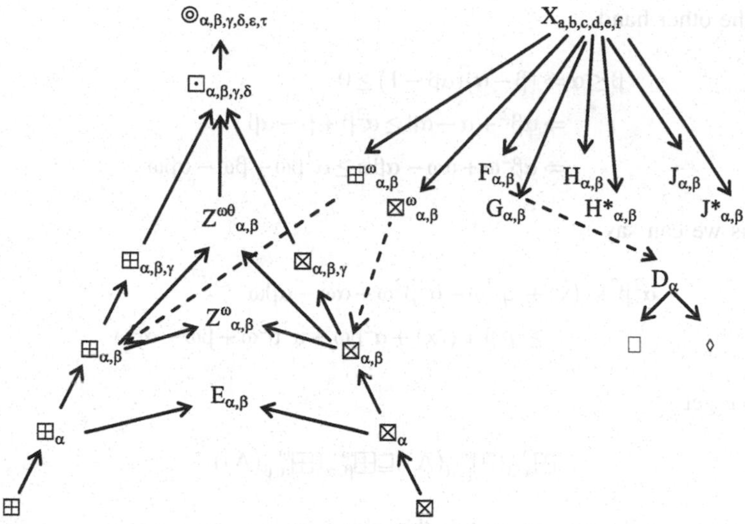

Fig. 8

From this definition, we get the following new diagram which is the extension of the last diagram of intuitionistic fuzzy operators on IFSs in Fig. 8.

Some fundamental properties of these operators are following.

Theorem 1 *Let X be a universal,* $A \in IFS\,(X)$ *and* $\alpha, \beta, \omega \in [0, 1]$ *then*

(a) *if* $\beta \leq \alpha$ *then* $\boxplus^{\omega}_{\alpha,\beta}(\boxplus^{\omega}_{\alpha,\beta}(A)) \sqsubseteq \boxplus^{\omega}_{\beta,\alpha}(\boxplus^{\omega}_{\beta,\alpha}(A))$

(b) *if* $\beta \leq \alpha$ *then* $\boxtimes^{\omega}_{\alpha,\beta}(\boxtimes^{\omega}_{\alpha,\beta}(A)) \sqsubseteq \boxtimes^{\omega}_{\beta,\alpha}(\boxtimes^{\omega}_{\beta,\alpha}(A))$

Proof

(a) If we use $\beta \leq \alpha$ then we get,

$$\begin{aligned}\beta \leq \alpha &\Rightarrow (\alpha-\beta)(\alpha+\beta+2\alpha\beta) \leq 0\\ &\Rightarrow \beta^2(1-2\alpha) \leq \alpha^2(1-2\beta)\\ &\Rightarrow \beta^2(1-2\alpha)\omega \leq \alpha^2(1-2\beta)\omega\end{aligned}$$

and with this inequality we can say

$$\begin{aligned}&\alpha\beta\mu_A(x)+\alpha\beta(1-\beta)\nu_A(x)+\beta(1-\alpha)(\alpha\beta\nu_A(x)+\beta\omega-\alpha\beta\omega)\\ &\leq \alpha\beta\mu_A(x)+\alpha\beta(1-\alpha)\nu_A(x)+\alpha(1-\beta)(\alpha\beta\nu_A(x)+\alpha\omega-\alpha\beta\omega)\end{aligned}$$

□

On the other hand

$$\begin{aligned}\beta \le \alpha &\Rightarrow (\beta-\alpha)(\alpha\beta-1)\ge 0\\ &\Rightarrow \alpha\beta^2+\alpha-\alpha\beta \ge \alpha^2\beta+\beta-\alpha\beta\\ &\Rightarrow \alpha\beta^2\omega+\alpha\omega-\alpha\beta\omega \ge \alpha^2\beta\omega+\beta\omega-\alpha\beta\omega\end{aligned}$$

with this we can say

$$\begin{aligned}&\alpha^2\beta^2\nu_A(x)+\alpha\beta^2\omega-\alpha^2\beta^2\omega+\alpha\omega-\alpha\beta\omega\\ &\qquad \ge \alpha^2\beta^2\nu_A(x)+\alpha^2\beta\omega-\alpha^2\beta^2\omega+\beta\omega-\alpha\beta\omega\end{aligned}$$

So we get

$$\boxplus^{\omega}_{\alpha,\beta}(\boxplus^{\omega}_{\beta,\alpha}(A))\sqsubseteq\boxplus^{\omega}_{\beta,\alpha}(\boxplus^{\omega}_{\alpha,\beta}(A))$$

We can show the property (b) as the same way.

Proposition 1 *Let X be a universal, A ∈ IFS (X) and* $\alpha, \beta \in [0, 1)$ *then*

(a) $\boxplus^{\frac{\beta}{1-\alpha}}_{1,\alpha}(A)=\boxplus_{\alpha,\beta}(A)$
(b) $\boxtimes^{\frac{\beta}{1-\alpha}}_{\alpha,1}(A)=\boxtimes_{\alpha,\beta}(A)$

Proof It is clear from definition. □

Definition 21 ([10]) Let X be a set and $A=\{<x, \mu_A(x), \nu_A(x)>: x\in X\} \in IFS(X)$, $\alpha, \beta, \omega, \theta \in [0, 1]$ .We define the following operator:

$$\begin{aligned}E^{\omega,\theta}_{\alpha,\beta}(A)=\{<x,\ &\beta((1-(1-\alpha)(1-\theta))\mu_A(x)+(1-\alpha)\theta\nu_A(x)+(1-\alpha)(1-\theta)\omega),\\ &\alpha((1-\beta)\theta\mu_A(x)+(1-(1-\beta)(1-\theta))\nu_A(x)+(1-\beta)(1-\theta)\omega)>: x\in X\}\end{aligned}$$

Proposition 2 ([10]) *Let X be a set and A ∈ IFS (X),* $\alpha, \beta, \omega, \theta \in [0, 1]$

$$E^{\omega,\theta}_{\alpha,\beta}(A^c)=E^{\omega,\theta}_{\beta,\alpha}(A)^c$$

Proof It is clear from definition. □

Proposition 3 ([10]) *Let X be a set and A ∈ IFS (X),* $\alpha, \beta, \omega, \theta \in [0, 1]$ *if* $\beta\le\alpha$ *then*

$$E^{\omega,\theta}_{\alpha,\beta}(A)\sqsubseteq E^{\omega,\theta}_{\beta,\alpha}(A)$$

Proof If we use $\beta \leq \alpha$ then

$$\begin{aligned}\beta \leq \alpha &\Rightarrow \beta(\theta(\mu_A(x)+\nu_A(x))+\omega(1-\theta)) \leq \alpha(\theta(\mu_A(x)+\nu_A(x))+\omega(1-\theta))\\ &\Rightarrow \beta(\theta(\mu_A(x)+\nu_A(x))+\omega(1-\theta))+\alpha\beta(\mu_A(x)+\theta\mu_A(x)-\theta\nu_A(x))\\ &\leq \alpha(\theta(\mu_A(x)+\nu_A(x))+\omega(1-\theta))+\alpha\beta(\mu_A(x)+\theta\mu_A(x)-\theta\nu_A(x))\end{aligned}$$

so we can say $E^{\omega,\theta}_{\alpha,\beta}(A) \sqsubseteq E^{\omega,\theta}_{\beta,\alpha}(A)$ □

Proposition 4 ([10]) *Let X be a set and A ∈ IFS (X),* $\alpha, \beta, \omega, \theta \in [0, 1]$ *if* $\omega \leq \theta$ *then*

$$E^{\omega,\theta}_{\alpha,\beta}(A) \sqsubseteq E^{\theta,\omega}_{\alpha,\beta}(A)$$

Proof It is clear from definition. □

Definition 22 Let X be a set, A ∈ IFS (X) and $\alpha, \beta \in [0, 1]$. We define the following operator:

$$B_{\alpha,\beta}(A) = \{\langle x, \beta(\mu_A(x) + (1-\alpha)\nu_A(x)), \alpha((1-\beta)\mu_A(x) + \nu_A(x))\rangle : x \in X\}$$

Definition 23 Let X be a set, A ∈ IFS (X) and $\alpha, \beta, \omega \in [0, 1]$. We define the following operator:

$$\boxminus_{\alpha,\beta}(A) = \{\langle x, \beta(\mu_A(x) + (1-\beta)\nu_A(x)), \alpha((1-\alpha)\mu_A(x) + \nu_A(x))\rangle : x \in X\}$$

Theorem 2 *Let X be a set, A ∈ IFS (X) and* $\alpha, \beta \in [0, 1]$.

$$B_{\alpha,\alpha}(A) = \boxminus_{\alpha,\alpha}(A)$$

Proof It is clear from definition. □

Theorem 3 *Let X be a set, A ∈ IFS (X) and* $\alpha, \beta, \omega \in [0, 1]$

(a) $\boxplus^{\omega}_{\alpha,\beta}(A^c) = \boxtimes^{\omega}_{\beta,\alpha}(A)^c$
(b) $\boxtimes^{\omega}_{\alpha,\beta}(A^c) = \boxplus^{\omega}_{\beta,\alpha}(A)^c$
(c) $\boxminus_{\alpha,\beta}(A^c) = \boxminus_{\beta,\alpha}(A)^c$

Proof

(a) From definition of this operators and complement of an intuitionistic fuzzy set we get those,
$\boxtimes^{\omega}_{\beta,\alpha}(A)^c = \{ < x, \beta((1-\alpha)\mu_A(x) + \nu_A(x)), \alpha(\beta\mu_A(x) + \omega - \omega\beta) > : x \in X\}$

and

$$\boxplus^{\omega}_{\alpha,\beta}(A^c)=\{\langle x,\ \beta(\nu_A(x)+(1-\alpha)\mu_A(x)),\ \alpha(\beta\mu_A(x)+\omega-\omega\beta)\rangle : x\in X\}$$

So, we get $\boxplus^{\omega}_{\alpha,\beta}(A^c)=\boxtimes^{\omega}_{\beta,\alpha}(A)^c$.

(b) If we use definitions then we get

$$\boxminus_{\alpha,\beta}(A^c)=\{\langle x,\ \beta(\nu_A(x)+(1-\beta)\mu_A(x)),\ \alpha((1-\alpha)\nu_A(x)+\mu_A(x))\rangle : x\in X\}$$

and

$$\boxminus_{\beta,\alpha}(A)^c=\{\langle x,\ \beta((1-\beta)\mu_A(x)+\nu_A(x)),\ \alpha(\mu_A(x)+(1-\alpha)\nu_A(x))\rangle : x\in X\}$$

□

Theorem 4 *Let X be a set and A IFS (X),* $\alpha,\ \beta\in[0,1]$. *If* $\alpha\geq\frac{1}{2}$, $\beta\leq\frac{1}{2}$ *then*

$$B_{\alpha\beta}(B_{\beta\alpha}(A))\sqsubseteq B_{\beta\alpha}(B_{\alpha\beta}(A))$$

Proof If we use $\alpha\geq\frac{1}{2}$ and $\beta\leq\frac{1}{2}$ then we get,

$$(1-2\alpha)\leq(1-2\beta)\Rightarrow\beta^2(1-2\alpha)(\mu_A(x)+\nu_A(x))\leq\alpha^2(1-2\beta)(\mu_A(x)+\nu_A(x))$$

So,

$$\begin{aligned}&\alpha\beta\mu_A(x)+\alpha\beta(1-\beta)\nu_A(x)+\beta^2(1-\alpha)^2\mu_A(x)+\beta^2(1-\alpha)\nu_A(x)\\&\leq\alpha\beta\mu_A(x)+\alpha\beta(1-\alpha)\nu_A(x)+\alpha^2(1-\beta)^2\mu_A(x)+\alpha^2(1-\beta)\nu_A(x)\end{aligned}$$

and

$$\begin{aligned}&\alpha^2(1-\beta)\mu_A(x)+\alpha^2(1-\beta)^2\nu_A(x)+\alpha\beta(1-\alpha)\mu_A(x)+\alpha\beta\nu_A(x)\\&\geq\beta^2(1-\alpha)\mu_A(x)+\beta^2(1-\alpha)^2\nu_A(x)+\alpha\beta(1-\beta)\mu_A(x)+\alpha\beta\nu_A(x)\end{aligned}$$

with these inequalities $B_{\alpha\beta}(B_{\beta\alpha}(A))\sqsubseteq B_{\beta\alpha}(B_{\alpha\beta}(A))$ □

Proposition 5 *Let X be a set and* $A\in IFS\ (X)$, $\alpha,\ \beta\in[0,1]$ *then*

$$B_{\alpha\beta}(A^c)=B_{\beta\alpha}(A)^c$$

Proof It is clear from definition. □

Proposition 6 *Let X be a set and* $A\in IFS\ (X)$, $\alpha,\ \beta,\ \omega\in[0,1]$ *then*

(a) $E^{\omega,0}_{\alpha,\beta}(A)=Z^{\omega}_{\alpha,\beta}(A)$
(b) $E^{\omega,1}_{\alpha,\beta}(A)=B_{\alpha,\beta}(A)$
(c) $E^{0,0}_{\alpha,\beta}(A)=G_{\alpha\beta,\alpha\beta}(A)$

(d) $E^{1,0}_{\alpha,\beta}(A) = E_{\alpha,\beta}(A)$
(e) $E^{0,0}_{1,0}(A) = \emptyset$
(f) $E^{0,0}_{0,1}(A) = X$

Proof Clear from definition. □

Proposition 7 *Let X be a set and A ∈ IFS (X),* α, β, ω ∈ [0, 1] *then*

(a) $E^{1,0}_{\alpha,1}(A) = \boxtimes_{\alpha}(A)$
(b) $E^{\omega,0}_{\alpha,1}(A) = \boxtimes_{\alpha,\omega(1-\alpha)}(A)$
(c) $E^{\omega,0}_{1,\beta}(A) = \boxplus_{\beta,\omega(1-\beta)}(A)$
(d) $E^{\omega,\theta}_{1,1}(A) = A$
(e) $E^{1,0}_{1,\beta}(A) = \boxplus_{\beta}(A)$
(f) $E^{\omega,1}_{\alpha,1}(A) = B_{\alpha,1}(A)$
(g) $E^{1,1}_{1,\beta}(A) = B_{1,\beta}(A)$

Proof Clear from definition. □

From these properties, as in Fig. 9, we get the following new diagram for intuitionistic fuzzy modal operators,

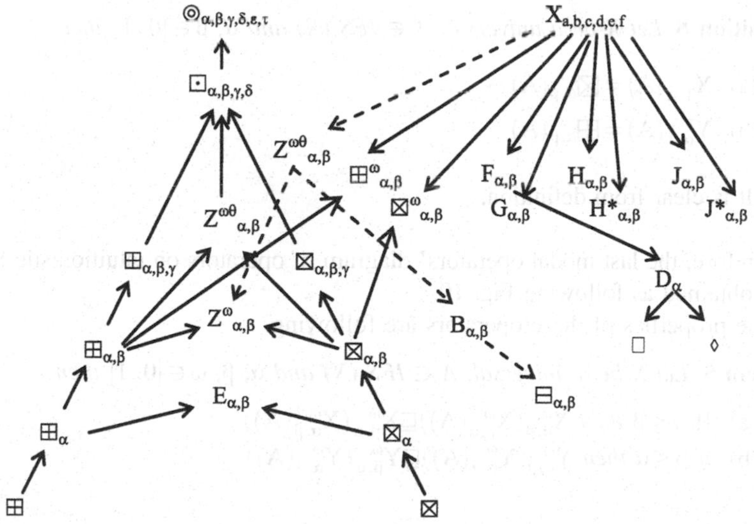

Fig. 9

3 The New Operators $X^{\omega}_{\alpha\beta}$ and $Y^{\omega}_{\alpha\beta}$

In this section, we give two new operators on intuitionistic fuzzy sets. They satisfy the properties of one type modal operators. So, these operators belong to family of one type operators.

Definition 24 Let X be a universal, $A \in IFS\ (X)$ and $\alpha, \beta, \omega \in [0, 1]$. We define the following operators,

(a) $X^{\omega}_{\alpha,\beta}(A) = \{\langle x, \beta(\mu_A(x) + (1-\alpha)\nu_A(x)) + \alpha(\omega - \omega\beta), \alpha\beta\nu_A(x)\rangle : x \in X\}$
(b) $Y^{\omega}_{\alpha,\beta}(A) = \{\langle x, \alpha\beta\mu_A(x), \alpha((1-\beta)\mu_A(x) + \nu_A(x)) + \beta(\omega - \omega\alpha)\rangle : x \in X\}$

It is clear that;

$$\begin{aligned}
&\beta(\mu_A(x) + (1-\alpha)\nu_A(x)) + \alpha(\omega - \omega\beta) + \alpha\beta\nu_A(x) \\
&= \beta(\mu_A(x) + \nu_A(x)) + \alpha\omega(1-\beta) \\
&\leq \beta + \alpha\omega - \alpha\beta\omega = \beta(1-\alpha\omega) + \alpha\omega \leq 1
\end{aligned}$$

and

$$\begin{aligned}
&\alpha\beta\mu_A(x) + \alpha((1-\beta)\mu_A(x) + \nu_A(x)) + \beta(\omega - \omega\alpha) \\
&= \alpha(\mu_A(x) + \nu_A(x)) + \beta\omega(1-\alpha) \\
&\leq \alpha + \beta\omega - \alpha\beta\omega = \alpha(1-\beta\omega) + \beta\omega \leq 1
\end{aligned}$$

Therefore we get $X^{\omega}_{\alpha,\beta}(A), Y^{\omega}_{\alpha,\beta}(A) \in IFS(X)$.

Proposition 8 *Let X be a universal,* $A \in IFS\ (X)$ *and* $\alpha, \beta \in [0, 1)$ *then*

(a) $X^{\frac{\beta}{1-\alpha}}_{1,\alpha}(A) = \boxtimes_{\alpha,\beta}(A)$
(b) $Y^{\frac{\beta}{1-\alpha}}_{\alpha,1}(A) = \boxplus_{\alpha,\beta}(A)$

Proof It is clear from definition. □

Therefore, the last modal operators' diagram of operators on intuitionistic fuzzy sets is obtained as following Fig. 10.

Some properties of these operators are following.

Theorem 5 *Let X be a universal,* $A \in IFS\ (X)$ *and* $\alpha, \beta, \omega \in [0, 1]$ *then*

(a) If $\alpha \leq \beta$ *then* $X^{\omega}_{\alpha,\beta}(X^{\omega}_{\beta,\alpha}(A)) \sqsubseteq X^{\omega}_{\beta,\alpha}(X^{\omega}_{\alpha,\beta}(A))$
(b) *if* $\beta \leq \alpha$ *then* $Y^{\omega}_{\alpha,\beta}(Y^{\omega}_{\beta,\alpha}(A)) \sqsubseteq Y^{\omega}_{\beta,\alpha}(Y^{\omega}_{\alpha,\beta}(A))$

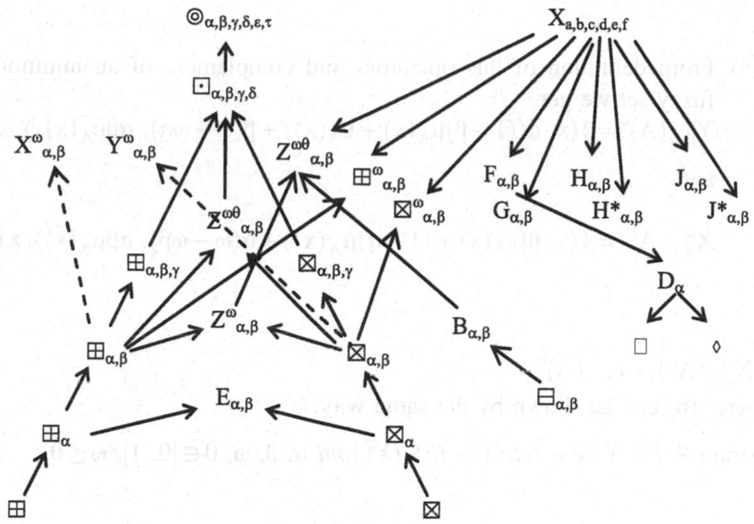

Fig. 10

Proof

(a) If we use $\alpha \le \beta$ then,

$$\begin{aligned} \alpha \le \beta &\Rightarrow (\alpha-\beta)(1-\alpha-\beta+\alpha\beta) \le 0 \\ &\Rightarrow \beta^2(1-\alpha)+\alpha \le \alpha^2(1-\beta)+\beta \\ &\Rightarrow \omega[\beta^2(1-\alpha)+\alpha] \le \omega[\alpha^2(1-\beta)+\beta] \end{aligned}$$

and with this inequality we can say

$$\begin{aligned} &\alpha\beta\mu_A(x)+\alpha\beta(1-\beta)\nu_A(x)+\beta^2(\omega-\alpha\omega)+\alpha\beta^2(1-\alpha)\nu_A(x)+\alpha\omega-\alpha\beta\omega \\ &\le \alpha\beta\mu_A(x)+\alpha\beta(1-\alpha)\nu_A(x)+\alpha^2(\omega-\beta\omega)+\alpha^2(1-\beta)(\beta\nu_A(x))+\beta\omega-\alpha\beta\omega \end{aligned}$$

□

On the other hand

$$\alpha^2\beta^2\nu_A(x) = \alpha^2\beta^2\nu_A(x)$$

So, $X^{\omega}_{\alpha,\beta}(X^{\omega}_{\beta,\alpha}(A)) \sqsubseteq X^{\omega}_{\beta,\alpha}(X^{\omega}_{\alpha,\beta}(A))$

We can show the property (b) same way.

Theorem 6 *Let X be a set, A ∈ IFS (X) and* $\alpha, \beta, \omega \in [0, 1]$.

(a) $X^{\omega}_{\alpha,\beta}(A^c) = Y^{\omega}_{\beta,\alpha}(A)^c$

(b) $Y^{\omega}_{\alpha,\beta}(A^c) = X^{\omega}_{\beta,\alpha}(A)^c$

Proof

(a) From definition of this operators and complement of an intuitionistic fuzzy set we get,
$Y^{\omega}_{\beta,\alpha}(A)^c = \{\langle x, \alpha((1-\beta)\mu_A(x)+\nu_A(x))+\beta(\omega-\omega\alpha), \alpha\beta\mu_A(x), \rangle : x \in X\}$
and

$$X^{\omega}_{\alpha,\beta}(A^c) = \{\langle x, \beta(\nu_A(x)+(1-\alpha)\mu_A(x))+\alpha(\omega-\omega\beta), \alpha\beta\mu_A(x)\rangle : x \in X\}$$

□

So, $X^{\omega}_{\alpha,\beta}(A^c) = Y^{\omega}_{\beta,\alpha}(A)^c$.
Property (b) can be shown by the same way.

Proposition 9 *Let X be a set, A ∈ IFS (X) and* $\alpha, \beta, \omega, \theta \in [0, 1]$, $\omega \leq \theta$.

(a) $X^{\omega}_{\alpha,\beta}(A) \sqsubseteq X^{\theta}_{\alpha,\beta}(A)$
(b) $Y^{\omega}_{\alpha,\beta}(A) \sqsubseteq Y^{\theta}_{\alpha,\beta}(A)$

Proof

(a) If we use $\omega \leq \theta$ then we can see,

$$\omega \leq \theta \Rightarrow \beta(1-\alpha)\omega \leq \beta(1-\alpha)\theta$$

and through this inequality,

$$\alpha((1-\beta)\mu_A(x)+\nu_A(x))+\beta(\omega-\omega\alpha) \leq \alpha((1-\beta)\mu_A(x)+\nu_A(x))+\beta(\theta-\theta\alpha)$$

□

So, $Y^{\omega}_{\alpha,\beta}(A) \sqsubseteq Y^{\theta}_{\alpha,\beta}(A)$

Theorem 7 *Let X be a set, A ∈ IFS (X) and* $\alpha, \beta, \omega \in [0, 1]$.

(a) $X^{\omega}_{\alpha,\beta}(\boxplus_\alpha A) \sqsubseteq X^{\omega}_{\alpha,\beta}(\boxtimes_\alpha A)$
(b) $Y^{\omega}_{\alpha,\beta}(\boxplus_\alpha A) \sqsubseteq Y^{\omega}_{\alpha,\beta}(\boxtimes_\alpha A)$

Proof

(a)

$$\begin{aligned}\beta(1-\alpha)^2 \leq \beta(1-\alpha) \Rightarrow \beta(\alpha\mu_A(x)+\alpha(1-\alpha)\nu_A(x)+(1-\alpha)^2)+\alpha(\omega-\omega\beta)\\ \leq \beta(\alpha\mu_A(x)+\alpha(1-\alpha)\nu_A(x)+(1-\alpha))+\alpha(\omega-\omega\beta)\end{aligned}$$

and

$$\alpha\beta - \alpha^2\beta \geq 0 \Rightarrow \alpha^2\beta\nu_A(x) + \alpha\beta - \alpha^2\beta \geq \alpha^2\beta\nu_A(x)$$

Therefore, $X^{\omega}_{\alpha,\beta}(\boxplus_\alpha A) \sqsubseteq X^{\omega}_{\alpha,\beta}(\boxtimes_\alpha A)$

Similarly, if we use $\alpha\beta(1-\alpha) \geq 0$ and $\alpha(1-\alpha) \geq \alpha(1-\beta)(1-\alpha)$ we get

$$Y^{\omega}_{\alpha,\beta}(\boxplus_\alpha A) \sqsubseteq Y^{\omega}_{\alpha,\beta}(\boxtimes_\alpha A)$$

References

1. Atanassov, K.T.: Intuitionistic fuzzy sets. Fuzzy Sets Syst. **20**, 87–96 (1986)
2. Atanassov, K.T.: Intuitionistic Fuzzy Sets. Phiysica-Verlag, Heidelberg (1999)
3. Atanassov, K.T.: Remark on two operations over intuitionistic fuzzy sets. Int. J. Uncertain. Fuzzyness Knowl. Syst. **9**(1), 71–75 (2001)
4. Atanassov, K.T.: The most general form of one type of intuitionistic fuzzy modal operators. NIFS **12**(2), 36–38 (2006)
5. Atanassov, K.T.: Some Properties of the operators from one type of intuitionistic fuzzy modal operators. Adv. Stud. Contemp. Math. **15**(1), 13–20 (2007)
6. Atanassov, K.T.: The most general form of one type of intuitionistic fuzzy modal operators, Part 2. NIFS **14**(1), 27–32 (2008)
7. Çuvalcoğlu, G.: Some Properties of $E_{\alpha,\beta}$ operator. Adv. Stud. Contemp. Math. **14**(2), 305–310 (2007)
8. Çuvalcoğlu, G.: On the diagram of one type modal operators on intuitionistic fuzzy sets: last expanding with $Z_{\alpha,\beta}^{\omega,\theta}$. Iran. J. Fuzzy Syst. **10**(1), 89–106 (2013)
9. Çuvalcoğlu, G.: The extension of modal operators' diagram with last operators. Notes on Intuitionistic Fuzzy Sets, **19**(3), 56–61 (2013)
10. Dencheva, K.: Extension of intuitionistic fuzzy modal operators ⊞ and ⊠. In: Proceedings of the Second International. IEEE Symposium: Intelligent Systems, vol. 3, pp. 21–22. Varna, 22–24 June 2004
11. Doycheva, B.: Inequalities with intuitionistic fuzzy topological and Gökhan Çuvalcoğlu's operators. NIFS **14**(1), 20–22 (2008)
12. Li, D., Shan, F., Cheng, C.: On Properties of four IFS operators. Fuzzy Sets Syst. **154**, 151–155 (2005)
13. Zadeh, L.A.: Fuzzy sets. Inf. Control **8**, 338–353 (1965)

Intuitionistic Fuzzy Relational Equations in BL−Algebras

Ketty Peeva

Abstract We investigate direct and inverse problem resolution for intuitionistic fuzzy relational equations in some BL−algebras, when the composition for the membership degrees is a sup $-t-$norm and for non-membership degrees is an inf $-s-$norm. Criterion for solvability of intuitionistic fuzzy relational equation is proposed and analytical expressions for maximal solution is given.

Keywords Intuitionistic fuzzy relations · Direct and inverse problems · BL−algebra

1 Introduction

Intuitionistic fuzzy sets (IFS) were introduced by K. Atanassov in 1983 [1]. After publishing his monograph [2], the interest on IFS was rapidly increasing with many publications in variety of areas. My interest began about 2000 (see [20]). My attention was focused on intuitionistic fuzzy relations—direct and inverse problems, their algorithmical and software resolution [16, 22–24, 27]. Intuitionistic fuzzy relations are studied for instance in [5, 9, 13, 18–20, 22–24, 26].

Direct and inverse problem resolution for intuitionistic fuzzy linear system of equations, when the composition is max − min for membership degrees and min − max for non-membership degrees is studied first in [20–23] and corresponding software is given in [16, 27].

In this chapter we present direct and inverse problem resolution for intuitionistic fuzzy relations in some BL−algebras, when the composition $*_{BL}$ for the membership degrees is a sup $-t-$norm and for non-membership degrees is an inf $-s-$norm:

$$A *_{BL} B = C.$$

K. Peeva (✉)
Faculty of Applied Mathematics and Informatics,
Technical University of Sofia, 8, Kl. Ohridski St., Sofia 1000, Bulgaria
e-mail: kgp@tu-sofia.bg

P. Angelov and S. Sotirov (eds.), *Imprecision and Uncertainty in Information Representation and Processing*, Studies in Fuzziness and Soft Computing 332,
DOI 10.1007/978-3-319-26302-1_6

Here A, B and C are finite intuitionistic fuzzy matrices. Two of them are given, one is unknown:

(i) If A and B are given, computing the unknown matrix C is called direct problem resolution.
(ii) If A and C are given, computing the unknown matrix B is called inverse problem resolution.

In Sect. 2 we introduce $t-$norms, $BL-$algebras, intuitionistic fuzzy sets, intuitionistic fuzzy relations and intuitionistic fuzzy matrices. Section 3 presents direct problem resolution for intuitionistic fuzzy matrices in $BL-$algebras with examples in Gödel algebra, Goguen algebra and Łukasiewicz algebra. Section 4 covers intuitionistic fuzzy relational equations in $BL-$algebras, finding maximal solution and establishing consistency of the equation, as well as suitable examples. Concluding section proposes ideas for next development.

Terminology for algebra, orders and lattices is given according to [12, 17], for fuzzy sets, fuzzy relations and for intuitionistic fuzzy sets—according to [2, 7, 10, 15, 22], for computational complexity and algorithms is as in [11].

2 Basic Notions

Partial order relation on a partially ordered set (poset) P is denoted by the symbol $\leq$. By a *greatest element* of a poset P we mean an element $b \in P$ such that $x \leq b$ for all $x \in P$. The *least element* of P is defined dually.

The tree well known couples of $t-$norms and $t-$conorms (or $s-$norms) are given in Table 1.

2.1 BL−Algebra

$BL-$algebra [14] is the algebraic structure:

$$BL = \langle L, \vee, \wedge, *, \rightarrow, 0, 1 \rangle,$$

Table 1 $t-$norms and $s-$norms

$t-$norm	Name	Expression	$s-$norm	Name	Expression
t_3	Minimum, Gödel $t-$norm	$t_3(x, y) = \min\{x, y\}$	s_3	Maximum, Gödel $t-$conorm	$s_3(x, y) = \max\{x, y\}$
t_2	Algebraic product	$t_2(x, y) = xy$	s_2	Probabilistic sum	$s_2(x, y) = x + y - xy$
t_1	Łukasiewicz $t-$norm	$t_1(x, y) = \max\{x + y - 1, 0\}$	s_1	Bounded sum	$s_1(x, y) = \min\{x + y, 1\}$

where $\vee, \wedge, *, \rightarrow$ are binary operations, 0, 1 are constants and:

(i) $L = \langle L, \vee, \wedge, 0, 1\rangle$ is a lattice with universal bounds 0 and 1;
(ii) $L = \langle L, *, 1\rangle$ is a commutative semigroup;
(iii) $*$ and $\rightarrow$ establish an adjoint couple:

$$z \leq (x \rightarrow y) \Leftrightarrow x * z \leq y, \forall x, y, z \in L.$$

(iv) for all $x, y \in L$

$$x * (x \rightarrow y) = x \wedge y \quad \text{and} \quad (x \rightarrow y) \vee (y \rightarrow x) = 1.$$

We suppose in next exposition that $L = [0, 1]$ and $x, y \in [0, 1]$.
The following algebras are examples for $BL-$algebras.

1. Gödel algebra

$$BL_G = \langle [0, 1], \vee, \wedge, \rightarrow_G, 0, 1\rangle,$$

where operations are

(1) *Maximum* or s_3-conorm:

$$\max\{x, y\} = x \vee y. \tag{1}$$

(2) *Minimum* or t_3-norm:

$$\min\{x, y\} = x \wedge y. \tag{2}$$

(3) The residuum $\rightarrow_G$ is

$$x \rightarrow_G y = \begin{cases} 1 \text{ if } x \leq y \\ y \text{ if } x > y \end{cases}. \tag{3}$$

(4) A supplementary operation is useful

$$x \varepsilon y = \begin{cases} y, & \text{if } x < y \\ 0, & \text{if } x \geq y \end{cases}. \tag{4}$$

It is not difficult to prove that (3) and (4) satisfy:

$$\neg (x \rightarrow_G y) = \neg x \varepsilon \neg y. \tag{5}$$

2. Product (Goguen) algebra

$$BL_P = \langle [0, 1], \vee, \wedge, \circ, \rightarrow_P, 0, 1\rangle,$$

where max and min are as (1) and (2), respectively, $\circ$ is the conventional real number multiplication (the t_2 norm, i. e, $t_2(x, y) = xy$) and the residuum $\rightarrow_P$ is

$$x \rightarrow_P y = \begin{cases} 1 \text{ if } & x \le y \\ \frac{y}{x} \text{ if } & x > y \end{cases}. \tag{6}$$

Here the supplementary useful operation is:

$$x\gamma y = \begin{cases} 0 \text{ if } & x \ge y \\ \frac{y-x}{1-x} \text{ if } & x < y \end{cases}. \tag{7}$$

It is not difficult to prove that (6) and (7) satisfy:

$$\neg (x \rightarrow_P y) = \neg x \, \gamma \, \neg y. \tag{8}$$

3. Łukasiewicz algebra

$$BL_L = \langle [0, 1], \vee, \wedge, \otimes, \rightarrow_L, 0, 1 \rangle,$$

where max and min are as (1) and (2), respectively, and

(1) $x \otimes y = 0 \vee (x + y - 1) \equiv t_1(x, y)$.
(2) The residuum $\rightarrow_L$ is

$$x \rightarrow_L y = 1 \wedge (1 - x + y). \tag{9}$$

(3) A supplementary operation is useful

$$x \delta y = 0 \vee (y - x). \tag{10}$$

It is not difficult to prove that (9) and (10) satisfy:

$$\neg (x \rightarrow_L y) = \neg x \, \delta \, \neg y. \tag{11}$$

2.2 *Intuitionistic Fuzzy Sets*

Let $E \neq \emptyset$ be a crisp set and $A \subseteq E$. An ***intuitionistic fuzzy set*** [2] $\hat{A}$ on E is described as

$$\hat{A} = \left\{ \langle x, \, \mu_A(x), \, \nu_A(x) \rangle \, | x \in E \right\},$$

where for each $x \in E$, $\mu_A : E \to [0,\ 1]$ defines the degree of membership and $\nu_A : E \to [0,\ 1]$ defines the degree of non-membership, respectively, of the elements $x \in E$ to $\hat{A}$ and for each $x \in E$ holds $0 \leq \mu_A(x) + \nu_A(x) \leq 1$.

The class of all intuitionistic fuzzy sets over E is denoted by $Int(E)$.

Remark In [2] IFS is defined with degree of membership $\mu_A : E \to [0,\ 1]$, degree of non-membership and $\nu_A : E \to [0,\ 1]$ and uncertainity degree $\pi_A : E \to [0, 1]$ of the elements $x \in E$ to $\hat{A}$ and for each $x \in E$ holds $\mu_A(x) + \nu_A(x) + \pi_A(x) = 1$, but in this chapter we study IFS defined only with degree of membership an degree of non-membership.

2.3 Intuitionistic Fuzzy Relations and Intuitionistic Fuzzy Matrices

An ***intutionistic fuzzy relation*** (IFR) between two nonempty crisp sets X and Y is an intutionistic fuzzy set on $X \times Y$, written $R \in Int(X \times Y)$. $X \times Y$ is called ***support*** of R. We write $R \subseteq X \times Y$ for the (conventional) fuzzy relation R between X and Y.

Any IFR $R \in Int(X \times Y)$ is given as follows:

$$R = \left\{ \langle (x,\ y),\ \mu_R(x,\ y),\ \nu_R(x,\ y) \rangle \,\middle|\, (x,\ y) \in X \times Y,\ \mu_R,\ \nu_R : X \times Y \to [0,\ 1] \right\},$$

$$0 \leq \mu_R(x, y) + \nu_R(x, y) \leq 1$$

for each $(x, y) \in X \times Y$.

The matrix $A = (\mu^A_{ij},\ \nu^A_{ij})_{m \times n}$ with $\mu^A_{ij},\ \nu^A_{ij} \in [0,\ 1]$ such that

$$0 \leq \mu^A_{ij} + \nu^A_{ij} \leq 1 \text{ for each } i,\ j, 1 \leq i \leq m,\ 1 \leq j \leq n,$$

is called an ***intuitionistic fuzzy matrix*** (IFM) of type $m \times n$.

When the IFR is over finite support, it is representable by IFM, written for convenience with the same letter. For instance, if the IFR $R \in Int(X \times Y)$ is over finite support, its representative matrix is stipulated to be the matrix $R = (\mu^R_{x_iy_j},\ \nu^R_{x_iy_j})_{m \times n}$ such that

$$\mu^R_{x_iy_j} = \mu_R(x_i, y_j), \quad \nu^R_{x_iy_j} = \nu_R(x_i, y_j).$$

According to this stipulation instead of IFRs we consider intuitionistic fuzzy matrices and operations with them.

3 Direct Problem

Two finite IFMs $A = (\mu_{ij}^A, \nu_{ij}^A)_{m\times p}$ and $B = (\mu_{ij}^B, \nu_{ij}^B)_{p\times n}$ are called ***conformable*** in this order, if the number of columns in A is equal to the number of rows in B.

Let A and B be finite conformable IFMs.

Definition 1 If $A = (\mu_{ij}^A, \nu_{ij}^A)_{m\times p}$ and $B = (\mu_{ij}^B, \nu_{ij}^B)_{p\times n}$ are finite IFMs, then the matrix $C = (\mu_{ij}^C, \nu_{ij}^C)_{m\times n}$ is called ***intuitionistic fuzzy s − t product*** of A and B in BL−algebra, written $C = A *_{BL} B$, if for each $i, j, 1 \le i \le m, \ 1 \le j \le n$ it holds:

$$\mu_{ij}^C = s_3{}_{k=1}^{p}\left(t_r(\mu_{ik}^A, \mu_{kj}^B)\right), \quad \nu_{ij}^C = t_3{}_{k=1}^{p}\left(s_r(\nu_{ik}^A, \nu_{kj}^B)\right), \tag{12}$$

where s_3 and t_3 are maximum s−norm and minimum t−norm respectively; for $r = 1, 2, 3$ we have the corresponding s_r norm and t_r norm, respectively, see Table 1.

Computing the product $C = A *_{BL} B$ is called ***direct problem resolution*** for $*_{BL}$ composition of the matrices A and B. The resulting product-matrix $C = A *_{BL} B$ is also IFM because of the duality principle.

Direct problem for $*_{BL}$ composition of matrices is solvable in polynomial time.

Example 1 Find:

1. Intuitionistic fuzzy Gödel product $C = A *_G B$;
2. Intuitionistic fuzzy Goguen product $C = A *_P B$;
3. Intuitionistic fuzzy Łukasiewicz product $C = A *_L B$,

if

$$A = \begin{pmatrix} \langle 0.5, 0.4\rangle & \langle 0.5, 0.5\rangle & \langle 0.9, 0.1\rangle & \langle 0.3, 0.6\rangle \\ \langle 0.8, 0.1\rangle & \langle 0.1, 0.9\rangle & \langle 0.7, 0.2\rangle & \langle 0.5, 0.5\rangle \\ \langle 0.7, 0.3\rangle & \langle 0.9, 0.1\rangle & \langle 0.3, 0.6\rangle & \langle 0.4, 0.5\rangle \end{pmatrix}, \tag{13}$$

$$B = \begin{pmatrix} \langle 0.7, 0.3\rangle & \langle 0.9, 0.\rangle & \langle 0.5, 0.5\rangle \\ \langle 0.9, 0.1\rangle & \langle 0.3, 0.6\rangle & \langle 0.2, 0.7\rangle \\ \langle 0.5, 0.5\rangle & \langle 0.6, 0.3\rangle & \langle 1., 0.\rangle \\ \langle 0.4, 0.5\rangle & \langle 0.7, 0.2\rangle & \langle 0.5, 0.5\rangle \end{pmatrix}. \tag{14}$$

1. ***Intuitionistic fuzzy Gödel product*** [20, 23] of A and B, written $C_G = A *_G B$, is defined by

$$\mu_{ij}^C = \bigvee_{k=1}^{p} (\mu_{ik}^A \wedge \mu_{kj}^B), \quad \nu_{ij}^C = \bigwedge_{k=1}^{p} (\nu_{ik}^A \vee \nu_{kj}^B). \tag{15}$$

for each $i, j, 1 \le i \le m, \ 1 \le j \le n$.

Expression (15) is obtained from (12) for $r = 3$.

$C_G = A *_G B$ can be computed by software given in [16, 27]—for the membership degrees $\mu_{ij}^C = \bigvee_{k=1}^{p} (\mu_{ik}^A \wedge \mu_{kj}^B)$ one can use the max − min product code and for the non-membership degrees $\nu_{ij}^{C} = \bigwedge_{k=1}^{p} (\nu_{ik}^{A} \vee \nu_{kj}^{B})$ the min − max product code is useful:

$$C_G = A *_G B = \begin{pmatrix} \langle 0.5,\ 0.4\rangle & \langle 0.6,\ 0.3\rangle & \langle 0.9,\ 0.1\rangle \\ \langle 0.7,\ 0.3\rangle & \langle 0.8,\ 0.1\rangle & \langle 0.7,\ 0.2\rangle \\ \langle 0.9,\ 0.1\rangle & \langle 0.7,\ 0.3\rangle & \langle 0.5,\ 0.5\rangle \end{pmatrix}. \tag{16}$$

If the matrices A and B are IFM, then the matrix $C = A *_G B$ is also IFM, because $(\max, \min, N_s)$ is a dual triple [15].

2. ***Intuitionistic fuzzy Goguen product*** of A and B, written $C_P = A *_P B$, if for each $i, j, 1 \le i \le m,\ \ 1 \le j \le n$

$$\mu_{ij}^{C} = \bigvee_{k=1}^{p} (\mu_{ik}^{A} \cdot \mu_{kj}^{B}), \quad \nu_{ij}^{C} = \bigwedge_{k=1}^{p} (\nu_{ik}^{A} \oplus \nu_{kj}^{B}). \tag{17}$$

In this case expression (17) is obtained from expression (12) for $r = 2$. The sign $\oplus$ is used for probabilistic sum, i.e. $s_2(x, y) = x + y - xy = x \oplus y$, see Table 1.

The result in this case for (13) and (14) is the max −product for the membership degrees $\mu_{ij}^{C} = \bigvee_{k=1}^{p} (\mu_{ik}^{A} \cdot \mu_{kj}^{B})$ and the min −probabilistic sum for non-membership degrees $\nu_{ij}^{C} = \bigwedge_{k=1}^{p} (\nu_{ik}^{A} \oplus \nu_{kj}^{B})$.

If the matrices A and B are IFM, then the matrix $C_P = A *_P B$ is also IFM, because $(\max, t_2, N_s)$ is a dual triple [15].

One can find software only for computing the membership degrees as max-product code in [16, 27].

$$C_P = A *_P B = \begin{pmatrix} \langle 0.45,\ 0.55\rangle & \langle 0.54,\ 0.37\rangle & \langle 0.9,\ 0.1\rangle \\ \langle 0.56,\ 0.37\rangle & \langle 0.72,\ 0.1\rangle & \langle 0.7,\ 0.2\rangle \\ \langle 0.81,\ 0.19\rangle & \langle 0.63,\ 0.3\rangle & \langle 0.35,\ 0.6\rangle \end{pmatrix}. \tag{18}$$

3. ***Intuitionistic fuzzy Łukasiewicz product*** of A and B, written $C_L = A *_L B$, if for each $i, j, 1 \le i \le m,\ \ 1 \le j \le n$

$$\mu_{ij}^{C} = \bigvee_{k=1}^{p} \left(t_1(\mu_{ik}^{A}, \mu_{kj}^{B})\right), \quad \nu_{ij}^{C} = \bigwedge_{k=1}^{p} \left(s_1(\nu_{ik}^{A}, \nu_{kj}^{B})\right). \tag{19}$$

In this case expression (19) is obtained from expression (12) for $r = 1$.

The numerical example results:

$$C_L = A *_L B = \begin{pmatrix} \langle 0.4,\ 0.6\rangle & \langle 0.5,\ 0.4\rangle & \langle 0.9,\ 0.1\rangle \\ \langle 0.5,\ 0.4\rangle & \langle 0.7,\ 0.1\rangle & \langle 0.7,\ 0.2\rangle \\ \langle 0.8,\ 0.2\rangle & \langle 0.6,\ 0.3\rangle & \langle 0.3,\ 0.6\rangle \end{pmatrix}. \tag{20}$$

The result in this case for (13) and (14) is the max −Łukasiewicz product for the membership degrees $\mu_{ij}^C = \bigvee_{k=1}^{p} \left(t_1(\mu_{ik}^A, \mu_{kj}^B) \right)$ and the min −bounded sum for non-membership degrees $\nu_{ij}^C = \bigwedge_{k=1}^{p} \left(s_1(\nu_{ik}^A, \nu_{kj}^B) \right)$.

There does not exist software neither for $*_L$ composition, nor for any of the membership or non-membership degrees.

If the matrices A and B are IFM, then the matrix $C_L = A *_L B$ is also IFM, because $(\max, t_1, N_s)$ is a dual triple [3, 8, 15].

Definition 2 If $A = (\mu_{ij}^A, \nu_{ij}^A)_{m\times p}$ and $B = (\mu_{ij}^B, \nu_{ij}^B)_{p\times n}$ are IFMs, then the matrix $C = (\mu_{ij}^C, \nu_{ij}^C)_{m\times n}$ is called ***intuitionistic fuzzy*** $t-\rightarrow_{BL}$ ***product*** of A and B in $BL-$algebra, written $C = A \rightarrow_{BL} B$, if for each $i,\ j,\ 1 \le i \le m,\ \ 1 \le j \le n$ it holds:

$$\mu_{ij}^C = t_3{}_{k=1}^{p} \left((\mu_{ik}^A \rightarrow_{BL} \mu_{kj}^B) \right), \quad \nu_{ij}^C = s_3{}_{k=1}^{p} \left((\nu_{ik}^A \overset{r}{\rightarrow}_{BL} \nu_{kj}^B) \right), \tag{21}$$

where

s_3 and t_3 are maximum $s-$norm and minimum $t-$norm respectively;
$\overset{r}{\rightarrow}_{BL} = \delta$ for $r = 1$, see (10);
$\overset{r}{\rightarrow}_{BL} = \gamma$ for $r = 2$, see (7);
$\overset{r}{\rightarrow}_{BL} = \varepsilon$ for $r = 3$, see (4).

Computing the product $C = A \rightarrow_{BL} B$ is called ***direct problem resolution*** for $\rightarrow_{BL}$ composition of the matrices A and B . The resulting product-matrix $C = A \rightarrow_{BL} B$ is also IFM because of the duality principle.

Direct problem for $\rightarrow_{BL}$ composition of matrices is solvable in polynomial time.

We give examples for obtaining intuitionistic $t-\rightarrow_{BL}$ product of matrices in various $BL-$algebras.

Example 2 1. ***Intuitionistic fuzzy Gödel*** $t-\rightarrow_G$ ***product*** of A and B, written $C = A \rightarrow_G B$, if $r = 3$ in (21):

$$\mu_{ij}^C = \bigwedge_{k=1}^{p} (\mu_{ik}^A \rightarrow_G \mu_{kj}^B), \quad \nu_{ij}^C = \bigvee_{k=1}^{p} (\nu_{ik}^A\ \varepsilon\ \nu_{kj}^B), \tag{22}$$

for each $i,\ j,\ 1 \le i \le m,\ \ 1 \le j \le n$.

If the matrices A and B are IFM, then the matrix $C = A \rightarrow_G B$ is also IFM, because $(\max, \min, N_s)$ is a dual triple [15] and (5) is valid.

As example, we implement (22) to compute $A^t \rightarrow_G C_G$, where A^t is the transposed of A (see (13)) and C_G is due to (16).

$$\hat{B}_G = A^t \rightarrow_G C_G = \begin{pmatrix} \langle 0.7, 0.3\rangle & \langle 1., 0.\rangle & \langle 0.5, 0.5\rangle \\ \langle 1, 0.\rangle & \langle 0.7, 0.3\rangle & \langle 0.5, 0.5\rangle \\ \langle 0.5, 0.4\rangle & \langle 0.6, 0.3\rangle & \langle 1., 0.\rangle \\ \langle 1., 0.\rangle & \langle 1., 0.\rangle & \langle 1., 0.\rangle \end{pmatrix}. \tag{23}$$

For computing (23) by software given in [16], one computes separately membership degrees, using the code for $\min -\alpha$ product and non-membership degrees by the code for $\max -\epsilon$ product—in [16] it does not exist special code for the $\rightarrow_G$ intuitionistic product.

2. ***Intuitionistic fuzzy Goguen t–*** $\rightarrow_P$ ***product*** of A and B, written $C = A \rightarrow_P B$, if $r = 2$ in (21):

$$\mu_{ij}^C = \bigwedge_{k=1}^{p} (\mu_{ik}^A \rightarrow_P \mu_{kj}^B), \quad \nu_{ij}^C = \bigvee_{k=1}^{p} (\nu_{ik}^A \, \gamma \, \nu_{kj}^B) \tag{24}$$

for each $i, j, 1 \le i \le m, \quad 1 \le j \le n.$

If the matrices A and B are IFM, then the matrix $C = A \rightarrow_P B$ is also IFM, because $(\max, ., N_s)$ is a dual triple [15] and (8) is valid.

There does not exist software for computing $C = A \rightarrow_P B$.

As example, we implement (24) to compute $A^t \rightarrow_P C_P$, where A^t is the transposed of A (see(13)) and C_P is due to (18).

$$\hat{B}_P = A^t \rightarrow_P C_P = \begin{pmatrix} \langle 0.7, 0.27\rangle & \langle 0.9., 0.\rangle & \langle 0.5, 0.3\rangle \\ \langle 0.9, 0.09\rangle & \langle 0.7, 0.2\rangle & \langle 0.3889, 0.5\rangle \\ \langle 0.5, 0.45\rangle & \langle 0.6, 0.27\rangle & \langle 1., 0.\rangle \\ \langle 1., 0.\rangle & \langle 1., 0.\rangle & \langle 0.875, 0.1\rangle \end{pmatrix}. \tag{25}$$

3. ***Intuitionistic fuzzy Łukasiewicz t–*** $\rightarrow_L$ ***product*** of A and B, written $C = A \rightarrow_L B$, if $r = 1$ in (21):

$$\mu_{ij}^C = \bigwedge_{k=1}^{p} (\mu_{ik}^A \rightarrow_L \mu_{kj}^B), \quad \nu_{ij}^C = \bigvee_{k=1}^{p} (\nu_{ik}^A \, \delta \, \nu_{kj}^B) \tag{26}$$

for each $i, j, 1 \le i \le m, \quad 1 \le j \le n.$

If the matrices A and B are IFM, then the matrix $C = A \rightarrow_L B$ is also IFM, because $(\max, t_1, N_s)$ is a dual triple [15] and (11) is valid.

There does not exist software for computing $C = A \rightarrow_L B$.

As example, we implement (26) to compute $A^t \rightarrow_L C_L$, where A^t is the transposed of A (see(13)) and C_L is due to (20).

$$\hat{B}_L = A^t \rightarrow_L C_L = \begin{pmatrix} \langle 0.7,\ 0.3\rangle & \langle 0.9.,\ 0.\rangle & \langle 0.6,\ 0.3\rangle \\ \langle 0.9,\ 0.1\rangle & \langle 0.7,\ 0.2\rangle & \langle 0.4,\ 0.5\rangle \\ \langle 0.5,\ 0.5\rangle & \langle 0.6,\ 0.3\rangle & \langle 1.,\ 0.\rangle \\ \langle 1.,\ 0.\rangle & \langle 1.,\ 0.\rangle & \langle 0.9,\ 0.1\rangle \end{pmatrix}. \tag{27}$$

Hence, if the matrices A and B are IFM, then the matrices $A *_{BL} B$ and $A \rightarrow_{BL} B$ are also IFM, because of the duality principle.

For finite conformable intuitionistic fuzzy matrices direct problem for $*_{BL}$ and $\rightarrow_{BL}$ compositions of matrices is solvable in polynomial time, but software is not developed.

4 Inverse Problem

Let A and B be conformable IFMs.

(i) The equation

$$C = A *_{BL} B, \tag{28}$$

where one of the matrices on the left side is unknown and the other two matrices are given, is called $*_{BL}$ ***intuitionistic fuzzy matrix equation***.

(ii) The equation

$$C = A \rightarrow_{BL} B, \tag{29}$$

where one of the matrices on the left side is unknown and the other two matrices are given, is called $\rightarrow_{BL}$ ***intuitionistic fuzzy matrix equation***.

In (28) and (29) $A = (a_{ij})_{m\times p}$ stands for the IFM of coefficients, $B = (b_{ij})_{p\times n}$ – for the IFM of unknowns, $C = (c_{ij})_{m\times n}$ is the right-hand side of the equation, a_{ij}, b_{ij}, $c_{ij} \in [\,0,\ 1\,]$ for each i and each j.

Solving (28) or (29) for the unknown matrix is called ***inverse problem resolution for intuitionistic fuzzy matrix equation in BL–algebra***. In this chapter we present inverse problem resolution for (28).

For $X = (\langle \mu_{ij}(x), \nu_{ij}(x)\rangle)_{p\times n}$ and $Y = (\langle \mu_{ij}(y), \nu_{ij}(y)\rangle)_{p\times n}$ the inequality

$$X \leq Y$$

means $\mu_{ij}(x) \leq \mu_{ij}(y)$ and $\nu_{ij}(x) \geq \nu_{ij}(y)$ for each $i = 1, \ldots, p, j = 1, \ldots, n$.

Definition 3 For the IRE $A *_{BL} B = C$:

(i) The matrix $X^0_{p\times m}$ with $x^0_{ij} \in [0, 1]$, when $1 \leq i \leq p$, $1 \leq j \leq n$, is called a ***solution*** of $A *_{BL} B = C$ if $A *_{BL} X^0 = C$ holds.

(ii) The set of all solutions of (28) is called ***complete solution set*** and it is denoted by $\mathbb{X}$. If $\mathbb{X} \neq \emptyset$ then (28) is called ***consistent***, otherwise it is called ***inconsistent***.

(iii) A solution $\check{X} \in \mathbb{X}$ is called a ***lower*** or ***minimal solution*** of (28) if for any $X \in \mathbb{X}$ the relation $X \leq \check{X}$ implies $X = \check{X}$, where $\leq$ denotes the partial order, induced in $\mathbb{X}$ by the order of [0, 1]. Dually, a solution $\hat{X} \in \mathbb{X}$ is called an ***upper*** or ***maximal*** solution of (28) if for any $X \in \mathbb{X}$ the relation $\hat{X} \leq X$ implies $X = \hat{X}$. When the upper solution is unique, it is called the ***greatest*** or ***maximum solution***. When the lower solution is unique, it is called the ***least*** or ***minimum solution***.

We present finding the greatest solution of intuitionistic fuzzy relational equation (28), we also give a criterion for its consistency.

Theorem 1 *Let A and C be finite IFMs, and let* $\mathbf{B}$ *be the set of all matrices B, such that* $A *_{BL} B = C$. *Then:*

(i) $\mathbf{B} \neq \emptyset \quad \Leftrightarrow \quad A^t \rightarrow_{BL} C \in \mathbf{B}$;

(ii) If the equation (28) is solvable for B then $A^t \rightarrow_{BL} C$ *is its greatest solution.*

(iii) There exists polynomial time algorithm for computing $A^t \rightarrow_{BL} C$.

Here A^t denotes the transpose of A.

For fuzzy relations (that are not intuitionistic), Theorem 1 is given in [3, 8].

Theorem 1 is valid for intuitionistic fuzzy relational equations $R *_{BL} Q = T$ and $R \rightarrow_{BL} Q = T$, when the relations are over finite universal sets.

Corollary 1 *The following statements are valid for the equation* $C = A *_{BL} B$:

(i) The Equation (28) is solvable iff $C = A *_{BL} (A^t \rightarrow_{BL} C)$ *holds;*

(ii) There exists polynomial time algorithm for establishing solvability of the equation (28) and for computing its greatest solution $\hat{B}_{BL} = A^t \rightarrow_{BL} C$.

If the matrices are fuzzy (but not intuitionistic), results for the greatest solution of a system of fuzzy linear equations are obtained in references: for max − min and max −product composition in [21, 22], for max −Łukasiewicz—in [25], and for the minimum solution for min − max composition—in [21].

Example 3 The greatest solution of IFRE (28), if A is given by (13):

(i) in Goguen algebra, when C is the IFM (16), is $\hat{B}_G = A^t \rightarrow_G C$, already computed in Example 2 (1), see (23);

(ii) in Product algebra, when C is the IFM (18), is $\hat{B}_P = A^t \rightarrow_P C$, already computed in Example 2 (2), see (25);

(iii) in Łukasiewicz algebra, when C is the IFM (20), $\hat{B}_L = A^t \rightarrow_L C$, already computed in Example 2 (3), see (27).

In particular, the results are valid for intuitionistic fuzzy linear systems of equations.

5 Conclusions

While finding the greatest solution of IFRE needs polynomial time, finding complete solution set is an open problem and supplementary—it has exponential time complexity [6, 21].

There do not exist methods and software for solving inverse problem for intuitionistic fuzzy relational equations.

Inverse problem resolution for (29) is an open problem.

References

1. Atanassov, K.: Intuitionistic fuzzy sets. VII ITKR's Session, Soa, June 1983 (Deposed in Central Sci. - Techn. Library of Bulg. Acad. of Sci., 1697/84) (in Bulgarian)
2. Atanassov, K.: Intuitionistic Fuzzy Sets. Physica-Verlag, Heidelberg (1999)
3. Bartl, Ed.: Minimal solutions of generalized fuzzy relational equations: Probabilistic algorithm based on greedy approach. Fuzzy Sets and Systems, (2014) in press, http://dx.doi.org/10.1016/j.fss.2014.02.012
4. Bartl, E., Belohlavek, R.: Sup-t-norm and inf-residuum are a single type of relational equations. Int. J. Gen. Syst. **40**, 599–609 (2011)
5. Dinda, B., Samanta, T.K.: Relations on intuitionistic fuzzy soft sets. Gen. Math. Notes **1**(2), 74–83. ISSN 2219-7184
6. Chen, L., Wang, P.: Fuzzy relational equations (I): the general and specialized solving algorithms. Soft Comput. **6**, 428–435 (2002)
7. De Baets, B.: Analytical solution methods for fuzzy relational equations. In: Dubois, D., Prade, H. (eds.) Handbooks of Fuzzy Sets Series: Fundamentals of Fuzzy Sets, vol. 1, pp. 291–340. Kluwer Academic Publishers (2000)
8. Belohlavek, R.: Sup-t-norm and inf-residuum are one type of relational product: unifying framework and consequences. Fuzzy Sets Syst. **197**, 45–58 (2012)
9. Burillo, P., Bustince, H.: Intuitionistic fuzzy relations (part I). Mathw. Soft Comput. **2**, 5–38 (1995)
10. Di Nola, A., Pedrycz, W., Sessa, S., Sanchez, E.: Fuzzy Relation Equations and Their Application to Knowledge Engineering. Kluwer Academic Press, Dordrecht (1989)
11. Garey, M.R., Johnson, D.S.: Computers and Intractability, A Guide to the Theory of NP-Completeness. Freeman, San Francisco (1979)
12. Grätzer, G.: General Lattice Theory. Akademie-Verlag, Berlin (1978)
13. Deschrijver, Gl., Kerre, E.E.: On the composition of intuitionistic fuzzy relations. Fuzzy Sets Syst. lbf 136 333–361 (2003)
14. Hájek, P.: Metamathematics of Fuzzy Logic. Kluwer Academic Publishers, Dordrecht (1998)
15. Klir, G., Clair, U.H. St., Bo, Y.: Fuzzy Set Theory Foundations and Applications. Prentice Hall PRT (1977)
16. Kyosev, Y., Peeva, K.: http://www.mathworks.com/matlabcentral, fuzzy-relational-calculus-toolbox (2004)
17. MacLane, S., Birkhoff, G.: Algebra. Macmillan, New York (1979)
18. Meenakshi, A.R., Gandhimathi, T.: Intuitionistic fuzzy relational equations. Adv. Fuzzy Math. **5**(3), 239–244 (2010)
19. Meenakshi, A.R., Gandhimathi, T.: System of intuitionistic fuzzy relational equations. Glob. J. Math. Sci.: Theory Pract. **4**(1), 49–55 (2012) ISSN 0974-3200
20. Peeva, K.: Resolution of composite intuitionistic fuzzy relational equations. Notes on IFS **61**, 15–24 (2000)

21. Peeva, K.: Resolution of fuzzy relational equations—method, algorithm and software with applications, information sciences **234**, 44–63 (2013) ISSN 0020-0255
22. Peeva, K., Kyosev, Y.: Fuzzy relational calculus-theory, applications and software (with CD-ROM). In the series Advances in Fuzzy Systems—Applications and Theory, vol. 22, World Scientific Publishing Company (2004)
23. Peeva, K., Kyosev, Y.: Software for direct and inverse problem resolution in intuitionistic fuzzy relational calculus. In: Second International Workshop on IFS and Generalized Nets, Warsaw, pp. 60–68, 26–27 July 2002; in Issues in Intuitionistic Fuzzy Sets and Generalized Nets, vol. 2, Warsaw School of Information Technology Press, pp. 117–125, ISBN 83-88311-75-1, Warszawa (2004)
24. Peeva, K., Kyosev, Y.: Solving problems in intuitionistic fuzzy relational calculus with fuzzy relational calculus toolbox. In: Eight International Conference on IFSs, Varna, pp. 37–43, 20–21 June 2004, NIFS 10 (2004)
25. Shivanian,´E.: An algorithm for finding solutions of fuzzy relation equations with max-Lukasiwicz composition. Mathw. Soft Comput. **17**, 15–26 (2010)
26. Wei, Zh.: Intuitionistic fuzzy relation equations. In: 2010 International Conference on Educational and Information Technology (ICEIT), 17–19 September 2010, vol. 2, pp. V2-249–V2-251, E-ISBN:978-1-4244-8035-7, Print ISBN:978-1-4244-8033-3
27. Zahariev, Z.: http://www.mathworks.com/matlabcentral, fuzzy-calculus-core-fc2ore (2010)

Intuitionistic Fuzzy Weakly Open Mappings

Biljana Krsteska

Abstract In this paper, we will introduce and characterize intuitionistic fuzzy weakly open mappings between intuitionistic fuzzy topological spaces. We will investigate their properties and relationships with other early defined classes of intuitionistic fuzzy mappings.

Keywords Intuitionistic fuzzy topology · Intuitionistic fuzzy weakly open mapping · Intuitionistic fuzzy weakly closed mapping

1 Introduction and Preliminaries

After the introduction of fuzzy sets by Zadeh [3], there have been numbers of generalizations of this fundamental concept. The notion of intuitionistic fuzzy sets introduced by Atanassov [1] is one among them. Using the notion of intuitionistic fuzzy sets, Coker [2] introduced the notion of intuitionistic fuzzy topological spaces. In this paper, we will introduce and characterize intuitionistic fuzzy weakly open mappings between intuitionistic fuzzy topological spaces and also we study these mappings in relation to some other types of already known mappings.

Throughout this paper, (X, τ), (Y, σ) and (Z, η), or simply X, Y and Z, are always means an intuitionistic fuzzy topological spaces [2].

2 Intuitionistic Fuzzy Weakly Open Mappings

Definition 2.1 A mapping f: $(X, \tau) \to (Y, \sigma)$ is said to be an intuitionistic fuzzy weakly open if $f(A) \leq \text{Int}f(\text{Cl}A)$ for each intuitionistic fuzzy open set A in X.

B. Krsteska (✉)
Faculty of Mathematics and Natural Sciences, Skopje, Macedonia
e-mail: madob2006@yahoo.com

P. Angelov and S. Sotirov (eds.), *Imprecision and Uncertainty in Information Representation and Processing*, Studies in Fuzziness and Soft Computing 332,
DOI 10.1007/978-3-319-26302-1_7

It is evident that, every intuitionistic fuzzy open mapping is an intuitionistic fuzzy weakly open. But the converse need not be true in general.

Example 2.1 Let $X=\{a,b,c\}$ and $\tau=\{0_X, A, 1_X\}$ and $\sigma=\{0_X, B, 1_X\}$, where $A=\langle x, (\frac{a}{0,5}, \frac{b}{0,3}, \frac{c}{0,6}), (\frac{a}{0,5}, \frac{b}{0,3}, \frac{c}{0,5})\rangle$ and $B=\langle x, (\frac{a}{0,2}, \frac{b}{0,4}, \frac{c}{0,3}), (\frac{a}{0,7}, \frac{b}{0,6}, \frac{c}{0,7})\rangle$. Then the identity function $id: (X,\tau) \to (X,\sigma)$ is an intuitionistic fuzzy weakly open mapping which is not an intuitionistic fuzzy open mapping.

Theorem 2.3 *Let* f: (X, τ) → (Y, σ) *be a surjective mapping. The following conditions are equivalent:*

(i) *f is an intuitionistic fuzzy weakly open;*
(ii) $f(\mathrm{Int}A)) \le \mathrm{Int}f(A)$ *for each intuitionistic fuzzy set A of X;*
(iii) $\mathrm{Int}f^{-1}(B) \le f^{-1}(\mathrm{Int}B)$ *for each intuitionistic fuzzy set B of Y;*
(iv) $f^{-1}(\mathrm{Cl}B) \le \mathrm{Cl}f^{-1}(B)$ *for each intuitionistic fuzzy set B of Y;*
(v) *f(A) is intuitionistic fuzzy open in Y, for each intuitionistic fuzzy $\theta-$ open set A in X;*

Proof

(i) ⇒ (ii) Let A be any intuitionistic fuzzy set of X and x(a,b) be an intuitionistic fuzzy point in IntA. Then, there exists an intuitionistic fuzzy open neighbourhood V of x(a,b) such that $V \le \mathrm{Cl}V \le A$. Then, we have $f(V) \le f(\mathrm{Cl}V) \le f(A)$. Since f is intuitionistic fuzzy weakly open, $f(V) \le \mathrm{Int}f(\mathrm{Cl}V) \le \mathrm{Int}f(A)$. It implies that f(x(a,b)) is an IFP in Intf(A). This shows that $x(a,b) \in f^{-1}(\mathrm{int}f(A))$. Thus $\mathrm{Int}_\theta A \le f^{-1}(\mathrm{Int}f(A))$, and so we obtain $f(\mathrm{int}A) \le \mathrm{Int}f(A)$.

(ii) ⇒ (i) Let G be an intuitionistic fuzzy open set in X. As $G \le \mathrm{Int}_\theta(\mathrm{Cl}G)$, $f(G) \le f(\mathrm{Int}_\theta(\mathrm{Cl}G)) \le \mathrm{Int}f(\mathrm{Cl}G)$. Hence, f is intuitionistic fuzzy weakly open.

(ii) ⇒ (iii) Let B be any intuitionistic fuzzy set of Y. Then by (ii), $f(\mathrm{Int}_\theta(f^{-1}(B))) \le \mathrm{Int}B$. Therefore, $\mathrm{Int}_\theta f^{-1}(B) \le f^{-1}(\mathrm{Int}B)$.

(iii) ⇒ (ii) This is obvious.

(iii) ⇒ (iv) Let B be any intuitionistic fuzzy subset of Y. By (iii), we have $1-\mathrm{Cl}_\theta f^{-1}(B) = \mathrm{Int}_\theta(1-f^{-1}(B)) = \mathrm{Int}_\theta f^{-1}(1-B) \le f^{-1}(\mathrm{Int}(1-B)) = f^{-1}(1-\mathrm{Cl}B) = 1 - f^{-1}(\mathrm{Cl}B)$. Therefore, we obtain $f^{-1}(\mathrm{Cl}B) \le \mathrm{Cl}_\theta f^{-1}(B)$.

(iv) ⇒ (iii) It is similar to (iii) ⇒ (iv).

(iv) ⇒ (v) Let A be an intuitionistic fuzzy $\theta-$open set in X. Then $1 - f(A)$ is an intuitionistic fuzzy set of Y and by (iv), $f^{-1}(\mathrm{Cl}(1-f(A))) \le \mathrm{Cl}_\theta f^{-1}(1-f(A))$. Therefore, $1-f^{-1}(\mathrm{Int}f(A)) \le \mathrm{Cl}_\theta(1-A) = 1-A$. Then, $A \le f^{-1}(\mathrm{Int}f(A))$ which implies $f(A) \le \mathrm{Int}f(A)$. Hence f(A) is intuitionistic fuzzy open in Y.

(v) ⇒ (vi) Let B be any intuitionistic fuzzy set in Y and A be an intuitionistic fuzzy $\theta-$closed set in X such that $f^{-1}(B) \le A$. Since $1 - A$ is intuitionistic fuzzy $\theta-$open in X, by (v), f(1 − A) is intuitionistic fuzzy open in Y. Let F = 1 − f(1 − A). Then F is intuitionistic fuzzy closed and also B ≤ F. Now we have $f^{-1}(F) = f^{-1}(1 - f(1 - A)) = 1 - f^{-1}(f(A)) \leq A$.

(vi) ⇒ (iv) Let B be any intuitionistic fuzzy set in Y. Then $A = Cl_{\theta}(f^{-1}(B))$ is intuitionistic fuzzy $\theta-$closed in X and $f^{-1}(B) \leq A$. Then there exists an intuitionistic fuzzy closed set F in Y containing B such that $f^{-1}(F) \leq A$. Since F is intuitionistic fuzzy closed, we obtain that $f^{-1}(ClB) \leq f^{-1}(F) \leq Cl_{\theta}f^{-1}(B)$. ■

Theorem 2.4 *Let* f: (X, τ) → (Y, σ) *be a mapping. Then the following statements are equivalent:*

(i) f is an intuitionistic fuzzy weakly open mapping;
(ii) for each intuitionistic fuzzy point x(a,b) in X and each intuitionistic fuzzy open set G of X containing x(a,b), there exists an intuitionistic fuzzy open set F containing f(x(a,b)) such that F ≤ f(ClG).

Proof

(i) ⇒ (ii) Let x(a,b) be an IFP in X and G be an intuitionistic fuzzy open set in X containing x(a,b). Since f is intuitionistic fuzzy weakly open, f(G) ≤ Intf (ClG). Let F = Intf(ClG). Hence F ≤ f(ClG), with F containing f(x(a,b)).

(ii) ⇒ (i) Let G be an intuitionistic fuzzy open set in X and let y(a,b) ∈ f(G) By (ii), F ≤ f(ClG) for some open set F in Y containing y(a,b). Hence we have, y(a,b)∈F ≤ Intf(ClG). This shows that f(G) ≤ Intf(ClG) and f is intuitionistic fuzzy weakly open. ■

Theorem 2.5 *Let* f: (X, τ) → (Y, σ) *be a bijective mapping. Then the following statements are equivalent:*

(i) *f is intuitionistic fuzzy weakly open;*
(ii) *Clf(A) ≤ f(ClA) for each intuitionistic fuzzy open set A in X;*
(iii) *Clf(IntB) ≤ f(B) for each intuitionistic fuzzy closed set B in X.*

Proof

(i) ⇒ (iii) Let B be an intuitionistic fuzzy closed set in X. Then we have f(1−B) = 1 − f(B) ≤ Intf(Cl(1 − B)) and so 1 − f(B) ≤ 1 − Clf(IntB). Hence Clf(IntB) ≤ f(B).

(iii) ⇒ (ii) Let A be an intuitionistic fuzzy open set in X. Since Cl(A) is an intuitionistic fuzzy closed set and A ≤ Int(ClA) by (iii), we have Clf (A) ≤ Clf(Int(ClA)) ≤ fCl(A).

(ii) ⇒ (iii) Similar to (iii) ⇒ (ii).
(iii) ≤ (i) Clear. ■

The proof of the following theorem is obvious and thus omitted.

Theorem 2.6 *For a mapping* f: (X, τ) → (Y, σ) *the following statements are equivalent:*

(i) *f is intuitionistic fuzzy weakly open;*
(ii) *for each intuitionistic fuzzy closed set B of X, f(IntB) ≤ Intf(B);*
(iii) *For each intuitionistic fuzzy open set A of X, f(Int(ClA)) ≤ Int(fCl(A));*
(iv) *for each intuitionistic fuzzy regular open set A of X, f(A) ≤ Intf(ClA);*
(v) *for every intuitionistic fuzzy preopen set A of X, f(A) ≤ Intf(ClA);*
(vi) *for every intuitionistic fuzzy θ − open set A of X, f(A) ≤ Intf(ClA).*

Theorem 2.7 *If* f: (X, τ) → (Y, σ) *is intuitionistic fuzzy weakly open mapping and Intf(ClA) ≤ f(A) for every intuitionistic fuzzy open set A of X, then f is intuitionistic fuzzy open mapping.*

Proof Let A be an intuitionistic fuzzy open set of X. Since f is intuitionistic fuzzy weakly open f(A) ≤ Intf(ClA). However, since Intf(ClA) ≤ f(A) for every intuitionistic fuzzy open set A of X, we obtain that f(A) = Intf(ClA) and therefore f(A) is intuitionistic fuzzy open set. Hence f is intuitionistic fuzzy open mapping. ■

Definition 2.2 A mapping f: (X, τ) → (Y, σ) is said to be intuitionistic fuzzy contra open (resp. intuitionistic fuzzy contra-closed) if f(A) is an intuitionistic fuzzy closed set (resp. intuitionistic fuzzy open) set of Y for each intuitionistic fuzzy open (resp. intuitionistic fuzzy closed) set A in X.

Theorem 2.8

(i) *If f: (X, τ) → (Y, σ) is intuitionistic fuzzy preopen and intuitionistic fuzzy contra open, then f is intuitionistic fuzzy weakly open mapping.*
(ii) *If f: (X, τ) → (Y, σ) is intuitionistic fuzzy contra closed, then f is an intuitionistic fuzzy weakly open mapping.*

Proof

(i) Let A be an intuitionistic fuzzy open set of X. Since f is intuitionistic fuzzy preopen f(U) ≤ Int(Clf(A)) and since f is intuitionistic fuzzy contra-open, f(U) is intuitionistic fuzzy closed. Therefore, f(A) ≤ Int(Clf (A)) = Intf(A) ≤ Intf(ClA) ≤ Intf(ClA).
(ii) Let A be an intuitionistic fuzzy open set of X. Then, we have f(A) ≤ f (ClA) ≤ Intf(ClA). ■

Remark 3.9 The converse of Theorem 2.8 does not hold. The mapping f defined on Example 2.1 is an intuitionistic weakly open mapping but it is not an intuitionistic fuzzy preopen mapping.

References

1. Atanassov, K.: Intuitionistic fuzzy sets. Fuzzy Sets Syst. **20**, 87–96 (1986)
2. Coker, D.: An introduction to intuitionistic fuzzy topological spaces. Fuzzy Sets Syst. **88**, 81–89 (1997)
3. Zadeh, L.A.: Fuzzy sets. Inf. Control **8**, 338–353 (1965)

Is 'Fuzzy Theory' An Appropriate Tool for Large Size Decision Problems?

Ranjit Biswas

Abstract This chapter presents a review work in brief of the work [11] which is on a recently unearthed domain of the intuitionistic fuzzy set theory of Atanassov [1-8]. The most useful soft computing set theories [17–23, 25–29, 31, 32] being used to solve the real life decision making problems are: fuzzy set theory, intuitionistic fuzzy set theory (vague sets are nothing but intuitionistic fuzzy sets, justified and reported by many authors), i–v fuzzy set theory, i-v intuitionistic fuzzy set theory, L-fuzzy set theory, type-2 fuzzy set theory, and also rough set theory, soft set theory, etc. While facing a decision making problem, the concerned decision maker in many cases choose one or more of these soft computing set theories by his own choice. Corresponding to each element x of all the universes involved in the decision problem, the value of μ(x) is proposed by the concerned decision maker by his best possible judgment. In real life situation, most of the decision making problems are of large size in the sense of the number of universes and the number of elements in the universes. For example, the populations in Big Data Statistics, be it R-Statistics or NR-Statistics [10], are all about big data; and decision analysis in many such cases involve the application of various soft-computing tools. But there arises a question: Is 'Fuzzy Theory' an appropriate tool for solving large size decision problems? In the work [11] a rigorous amount of mathematical analysis, logical analysis and justifications have been made to answer this question, introducing the 'Theory of CIFS' (Cognitive Intuitionistic Fuzzy System). In this chapter we revisit the mathematical analysis of [11] in brief, and discuss only the important issues of the 'Theory of CIFS' presented in [11]. Many of the decision problems are solved in computers using fuzzy numbers. It is observed that the existing notion of triangular fuzzy numbers and trapezoidal fuzzy numbers are having major drawbacks to the decision makers while solving problems using computer programs or softwares, the issue which is also discussed in this chapter.

R. Biswas (✉)
Department of Computer Science and Engineering, Faculty of Engineering and Technology, Jamia Hamdard University, Hamdard Nagar, New Delhi 110062, India
e-mail: ranjitbiswas@yahoo.com

P. Angelov and S. Sotirov (eds.), *Imprecision and Uncertainty in Information Representation and Processing*, Studies in Fuzziness and Soft Computing 332,
DOI 10.1007/978-3-319-26302-1_8

Keywords CIFS • Atanassov trio functions • Atanassov initialization • Atanassov constraint • Atanassov trio bags • *h*-bag • *m*-bag • *n*-bag • Atanassov processing time (APT) • CESFM • T-fuzzy number • Z-fuzzy number • T-intuitionistic fuzzy number • Z-intuitionistic fuzzy number

1 Introduction

The 'Theory of CIFS' (Cognitive Intuitionistic Fuzzy System) introduced in [11] is initiated with the most important question (but yet an open unsolved problem) in the theory of soft-computing which is as mentioned below:

How Does the Cognition System of a Human or of an Animal or Bird (or of any living thing which has brain) Evaluate the Membership Value $\mu(x)$?

The work in [11] is based on philosophical as well as logical views on the subject of decoding the 'progress' of decision making process in the Human/Animal cognition systems while evaluating the membership value μ(x) in a fuzzy set or in an intuitionistic fuzzy set or in any such soft computing set model or in a crisp set. By 'cognition system' it is meant the cognition system of a human being or of a living animal or of a bird or of any living thing which has brain (ignoring the machines, robots, or software which have artificial intelligence). While a hungry lion finds his food like one cow or one buffalo or one deer (or any other animal of his own food list) in his forest, he decides a lot by his best possible judgment on a number of significant parameters before he starts to chase and also even during the real time period of his chasing. Even in many situations he decides whether it is appropriate to chase, or even after chasing he decides every moment whether to give up chasing or to continue chasing without any problem of his own security, etc. No doubt that he takes these real time decisions by his best possible judgment using his own logic/theory, which is not known to us. But whatever be the different type of logic/theory be used by different kind of decision makers in various decision problems, the kernel of their brain executes a unique common logic of CIFS, irrespective of their intellectual capabilities. This was fact during stone age period of earth, and will remain so for ever on this earth.

In this chapter we revisit the 'Theory of CIFS' [11] in brief, and discuss some of the important issues of it. At the end we identify major demerits of the existing notion of triangular fuzzy numbers and trapezoidal fuzzy numbers due to which computational difficulties are being faced by the decision makers while solving their problems in computers using softwares.

2 Theory of CIFS: A Revisit in Brief

The **Theory of CIFS (Cognitive Intuitionistic Fuzzy System)** in [11] says that a crisp decision maker or a fuzzy decision maker (or any soft decision maker) can not decide on any decision making issue without using intuitionistic fuzzy set (IFS) theory, but he does not necessarily need to have any knowledge of intuitionistic fuzzy set theory. The permanent residence of the 'Theory of IFS' inside the brain (CPU) of every living thing (i.e. every decision maker) is a hidden truth, not by choice of the concerned living thing. In fact the 'Theory of IFS' is a permanent and hidden resident inside the kernel (i.e. at the lowest level) of the processor/brain of every cognition system (be of human or of animal or of bird or of any living thing) in the form of like a **'in-built system-software'** in the Operating System. This software like system automatically gets executed at the lowest level based upon the platform of intuitionistic fuzzy theory while evaluating any membership value μ(x) for a fuzzy set or for any soft-computing set or crisp set. The evaluated μ(x) is always the output at higher level. Although the execution happens in a hidden way at the lowest level (like execution of a machine language program in CPU) but it continuously outputs to update the estimated value of μ(x) at higher level in the cognition system till some amount of time. But for this, it does not require that a fuzzy decision maker or a crisp ordinary decision maker must be aware or knowledgeable about IFS Theory. Consider the case of a FORTRAN programmer who chooses the tool 'FORTRAN language' by his own choice and executes his program written by him in FORTRAN language corresponding to a given engineering problem. But for this, it does not require that the programmer must be aware or knowledgeable about machine language programming!. The analogous fact is true for a fuzzy decision maker too, who estimates μ(x) using the domain of his fuzzy knowledge whereas at the lowest level inside his cognition system the exact execution happens under intuitionistic fuzzy systems only, the theory which is established in [9, 11].

The fuzzy sets are a special case of intuitionistic fuzzy sets, but the existing concept that "the intuitionistic fuzzy sets are higher order fuzzy sets" is **incorrect** (an example of similar **incorrect** concept can be imagined if somebody says that "fuzzy sets can be viewed as higher order crisp sets"!). It is fact that the Theory of IFS is the most appropriate model for translation of imprecise objects while the fuzzy sets are 'lower order' or 'lower dimensional' intuitionistic fuzzy sets as special case. It is rigorously justified in [11] with examples that it may not be an appropriate choice to use fuzzy theory if the problem under study involves estimation of membership values for large number of elements. Unfortunately, most of the real life problems around us consist of many universes where each universe is having many elements. However, two interesting examples of 'decision making problems' with solutions are presented in [11] out of which one example shows the dominance of the application potential of intuitionistic fuzzy set theory over fuzzy set theory, and the other shows the converse i.e. the dominance of the application potential of fuzzy set theory over intuitionistic fuzzy set theory in some cases (where decision makers are pre-selected and most intellectual in the context of the subject pertaining to the concerned problem).

The following two hypothesis are hidden facts in fuzzy computing or in any soft computing process, which have been established in the Theory of CIFS [11] with a detail analysis:-

Fact-1:
A decision maker (intelligent agent) can never use or apply 'fuzzy theory' or any soft-computing set theory without intuitionistic fuzzy system.

Fact-2:
The Fact-1 does not necessarily require that a fuzzy decision maker (or a crisp ordinary decision maker or a decision maker with any other soft theory models or a decision maker like animal/bird which has brain, etc.) must be aware or knowledgeable about IFS Theory!

The Theory of CIFS was initiated in [11] with the fundamental issues like: How the estimation process of the "membership value μ(x) of an element x to belong to a fuzzy set A or an IFS A" is initiated in the cognition system at time $t = 0$ and completed after time $t = T\,(>0)$; How does in reality the 'progress' of decision making process for μ(x) actually happen inside the brain with respect to the variable 'time'.

Suppose that the complete processing time taken by the decision maker to come to his final judgment about μ(x) is T (>0) unit of time. In [11] we designate this time of processing for evaluating the membership value μ(x) as **"Atanassov Processing Time"** (APT) for the element x corresponding to this decision maker, and denoted by the notation APT(x) = T. Thus, the value of μ(x) is proposed by the decision maker for which the time-cost is T (>0), and hence in fuzzy theory one can compute $\upsilon(x) = 1 - \mu(x)$ by doing an arithmetic just, without any further cost of time towards decision process. There is in fact no element of soft-computing in calculating the value of υ(x) in fuzzy theory.

In [11] the Atanassov Trio Functions and Atanassov Constraint are defined as below:

Let R^* be the set of all non-negative real numbers. Consider a pre-fixed fuzzy decision maker. For any given element x of the set X to belong to the fuzzy set A of X, the membership value μ(x) is the final output of a hidden "cognitive intuitionistic fuzzy system" in the brain of the fuzzy decision maker where the following three functions are co-active:-

(i) **h(x, t)** called by 'Hesitation Function' whose domain is $X \times R^*$ and range is [0,1]. For a fixed element x of the set X, h(x, t) is a non-increasing continuous function of time t.
(ii) **m(x, t)** called by 'Membership Function' whose domain is $X \times R^*$ and range is [0,1]. For a fixed element x of the set X, m(x, t) is a non-decreasing continuous function of time t.
(iii) **n(x, t)** called by 'Non-membership Function' whose domain is $X \times R^*$ and range is [0,1]. For a fixed element x of the set X, n(x, t) is a non-decreasing continuous function of time t.

These three functions <m(x, t), n(x, t), h(x, t)> are called **Atanassov Trio Functions (AT functions).**

These functions are subject to the constraint:

$$h(x,t) + m(x,t) + n(x,t) = 1 \quad \text{for any time t.} \quad (\textbf{Atanassov Constraint}).$$

Each of the AT functions is to be basically treated as a function of time t, because it is always considered once an element x be picked up for evaluation by the decision maker. For a fixed decision maker (intelligent agent), corresponding to every element x of X to belong to the fuzzy set A, there exists Atanassov Trio Functions i.e. a set of three AT functions. The value μ(x) in the fuzzy set A comes at the instant t = T from the function m(x, t). This function m(x, t) finally converges at the value μ(x) after a course of sufficient growth for a total T amount of time. The function m(x, t) gets feeding from h(x, t) in a continuous manner starting from time t = 0 till time t = T. None else feeds m(x, t).

The membership value μ(x) for an element x in a fuzzy set A (or in an intuitionistic fuzzy set or in a soft computing set) can never be derived without the activation of AT functions in the cognition system, and this happens to any brain of human being or animal or of any living thing, irrespective of his education or knowledge. This was a fact in the stone age of the earth too, and will continue to remain as a fact for ever. A crisp decision maker or a fuzzy decision maker or any soft computing decision maker does not need to have knowledge about 'Intuitionistic Fuzzy Set Theory'.

At time t = 0 i.e. at the starting instant of time for evaluating the membership value μ(x), any decision process in the cognition system starts with AT functions with the following initial values:-

$$\mathbf{h(x,0) = 1, \quad with\ m(x,0) = 0\ and\ n(x,0) = 0.}$$

The clock starts from time t = 0 and the whistle blows from this initialization only. This initialization <0, 0, 1> is called by **'Atanassov Initialization'**.

It is important to understand that Atanassov Initialization is not initialized by any choice of the decision maker or by any decision of the decision maker or by any prior information from the kernel of the cognition system to the outer-sense of the decision maker. It is never initialized by the decision maker himself, but it gets automatically initialized at the kernel during the execution of any decision making process. By decision maker, we shall mean here a human or an animal or a bird or any living thing which has a brain (we exclude the cases of intelligent robots or intelligent machines or intelligent software which have artificial intelligence).

During the progress of decision making process with respect to the variable 'time' in the brain while evaluating the membership value μ(x), imagine that the values of AT functions are stored and updated continuously, with respect to time, in the three bags (imaginary bags): h-bag, m-bag and n-bag. The updating happens like in computer memory, always replacing their previous values. These three bags are called by **Atanassov Trio Bags** (see Fig. 1).

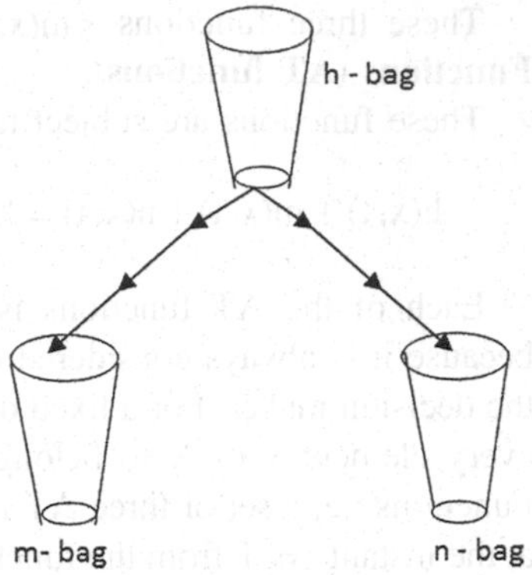

Fig. 1 Atanassov Trio bags

It is obvious that at time t = 0 each of the Atanassov Trio Bags contains the value corresponding to Atanassov Initialization, not else. Immediately after that, the m-bag and n-bag start getting credited with zero or more amount of values continuously from the h-bag, subject to fulfillment of Atanassov Constraint at every instant of time t. But there never happens a reverse flow, i.e. the h-bag does never get credited from any or both of m-bag and n-bag.

While evaluating the membership value μ(x) of an element x, the Atanassov Initialization happens at time t = 0 at the human cognition system (or at the cognition system of the animal or bird whoever be the decision maker). At the very next instant of time, i.e. from time t > 0, the following actions happens simultaneously to the AT functions subject to fulfillment of Atanassov constraint (assuming that the transaction time from h-bag to any bag is always nil):

(i) h(x, t) starts reducing (at least non-increasing), and
(ii) m(x, t) as well as n(x, t) start increasing (non-decreasing).

After certain amount of time, say after t = T (>0) the processing of the decision making process stops (converges) at the following state, say:

$$h(x, T) = \pi(x), \quad \text{with } m(x, T) = \mu(x) \text{ and } n(x, T) = \vartheta(x)$$

where $\pi(x) + \mu(x) + \vartheta(x) = 1$, and after which there is no further updation happens to the values of AT functions in the cognition system (Fig. 2).

Thus at the end of the convergence process at Tth instant of time where T is the value of APT(x), the following results outcome:-

$$\underset{t \to T}{Lt}\, m(x,t) = \mu(x), \quad \underset{t \to T}{Lt}\, n(x,t) = \vartheta(x) \quad \text{and} \quad \underset{t \to T}{Lt}\, h(x,t) = \pi(x),$$

such that $\mu(x) + \vartheta(x) + \pi(x) = 1$.

It is fact that the cognition system of a decision maker (fuzzy decision maker or intuitionistic fuzzy decision maker or crisp decision maker) can not evaluate the membership value μ(x) of an element x without initiating from the Atanassov

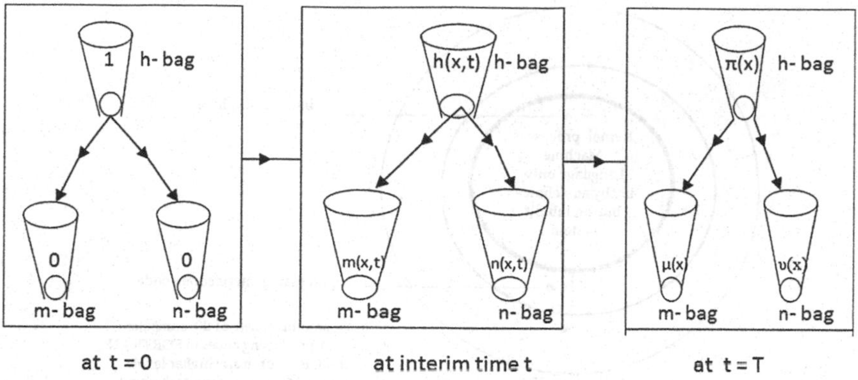

Fig. 2 Evaluation of μ(x) starting from 'Atanassov initialization'

Fig. 3 Cognition system of a decision maker in CIFS, be it a human or animal or bird or any living thing which has brain

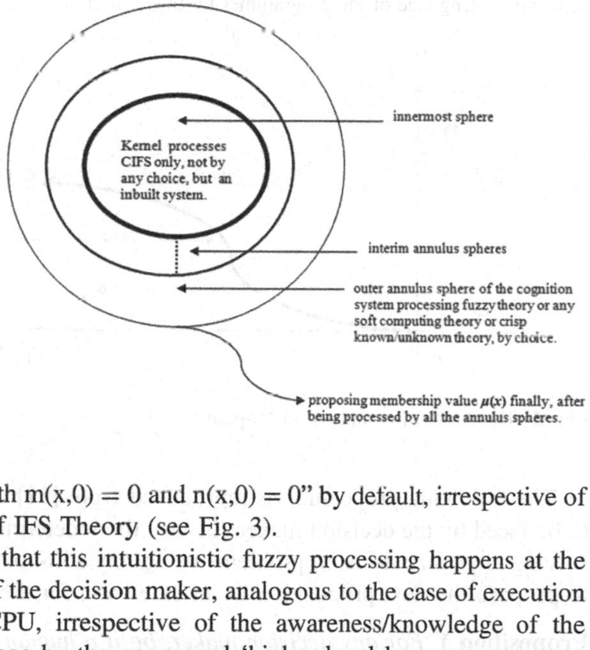

Initialization "h(x,0) = 1 with m(x,0) = 0 and n(x,0) = 0" by default, irrespective of his awareness/knowledge of IFS Theory (see Fig. 3).

It is because of the fact that this intuitionistic fuzzy processing happens at the kernel of the brain (CPU) of the decision maker, analogous to the case of execution of FORTRAN codes in CPU, irrespective of the awareness/knowledge of the concept of Machine Language by the concerned 'higher level language programmer' (see Fig. 4). Here the decision maker may be a fuzzy decision maker or any kind of decision maker (who may be a layman of IFS theory or of Fuzzy theory, or who could be even an animal or a living thing having brain).

The four variable parameters m, n, h and t could be viewed to form a 4-dimensional hyperspace in the Theory of CIFS. Since our interest is on the trio m, n and h only, we consider 3-D geometry with three mutually perpendicular axes called by m-axis, n-axis and h-axis, forming a 3-D mnh-space.

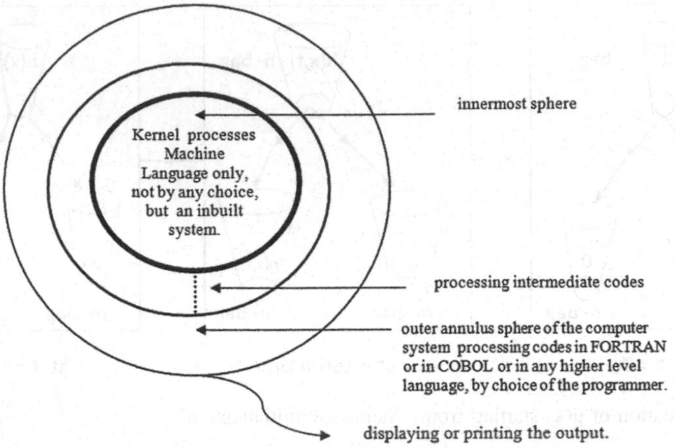

Fig. 4 CPU of a computer with common machine language at the kernel irrespective of any higher level language of the programmer by his own choice

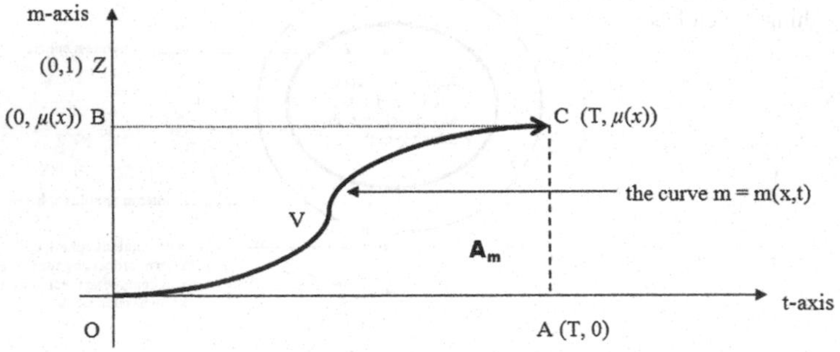

Fig. 5 The curve m = m(x, t) on tm-plane

The following proposition is reproduced from [11] to justify the weak situation to be faced by the decision makers in case the indeterministic part be ignored while solving large size decision problems using any soft-computing set theory, and then to propose the next proposition as an extension for the CIFS.

Proposition 1 *For any decision maker, be it a human or an animal or any living thing which has brain, it is impossible that his brain (kernel of his cognitive system) does always have the indeterministic component (i.e. the hesitation component or undecided component) h(x, t) to be nil for the element x of the universe X, while going to propose the corresponding membership value* $\mu(x)$.

Proof Suppose that APT(x) = T.

Consider the 2-D curve m = m(x, t) on tm-plane (as shown in Fig. 5). Suppose that A_m is the area under the curve m = m(x, t) bounded by the lines t-axis, m-axis

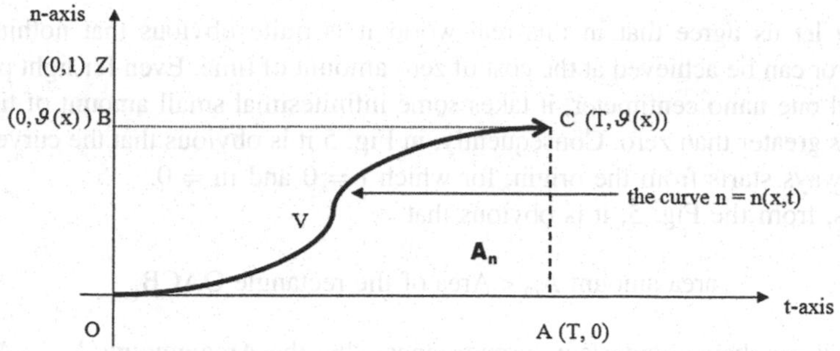

Fig. 6 The curve n = n(x, t) on tn-plane

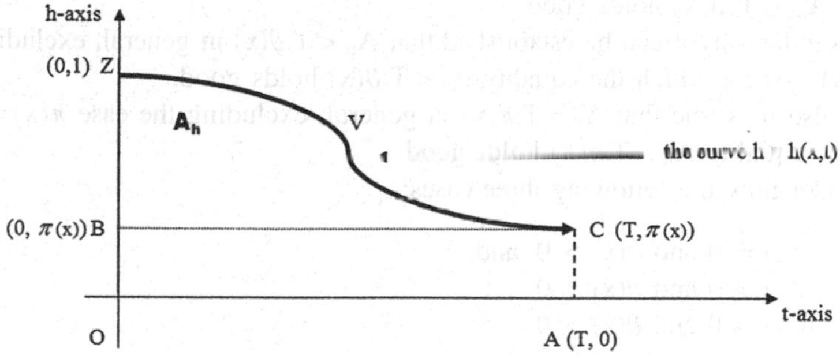

Fig. 7 The curve h = h(x, t) on th-plane

and the line t = T. Clearly, corresponding to a given decision maker the quantity A_m is not a function of t but of the parameter x. □

Now consider the 2-D curve n = n(x, t) on tn-plane (as shown in Fig. 6). Suppose that A_n is the area under the curve n = n(x, t) bounded by the lines t-axis, n-axis and the line t = T.

Also consider the 2-D curve h = h(x, t) on th-plane (as shown in Fig. 7).

Suppose that A_h is the area under the curve h = h(x, t) bounded by the lines t-axis, h-axis and the line t = T.

Now consider the Atanassov Constraint

$$m(x,t)+n(x,t)+h(x,t)=1.$$

Integrating with respect to time t we get,

$$\int_0^T m(x,t)dt+\int_0^T n(x,t)dt+\int_0^T h(x,t)dt=\mathrm{T} \tag{1}$$

$$\text{or,}\quad A_m+A_n+A_h=T$$

Now let us agree that in this real world it is quite obvious that nothing can happen or can be achieved at the cost of zero amount of time. Even for light particle to travel one nano centimeter, it takes some infinitesimal small amount of time Δt which is greater than zero. Consequently, in Fig. 5 it is obvious that the curve OVC does always starts from the origin, for which $t = 0$ and $m = 0$.

Thus, from the Fig. 5, it is obvious that

$$\text{area amount } A_m < \text{Area of the rectangle OACB.}$$

(There is no chance under any circumstances that the Area amount A_m = Area of the rectangle OACB, unless both are equal to zero).

Therefore, $A_m < T.\mu(x)$ in general, excluding the case $\mu(x) = 0$ for which the equality $A_m < T.\mu(x)$ holds good.

In a similar way it can be established that $A_n < T.\vartheta(x)$ in general, excluding the case $\vartheta(x) = 0$ for which the equality $A_n < T.\vartheta(x)$ holds good.

And also it is true that $A_h > T.\pi(x)$ in general, excluding the case $\pi(x) = 1$ for which the equality $A_h > T.\pi(x)$ holds good.

Consider now the following three cases:-

Case(1) $\mu(x) = 0$ and $\vartheta(x) > 0$. and
Case(2) $\mu(x) > 0$ and $\vartheta(x) = 0$
Case(3) $\mu(x) > 0$ and $\vartheta(x) > 0$

It is obvious that for all of these three cases

$$A_m + A_n < (T.\mu(x) + T.\vartheta(x)) \tag{2}$$

Now let us prove our proposition **by contradiction.**

For this, let us suppose that:

For any decision maker, be it a human or an animal or any living thing, it is ***possible*** *that his brain (kernel of his cognitive system) does always have the indeterministic component (i.e. the hesitation component or undecided component) h(x, t) to be nil while going to propose the membership value $\mu(x)$.*

Therefore, $h(x, t) = 0$ for every $t \in [0,T]$.

From (1), $T = A_m + A_n$

Therefore, $T < (T.\mu(x) + T.\vartheta(x))$, using (2).

This means that $\mu(x) + \vartheta(x) > 1$, which is not possible in any soft-computing set theory (for instance, not possible in Fuzzy Set theory). Hence the Proposition.

Proposition 2 *For any decision maker in the Theory of CIFS, be it a human or an animal or any living thing which has brain, it is impossible that his brain (kernel of his cognitive system) does always have the indeterministic component (i.e. the hesitation component or undecided component) h(x, t) to be nil for the element x of the universe X, during the progress of decision making process for evaluating $\mu(x)$.*

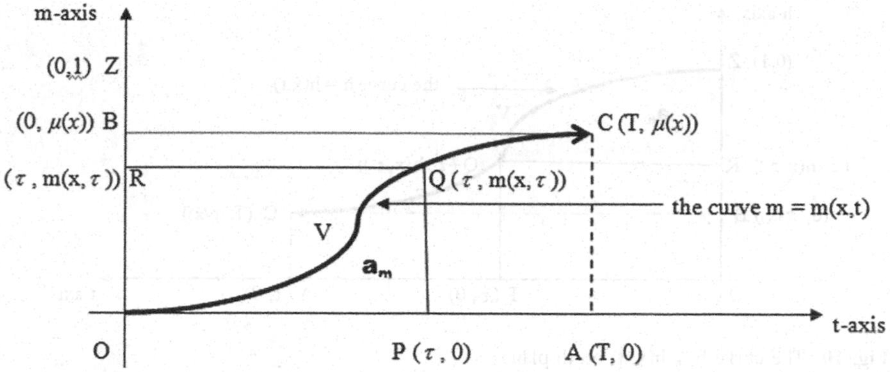

Fig. 8 The curve m = m(x, t) on tm-plane

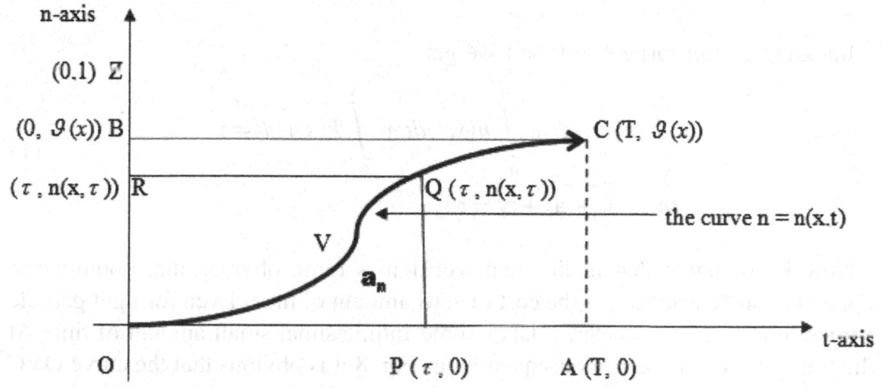

Fig. 9 The curve n = n(x, t) on tn-plane

Proof Suppose that APT(x) = T. Consider any arbitrary time $\tau < T$.

Consider the 2-D curve m = m(x, t) on tm-plane (as shown in Fig. 8). Suppose that a_m is the area under the curve m = m(x, t) bounded by the lines t-axis, m-axis and the line $t = \tau$. Clearly, assuming the amount τ now fixed here, the quantity a_m is not a function of t but of the parameter x. □

Now consider the 2-D curve n = n(x, t) on tn-plane (as shown in Fig. 9). Suppose that a_n is the area under the curve n = n(x, t) bounded by the lines t-axis, n-axis and the line $t = \tau$.

Also consider the 2-D curve h = h(x, t) on th-plane (as shown in Fig. 10).

Suppose that a_h is the area under the curve h = h(x, t) bounded by the lines t-axis, h-axis and the line $t = \tau$.

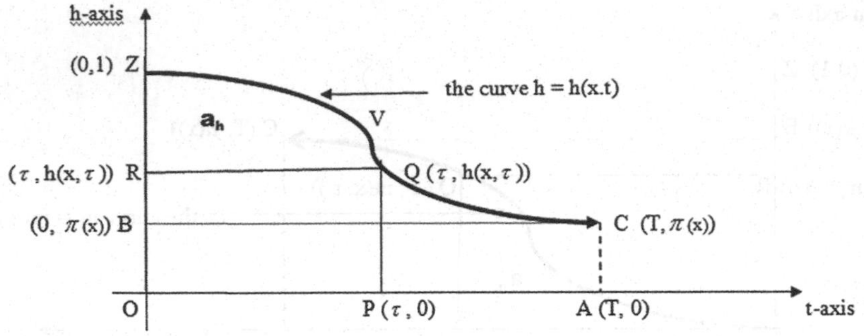

Fig. 10 The curve h = h(x, t) on th-plane

Now consider the Atanassov Constraint

$$m(x,t) + n(x,t) + h(x,t) = 1.$$

Integrating with respect to time t we get,

$$\int_0^{\tau} m(x,t)dt + \int_0^{\tau} n(x,t)dt + \int_0^{\tau} h(x,t)dt = \tau \tag{1}$$

$$\text{or,}\quad a_m + a_n + a_h = \tau$$

Now let us agree that in this real world it is quite obvious that nothing can happen or can be achieved at the cost of zero amount of time. Even for light particle to travel one nano centimeter, it takes some infinitesimal small amount of time Δt which is greater than zero. Consequently, in Fig. 8 it is obvious that the curve OVC does always starts from the origin, for which t = 0 and m = 0.

Thus, from the Fig. 8, it is obvious that

$$\text{area amount } a_m < \text{Area of the rectangle OPQR.}$$

(There is no chance under any circumstances that the Area amount a_m = Area of the rectangle OPQR, unless both are equal to zero).

Therefore, $a_m < \tau.\ m(x, \tau)$ in general, excluding the case $m(x,\tau) = 0$ for which the equality $a_m = \tau.\ m(x, \tau)$ holds good.

In a similar way it can be established that $a_n < \tau.\ n(x, \tau)$ in general, excluding the case $n(x, \tau) = 0$ for which the equality $a_n = \tau.\ n(x,\tau)$ holds good.

Similarly, it is true that $a_h > \tau.\ h(x, \tau)$ in general, excluding the case $h(x,\tau) = 1$ for which the equality $a_h = \tau.\ h(x, \tau)$ holds good.

Consider now the following two cases:-

Case(1): $m(x, \tau) = 0$ and $n(x, \tau) > 0$. and
Case(2): $m(x, \tau) > 0$ and $n(x, \tau) = 0$. and
Case(3): $m(x, \tau) > 0$ and $n(x, \tau) > 0$.

It is obvious that for all of these three cases

$$a_m + a_n < (\tau.m(x, \tau) + \tau.n(x, \tau)) \quad (2)$$

Now let us prove our proposition **by contradiction.**

For this, let us suppose that:

For any decision maker, be it a human or an animal or any living thing, it is ***possible*** *that his brain (kernel of his cognitive system) does always have the indeterministic component (i.e. the hesitation component or undecided component) h(x, t) to be nil during the progress of decision making process for evaluating μ(x).*

Therefore, $h(x, t) = 0$ for every $t \in [0, \tau]$.

From (1), $\tau = a_m + a_n$

Therefore, $\tau < (\tau.\ m(x, \tau) + \tau.\ n(x, \tau))$, using (2).

This means that $m(x, \tau) + n(x, \tau) > 1$, which is not possible in the Theory of CIFS. Hence the Proposition.

3 Is 'Fuzzy Theory' an Appropriate Tool for Large Size Decision Problems?

It is observed that $0 \leq \pi(x) \leq 1$ for every x of the universe X, whoever be the decision maker. As a special case, it may happen for one or few elements in the IFS A that $\pi(x) = 0$. But in ground reality, for a decision maker by the best possible processing in his cognition system, the data '$\pi(x) = 0$' can not be true in general for all and across all the elements x of any universe X while proposing an IFS A of X (as justified further in **Proposition 1 and 2**). Even if it be true for one or few or many elements, it is illogical to believe that it is true for all and across all the elements of any universe X while proposing an IFS A. Consequently, it is a rare case that an intuitionistic fuzzy set eventually becomes equivalent to a fuzzy set. It is not a feasible case that a fuzzy decision maker can ignore $\pi(x)$ by his own decision if it is not zero.

Further to that, any real life soft-computing problem on this earth usually occurs involving more than one universe. There could be r number of universes viz. X_1, X_2, ……, X_r in a given problem under consideration by a decision maker. And in that case it is extremely illogical to believe that '$\pi(x) = 0$' is true for all the elements of all the r universes. Neither any real logical system(s) nor the Nature can force a decision maker (human being or animal or any living object having a brain) either to stand strictly at the decision: "$\pi(x) = 0$ for every x of every X", or "to abandon his decision process otherwise".

Any decision process for deciding the membership value μ(x) starts with Atanassov's initialization <0, 0, 1> and then after certain amount of time T (called by Atanassov Processing Time) it converges to the trio $<\mu(x), \vartheta(x), \pi(x)>$ without any further updation of the AT functions. In general in most of the cases, $\pi(x) \neq$ NIL.

Even if $\pi(x)$ = NIL for one element x, it is a very rare situation that $\pi(x)$ will be Nil for all the elements x of X in the IFS proposed by the decision maker. It is a very very special case, the following justification will support further to this hypothesis:-

Suppose that, to solve an ill-defined problem we have to consider 20 number of universes, while in each universe there is at least 200 elements. Suppose that, to solve this problem an intelligent decision maker (say, a soft computing human expert) needs to consider more than 30 fuzzy sets of each universe. Clearly, he has to propose membership values by his best possible judgment for more than 1,20,000 elements. For deciding the membership value for each of these 1,20,000 elements the cognition system of the decision maker by default begins with Atanassov's initialization $<0, 0, 1>$ and after certain time converges to the decision about $\mu(x)$ for the element. Can we presume that for each of these 1,20,000 elements his convergence process starting from the Atanassov's initialization trio $<0, 0, 1>$ will stop at the trio $< \mu(x), \vartheta(x), 0 >$ with $\pi(x) = 0$ for each and every x? Can we presume that there is not a single element x out of 1,20,000 elements for which the convergence process ends with some non-zero amount of $\pi(x)$?

This surely justifies that it may not be appropriate to use fuzzy theory if the problem under study involves the estimation of membership values for large number of elements of one or more universes. For instance the populations in Big Data Statistics [10, 16], be it R-Statistics or NR-Statistics, are all about big data expanding in 4Vs very fast; and decision analysis in many such cases involve the application of various soft-computing tools, but it is most important to have excellent results only.

In our everyday life, every human being plays the role of a decision maker at every moment of time (ignoring his sleeping period at night). He is compelled to decide every day for large number of imprecise problems of various nature. But one can not be always an excellent and outstanding decision maker for all the unknown (or known), unpredictable, homogeneous/heterogeneous, precise/imprecise problems he faces every day without 'any element of hesitation' at all.

Consequently, it is well justified in [11] that for a large or moderate size soft computing problem, it may not be appropriate to use the tool 'Fuzzy Theory' in order to get excellent results.

However, there are also a number of real life cases where only the best/excellent decision makers (in most of the cases being pre-selected or pre-choosen) are allowed to take decisions who can do the job and are supposed to do the job **'without any hesitation'** on any issue of the problem under consideration, i.e. the outcome '$\pi(x) = 0$' is to happen to be true everywhere during the execution of the problem-solving by them. The example of CESFM presented in [11, 14] is a very ideal example to understand the situation where fuzzy theory is more appropriate tool than intuitionistic fuzzy theory in some special cases.

4 Major Failures of the Existing Notion of Fuzzy Numbers to the Decision Makers

While doing computations to solve real life ill-posed problems in computers using a program or a software, the decision makers (including academicians, scientists and engineers) face a serious problem due to the failures of the role of the existing popular notion of fuzzy numbers [24]: Triangular Fuzzy Numbers and Trapezoidal Fuzzy Numbers. This failure are frequently observed by the scientists, mathematicians, statisticians, engineers and academicians while they are in many areas of simple or complex computations. For solving a very precise and well posed problem in a computer, a decision makers may use crisp numbers only (say, real numbers) in modeling the problem into a Linear Programming Problem or a Non-linear Programming Problem or a Game Theory Problem or a Statistical Problem, etc. In that case he does not face any problem in memory utilization. A positive real number x can be stored in memory in consecutive four bytes reserved for x. The addition, subtraction, multiplication, division of two real numbers x and y are again real numbers and hence can be stored in memory in consecutive four bytes reserved for z. It is because of the reason that the set R of real numbers forms a division algebra, and does also form a region [13, 15]. Consequently, decision makers can fluently use real numbers, can fluently store them and their results in the space of z. But this is not possible [12] if a decision maker solves an ill defined problem using Triangular Fuzzy Numbers and/or Trapezoidal Fuzzy Numbers [24]. Both Triangular Fuzzy Numbers and Trapezoidal Fuzzy Numbers are paralyzed with exactly analogous type of drawbacks. For example, consider the triangular fuzzy numbers a = <1, 2, 3> and b = <4, 5, 6> and see carefully that a * b is not a triangular fuzzy number!. Thus after the multiplication operation on triangular fuzzy numbers a and b be performed, the result is loosing the data-structure of the two operands! To store the triangular fuzzy numbers a or b in memory we need six bytes, but how to store the simple multiplication result a * b which is not a triangular fuzzy number!.

Similarly it can be noticed that division of two triangular fuzzy numbers is not a triangular fuzzy number. These are the major drawbacks of triangular fuzzy numbers. The main problem is that what kind of ADT (Abstract Data Type) is to be considered in fuzzy arithmetic computing? What kind of storage-structure is to be considered? In computer science, a programmer knows that if a, b, c are real(float) numbers then x = a * b + c is also real, and hence the programmers reserves in memory the number of bytes for x accordingly. If a, b, c, are three dimensional vectors then $x = a \times b + c$ is also a three dimensional vector where X denotes cross-product. But if a, b, c are triangular fuzzy numbers, then x = a * b + c is not a triangular fuzzy number (although it is a fuzzy number). This serious problem is faced by the fuzzy mathematicians, fuzzy statisticians, and fuzzy experts of all other fields. To solve this crisis, the notion of T-fuzzy numbers and Z-fuzzy numbers are introduced in this section, but retaining compatibility with the Tokunaba and Yasunobu's model [30] of fuzzy computer, probably the first ever attempt to model fuzzy computers.

4.1 T-Fuzzy Numbers and Z-Fuzzy Numbers

In this section the existing concept of triangular fuzzy numbers and trapezoidal fuzzy numbers is updated by introducing T-fuzzy numbers (Triangle Type Fuzzy Numbers) and Z-fuzzy numbers (Trapezoidal Type Fuzzy Number) respectively. The 'triangular fuzzy numbers' and 'Triangle Type Fuzzy Numbers' are not same. Similarly the 'trapezoidal fuzzy numbers' and 'Trapezoidal Type Fuzzy Numbers' are not same. The existing notion of triangular fuzzy numbers is a special case of T-fuzzy numbers and the existing notion of trapezoidal fuzzy numbers is a special case of Z-fuzzy numbers.

T-Fuzzy Number

A triangle type fuzzy number (T-fuzzy number) $\tilde{a}$ is of the form

$$\tilde{a} = < a_1, a_2, a_3, l(x), r(x) >$$

where the functions $l(x)$: $[a_1, a_2] \rightarrow [0, 1]$ and $r(x)$: $[a_2, a_3] \rightarrow [0, 1]$ are inversible (bijective), fuzzy convex and satisfy the following conditions:

(i) $l(a_1) = 0$
(ii) $l(a_2) = 1 = r(a_2)$
(iii) $r(a_3) = 0$

The curve $y = l(x)$ is called the *left boundary*, and the curve $y = r(x)$ is called the *right boundary* of the T-fuzzy number $\tilde{a}$.

Thus the membership function of a T-fuzzy number $\tilde{a} = <a_1, a_2, a_3, l(x), r(x)>$ will be as follows:-

$$\mu_{\tilde{a}}(x) = \begin{cases} 0 & \text{if} \quad x \le a_1 \\ l(x) & \text{if} \quad a_1 \le x \le a_2 \\ r(x) & \text{if} \quad a_2 \le x \le a_3 \\ 0 & \text{if} \quad x \ge a_3 \end{cases}$$

Z-Fuzzy Number

A trapezoidal type fuzzy number (Z-fuzzy number) $\tilde{a}$ is of the form

$$\tilde{a} = < a_1, a_2, a_3, a_4, l(x), m(x), r(x) >,$$

where the functions $l(x)$: $[a_1, a_2] \rightarrow [0, 1]$, and $r(x)$: $[a_3, a_4] \rightarrow [0, 1]$ are inversible (bijective), fuzzy convex and satisfy the following conditions:

(i) $l(a_1) = 0$
(ii) $l(a_2) = 1 = r(a_3)$
(iii) $r(a_4) = 0$ and
(iv) $m(x)$: $[a_2, a_3] \rightarrow \{1\}$ is a constant function.

The curve y = l(x) is called the *left boundary*, the curve y = r(x) is called the *right boundary*, and the curve (straight line) y = m(x) is called the *roof-line* of the Z-fuzzy number ã. Thus the membership function of a Z-fuzzy number ã = <a_1, a_2, a_3, a_4, l(x), m(x), r(x)> will be as follows:-

$$\mu_{\tilde{a}}(x) = \begin{cases} 0 & \text{if} \quad x \le a_1 \\ l(x) & \text{if} \quad a_1 \le x \le a_2 \\ 1 & \text{if} \quad a_2 \le x \le a_3 \\ r(x) & \text{if} \quad a_3 \le x \le a_4 \\ 0 & \text{if} \quad x \ge a_4 \end{cases}$$

In the next part we define the various arithmetic operations over T-fuzzy numbers and Z-fuzzy numbers.

4.2 *Arithmetic of T-Fuzzy Numbers and Z-Fuzzy Numbers*

Clearly, the existing concept of triangular fuzzy numbers is a special case of T-fuzzy numbers. Many authors studied arithmetic of fuzzy numbers and applied in wide variety of fields viz. Fuzzy Linear Programming, Fuzzy Optimization, Fuzzy DBMS, Fuzzy Searching Techniques in AI, Neuro-fuzzy Systems, Fuzzy Pattern Recognition etc. to list a few only. The work of Kaufmann and Gupta [24] is interesting. They noted that fuzzy numbers can be treated as a generalization of the concept of the confidence interval. Let ã = (a_1, a_2, a_3) be a triangular fuzzy number, and suppose that for a given level α of presumption $\mu_{\tilde{a}}(x) \ge \alpha$ is true in the interval [a_l, a_u] and not true outside, where $[a_l, a_u] \subseteq [a_1, a_3]$. Then we say that this fuzzy number ã has the confidence interval [a_l, a_u] corresponding to the level of presumption α. For adding two triangular fuzzy numbers, we simply add the confidence intervals of them corresponding to the common values of α. In a similar way the other operations like Subtraction, Multiplication, Division, Scalar Multiplication etc. can be performed. To study the arithmetic with T-fuzzy numbers, we will follow an analogous art of Kaufmann and Gupta [24]. In the next part of this section we present the operations over T-fuzzy numbers. The operations over Z-fuzzy numbers are to be carried out in a similar manner. We use the notations +, −, *, / for the operations of Addition, Subtraction, Multiplication and Division respectively of two T-fuzzy numbers.

4.2.1 Addition of two T-Fuzzy Numbers

Consider two T-fuzzy numbers A and B given by

$$A = < a_1, a_2, a_3, l_A(x), r_A(x) >, \quad \text{and}$$
$$B = < b_1, b_2, b_3, l_B(x), r_B(x) >.$$

For a level of presumption α, suppose that the confidence intervals of A and B are

$$A_\alpha = [A_1^\alpha, A_2^\alpha] \quad \text{and} \quad B_\alpha[B_1^\alpha, B_2^\alpha].$$

If C = A + B, (using the symbol '+' to denote the operation of addition of two T-fuzzy numbers), then using Kaufmann and Gupta's style we have

$$C_\alpha = A_\alpha \oplus B_\alpha = [A_1^\alpha + B_1^\alpha, A_2^\alpha + B_2^\alpha].$$

where ⊕ denotes the symbol for interval-addition.

We will now establish the following proposition:

Proposition 3 *The fuzzy number C is also a T-fuzzy number.*

Proof The equations of the two boundaries of the T-fuzzy number A could be viewed as

$$\alpha = l_A(A_1^\alpha) \quad \text{and} \quad \alpha = r_A(A_2^\alpha).$$

□

This gives $A_\alpha = [l_A^{-1}(\alpha), r_A^{-1}(\alpha)]$.
Similarly $B_\alpha = [l_B^{-1}(\alpha), r_B^{-1}(\alpha)]$.
Therefore, $C_\alpha = [C_1^\alpha, C_2^\alpha] = [l_A^{-1}(\alpha) + l_B^{-1}(\alpha), r_A^{-1}(\alpha) + r_B^{-1}(\alpha)]$.

The left boundary of the T-fuzzy number C is given by the curve $\alpha = l_C(x)$ where α is the solution of the equation $x = l_A^{-1}(\alpha) + l_B^{-1}(\alpha)$; and the right boundary is given by the curve $\alpha = r_C(x)$, where α is the solution of the equation $x = r_A^{-1}(\alpha) + r_B^{-1}(\alpha)$.

We will now verify that

(i) l_c, r_c are inversible functions, and
(ii) l_c, r_c are fuzzy convex.

Consider the function $\alpha = l_C(x)$.

We see that $l_C^{-1}(\alpha) = l_A^{-1}(\alpha) + l_B^{-1}(\alpha)$, which is unique. This shows that l_c is an inversible function. Similarly we can argue for r_c.

To prove that l_c is fuzzy convex, it is sufficient if we prove that l_c is an increasing function.

For this, take $x_2 > x_1$ and suppose that $l_C(x_2) = \gamma_2$ and $l_C(x_1) = \gamma_1$.

Now, $\quad l_C^{-1}(\gamma_2) > l_C^{-1}(\gamma_1)$
or, $\quad l_A^{-1}(\gamma_2) + l_B^{-1}(\gamma_2) > l_A^{-1}(\gamma_1) + l_B^{-1}(\gamma_1)$

which implies that $\gamma_2 > \gamma_1$.

(Because if $\gamma_2 > \gamma_1$ is not true, then we must have $\gamma_1 \geq \gamma_2$; and in that case we must have $l_A^{-1}(\gamma_1) + l_B^{-1}(\gamma_1) \geq l_A^{-1}(\gamma_2) + l_B^{-1}(\gamma_2)$ which is not true). Hence l_c is an increasing function.

In a similar way we can prove that r_c is a decreasing function.

Now, to find out the values of c_1, c_2 and c_3 we solve the following equations respectively for x:

(i) $l_C(x) = 0$, (ii) $l_C(x) = 1$, [or, the equation $r_C(x) = 1$], and (iii) $r_C(x) = 0$.

The equation $l_C(x) = 0$ gives

$$\begin{aligned} c_1 &= l_C^{-1}(0) \\ &= l_A^{-1}(0) + l_B^{-1}(0) \\ &= a_1 + b_1 \end{aligned}$$

Similarly the equation $r_C(x) = 0$ gives $c_3 = a_3 + b_3$.

The equation $l_C(x) = 1$ gives

$$\begin{aligned} c_2 &= l_C^{-1}(1) \\ &= l_A^{-1}(1) + l_B^{-1}(1) \\ &= a_2 + b_2 \end{aligned}$$

The equation $r_C(x) = 1$ too gives the same value for c_2.

Thus we get that, addition of two T-fuzzy numbers $A = <a_1, a_2, a_3, l_A(x), r_A(x)>$ and $B = <b_1, b_2, b_3, l_B(x), r_B(x)>$ is the T-fuzzy number C given by

$$\begin{aligned} C &= <a_1, a_2, a_3, l_A, r_A> + <b_1, b_2, b_3, l_B, r_B> \\ &= <c_1, c_2, c_3, l_C, r_C> \end{aligned}$$

where

(i) $c_i = a_i + b_i$, for i = 1, 2, 3
(ii) $l_C(x)$ is the solution of the equation $x = l_A^{-1}(\alpha) + l_B^{-1}(\alpha)$ for unknown α, and
(iii) $r_C(x)$ is the solution of the equation $x = r_A^{-1}(\alpha) + r_B^{-1}(\alpha)$ for unknown α.

It is obvious that this addition operation is commutative.

In a similar way we see that the following results are true.

4.2.2 Subtraction of Two T-Fuzzy Numbers

Subtraction of the T-fuzzy number $A = \langle a_1, a_2, a_3, l_A(x), r_A(x)\rangle$ from the T-fuzzy number $B = \langle b_1, b_2, b_3, l_B(x), r_B(x)\rangle$ yields the T-fuzzy number C given by

$$C = \langle a_1, a_2, a_3, l_A, r_A\rangle - \langle b_1, b_2, b_3, l_B, r_B\rangle = \langle c_1, c_2, c_3, l_C, r_C\rangle$$

where

(i) $c_i = a_i - b_i$, for i = 1, 2, 3
(ii) $l_C(x)$ is the solution of the equation $x = l_A^{-1}(\alpha) + l_B^{-1}(\alpha)$ for unknown α, and
(iii) $r_C(x)$ is the solution of the equation $x = r_A^{-1}(\alpha) + r_B^{-1}(\alpha)$ for unknown α.

4.2.3 Multiplication of Two T-Fuzzy Numbers

Multiplication of two T-fuzzy numbers $A = \langle a_1, a_2, a_3, l_A(x), r_A(x)\rangle$ and $B = \langle b_1, b_2, b_3, l_B(x), r_B(x)\rangle$ denoted by A * B is the T-fuzzy number C given by

$$C = \langle a_1, a_2, a_3, l_A, r_A\rangle * \langle b_1, b_2, b_3, l_B, r_B\rangle = \langle c_1, c_2, c_3, l_C, r_C\rangle.$$

where

(i) $c_i = a_i \,.\, b_i$, for i = 1, 2, 3
(ii) $l_C(x)$ is the solution of the equation $x = l_A^{-1}(\alpha)\,.\,l_B^{-1}(\alpha)$ for α, and
(iii) $r_C(x)$ is the solution of the equation $x = r_A^{-1}(\alpha)\,.\,r_B^{-1}(\alpha)$ for α.

It is obvious that this multiplication operation is commutative.

4.2.4 Division of Two T-Fuzzy Numbers

Division of the T-fuzzy number $A = \langle a_1, a_2, a_3, l_A(x), r_A(x)\rangle$ by the T-fuzzy number $B = \langle b_1, b_2, b_3, l_B(x), r_B(x)\rangle$ denoted by A/B is the T-fuzzy number C given by

$$C = \langle a_1, a_2, a_3, l_A, r_A\rangle / \langle b_1, b_2, b_3, l_B, r_B\rangle$$
$$= \langle c_1, c_2, c_3, l_C, r_C\rangle.$$

where

(i) $c_1 = a_1/b_3$, $c_2 = a_2/b_2$, $c_3 = a_3/b_1$.
(ii) $l_C(x)$ is the solution of the equation $x = l_A^{-1}(\alpha)/r_B^{-1}(\alpha)$ for α, and
(iii) $r_C(x)$ is the solution of the equation $x = r_A^{-1}(\alpha)/l_B^{-1}(\alpha)$ for α.

4.2.5 Scalar Multiplication

For any scalar k, the scalar multiplication of a T-fuzzy number $A = <a_1, a_2, a_3, l_A(x), r_A(x)>$ with k results in the T-fuzzy number C given by $C = k. < a_1, a_2, a_3, l_A, r_A > = < c_1, c_2, c_3, l_C, r_C >$ where

(i) $c_i = k.a_i$ for i = 1, 2, 3
(ii) $l_C(x)$ is the solution of the equation $x = k.\,l_A^{-1}(\alpha)$ for α, and
(iii) $r_C(x)$ is the solution of the equation $x = k.\,r_A^{-1}(\alpha)$ for α.

The membership function of the T-fuzzy number C will be

$$\mu_C(x) = \begin{cases} 0 & \text{if } x \le ka_1 \\ l(x/k) & \text{if } ka_1 \le x \le ka_2 \\ r(x/k) & \text{if } ka_2 \le x \le ka_3 \\ 0 & \text{if } x \ge ka_3 \end{cases}$$

5 Conclusion

This chapter presents a very brief review exercise of the work of [11] entitled: "Is 'Fuzzy Theory' An Appropriate Tool For Large Size Problems?", in which the Theory of CISF is introduced. Theory of CISF is basically on the subject of decoding the 'progress' of decision making process in the Human/Animal cognition systems while evaluating the membership value μ(x) in a fuzzy set or in an intuitionistic fuzzy set or in any such soft computing set model or in a crisp set. The theory of CIFS in [9, 11] explains and well justifies that it may not be an appropriate decision to apply fuzzy set theory if the problem under consideration involves the estimation of membership values for a large number of elements. The two hidden facts about fuzzy set theory (and, about any soft computing set theory) established in [11] are:

Fact-1:
A decision maker (intelligent agent) can never use or apply 'fuzzy theory' or any soft-computing set theory without intuitionistic fuzzy system.
Fact-2:

The Fact-1 does not necessarily require that a fuzzy decision maker (or a crisp ordinary decision maker or a decision maker with any other soft theory models or a decision maker like animal/bird which has brain, etc.) must be aware or knowledgeable about IFS Theory!

It has been philosophically and logically justified that whenever fuzzy theory or any soft computing set theory be applied to any real problem, it happens by the mandatory application of intuitionistic fuzzy system inside the brain (CPU), but it is fact that the decision maker (human being or animal or any living thing having a brain or processor element) need not be aware of IFS Theory!. The decision maker is, neither knowingly nor un-knowingly, applying the IFS theory during the progress of any decision making process by any theory/logic of his own choice; because of the hidden truth that the cognition system has an in-built system software type which spontaneously processes IF philosophy at the kernel.

It is justified in [11] that although fuzzy sets are a special case of intuitionistic fuzzy sets, but the existing concept that 'the intuitionistic fuzzy sets are higher order fuzzy sets" is **incorrect** (a similar **incorrect** concept can be imagined if one says that 'fuzzy sets can be viewed as higher order crisp sets'). Rather, the fact is that the IFSs are the most appropriate optimal model for translation of imprecise objects while the fuzzy sets are 'lower order' or 'lower dimensional' intuitionistic fuzzy sets.

Decision making activities of ill-defined problems are a routine work at every moment for every living agent. The same was true during the period prior to the discovery of crisp set theory, even starting from the stone age too. Decoding the 'progress' of decision making process in the human cognition systems (or, in the cognition system of any living animal which has a brain or a processor element, be it of a fuzzy decision maker or an intuitionistic fuzzy decision maker or any intelligent decision maker) while evaluating the membership value μ(x) to construct a fuzzy set or an IFS or any such soft computing set model, it is observed that the exact algorithm processed (analogous to the execution of machine language program corresponding to any higher level language program) is absolutely nothing but **intuitionistic fuzzy** only, which is not by any choice of the decision maker but by in-built **CIFS.**

However, the intuitionistic fuzzy processing in the cognition system of the membership value μ(x) as a special case many times may converge at **fuzzy** or at the **crisp** output for one or more elements of the universe of discourse.

The Atanassov 'Theory of IFS' is purely a choice of the decision maker. The decision maker must be knowledgeable about the 'Theory of IFS' if he wants to use it for solving any ill-defined problem. But the 'Theory of CIFS' is not and never a choice of the decision maker. It is an in-built in the cognition system of every decision maker, irrespective of his any knowledge about intuitionistic fuzzy set theory. Whoever be the decision maker, be it a human or an animal or a bird or any living thing which has brain, the 'Theory of CIFS' is automatically and mandatorily followed and finally executed inside the kernel of the cognition system, irrespective of his intellectual capability, irrespective of his any knowledge about intuitionistic fuzzy set theory.

Pattern Recognition or Object Recognition is one of the earliest and probably the most important and most executed Decision Making Problem on this earth. This problem is being solved by every human being, every living animal and every bird as a routine exercise probably very large (if not infinite) number of times every day in his real life environment. This problem has converted every human being, every living animal and every bird into a decision maker. Any problem of pattern recognition or object recognition is an impossible task without CISF. When your eyes are open, you see something and can (or can not) recognize it. Every moment your brain is recognizing something i.e. Every moment your brain is solving some kind of pattern recognition problem. For example, at some real instant of time when you see a chair by your eyes, you immediately (in time $t > 0$) recognize it and say that it is 'chair' and/or you also may say that it is a furniture, etc. rejecting all other infinite number of possibilities like: it is a dog, it is a book, it is a banana, car, tiger, mango, a number 54, a building, table, laptop, etc. When you see a mango by your eyes, you immediately (in time $t > 0$) recognize it and say that it is 'mango', rejecting all other infinite number of possibilities like: it is a dog, it is a book, it is a banana, car, tiger, a number 54, a building, table, laptop, etc.

While a hungry tiger chases a buffalo in a jungle, he decides a lot by his own logic (the logic which is unknown to us). The tiger does not know fuzzy logic or intuitionistic fuzzy logic or type-2 fuzzy logic, etc. But he has his own logic by which he decides and very rightly decides about many issues like:

(i) which buffalo to chase now (out of thousands buffalos available in his proximity). It is also fact that in many occasions he does not choose the buffalo of his nearest proximity due to some reason, or sometimes he decides to choose to chase a baby buffalo because he does also optimize the chance of his success in the real time scenario.
(ii) even sometimes after chasing a particular buffalo for about 300 m, he decides to give up his run (leading to failure to get his food in this attempt), or sometimes he decides to shift his target to another buffalo.

All these are real time decision oriented activities done by his best possible judgement, by his own logic which is unknown to us. This tiger may be illiterate according to our rich literature or rich logic, but surely he is literate by his own logic, by his own literature which are unknown to us. Whatever be the logic or literature being practiced by this tiger, the kernel always executes the algorithms of CIFS being initiated by Atanassov Initialization, not by any choice of the tiger. The different logic or literature used by different decision makers are analogous to higher level language operating in the outer annulus sphere (see Fig. 3) of the cognition system, but the kernel of every decision maker (be it human being, animal/bird, or any living thing which has brain) functions by a common machine language of CIFS irrespective of all kind of the knowledge of the decision maker which resides at the outer annulus sphere. This logic is well analysed in [11] in details, justifying that the soft-computing solution of any problem of Object Recognition can be well solved using the theory of IFS, but can not be so well solved if fuzzy theory be applied. However by another interesting example of CESFM on football sports explained in [11], it is shown that for a given problem if the decision makers of excellent talent be allowed to give their best possible judgment to the issues (i.e. if they are the best available decision makers on the subject under consideration), then fuzzy set theory will be more appropriate than intuitionistic fuzzy set theory. In the theory of CIFS in [11] it is well justified with several examples that in most of the cases of real life problems Intuitionistic Fuzzy

Set Theory of Prof. Atanassov will be the more appropriate tool for applications compared to the Fuzzy Set Theory of Prof. Zadeh. But in the Theory of CESFM in football sports the decisions are to be taken by the FIFA Referees of best qualities and of best intellectual capabilities (on the subject) of the world who are expected to have no element of hesitation while proposing membership values. Hence this is a particular case of interest where Fuzzy Set Theory is a better tool for the soft-computing CESFM method compared to the Intuitionistic Fuzzy Set Theory.

It has been noticed well that multiplication or division (and many other operations) of two triangular fuzzy numbers (trapezoidal fuzzy numbers) does not yield a fuzzy number which is a triangular fuzzy number (trapezoidal fuzzy numbers). This leads to a serious architectureus problem to the computer scientists while attempting to design fuzzy computers. In particular the programmers, while solving ill-defined engineering problems or fuzzy optimization problems or any soft-computing problems where multiplication/division operations are involved, have been facing problem regarding distortion of data structures of the output values. This problem is overcome by defining T-fuzzy numbers and Z-fuzzy numbers, just by the way they are constructed. Various operations are defined on T-fuzzy numbers and Z-fuzzy numbers compatible with the Tokunaga and Yasunobu's model [30] of fuzzy computer.

If a, b are two T-fuzzy numbers (Z-fuzzy numbers) then the following are now true:

(i) $a + b$ is also a T-fuzzy number (Z-fuzzy numbers)
(ii) $a - b$ is also a T-fuzzy number (Z-fuzzy numbers)
(iii) $a * b$ is also a T-fuzzy number (Z-fuzzy numbers)
(iv) a/b is also a T-fuzzy number (Z-fuzzy numbers)
(v) a^2 is also a T-fuzzy number (Z-fuzzy numbers)
(vi) a^n is also a T-fuzzy number (Z-fuzzy numbers)
(vii) $1/a$ is also a T-fuzzy number (Z-fuzzy numbers) etc.

Consequently, with these revised notion of the fuzzy numbers: T-fuzzy numbers and Z-fuzzy numbers, a possible way could now be discovered by rigorous future research work to define methods of fuzzy computing like sqrt(n), e^n, log n, etc. and fuzzy trigonometrical terms like sin a, cos a, etc. where n and a are fuzzy numbers. The Tokunaga and Yasunobu's model [30] of fuzzy computer will be able to compute fuzzy arithmetic expression (which is in infix notation) of T-fuzzy numbers in two steps:-

Step-1. It will convert the infix form of fuzzy arithmetic expression into postfix fuzzy arithmetic expression.

Step-2. It will evaluate the fuzzy postfix expression to give the result which is a T-fuzzy number.

In each step, the fuzzy stack will be the main tool to accomplish the task. But there is no literature reported so far on fuzzy stacks, possibly because of the drawback of existing notion of triangular fuzzy numbers. Anyway, in the future research work the fuzzy experts will concentrate upon the problem how to define the notion of fuzzy stacks which can be used to evaluate fuzzy arithmetic expressions of T-fuzzy numbers. Attempt may be made to define fuzzy queues and many other fuzzy data-structures, their applications in fuzzy computing. The notion of T-intuitionistic fuzzy numbers and Z-intuitionistic fuzzy numbers can also be introduced in a similar way. Attempt will also be made to explore whether and how the set of all T-fuzzy numbers (Z-fuzzy numbers) forms a region [13, 15] or at least forms a division algebra, whether and how the set of all T-intuitionistic fuzzy numbers (Z-intuitionistic fuzzy numbers) forms a region [13, 15] or at least forms a division algebra. Otherwise the use of fuzzy numbers and/or intuitionistic fuzzy numbers will not be fruitful for any kind of complex computation for solving any mathematical or engineering or optimization or statistical or decision making problem. Presently it is an important unsolved problem.

References

1. Atanassov, K.T.: Intuitionistic fuzzy sets. Fuzzy Sets Syst. **20**, 87–96 (1986)
2. Atanassov, K.T.: More on intuitionistic fuzzy sets. Fuzzy Sets Syst. **33**, 37–45 (1989)
3. Atanassov, K.T.: New operations defined over the intuitionistic fuzzy sets. Fuzzy Sets Syst. **6**, 137–142 (1994)
4. Atanassov, K.T.: Operators over interval valued intuitionistic fuzzy sets. Fuzzy Sets Syst. **64**, 159–174 (1994)
5. Atanassov, K.T.: Intuitionistic Fuzzy Sets: Theory and Applications. Springer, Heidelberg (1999)
6. Atanassov, K.T.: On Intuitionistic Fuzzy Sets Theory. Springer, Berlin (2012)
7. Atanassov, K.T., Gargov, G.: Interval-valued intuitionistic fuzzy sets. Fuzzy Sets Syst. **31**, 343–349 (1989)
8. Atanassov, K., Pasi, G., Yager, R.R.: Intuitionistic fuzzy interpretations of multi-criteria multi-person and multi-measurement tool decision making. Int. J. Syst. Sci. **36**, 859–868 (2005)
9. Biswas, R.: Decoding the 'progress' of decision making process in the human/animal cognition systems while evaluating the membership value μ(x). Issues Intuitionistic Fuzzy Sets Generalized Nets **10**, 21–53 (2013)
10. Biswas, R.: Introducing soft statistical measures. J. Fuzzy Math. **22**(4), 819–851 (2014)
11. Biswas, R.: Is 'Fuzzy Theory' An Appropriate Tool For Large Size Problems?. In: the series of SpringerBriefs in Computational Intelligence. Springer, Heidelberg (2015)
12. Biswas, R.: Fuzzy numbers redefined. Information **15**(4), 1369–1380 (2012)
13. Biswas, R.: Region algebra, theory of objects and theory of numbers. Int. J. Algebra **6**(8), 1371–1417 (2012)
14. Biswas, R.: "THEORY OF CESFM": a proposal to FIFA & IFAB for a new 'continuous evaluation fuzzy method' of deciding the WINNER of a football match that would have otherwise been drawn or tied after 90 minutes of play. J Fuzzy Math. **23**(4). (in Press)
15. Biswas, R.: "Theory of Numbers" of a Complete Region. Notes on Number Theory and Discrete Mathematics. **21**(3), 1–21 (2015)
16. Biswas, R.: "Atrain Distributed System" (ADS): an infinitely scalable architecture for processing big data of any 4Vs. In: Acharjya, D.P., Dehuri, S., Sanyal, S. (ed.) Computational Intelligence for Big Data Analysis Frontier Advances and Applications, Part-1, pp. 1–53 Springer International Publishing, Switzerland (2015)

17. Bouchon-Meunier, B., Yager, R.R., Zadeh, L.A.: Fuzzy Logic and Soft Computing. World Scientific, Singapore (1995)
18. Dubois, D., Prade, H.: Fuzzy Sets and Systems: Theory and Applications. Academic Press, New York (1990)
19. Dubois, D., Prade, H.: Twofold fuzzy sets and rough sets: some issues in knowledge representation. Fuzzy Sets Syst. **23**, 3–18 (1987)
20. Gau, W.L., Buehrer, D.J.: Vague sets. IEEE Trans. Syst. Man Cybern. **23**(2), 610–614 (1993)
21. Goguen, J.A.: L-fuzzy sets. J. Math. Anal. Appl. **18**, 145–174 (1967)
22. Gorzalzany, M.B.: A method of inference in approximate reasoning based on interval-valued fuzzy sets. Fuzzy Sets Syst. **21**, 1–17 (1987)
23. Kaufmann, A.: Introduction to the Theory of Fuzzy Subsets. Academic Press, New York (1975)
24. Kaufmann, A., Gupta, M.M.: Introduction to Fuzzy Arithmetic Theory and Application. Van Nostrand Reinhold, New York (1991)
25. Klir, G.K., Yuan, B.: Fuzzy Sets and Fuzzy Logic, Theory and Applications. Prentice Hall, New Jersey (1995)
26. Mizumoto, M., Tanaka, K.: Some properties of fuzzy set of type 2. Inf. Control **31**, 321–340 (1976)
27. Mololodtsov, D.: Soft set theory-first results. Comput. Math. Appl. **37**(4/5), 19–31 (1999)
28. Novak, V.: Fuzzy Sets and Their Applications. Adam Hilger, Bristol (1986)
29. Pawlak, Z.: Rough Sets. Int. J. Inf. Comput. Sci. **11**, 341–356 (1982)
30. Tokunaga, H., Yasunobu, S.: The Fuzzy Computer, in Applied Research in Fuzzy Technology (Results of the Laboratory for International Fuzzy Engineering (LIFE) Series: International Series in Intelligent Technologies, vol. 1. Springer, Berlin Heidelberg, New York (1994)
31. Zadeh, L.A.: Fuzzy sets. Inf Control **8**, 338–353 (1965)
32. Zimmermann, H.J.: Fuzzy Set Theory and Its Applications. Kluwer Academic Publishers, Boston/Dordrecht/London (1991)

Properties and Applications of Pythagorean Fuzzy Sets

Ronald R. Yager

Abstract We introduce the concept of Pythagorean fuzzy subsets and discuss its relationship with intuitionistic fuzzy subsets. We focus on the negation and its relationship to the Pythagorean theorem. We describe some of the basic set operations on Pythagorean fuzzy subsets. We look at the relationship between Pythagorean membership grades and complex numbers. We consider the problem of multi-criteria decision making with satisfactions expressed as Pythagorean membership grades. We look at the use of the geometric mean and ordered weighted geometric (OWG) operator for aggregating criteria satisfaction. We provide a method for comparing alternatives whose degrees of satisfaction to the decision criteria are expressed as Pythagorean membership grades.

Keywords Intuitionistic fuzzy sets • Non-standard membership grades • Pythagorean theorem • Complex numbers • Multi-criteria aggregation • Geometric mean

1 Introduction

Atanassov introduced the idea of intuitionistic fuzzy sets [1]. A considerable body of research has been devoted to these sets [2]. Intuitionistic fuzzy sets extend the representational capability of fuzzy sets from being able to represent partial membership to additional being able to represent lack of commitment or uncertainty in providing the membership grade. They are an example of what are referred to as non-standard fuzzy sets. In a standard fuzzy subset A one provides a membership grade A(x) ∈ [0, 1], indicating the degree of support for the membership of x in A. Implicit in this situation is the assumption that the degree of support against membership of x in A is valued as the negation of A(x), typically taken as 1 − A(x).

R.R. Yager (✉)
Machine Intelligence Institute, Iona College, New Rochelle, NY 10801, USA
e-mail: yager@panix.com

P. Angelov and S. Sotirov (eds.), *Imprecision and Uncertainty in Information Representation and Processing*, Studies in Fuzziness and Soft Computing 332,
DOI 10.1007/978-3-319-26302-1_9

Instead of accepting this implicit assumption for the support against membership intuitionistic fuzzy allows for a separate specification of this value, being only constrained by the requirement that the sum of the supports for and against doesn't exceed one. In adding this capability Atanassov has allowed the providers of membership grades to be uncertain or hesitant in providing their membership grades, thus if $A^+(x)$ and $A^-(x)$ are the degrees of support for and against membership then the value $1 - (A^+(x) + A^-(x))$ is the amount of uncommitted or uncertain membership.

Implicit in the use of intuitionistic fuzzy sets is the acceptance of the linear form of logical negation, $Neg(a) = 1 - a$. As shown by Sugeno and Yager there are other possible formations within the field of fuzzy sets for the modeling of negation [3]. One notable example is the negation $Neg(a) = (1 - a^2)^{1/2}$ which Yager has referred to as the Pythagorean negation. Using this formation for the negation instead of the linear formulation Yager [4–6] provided a related class of non-standard fuzzy sets that he referred to as Pythagorean fuzzy sets. Some researchers have begun using these sets in some applications [7, 8]. Here we look in more detail at the Pythagorean fuzzy sets. In addition to discussing the basic properties of these sets we provide a formulation in terms of complex numbers [9]. We consider the problem of multi-criteria decision making when the degrees of satisfaction are expressed as Pythagorean membership grades. We also provide a formation for comparing Pythagorean membership grades.

2 Pythagorean Membership Grades

In [4–6] Yager introduced a new class of fuzzy sets called Pythagorean fuzzy sets, PFS, which are closely related to Atanassov's intuitionistic fuzzy sets [1, 2]. We shall refer to the membership grades associated with these sets as Pythagorean membership grades, PMG's. In the following we describe the Pythagorean membership grades.

One way of expressing Pythagorean membership grades is by giving a pair of values r(x) and d(x) for each $x \in X$. Here $r(x) \in [0, 1]$ is called the strength of commitment at x and $d(x) \in [0, 1]$ is called the direction of commitment. Here r(x) and d(x) are associated with a pair of membership grades $A_Y(x)$ and $A_N(x)$ indicating respectively the support for membership of x in A and the support against membership of x in A. As we shall see $A_Y(x)$ and $A_N(x)$ are related using the Pythagorean complement with respect to r(x). In particular the value of $A_Y(x)$ and $A_N(x)$ are defined from r(x) and d(x) as

$$A_Y(x) = r(x)Cos(\theta(x))$$
$$A_N(x) = r(x)Sin(\theta(x))$$

where $\theta(x) = (1 - d(x))\frac{\pi}{2}$. Here we see $\theta(x)$ is expressed as radians and $\theta(x) \in [0, \frac{\pi}{2}]$. Thus we see the closer d(x) to 1, the closer $\theta(x)$ to 0, the more the commitment r(x) is supporting membership of x in A.

We now show that $A_Y(x)$ and $A_N(x)$ are Pythagorean complements with respect to r(x).

$$A_Y^2(x) + A_N^2(x) = r^2(x)\mathrm{Cos}^2(\theta(x)) + r^2(x)\sin^2(\theta(x))$$

and since it is known from the Pythagorean theorem that $\mathrm{Cos}^2(\theta) + \mathrm{Sin}^2(\theta) = 1$ then we have that $A_Y^2(x) + A_N^2(x) = r^2(x)$ and hence $A_Y^2(x) = r^2(x) - A_N^2(x)$ and $A_N^2(x) = r^2(x) - A_Y^2(x)$.

Thus A_Y and A_N are Pythagorean complements with respect to r(x).

Pythagorean membership grades allow some lack of commitment in addition to imprecision in assigning membership. We see that r(x), which is a value in the unit interval, is the strength of commitment about membership at point x, the larger r(x) the stronger the commitment. Let us know understand the meaning of the value d (x), the direction of the strength. We recall that $\theta(x) = (1 - d(x))\frac{\pi}{2}$. In the case when $d(x) = 1$, then $\theta(x) = 0$ and $\mathrm{Cos}(\theta(x)) = 1$ and $\mathrm{Sin}(\theta(x)) = 0$. Thus $A_Y(x) = r(x)$ and $A_N(x) = 0$. On the other hand if $d(x) = 0$ then $\theta(x) = \pi/2$ and we get $A_Y(x) = 0$ and $A_N(x) = 1$. Thus we see that d(x) is essentially indicating on a scale of 1 to 0 how fully the strength r(x) is pointing to membership. If $d(x) = 1$ the direction of r(x) is completely to membership while $d(x) = 0$ the direction of the strength is completely to non-membership. Intermediate values of d(x) indicate partial support to membership and non-membership.

Here we note that the Pythagorean membership grade can be expressed either by providing r(x) and d(x) or by r(x) and $\theta(x)$ were we express θ as radians in the range $[0, \frac{\pi}{2}]$.

Thus we see that the Pythagorean membership grade provides a type of imprecise membership grades, generally referred to as type 2. These membership grades, $A_Y(x)$ and $A_N(x)$, are related by the Pythagorean complement with respect to strength of commitment, $A_Y^2(x) + A_N^2(x) = r^2(x)$. Furthermore we have $\mathrm{Cos}(\theta(x)) = \frac{A_Y(x)}{r(x)}$ and hence $\theta(x) = \mathrm{Arccos}(\frac{A_Y(x)}{r(x)})$.

We note that more generally a Pythagorean membership grade A(x) is a pair of values (a, b) such that $a, b \in [0, 1]$ and $a^2 + b^2 \leq 1$. Here $a = A_Y(x)$, the degree of support for membership of x is A and $b = A_N(x)$ the degree of support against membership of x in A. We see that for this pair $a^2 + b^2 = r^2$. Thus a Pythagorean membership grade is a point on a circle of radius r. We also recall that any point (a, b) on a circle of radius $r^2 = a^2 + b^2$ can be expressed as (r Cos(θ), r Sin(θ)). Thus we see that $\mathrm{Cos}(\theta) = \frac{a}{r}$ and $\mathrm{Sin}(\theta) = \frac{b}{r}$ hence $\theta = \mathrm{arc\ Cos}(a/b)$ thus $d = \frac{\pi - 2\theta}{\pi}$. Thus the point (a, b) has strength of commitment and direction of commitment pair $r = (a^2 + b^2)^{1/2}$ and $d = \frac{\pi - 2\theta}{\pi}$. We emphasize that since we require that a and $b \in [0, 1]$ then $\theta \in [0, \frac{\pi}{2}]$, a Pythagorean membership grade is a point in the upper right quadrant.

Another example of non-standard fuzzy subset is the intuitionistic fuzzy subsets introduced by Atanassov. An intuitionistic membership grade $F(x) = (A^+(x), A^-(x))$ is also a pair (a, b) such that a, b $\in$ [0, 1]. Here $A^+(x)$, *a*, indicates the amount of guaranteed membership of x in A and $A^-(x)$, *b*, indicates the guaranteed non-membership in A however here we require that $a + b \leq 1$. The expression Hes $(x) = 1 - (A^+(x) + A^-(x))$ is called the hesitancy of x. It is a reflection of lack of commitment or uncertainty associated with the membership grade at x. We shall find it convenient to denote $S(x) = 1 - \text{Hes}(x) = A^+(x) + A^-1(x)$. It is a kind of total commitment.

Thus while both intuitionistic and Pythagorean membership allow for the representation of uncertain membership in grades in terms of pairs of values $(A^+(x), A^-(x))$ and $(A_Y(x), A_N(y))$ there are some important differences between these two representations. The first is that $A^+(x) + A^-(x) \leq 1$ while $A_Y^2(x) + A_N^2(x) \leq 1$.

We observe that for a and b $\in$ [0, 1] then $a^2 \leq a$ and $b^2 \leq b$ from this we observe that if $a + b \leq 1$ then $a^2 + b^2 \leq 1$. From this we can conclude the following theorem.

Theorem *The set of Pythagorean membership grades is greater than the set of intuitionistic membership grades.*

We see this as follows. First we note that every point (a, b) that is an intuitionistic membership grade is also a Pythagorean membership grade. We first observe that for any a and b $\in$ [0, 1] then $a^2 \leq a$ and $b^2 \leq b$ from this we observe that if $a + b \leq 1$ then $a^2 + b^2 \leq 1$. Secondly there are Pythagorean membership grades that not intuitionistic membership grades. Consider now the point $(\frac{\sqrt{3}}{2}, \frac{1}{2})$. We see that $\left(\frac{\sqrt{3}}{2}\right)^2 + \left(\frac{1}{2}\right)^2 = \frac{3}{4} + \frac{1}{4} = 1$ thus this is a Pythagorean membership grade. However since $\frac{\sqrt{3}}{2} = \frac{1.72}{2} = 0.866$ then $0.5 + 0.866 > 1$ this is not an intuitionistic membership grade.

This result can be clearly seen from Fig. 1. Here we observe that intuitionistic membership grades are all points under the line $x + y \leq 1$ and the Pythagorean membership grades are all points with $x^2 + y^2 \leq 1$. We see then that the Pythagorean membership grades allow for the representation on a larger body of non-standard membership grades then intuitionistic membership grades.

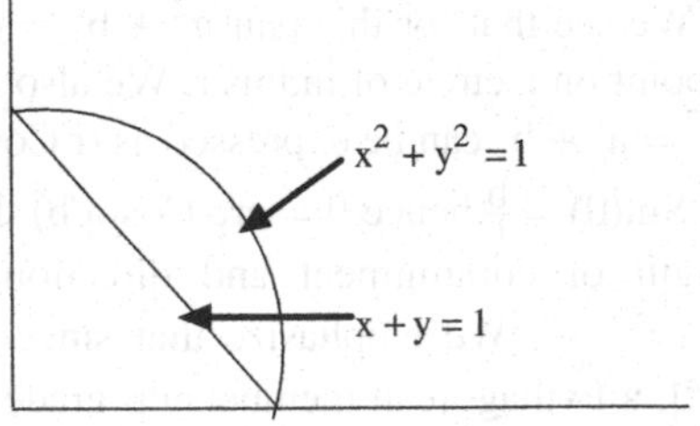

Fig. 1 Comparison of space of Pythagorean and intuitionistic membership grades

3 On the Negation Operation

Another distinction between Pythagorean and Intuitionistic fuzzy sets relates to their definitions of complement or negation. Before introducing the negation of Pythagorean fuzzy sets we need to say something about the complement operator [3]. A complement C operator is a mapping C: [0, 1] → [0, 1] that satisfies

(1) **Boundary Conditions**: C(0) = 0 and C(1) = 0
(2) **Monotonicity**: For all a, b ∈ [0, 1] if a ≤ b then C(a) ≥ C(b)
(3) **Continuity**
(4) **Involution**: C(C(a)) = a

We recall that the linear function C(a) = 1 − a is the classic example of a complement operator.

Yager [10, 11] introduced a family of complement operators. The Yager class of complements is defined by

$$C(a) = (1 - a^P)^{1/P}$$

where P ∈ (0, ∞). We observe that for P = 1 we get the classic linear complement C(a) = 1 − a. If p = 2 then we get

$$C(a) = (1 - a^2)^{1/2}$$

we note here

$$(C(a))^2 + a^2 = 1.$$

We shall refer to this as the Pythagorean complement.

We know introduce the related idea of complements with respect to r, where r ∈ [0, 1]. We define $C_{[r]}$: [0, r] → [0, r] as a complement with respect to r if

(1) $C_{[r]}(0) = r$, $C_{[r]}(r) = 0$
(2) $C_{[r]}$ is monotonic
(3) $C_{[r]}$ is continuous
(4) $C_{[r]}(C_{[r]}(a)) = a$ Involution

We note that the Yager class of complements is easily extended to be complements with respect to r,

$$C_{[r]}(a) = (r^P - a^P)^{1/P}$$

We note here that $(C_{[r]}(a))^p = r^P - a^P$ and hence $(C_{[r]}(a))^p + a^P = r^P$. It can be shown these satisfy the required conditions. Two important complements with respect to r are the linear complement $C_{[r]}(a) = r - a$ and the Pythagorean complement with respect to r, $C_{[r]}(a) = (r^2 - a^2)^{1/2}$.

We now turn to the idea of set complement. Assume A is an intuitionistic fuzzy set with intuitionistic membership grades, $<A^+(x), A^-(x)>$ where $A^+(x) + A^-(x) \leq 1$. We recall an intuitionistic fuzzy set A has complement $\bar{A}$ with membership grades [1, 2]

$$\bar{A}(x) = <\bar{A}^+(x), \bar{A}^-(x)> = <A^-(x), A^+(x)>.$$

We have simply interchanged values of degree of support for with of degree of support against. A more fundamental understanding of this operation can had recalling that the strength of commitment $S(x) = A^+(x) + A^-(x)$. Here we see $\bar{A}^+(x) = A^-(x) = S(x) - A^+(x)$ and $\bar{A}^-(x) = A^+(x) = S(x) - A^-(x)$. Using this we can express

$$\bar{A}(x) = <\bar{A}^+(x), \bar{A}^-(x)> = <S(x) - A^+(x), S(x) - A^-(x)>$$

Here then we have that $\bar{A}^+(x)$ is the linear complement of $A^+(x)$ with respect to S (x) and $\bar{A}^-(x)$ is the linear complement of $A^-(x)$ with respect to S(x).

In the case of the Pythagorean fuzzy sets we define the complement in analogous manner using the Pythagorean complement with respect to the commitment r(x) [4–6]. Assume a Pythagorean membership grade $A(x) = <A_Y(x), A_N(x)>$ we define its complement $\bar{A}(x) = <\bar{A}_Y(x), \bar{A}_N(x)>$ such that $\bar{A}_Y(x) = (r^2(x) - A_Y^2(x))^{1/2}$ and $\bar{A}_N(x) = (r^2(x) - A_N^2(x))^{1/2}$ where $r^2(x) = A_Y^2(x) + A_N^2(x)$. Here then $\bar{A}_Y(x)$ is the Pythagorean complement of $A_Y(x)$ with respect r(x) and $\bar{A}_N(x)$ is the Pythagorean complement of $A_N(x)$ with respect r(x). We easily see that $\bar{A}_Y(x) = (r^2(x) - A_Y^2(x))^{1/2} = (A_N^2(x))^{1/2} = A_N(x)$ and $\bar{A}_N(x) = (r^2(x) - A_N^2(x))^{1/2} = (A_Y^2(x))^{1/2} = A_Y(x)$. Thus here again we have $\bar{A}_Y(x) = A_N(x)$ and $\bar{A}_N(x) = A_Y(x)$.

We recall that if A and B are two intuitionistic fuzzy sets with intuitionistic membership grades $A(x) = <A^+(x), A^-(x)>$ and $B(x) = <B^+(x), B^-(x)>$ then as suggested by Atanassov [2] we say $A \subseteq B$ if $A^+(x) \leq B^+(x)$ and $A^-(x) \geq B^-(x)$ for all x. We see that since $\bar{A}(x) = = <A^-(x), A^+(x)>$ and $\bar{B}(x) = <B^-(x), B^+(x)>$ then if $A \subseteq B$ we have $\bar{B} \subseteq \bar{A}$.

In an analogous manner if E and F are two Pythagorean fuzzy sets with Pythagorean membership grades $E(x) = <E_Y(x), E_N(x)>$ and $F(x) = <F_Y(x), F_N(x)>$ we say $E \subseteq F$ if $E_Y(x) \leq F_Y(x)$ and $E_N(x) \geq F_N(x)$ for all x. Since $\bar{E}(x) = <E_N(x), E_Y(x)>$ and $\bar{F}(x) = <F_N(x), F_Y(x)>$ then if $E \subseteq F$ we have $\bar{F} \subseteq \bar{E}$.

4 Basic Set Operations for Pythagorean Fuzzy Sets

We now turn to the basic operations needed for combining Pythagorean fuzzy sets Assume A_1 and A_2 are two fuzzy subsets of X with Pythagorean memberships grades. For simplicity we denote for $A_1(x) = (a_1, b_1)$ and $A_2(x) = (a_2, b_2)$. Here we have $a_1^2 + b_1^2 = r_1^2 \leq 1$ and $a_2^2 + b_2^2 = r_2^2 \leq 1$. Consider now the intersection, $D = A_1 \cap A_2$. We shall D define $D(x) = (d_1, d_2)$ where $d_1 = \text{Min}(a_1, a_2)$ and $d_2 = \text{Max}(b_1, b_2)$. In order for D to be a Pythagorean fuzzy subset we must have that $d_1^2 + d_2^2 \leq 1$. We see since $d_2 = \text{Max}(b_1, b_2)$ then

$$d_2^2 = \text{Max}\left(b_1^2, b_2^2\right) = \text{Max}\left(r_1^2 - a_1^2, r_2^2 - a_2^2\right) \leq \text{Max}\left(r_1^2 - \text{Min}\left(a_1^2, a_2^2\right), r_2^2 - \text{Min}\left(a_1^2, a_2^2\right)\right)$$
$$d_2^2 \leq \text{Max}\left(1 - \text{Min}\left(a_1^2, a_2^2\right), 1 - \text{Min}\left(a_1^2, a_2^2\right)\right) \leq 1 - \text{Min}\left(a_1^2, a_2^2\right)$$

From this we see that $d_1^2 + d_2^2 \leq \text{Min}(a_1^2, a_2^2) + 1 - \text{Min}(a_1^2, a_2^2) \leq 1$. Thus we see that this satisfies the requirement of being a Pythagorean membership grade.

We now define $E = A_1 \cup A_2$ so that $E(x) = (e_1, e_2)$ where $e_1 = \text{Max}(a_1, a_2)$ and $e_2 = \text{Min}(b_1, b_2)$. In a manner analogous to the preceding we can show that (e_1, e_2) is a Pythagorean membership grade.

Thus, as in the case of the intuitionistic fuzzy sets we can define the set operations of intersection and union using the Max and Min operators. If A_1 and A_2 are two Pythagorean fuzzy sets with membership grades $A_1(x) = (a_1, b_1)$ and $A_2(x) = (a_2, b_2)$ then $D = A_1 \cap A_2$ and $E = A_2 \cup A_2$ are Pythagorean fuzzy sets such that

$$d(x) = (d_1, d_2) = (\text{Min}(a_1, a_2),\ \text{Max}(b_1, b_2))$$
$$e(x) = (e_1, e_2) = (\text{Max}(a_1, a_2),\ \text{Min}(b_1, b_2))$$

We further define the complement of A_1, $\bar{A}$, such that $\bar{A}(x) = ((r_1^2 - a_1^2)^{0.5}, (r_2^2 - b_1^2)^{0.5})$. where $r_1^2 = a_1^2 + b_1^2$ and hence $\bar{A}(x) = (b_1, a_1)$.

We now look at the more general question of aggregation of Pythagorean fuzzy sets.

Definition A function Agg: $[0, 1]^q \rightarrow [0, 1]$ is called an aggregation function [12, 13] if

(1) Agg(0, ..., 0) = 0
(2) Agg(1, ... 1) = 1
(3) $\text{Agg}(a_1, \ldots, a_q) \geq \text{Agg}(b_1, \ldots, b_q)$ if $a_j \geq b_j$ for all j

Conditions one and two are referred to as boundary conditions and condition three is a monotonicity requirement.

We now define the dual of an aggregation operator [12, 14].

Definition Let Agg be any aggregation operator defined on the unit interval I, Agg: $I^q \to I$, we define the dual of Agg, $\widetilde{Agg}$ as

$$\widetilde{Agg}(x_1, \ldots, x_q) = Neg(Agg(Neg(x_1), \ldots, Neg(x_q)))$$

where Neg is a complement operator.

Definition Assume A_1, ..., A_q are collection of PFS, Pythagorean Fuzzy Sets, with membership grades $A_j(x) = (A_{jY}(x), A_{jN}(x))$. We define E = Agg($A_1$, ..., A_q) as a PFS with Pythagorean membership grades $E(x) = (E_Y(x), E_N(x))$ such that

$$E_Y(x) = Agg(A_{1Y}(x), A_{2Y}(x), \ldots, A_{qY}(x))$$
$$E_N(x) = \widetilde{Agg}(A_{1N}(x), A_{2N}(x), \ldots, A_{qN}(x))$$

In the case of the Pythagorean membership grade $Neg(a) = (1 - a^2)^{1/2}$.

For E to be a PFS we require that $E_Y(x)$ and $E_N(x)$ satisfy $E_Y^2(x) + E_N^2(x) \leq 1$. Thus if $E_Y^2(x) + E_N^2(x) \leq 1$ the operator of Agg is closed, it maps the collection of PFS into a PFS.

We note that Agg is monotonic, $Agg(x_1, \ldots, x_q) \geq Agg(x_1, \ldots, y_n)$ if $x_j \geq y_j$ We now prove the following theorem using the monotonicity.

Theorem If Agg is monotonic then it is always the case that $E_Y^2(x) + E_N^2(x) \leq 1$.

Proof In the following for notational convince we shall denote the pair ($A_{jY}(x)$, $A_{jN}(x)$) as (a_j, b_j) where each pair (a_j, b_j) satisfies $a_j^2 + b_j^2 = r_j^2 \leq 1$

$$E_Y(x) = Agg(a_1, \ldots, a_q)$$
$$E_N(x) = \widetilde{Agg}(b_1, \ldots, b_q) = (1 - (Agg((1 - b_1^2)^{1/2}, (1 - b_2^2)^{1/2}, \ldots, (1 - b_2^2))^2)^{1/2}$$

We now recall that since $a_j^2 + b_j^2 \leq 1$ then $1 - b_j^2 \geq a_j^2$. From this we observe that

$$E_Y(x) = Agg(a_1, \ldots, a_q) \leq Agg((1 - b_1^2)^{1/2}, \ldots, (1 - b_q^2)^{1/2})$$

We now observe that

$$E_Y^2(x) + E_N^2(x) \leq (Agg((1 - b_1^2)^{1/2}, (1 - b_q^2)^{1/2}))^2 + ((1 - (Agg((1 - b_1^2)^{1/2}, \ldots, (1 - b_q^2))^2)^{1/2})^2$$
$$E_Y^2(x) + E_N^2(x) \leq (Agg((1 - b_1^2)^{1/2}, (1 - b_q^2)^{1/2}))^2 + 1 - (Agg((1 - b_1^2)^{1/2}, (1 - b_q^2)^{1/2}))^2$$
$$E_Y^2(x) + E_N^2(x) \leq 1$$

Thus we see that if $A_j(x)$ are a collection of PFS, Pythagorean fuzzy sets, then if we define E = Agg(A_1, ..., A_q> where such that $E_Y(x) = Agg_j(A_{jY}(x))$ and $E_N(x) = \widetilde{Agg}_j(A_{jN}(x)$ then E is a PFS with $E(x) = (E_Y(x), E_N(x))$. Thus the operation Agg is closed.

In [12] authors consider a number of important classes aggregation operators. We briefly look at these for the case of aggregation Pythagorean fuzzy subsets. Two important classes of aggregation functions are conjunctive and disjunctive operators. An aggregation function is said to be a conjunctive type operator if $\mathrm{Agg}(a_1, \ldots, a_n) \leq \mathrm{Min}(a_1, \ldots, a_n)$ and is called a disjunctive type operator if $\mathrm{Agg}(a_1, \ldots, a_n) \geq \mathrm{Max}(a_1, \ldots, a_n)$. The conjunctive operator generalizes the set intersection, *and*, operator while the disjunctive operator generalizes the set union, *or*, operator. It can be shown that if Agg is a conjunctive type operator then $\widetilde{\mathrm{Agg}}$ is disjunctive and if Agg is disjunctive then $\widetilde{\mathrm{Agg}}$ is conjunctive.

An special important type of conjunctive operators are t-norms and a related special important type of disjunctive operators are t-conorms [15]. We recall that a t-norm is defined as a binary aggregation operator that has one as an identity, T(a, 1) = a, and is associative T(a, T(b, c)) = T(T(a, b, c)). A conorm is also associative and has zero as its identity, S(a, 0) = a. It can be shown [12] that if Agg is a t-norm then $\widetilde{\mathrm{Agg}}$ is a t-conorm and also if Agg is a t-conorm then $\widetilde{\mathrm{Agg}}$ is a t-norm.

Another important class of Agg operators are mean type aggregation operators. These operators are defined by their satisfaction of the condition $\mathrm{Min}_j[a_j] \leq \mathrm{Agg}(a_1, \ldots, a_q) \leq \mathrm{Max}_j[a_j]$. It can be shown that if Agg is a mean type operator then $\widetilde{\mathrm{Agg}}$ is also mean type operator.

We now consider the following mean aggregation operator

$$\mathrm{Agg}(b_1, \ldots, b_n) = (\Sigma_j w_j b_j^2)^{1/2}$$

where $\Sigma_j w_j = 1$ and $w_j \in [0, 1]$. This is an example of a class of mean operators called weighted power means [12]. We can show its dual

$$\widetilde{\mathrm{Agg}}(b_1, \ldots, b_n) = \mathrm{Neg}(\mathrm{Agg}_j(\mathrm{Neg}(b_j)) = (\Sigma_j w_j b_j^2)^{1/2}$$

Thus $\mathrm{Agg}(a_1, \ldots, a_n) = (\Sigma_j w_j a_j^2)^{1/2}$ is self dual $\mathrm{Agg}(a_1, \ldots, a_n) = \widetilde{Agg}(a_1, \ldots, a_n)$.

Another example of mean of is the geometric mean $\mathrm{Agg}(b_1, \ldots, b_n) = \prod_{j=1}^{n} b_j^{w_j}$. It can be easily shown [4–6] that its dual is

$$\widetilde{\mathrm{Agg}}(b_1, \ldots, b_n) = (1 - \mathrm{Agg}((1 - b_1^2), \ldots\ldots, (1 - b_n^2)))^{1/2}$$

5 Pythagorean Membership Grades and Complex Numbers

In [4] we showed that for some purposes these types of Pythagorean membership grades could be effectively expressed using a complex number to represent the membership grade. We note in [16] Dick and Yager explored this relationship in considerable detail.

In anticipation of investigating this relationship between Pythagorean membership grades and complex numbered membership grades we review some ideas about complex numbers [9].

A complex number z is an ordered pair (x, y) interpreted as $z = x + iy$ where $i = \sqrt{-1}$, the so-called imaginary number. One can view a complex number as a point in a plane as shown in Fig. 2.

Fundamental to the manipulation of complex numbers is the Euler formula. For any real number x

$$e^{ix} = \text{Cos}(x) + i\ \text{Sin}(x)$$

where x is interpreted as radians. We recall 2π radians is 360°. A useful formula when using the Euler formula is the Pythagorean theorem, $\text{Cos}^2(x) + \text{Sin}^2(x) = 1$.

Assume any complex number $z = a + ib$, and let $r = \sqrt{a^2 + b^2} = |z|$. Consider now a value θ such that $\text{Cos}(\theta) = \frac{a}{\sqrt{a^2+b^2}}$. We see that from the Pythagorean theorem

$$\text{Sin}^2(\theta) = 1 - \text{Cos}^2(\theta) = \frac{a^2+b^2}{a^2+b^2} - \frac{a^2}{a^2+b^2} = \frac{b^2}{a^2+b^2}$$

Hence $\text{Sin}(\theta) = \frac{b}{\sqrt{a^2+b^2}}$. Here now we see that from $\text{Cos}(\theta) = \frac{a}{|z|}$ we get $a = |z|\ \text{Cos}(\theta)$ and from $\text{Sin}(\theta) = \frac{b}{|z|}$ we get $b = |z|\ \text{Sin}(\theta)$. Using this we can express the complex number $z = a + ib$ as $z = |z|\ \text{Cos}(\theta) + i|z|\ \text{Sin}(\theta) = |z|\ (\text{Cos}(\theta) + i\ \text{Sin}(\theta))$. Using the Euler formula we get

$$z = |z|e^{i\theta} = |z|(\text{Cos}(\theta) + i\ \text{Sin}(\theta)).$$

We see that any complex number $z = a + ib$ can be is alternatively expressed as $z = |z|e^{\,i\theta}$ where $|z| = (a^2 + b^2)^{1/2}$ and $\theta = \text{ArcCos}(\frac{a}{(a^2+b^2)^{1/2}})$, the angle whose cosine is $\frac{a}{z}$. We note $\theta = \text{ArcSin}(\frac{b}{(a^2+b^2)^{1/2}})$. The form $z = |z|e^{i\theta}$ is called the polar

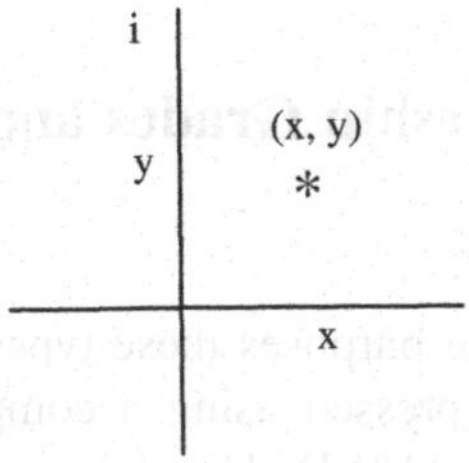

Fig. 2 Point in complex plane

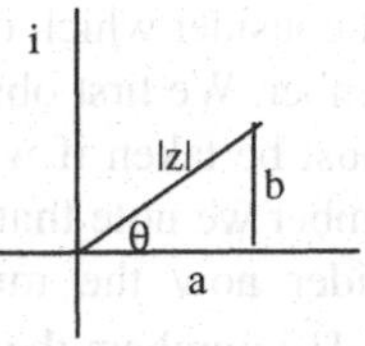

Fig. 3 Geometric perspective

representation of the complex number z. Geometrically we see this relationship in Fig. 3.

The polar representation greatly simplifies many operators involving complex numbers, especially multiplication. Consider the multiplication of two complex numbers, $z_1 = a_1 + ib_1$, and $z_2 = a_2 + ib_2$. We easily see that $z_1z_2 = (a_1a_2 - b_1b_2) + i(a_1b_2 + a_2b_1)$. Performing this using the polar representation is much simpler. In this case with $z_1 = r_1\, e^{i\theta_1}$ and $z_1 = r_1\, e^{i\theta_2}$. We get $z_1z_2 = r_1r_2\, e^{i(\theta_1 + \theta_2)}$. Another operation that is easy to perform is the polar domain is raising a complex number to a power. Here if $z = a + ib = re^{i\theta}$ then $(z)^m = r^m e^{im\theta} = r_2\, e^{i\theta_2}$. We note that $r_2 = r^m$ and $\theta_2 = m\theta$, the term θ multiplied by m. Thus we have

$$z^m = r^m(\mathrm{Cos}(m^\theta) + i\mathrm{Sin}(m\theta)).$$

We see if $m = -1$ then $z^{-1} = \frac{1}{z} = \frac{1}{r}(\mathrm{Cos}(-\theta) + i\mathrm{Sin}(-\theta))$. Recalling that $\mathrm{Cos}(-\theta) = \mathrm{Cos}(\theta)$ and $\mathrm{Sin}(-\theta) = -\mathrm{Sin}(\theta)$ then $z^{-1} = \frac{1}{r}(\mathrm{Cos}(\theta) - i\ \mathrm{Sin}(\theta))$

We also observe that if $z_1 = r_1\, e^{i\theta_1}$ and $z_1 = r_1\, e^{i\theta_2}$ then

$$\frac{z_1}{z_2} = \frac{r_1}{r_2} e^{i(\theta_1 - \theta_2)} = \frac{r_1}{r_2}(\mathrm{Cos}(\theta_1 - \theta_2) + i\ \mathrm{Sin}(\theta_1 - \theta_2))$$

We recall the conjugate of $z = a + ib$ is $\bar{z} = a - ib$. If $z = r(\mathrm{Cos}(\theta) + i\ \mathrm{Sin}(\theta)) = re^{i\theta}$ then $\bar{z} = r(\mathrm{Cos}(\theta) - i\ \mathrm{Sin}(\theta))$. Recalling that $\mathrm{Cos}(-\theta) = \mathrm{Cos}(\theta)$ and $\mathrm{Sin}(-\theta) = -\mathrm{Sin}(\theta)$ we see that $\bar{z} = r(\mathrm{Cos}(-\theta) + i\ \mathrm{Sin}(-\theta)) = re^{-i\theta}$. From this we see $z\,\bar{z} = rre^{+i\theta}e^{-i\theta} = r^2 = a^2 + b^2$.

Returning to our discussion of Pythagorean membership grades $(a(x), b(x)) = (r(x)\ \mathrm{Cos}(\theta(x)), r(x)\ \mathrm{Sin}(\theta(x)))$. We see that formally we can view these as complex numbers of the form $z(x) = r(x)e^{i\theta(x)}$. However we note that not all complex numbers of the form $z = re^{i\theta}$ are interpretable as Pythagorean membership grades. As we noted the requirement for a pair $(r\ \mathrm{Cos}(\theta), r\ \mathrm{Sin}(\theta))$ to be a Pythagorean membership grade is that $r\ \mathrm{Cos}(\theta)$ and $r\ \mathrm{Sin}(\theta)$ be in the unit interval and $r^2\ \mathrm{Cos}^2(\theta) + r^2\ \mathrm{Sin}^2(\theta) \leq 1$. These conditions require that $r \in [0, 1]$ and $\theta \in [0, \frac{\pi}{2}]$. So complex numbers $z = re^{i\theta}$ having the properties $r \in [0, 1]$ and $\theta \in [0, \frac{\pi}{2}]$ are examples of Pythagorean membership grades, we shall refer to these as **Π-i numbers**.

In the preceding we described a number of operations on complex numbers we now must consider which of these operations are useful in the domain of Π-i numbers. In particular, we must consider which operations allow us to start with Π-i numbers and end with a Π-i number. We first observe that the conjugate $\bar{z} = re^{-i\theta}$ is **not** a Π-i number, thus care must be taken if we use conjugation.

Assume $z = re^{i\theta}$ is a Π-i number we note that if $\alpha \in [0, 1]$ is a real number then $\alpha re^{i\theta}$ is a Π-i number. Consider now the multiplication of Π-i numbers. Let $z_1 = r_1e_1^{i\theta}$ and $z_2 = r_2e_2^{i\theta}$ be two Π-i numbers their product is $z = z_1z_2 = r_1\, r_2e^{i\theta_1 + \theta_2}$ is a Π-i number if $\theta_1 + \theta_2 \leq \frac{\pi}{2}$. Consider division with Π-i numbers $z = \frac{z_1}{z_2} = \frac{r_1}{r_2} e^{i(\theta_1 - \theta_2)}$. Here we see that z is a Π-i number if $r_2 \geq r_1$ and $\theta_1 \geq \theta_2$. The condition $r_2 \geq r_1$ is expected but the condition $\theta_1 \geq \theta_2$ is interesting. Assume $z = re^{i\theta}$ is a Π-i number and $m \in [0, 1]$ we see that $z^m = r^m\, e^{i\theta m}$ and since $r \leq r^m \leq 1$ and $0 \leq \theta m \leq \theta$ then z^m is Π-i number. Consider now the operation

$$z = (z_1z_2 \ldots z_n)^m = z_1^m z_2^m \ldots\ldots z_n^m = r_1^m r_2^m \ldots\ldots r_n^m e^{im(\theta_1 + \theta_2 + \ldots + \theta_n)}$$

We see that while $r_1^m r_2^m \ldots\ldots r_n^m \in [0, 1]$ if $m \in [0, 1]$ to be certain that $m \sum_{j=1}^{n} \theta_j \leq \frac{\pi}{2}$ we must have $m \leq 1/n$.

Consider now a more complex operation. Let $m_j \in [0, 1]$ for $j = 1$ to n and consider

$$z = z_1^{m_1} z_2^{m_2} \ldots\ldots z_n^{m_n} = r_1^{m_1} r_2^{m_2} \ldots\ldots r_n^{m_2} e^{i(m_1\theta_1 + m_2\theta_2 + \ldots + m_n\theta_n)}$$

We see that while $r_1^{m_1} r_2^{m_2} \ldots\ldots r_n^{m_2} \in [0, 1]$ to be sure that $\sum_{j=1}^{n} w_j\theta_j \leq \frac{\pi}{2}$ and in turn to be sure that z is Π-i number we require that $\sum_{j=1}^{n} w_j \leq 1$.

Let us look at the product and try to understand its semantics. We see $z = z_1z_2 = r_1\, r_2e^{i(\theta 1 + \theta 2)} = r\, e^{i\theta}$. We see that r is generally smaller then either r_1 or r_2, this a kind of reduction in strength of commitment, an increase in uncertainty. On the other hand θ is larger then either θ_1 or θ_2. The larger θ the more of the committed value r is pushed toward non-membership. One clear effect of this operation is reduction of support for membership. This seems to be somewhat in the spirit of and "anding" or conjunction operation.

One can consider some variation of the product of Π-i numbers that will always assure it is a Π-i number by defining $z = z_1 \otimes z_2 = r_1r_2e^{i((\theta_1 + \theta_2) \wedge \frac{\pi}{2})}$ here ∧ is the minimum operation.

6 Geometric Aggregation of Multiple Criteria

We now consider the issue of multiple-criteria decision-making. Assume we have a finite set X of alternatives and a collection of q criteria that we desire to be satisfied. We denote these criteria C_j, for $j = 1$ to q. Furthermore we let $C_j(x)$ indicate the degree of satisfaction of criteria C_j by alternative x. The problem here is to select the alternative that best satisfies the collection of criteria. One approach is to aggregate the satisfactions to the individual criteria by each alternative and then select the alternative with the maximum aggregated satisfaction. In [17–20] the authors discuss the use of the geometric mean to provide an aggregation of these criteria satisfactions for each x. They also associate with each criteria C_j an importance weight $w_j \in [0, 1]$ where $\sum_{j=1}^{q} w_j = 1$. Using this information they calculate the overall by alternative x, $C(x) = \prod_{j=1}^{q} C_j(x)^{w_j}$, it is the geometric mean [12].

In these works the authors assume $C_j(x) \in [0, 1]$ here we shall extend these ideas to the case where the $C_j(x)$ are Pythagorean membership grades. Here then $C_j(x) = [C_{Yj}(x), C_{Nj}(x)] = [r_j(x)\ \mathrm{Cos}(\theta_j(x), r_j(x)\ \mathrm{Sin}(\theta_j(x))]$ where $C_{Yj}(x) \in [0, 1]$ indicates the degree of support for satisfaction of C_j by x and $C_{Nj}(x) \in [0, 1]$ indicates the degree of support against satisfaction of C_j by x. Here we shall find it convenient to represent $C_j(x)$ as $r_j(x)e^{i\theta_j(x)}$. The use of Pythagorean degrees of satisfaction allows for the inclusion of imprecision and lack of commitment in modeling of the criteria satisfactions.

Our problem is to calculate $C(x) = \prod_{j=1}^{q} C_j(x)^{w_j}$ where $C_j(x) = r_j(x)e^{i\theta_j(x)}$. Here we have

$$C(x) = \prod_{j=1}^{q} C_j(x)^{w_j} = \prod_{j=1}^{q} (r_j(x)e^{i\theta_j(x)})^{w_i} = \prod_{j=1}^{q} (r_j(x))^{w_j} e^{i\sum w_j\theta_j(x)}$$

Denoting $r(x) = \prod_{j=1}^{q} (r_j(x))^{w_j}$ and $\theta(x) = \sum_{j=1}^{q} w_j\theta_j(x)$ we have $C(x) = r(x)e^{i\theta(x)}$. We see since each $r_j(x) \in [0, 1]$ and each $w_j \in [0, 1]$ that $r(x) = \prod_{j=1}^{q} (r_j(x))^{w_j} \in [0, 1]$. In addition since each $\theta_j(x) \in [0, \frac{\pi}{2}]$ and the w_j also satisfy $\sum_{j=1}^{q} w_j = 1$ then we have $\theta(x) \in [0, \frac{\pi}{2}]$. Thus we see that $C(x) = r(x)e^{i\theta(x)}$ is a Π-i number.

7 Comparing Pythagorean Membership Grades

As we have just seen the result of the aggregation of the criteria satisfactions is the association of a Pythagorean membership grade (a(x), b(x)) with each alternative x. The next question we are faced with is selecting the best alternative. Here we shall suggest a method for comparing Pythagorean membership grades. Let P indicate a generic Pythagorean membership grade, now we shall introduce a function that associates with P a value in the unit interval so that the bigger this value the more preferred the alternative. If $\mathcal{P}$ is the set of Pythagorean membership grades we want a function F: $\mathcal{P} \rightarrow [0, 1]$.

Let us recall then that there are two basic representations of a Pythagorean membership grade. The first is (a, b) here $a \in [0, 1]$ and $b \in [0, 1]$ and $a^2 + b^2 \leq 1$. The second is the polar co-ordinates (r, θ). Their relationship is that $r^2 = a^2 + b^2$ and $a = r\cos(\theta)$ and $b = r\sin(\theta)$. Here $r \in [0, 1]$ and $\theta \in [0, \pi/2]$. Closely related to this polar representation is a representation (r, d) with $d \in [0, 1]$ and $d = 1 - \frac{2\theta}{\pi}$. Thus when $\theta = 0$ we have $d = 1$ and when $\theta = \pi/2$ we have $d = 0$.

To obtain the desired function F we shall use fuzzy function modeling [21]. Consider Fig. 4, which describes the space of Pythagorean membership grades. What is clear is that at point A we want our function F to take its largest value of 1, as this corresponds to the point where an alternative fully satisfies the criteria. Point B corresponds to a membership grade indicating that the criteria are completely unsatisfied by the alternative. Here we want our function to take on the lowest value of zero. Finally the point C is a place where we neither have support for or against the satisfaction of the criteria. Here we will let our function take a neutral value of 0.5. We further note that the point A corresponds to the case where r is one and d is one (θ is zero). The point B corresponds to the case where r is one and d is zero (θ is π/2). The point C corresponds to the case where r is zero.

Using the above we define our function F using a fuzzy rule base with three rules [21]

If r is *close to one* and d is *close to one* then F is 1.

If *r is close to one* and d is *close to zero* then F is 0

If is r *close to zero* then F is 0.5

We represent *close to one* for r as a fuzzy subset E_1 on unit interval where $E_1(r) = r$. We represent *close to zero* for r as a fuzzy subset E_2 on unit interval where $E_2(r) = 1 - r$. We represent *close to one* for d as a fuzzy subset D_1 on unit interval where $D_1(r) = d$. We represent *close to zero* for d as a fuzzy subset D_2 on unit interval where $D_2(r) = 1 - d$.

Using the Takagi-Sugeno [22] approach for building functions from fuzzy rule bases we get

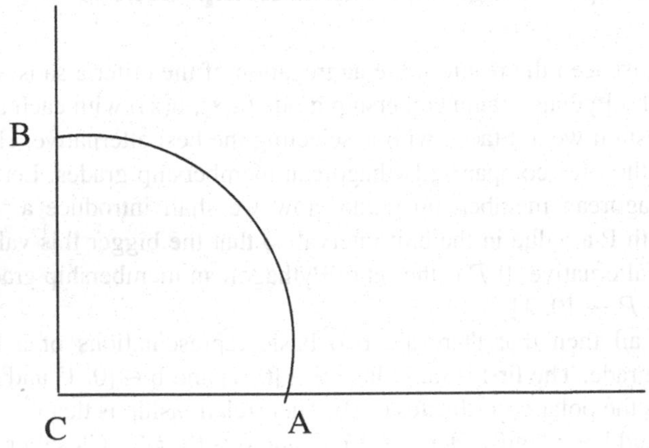

Fig. 4 Function points

$$F(r,d)=\frac{(1)E_1(r)D_1(d)+(0)E_1(r)D_2(d)+(0.5)E_2(r)}{E_1(r)D_1(d)+E_1(r)D_2(d)+E_2(r)}=\frac{(d)(r)+0.5(1-r)}{dr+(r)(1-d)+(1-r)}$$

We easily see that d · r + r(1 − d) + (1 − r) = 1 thus we get for our function

$$F(r,a)=d\,r+\frac{1}{2}(1-r)=\frac{1}{2}+r\left(d-\frac{1}{2}\right)$$

Since $d = 1 - \frac{2\theta}{\pi}$ we can also express the function as

$$F(r,\theta)=\frac{1}{2}+r\left(\frac{1}{2}-\frac{2\theta}{\pi}\right)$$

Let us look this for some particular instances. We see if r = 1 and θ = 0, point A, we get F(r, θ) = 1. We see if r = 1 and θ = π/2, point B, then F(r, θ) = 0. We see if r = 0, point C, then F(r, θ) = 0.5. This function satisfies the three rules. Let us look further at the performance of this function. If we let r be fixed then we see that

$$\frac{dF(r,\theta)}{d\theta}=-\frac{2r}{\pi},$$

it decreases as θ increases. Thus on a fixed arc of radius r we see that F decreases as we go from θ = 0 to θ = π/2.

Consider now the case of a fixed value for θ and let us see what happens when r changes. Here $\frac{dF(r,\theta)}{dr}=\frac{1}{2}-\frac{2\theta}{\pi}$. We see that for θ < π/4 F increases as r increases. For θ > π/4 then F decreases as r increases. Finally if θ = π/4 then F(r, π/4) = 0.5. It remains the same for all r.

We also observe then that if θ = 0, then $F(r,\theta)=\frac{1}{2}(1+r)$. Thus we see that as r goes for 0 to one, F(r θ) goes from 0.5 to 1. Similarly if θ = π/2 then $F(r,\theta)=\frac{1}{2}-\frac{r}{2}$. Here the as r goes from zero to one F goes from 0.5 to zero.

Thus we see if x and y are two alternatives such that their overall satisfaction to the multiple criteria are be expressed Pythagorean membership grades (r(x), θ(x)) and (r(y), θ(y)) if we calculate F(r(x), θ(x)) and F(r(y), θ(y)) the bigger of these corresponds to the preferred alternative.

8 Aggregation Using a Ordered Weighted Geometric Operator

An alternative approach for aggregation of the individual criteria discussed in [17–19] is the Ordered Weighted Geometric (OWG) operator which is based on the OWA operator introduced by Yager [23, 24]. We note as discussed in [19] as well as in [24] the use of ordered weighted type aggregation provides the capability to model various different types of user specified aggregation imperatives.

Here again we assume a collection of q criteria that we desire to be satisfied, C_k, for k = 1 to q. Again we let $C_k(x)$ indicate the degree of satisfaction of criteria C_k by alternative x. Here we also have a set of weights $\alpha_j \in [0, 1]$ such that $\sum_{j=1}^{q} \alpha_j = 1$. However the weight α_j rather then being associated with the criteria C_j is associated with the criteria with the jth largest satisfaction. In [17, 19] the authors assumed $C_k(x) \in [0, 1]$. From these $C_k(x)$ one obtains an index function, *ind*, so that ind(j) is the index of the jth largest $C_k(x)$. Using the ind(j) we obtain the OWG aggregation of the $C_k(x)$ as $C(x) = \prod_{j=1}^{q} C_{ind(j)}(x)^{\alpha_j}$. Here we emphasize the $C_{ind(j)}$ is the jth most satisfied criteria by alternative x and $C_{ind(j)}(x)$ is its degree of satisfaction by x. Here then we have ordered the criteria by there satisfaction under x to obtain *ind*.

Here we shall extend the use of the OWG operator to the case where the $C_k(x)$ are Pythagorean satisfaction grades. Here then $C_k(x) = [C_{Yk}(x), C_{Nk}(x)] = [r_k(x) \operatorname{Cos}(\theta_k(x), r_k(x) \operatorname{Sin}(\theta_k(x))]$ where $C_{Yk}(x) \in [0, 1]$ indicates the degree of support for satisfaction of C_k by x and $C_{Nk}(x) \in [0, 1]$ indicates the degree of support against satisfaction of C_k by x. Here again we shall find it convenient to represent $C_k(x)$ as $r_k(x)e^{i\theta_k(x)}$. While the situation appears very similar to the earlier situation, where we extended the geometric mean to the case of Pythagorean membership grades, there is one substantial difference. Here in the case of the OWG we must order the satisfactions, the $C_k(x)$. Since the $C_k(x)$, $r_k(x)e^{i\theta_k(x)}$, are not scalar numbers there is not an implicit ordering of the $C_k(x)$. So as to obtain the requisite ordering we shall use the function $F(r, \theta)$ to induce an ordering on the $C_k(x)$. In particular for each $C_k(x) = r_k(x)e^{i\theta_k(x)}$ we calculate a value $P(k) = F(r_k(x), \theta_k(x)) = \frac{1}{2} + r_k(x)(\frac{1}{2} - \frac{2\theta_k(x)}{\pi})$. Using the P(k)'s we obtain the function *ind* such ind(j) is the index of the criteria with jth largest value for P(k). We note that this is in the spirit of the idea of induced ordering introduced in [25, 26]. Once having the function ind(j) we are in a position to calculate $C(x) = \prod_{j=1}^{q} C_{ind(j)}(x)^{\alpha_j}$ as we did in the earlier case of geometric mean with Pythagorean satisfactions.

9 Conclusion

We introduced the idea of Pythagorean fuzzy subsets and discussed its relationship with intuitionistic fuzzy subsets. We focused on the negation and its relationship to the Pythagorean theorem. We described some of the basic set operations on Pythagorean fuzzy subsets. We looked at the relationship between Pythagorean membership grades and complex numbers. We considered the problem of multi-criteria decision making with satisfactions expressed as Pythagorean membership grades. We looked at the use of the geometric mean and ordered weighted geometric (OWG) operator for aggregating criteria satisfaction. We provided a method for comparing alternatives whose degrees of satisfaction to the decision criteria are expressed as Pythagorean membership grades.

References

1. Atanassov, K.T.: Intuitionistic fuzzy sets. Fuzzy Sets Syst. **20**, 87–96 (1986)
2. Atanassov, K.T.: On Intuitionistic Fuzzy Sets Theory. Springer, Heidelberg (2012)
3. Klir, G.J., Yuan, B.: Fuzzy Sets and Fuzzy Logic: Theory and Applications. Prentice Hall, Upper Saddle River (1995)
4. Yager, R.R., Abbasov, A.M.: Pythagorean membership grades, complex numbers and decision-making. Int. J. Intell. Syst. **28**, 436–452 (2013)
5. Yager, R.R.: Pythagorean fuzzy subsets. In: Proceedings of the Joint IFSA Congress and NAFIPS Meeting, pp. 57–61. Edmonton, Canada (2013)
6. Yager, R.R.: Pythagorean membership grades in multi-criteria decision making. IEEE Trans. Fuzzy Syst. **22**, 958–965 (2014)
7. Reformat, M.Z., Yager, R.R.: Suggesting recommendations using pythagorean fuzzy sets with an application to Netflix movie data. In: Laurent, A., Strauss, O., Bouchon-Meunier, B., Yager, R. (eds.) Information Processing and Management of Uncertainty in Knowledge-Based Systems: Proceedings of the 15th IPMU International Conference, Part I, pp. 546–587. Montpellier, France (2014)
8. Beliakov, G., James, S.: Averaging aggregation functions for preferences expressed as Pythagorean membership grades and fuzzy orthopairs. In: IEEE International Conference on Fuzzy Systems, pp. 298–305. Beijing, China (2014)
9. Churchill, R.V.: Complex Variables and Applications. McGraw-Hill, New York (1960)
10. Yager, R.R.: On the measure of fuzziness and negation part I: membership in the unit interval. Int. J. Gen. Syst. **5**, 221–229 (1979)
11. Yager, R.R.: On the measure of fuzziness and negation part II: lattices. Inf. Control **44**, 236–260 (1980)
12. Beliakov, G., Pradera, A., Calvo, T.: Aggregation Functions: A Guide for Practitioners. Springer, Heidelberg (2007)
13. Mesiar, R., Kolesarova, A., Calvo, T., Komornikova, M.: A review of aggregation functions. In: Bustince, H., Herrera, F., Montero, J. (eds.) Fuzzy Sets and Their Extensions: Representation, Aggregation and Models, pp. 121–144. Springer, Heidelberg (2008)
14. Grabisch, M., Marichal, J.-L., Mesiar, R., Pap, E.: Aggregation Functions. Cambridge University Press, Cambridge (2009)
15. Klement, E.P., Mesiar, R., Pap, E.: Triangular Norms. Kluwer, Dordrecht (2000)
16. Dick, S., Yager, R.R., Yazdanbakhsh, O.: On the properties of pythagorean and complex fuzzy sets. IEEE Transactions on Fuzzy Systems, (To Appear)
17. Chiclana, F., Herrera, F., Herrera-Viedma, E.: The ordered weighted geometric operator: properties and applications. In: Proceedings of 8th International Conference on Information Processing and Management of Uncertainty in Knowledge-based systems, pp. 985–991. Madrid (2000)
18. Xu, Z.S., Da, Q.L.: The ordered weighted geometric averaging operator. Int. J. Intell. Syst. **17**, 709–716 (2002)
19. Herrera, F., Herrera-Viedma, E., Chiclana, F.: A study of the origins and uses of the ordered weighted geometric operator in multicriteria decision making. Int. J. Intell. Syst. **18**, 689–707 (2003)

20. Yager, R.R., Xu, Z.: The continuous ordered weighted geometric operator and its application to decision making. Fuzzy Sets Syst. **157**, 1393–1402 (2006)
21. Yager, R.R., Filev, D.P.: Essentials of Fuzzy Modeling and Control. Wiley, New York (1994)
22. Takagi, T., Sugeno, M.: Fuzzy identification of systems and its application to modeling and control. IEEE Trans. Syst. Man Cybern. **15**, 116–132 (1985)
23. Xu, Z., Yager, R.R.: Some geometric aggregation operators based on intuitionistic fuzzy sets. Int. J. Gen Syst **35**, 417–433 (2006)
24. Yager, R.R.: On ordered weighted averaging aggregation operators in multi-criteria decision making. IEEE Trans. Syst. Man Cybern. **18**, 183–190 (1988)
25. Yager, R.R., Filev, D.P.: Induced ordered weighted averaging operators. IEEE Trans. Syst. Man Cybern. **29**, 141–150 (1999)
26. Yager, R.R.: On induced aggregation operators. In: Proceedings of the Eurofuse Workshop on Preference Modeling and Applications, pp. 1–9. Granada (2001)

Additive Generators Based on Generalized Arithmetic Operators in Interval-Valued Fuzzy and Atanassov's Intuitionistic Fuzzy Set Theory

Glad Deschrijver and Etienne E. Kerre

Abstract In this paper we investigate additive generators in Atanassov's intuitionistic fuzzy and interval-valued fuzzy set theory. Starting from generalized arithmetic operators satisfying some axioms we define additive generators and we characterize continuous generators which map exact elements to exact elements in terms of generators on the unit interval. We give a necessary and sufficient condition under which a generator actually generates a t-norm and we show that the generated t-norm belongs to particular classes of t-norms depending on the arithmetic operators involved in the definition of the generator.

Keywords Atanassov's intuitionistic fuzzy set · Interval-valued fuzzy set · Additive generator · t-norm

1 Introduction

Triangular norms on $([0, 1], \leq)$ were introduced in [1] and play an important role in fuzzy set theory (see e.g. [2–4] for more details). Additive generators are very useful in the construction of t-norms: any generator on $([0, 1], \leq)$ can be used to generate a t-norm. Generators play also an important role in the representation of continuous Archimedean t-norms on $([0, 1], \leq)$. Moreover, some properties of t-norms which have a generator can be related to properties of their generator. See e.g. [4–8] for more information about generators on the unit interval.

Dedicated to Prof. K. Atanassov on the occasion of his 60th anniversary.

G. Deschrijver · E.E. Kerre (✉)
Fuzziness and Uncertainty Modelling Research Unit, Department of Mathematics, Computer Science and Statistics, Ghent University, Krijgslaan 281 (S9), B-9000 Gent, Belgium
e-mail: Etienne.Kerre@UGent.be

G. Deschrijver
e-mail: Glad.Deschrijver@UGent.be

P. Angelov and S. Sotirov (eds.), *Imprecision and Uncertainty in Information Representation and Processing*, Studies in Fuzziness and Soft Computing 332,
DOI 10.1007/978-3-319-26302-1_10

Interval-valued fuzzy set theory [9, 10] is an extension of fuzzy set theory in which to each element of the universe a closed subinterval of the unit interval is assigned which approximates the unknown membership degree. Another extension of fuzzy set theory is intuitionistic fuzzy set theory introduced by Atanassov [11–13]. In [14] it is shown that the underlying lattice of Atanassov's intuitionistic fuzzy set theory is isomorphic to the underlying lattice of interval-valued fuzzy set theory and that both can be seen as L-fuzzy sets in the sense of Goguen [15] w.r.t. a special lattice $\mathcal{L}^I$. In [16] we introduced additive and multiplicative generators on $\mathcal{L}^I$ based on a special kind of addition introduced in [17]. In [18] another addition was introduced and many more additions can be introduced. Therefore, in this paper we will investigate additive generators on $\mathcal{L}^I$ independently of the addition.

2 The Lattice $\mathcal{L}^I$

Definition 1 We define $\mathcal{L}^I = (L^I, \leq_{L^I})$, where

$$L^I = \{[x_1, x_2] \mid (x_1, x_2) \in [0, 1]^2 \text{ and } x_1 \leq x_2\},$$

$$[x_1, x_2] \leq_{L^I} [y_1, y_2] \iff (x_1 \leq y_1 \text{ and } x_2 \leq y_2), \text{ for all } [x_1, x_2], [y_1, y_2] \text{ in } L^I.$$

Similarly as Lemma 2.1 in [14] it can be shown that $\mathcal{L}^I$ is a complete lattice.

Definition 2 [9, 10] An interval-valued fuzzy set on U is a mapping $A : U \to L^I$.

Definition 3 [11–13] An intuitionistic fuzzy set on U is a set

$$A = \{(u, \mu_A(u), \nu_A(u)) \mid u \in U\},$$

where $\mu_A(u) \in [0, 1]$ denotes the membership degree and $\nu_A(u) \in [0, 1]$ the non-membership degree of u in A and where for all $u \in U$,

$$\mu_A(u) + \nu_A(u) \leq 1.$$

An intuitionistic fuzzy set A on U can be represented by the L-fuzzy set A given by

$$A : U \to L^I : u \mapsto [\mu_A(u), 1 - \nu_A(u)],$$

In Fig. 1 the set L^I is shown. Note that to each element $x = [x_1, x_2]$ of L^I corresponds a point $(x_1, x_2) \in \mathbb{R}^2$.

In the sequel, if $x \in L^I$, then we denote its bounds by x_1 and x_2, i.e. $x = [x_1, x_2]$. The length $x_2 - x_1$ of the interval $x \in L^I$ is called the degree of uncertainty and is denoted by x_π. The smallest and the largest element of $\mathcal{L}^I$ are given by $0_{\mathcal{L}^I} = [0, 0]$ and $1_{\mathcal{L}^I} = [1, 1]$. Note that, for x, y in L^I, $x <_{L^I} y$ is equivalent to $x \leq_{L^I} y$ and

Fig. 1 The *grey area* is L^I

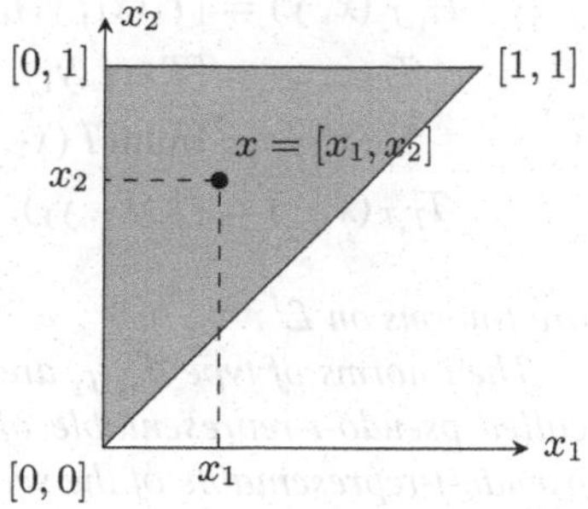

$x \neq y$, i.e. either $x_1 < y_1$ and $x_2 \leq y_2$, or $x_1 \leq y_1$ and $x_2 < y_2$. We define the relation $\ll_{L^I}$ by $x \ll_{L^I} y \iff x_1 < y_1$ and $x_2 < y_2$, for x, y in L^I. We define for further usage the sets

$$
\begin{aligned}
D &= \{[x, x] \mid x \in [0, 1]\},\\
\bar{L}^I &= \{[x_1, x_2] \mid (x_1, x_2) \in \mathbb{R}^2 \text{ and } x_1 \leq x_2\},\\
\bar{D} &= \{[x, x] \mid x \in \mathbb{R}\};\\
\bar{L}^I_+ &= \{[x_1, x_2] \mid (x_1, x_2) \in [0, +\infty[^2 \text{ and } x_1 \leq x_2\},\\
\bar{D}_+ &= \{[x, x] \mid x \in [0, +\infty[\},\\
\bar{L}^I_{+,0} &= \{[x_1, x_2] \mid (x_1, x_2) \in]0, +\infty[^2 \text{ and } x_1 \leq x_2\},\\
\bar{L}^I_{\infty,+} &= \{[x_1, x_2] \mid (x_1, x_2) \in [0, +\infty]^2 \text{ and } x_1 \leq x_2\},\\
\bar{D}_{\infty,+} &= \{[x, x] \mid x \in [0, +\infty]\}.
\end{aligned}
$$

Theorem 1 (Characterization of supremum in $\mathcal{L}^I$) [19] *Let A be an arbitrary non-empty subset of L^I and $a \in L^I$. Then $a = \sup A$ if and only if*

$$
\begin{aligned}
&(\forall x \in A)(x \leq_{L^I} a)\\
&\text{and } (\forall \epsilon_1 > 0)(\exists z \in A)(z_1 > a_1 - \epsilon_1)\\
&\text{and } (\forall \epsilon_2 > 0)(\exists z \in A)(z_2 > a_2 - \epsilon_2).
\end{aligned}
$$

Definition 4 A t-norm on $\mathcal{L}^I$ is a commutative, associative, increasing mapping $\mathcal{T} : (L^I)^2 \rightarrow L^I$ which satisfies $\mathcal{T}(1_{\mathcal{L}^I}, x) = x$, for all $x \in L^I$.

A t-conorm on $\mathcal{L}^I$ is a commutative, associative, increasing mapping $\mathcal{S} : (L^I)^2 \rightarrow L^I$ which satisfies $\mathcal{S}(0_{\mathcal{L}^I}, x) = x$, for all $x \in L^I$.

Theorem 2 *[19–21] Let T_1, T_2 and T be t-norms on $([0, 1], \leq)$ for which $T_1(x, y) \leq T_2(x, y)$, for all x, y in $[0, 1]$, and let $t \in [0, 1]$. The functions $\mathcal{T}_{T_1,T_2}$, $\mathcal{T}_T$, $\mathcal{T}'_T$ and $\mathcal{T}_{T,t}$ defined by, for all x, y in L^I,*

$$
\begin{aligned}
\mathcal{T}_{T_1,T_2}(x,y) &= [T_1(x_1,y_1), T_2(x_2,y_2)],\\
\mathcal{T}_T(x,y) &= [T(x_1,y_1), \max(T(x_1,y_2), T(x_2,y_1))],\\
\mathcal{T}'_T(x,y) &= [\min(T(x_1,y_2), T(x_2,y_1)), T(x_2,y_2)],\\
\mathcal{T}_{T,T}(x,y) &= [T(x_1,y_1), \max(T(t, T(x_1,y_1), T(x_1,y_2), T(x_2,y_1))],
\end{aligned}
$$

are t-norms on $\mathcal{L}^I$.

The t-norms of type $\mathcal{T}_{T_1,T_2}$ are called t-representable, the t-norms of type $\mathcal{T}_T$ are called pseudo-t-representable of the first kind, the t-norms of type $\mathcal{T}'_T$ are called pseudo-t-representable of the second kind.

If for a mapping f on $[0,1]$ and a mapping F on L^I it holds that $F(D) \subseteq \bar{D}$, and $F([a,a]) = [f(a), f(a)]$, for all $a \in L^I$, then we say that F is a natural extension of f to L^I. E.g. $\mathcal{T}_{T,T}$, $\mathcal{T}_T$, $\mathcal{T}_{T,t}$ and $\mathcal{T}'_T$ are all natural extensions of T to L^I.

Example 1 The following are well-known t-norms and t-conorms on the unit interval: for all x, y in $[0,1]$,

$$
\begin{aligned}
T_L(x,y) &= \max(0, x+y-1),\\
T_P(x,y) &= xy,\\
T_D(x,y) &= \begin{cases} \min(x,y), & \text{if } \max(x,y) = 1,\\ 0, & \text{else,} \end{cases}\\
S_L(x,y) &= \min(1, x+y).
\end{aligned}
$$

Using these t-norms and the above constructions, we can construct t-norms on $\mathcal{L}^I$. For example, using T_L we obtain, for all x, y in L^I,

$$
\begin{aligned}
\mathcal{T}_{T_L,T_L}(x,y) &= [\max(0, x_1+y_1-1), \max(0, x_2+y_2-1)],\\
\mathcal{T}_{T_L}(x,y) &= [\max(0, x_1+y_1-1), \max(0, x_1+y_2-1, x_2+y_1-1)].
\end{aligned}
$$

3 Arithmetic Operators on $\bar{L}^I$

We start from two arithmetic operators $\oplus : (\bar{L}^I)^2 \to \bar{L}^I$ and $\otimes : (\bar{L}^I_+)^2 \to \bar{L}^I$ satisfying the following properties,

(ADD-1) $\oplus$ is commutative,
(ADD-2) $\oplus$ is associative,
(ADD-3) $\oplus$ is increasing,
(ADD-4) $0_{\mathcal{L}^I} \oplus a = a$, for all $a \in \bar{L}^I$,
(ADD-5) $[\alpha,\alpha] \oplus [\beta,\beta] = [\alpha+\beta, \alpha+\beta]$, for all α, β in $[0, +\infty[$,
(MUL-1) $\otimes$ is commutative,
(MUL-2) $\otimes$ is associative,

(MUL-3) $\otimes$ is increasing,
(MUL-4) $1_{\mathcal{L}^I} \otimes a = a$, for all $a \in \bar{L}^I_+$,
(MUL-5) $[\alpha, \alpha] \otimes [\beta, \beta] = [\alpha\beta, \alpha\beta]$, for all α, β in $[0, +\infty[$.

The conditions (ADD-1)–(ADD-4) and (MUL-1)–(MUL-4) are natural conditions for any addition and multiplication operators. The conditions (ADD-5) and (MUL-5) ensure that these operators are natural extensions of the addition and multiplication of real numbers to $\bar{L}^I$.

Sometimes we will assume that $\oplus$ and $\otimes$ satisfy the following alternative conditions instead of (ADD-5) and (MUL-5):

(ADD-5') $[\alpha, \alpha] \oplus b = [\alpha + b_1, \alpha + b_2]$, for all $\alpha \in [0, +\infty[$ and $b \in \bar{L}^I$,
(MUL-5') $[\alpha, \alpha] \otimes b = [\alpha b_1, \alpha b_2]$, for all $\alpha \in [0, +\infty[$ and $b \in \bar{L}^I_+$.

Note that from (ADD-3) and (ADD-4) it follows that, for all a, b in $\bar{L}^I$, $a \oplus b \geq_{L^I} a$, if $b \geq_{L^I} 0_{\mathcal{L}^I}$. Similarly, we find that $a \otimes b \geq_{L^I} a$, if $b \geq_{L^I} 1_{\mathcal{L}^I}$, for all a, b in $\bar{L}^I_+$.

Define the mapping $\ominus$ by, for all x, y in $\bar{L}^I$,

$$1_{\mathcal{L}^I} \ominus x = [1 - x_2, 1 - x_1], \tag{1}$$

$$x \ominus y = 1_{\mathcal{L}^I} \ominus ((1_{\mathcal{L}^I} \ominus x) \oplus y). \tag{2}$$

Define finally the mapping $\oslash$ by, for all x, y in $\bar{L}^I_{+,0}$,

$$1_{\mathcal{L}^I} \oslash x = \left[\frac{1}{x_2}, \frac{1}{x_1}\right], \tag{3}$$

$$x \oslash y = 1_{\mathcal{L}^I} \oslash ((1_{\mathcal{L}^I} \oslash x) \otimes y). \tag{4}$$

We recall some properties that we need later on; other properties can be found in [22].

Theorem 3 [22] *The mapping $\ominus$ satisfies the following properties, for all α, β in $\mathbb{R}$ and a, b, c in $\bar{L}^I$,*

(i) $[\alpha, \alpha] \ominus [\beta, \beta] = [\alpha - \beta, \alpha - \beta]$,
(ii) $a \ominus (b \ominus c) = ((1_{\mathcal{L}^I} \ominus b) \oplus c) \ominus (1_{\mathcal{L}^I} \ominus a)$.

If $\oplus$ satisfies (ADD-5'), then

(iii) $[\alpha, \alpha] \ominus b = [\alpha - b_2, \alpha - b_1]$,
(iv) $(a \oplus b) \ominus [\alpha, \alpha] = a \oplus (b \ominus [\alpha, \alpha])$.

The mapping $\oslash$ satisfies the following properties, for all α, β in $]0, +\infty[$ and a, b, c in $\bar{L}^I_{+,0}$,

(v) $[\alpha, \alpha] \oslash [\beta, \beta] = \left[\frac{\alpha}{\beta}, \frac{\alpha}{\beta}\right]$,
(vi) $a \oslash (b \oslash c) = ((1_{\mathcal{L}^I} \oslash b) \otimes c) \oslash (1_{\mathcal{L}^I} \oslash a)$.

If $\otimes$ satisfies (MUL-5'), then

(vii) $[\alpha, \alpha] \oslash b = \left[\frac{\alpha}{b_2}, \frac{\alpha}{b_1}\right]$,
(viii) $(a \otimes b) \oslash [\alpha, \alpha] = a \otimes (b \oslash [\alpha, \alpha])$.

Example 2 We give some examples of arithmetic operators satisfying the conditions (ADD-1)–(ADD-4) and (MUL-1)–(MUL-4).

- In the interval calculus (see e.g. [23]) the following operators are defined: for all x, y in $\bar{L}^I$,

$$x \oplus y = [x_1 + y_1, x_2 + y_2],$$
$$x \ominus y = [x_1 - y_2, x_2 - y_1],$$
$$x \otimes y = [x_1 y_1, x_2 y_2], \quad \text{if } x, y \text{ in } \bar{L}^I_+,$$
$$x \oslash y = \left[\frac{x_1}{y_2}, \frac{x_2}{y_1}\right], \quad \text{if } x, y \text{ in } \bar{L}^I_{+,0}.$$

It is easy to see that these operators satisfy (ADD-1)–(ADD-4), (MUL-1)–(MUL-4), (1), (2), (3) and (4).

- In [17] the following operators are defined: for all x, y in $\bar{L}^I$,

$$x \oplus_{\mathcal{L}^I} y = [\min(x_1 + y_2, x_2 + y_1), x_2 + y_2],$$
$$x \ominus_{\mathcal{L}^I} y = [x_1 - y_2, \max(x_1 - y_1, x_2 - y_2)],$$
$$x \otimes_{\mathcal{L}^I} y = [x_1 y_1, \max(x_1 y_2, x_2 y_1)], \quad \text{if } x, y \text{ in } \bar{L}^I_+,$$
$$x \oslash_{\mathcal{L}^I} y = \left[\min\left(\frac{x_1}{y_1}, \frac{x_2}{y_2}\right), \frac{x_2}{y_1}\right], \quad \text{if } x, y \text{ in } \bar{L}^I_{+,0}.$$

It was proven in [17] that these operators satisfy (ADD-1)–(ADD-4), (MUL-1)–(MUL-4), (1), (2), (3) and (4).

- In [18] the following operators are defined for all $t \in]0, 1]$: for all x, y in $\bar{L}^I$,

$$x \oplus^t_{\mathcal{L}^I} y = [\min(1 - t + x_1 + y_1, x_1 + y_2, x_2 + y_1), x_2 + y_2],$$
$$x \ominus^t_{\mathcal{L}^I} y = [x_1 - y_2, \max(t + x_2 - y_1 - 1, x_1 - y_1, x_2 - y_2)],$$
$$x \otimes^t_{\mathcal{L}^I} y = [x_1 y_1, \max(t x_2 y_2, x_1 y_2, x_2 y_1)], \quad \text{if } x, y \text{ in } \bar{L}^I_+,$$
$$x \oslash^t_{\mathcal{L}^I} y = \left[\min\left(\frac{x_1}{t y_2}, \frac{x_1}{y_1}, \frac{x_2}{y_2}\right), \frac{x_2}{y_1}\right], \quad \text{if } x, y \text{ in } \bar{L}^I_{+,0}.$$

It was proven in [17] that these operators satisfy (ADD-1)–(ADD-4), (MUL-1)–(MUL-4), (1) and (2). In [22] it is shown that these operators satisfy (3) and (4).

- Define the following operators, for all x, y in $\bar{L}^I$,

$$x \oplus'_{\mathcal{L}^I} y = [x_1 + y_1, \max(x_1 + y_2, x_2 + y_1)],$$
$$x \ominus'_{\mathcal{L}^I} y = [\min(x_1 - y_1, x_2 - y_2), x_2 - y_1],$$
$$x \otimes'_{\mathcal{L}^I} y = [\min(x_1 y_2, x_2 y_1), x_2 y_2], \quad \text{if } x, y \text{ in } \bar{L}^I_+,$$
$$x \oslash'_{\mathcal{L}^I} y = \left[\frac{x_1}{y_2}, \max\left(\frac{x_1}{y_1}, \frac{x_2}{y_2}\right)\right], \quad \text{if } x, y \text{ in } \bar{L}^I_{+,0}.$$

It is easy to verify that these operators satisfy (ADD-1)–(ADD-4), (MUL-1)–(MUL-4), (1), (2), (3) and (4).

4 The Arithmetic Operators and Triangular Norms and Conorms on $\mathcal{L}^I$

Theorem 4 [22] *The mapping* $\mathcal{S}_\oplus : (L^I)^2 \to L^I$ *defined by, for all* x, y *in* L^I,

$$\mathcal{S}_\oplus(x, y) = \inf(1_{\mathcal{L}^I}, x \oplus y), \tag{5}$$

is a t-conorm on $\mathcal{L}^I$ *if and only if* $\oplus$ *satisfies the following condition:*

$$\begin{aligned} &(\forall (x, y, z) \in (L^I)^3) \\ &\Big(\big((\inf(1_{\mathcal{L}^I}, x \oplus y) \oplus z)_1 < 1 \text{ and } (x \oplus y)_2 > 1\big) \\ &\implies (\inf(1_{\mathcal{L}^I}, x \oplus y) \oplus z)_1 = (x \oplus \inf(1_{\mathcal{L}^I}, y \oplus z))_1\Big). \end{aligned} \tag{6}$$

Furthermore $\mathcal{S}_\oplus$ *is a natural extension of* S_L *to* L^I.

Theorem 4 shows that in order to check whether the mapping $\mathcal{S}_\oplus$ given by (5) is a t-conorm, it is sufficient to check the associativity for all x, y, z in L^I such that $(\inf(1_{\mathcal{L}^I}, x \oplus y) \oplus z)_1 < 1$ and $(x \oplus y)_2 > 1$.

Theorem 5 [22] *The mapping* $\mathcal{T}_\oplus : (L^I)^2 \to L^I$ *defined by, for all* x, y *in* L^I,

$$\mathcal{T}_\oplus(x, y) = \sup(0_{\mathcal{L}^I}, x \ominus (1_{\mathcal{L}^I} \ominus y)), \tag{7}$$

is a t-norm on $\mathcal{L}^I$ *if and only if* $\oplus$ *satisfies* (6). *Furthermore,* $\mathcal{T}_\oplus$ *is a natural extension of* T_L *to* L^I.

The following theorem gives a simpler sufficient condition so that $\mathcal{S}_\oplus$ is a t-conorm and $\mathcal{T}_\oplus$ is a t-norm on $\mathcal{L}^I$.

Theorem 6 [22] *Assume that* $\oplus$ *satisfies the following condition:*

$$\begin{aligned} &(\forall (x, y) \in \bar{L}^I_+ \times L^I) \\ &\Big(\big(([x_1, 1] \oplus y)_1 < 1 \text{ and } x_2 \in]1, 2]\big) \implies ([x_1, 1] \oplus y)_1 = (x \oplus y)_1\Big). \end{aligned} \tag{8}$$

Then the mappings $\mathcal{T}_\oplus, \mathcal{S}_\oplus : (L^I)^2 \to L^I$ *defined by, for all* x, y *in* L^I,

$$\mathcal{T}_\oplus(x, y) = \sup(0_{\mathcal{L}^I}, x \ominus (1_{\mathcal{L}^I} \ominus y)),$$
$$\mathcal{S}_\oplus(x, y) = \inf(1_{\mathcal{L}^I}, x \oplus y),$$

are a t-norm and a t-conorm on $\mathcal{L}^I$ *respectively. Furthermore* $\mathcal{T}_\oplus$ *is a natural extension of* T_L *to* L^I, *and* $\mathcal{S}_\oplus$ *is a natural extension of* S_L *to* L^I.

Theorem 7 [22] *The mapping* $\mathcal{T}_\otimes : (L^I)^2 \to L^I$ *defined by, for all* x, y *in* L^I,

$$\mathcal{T}_\otimes(x, y) = x \otimes y,$$

is a t-norm on $\mathcal{L}^I$. *Furthermore* $\mathcal{T}_\otimes$ *is a natural extension of* T_P *to* L^I.

In the following theorem an alternative way of extending the Łukasiewicz t-norm on the unit interval to $\mathcal{L}^I$ using the arithmetic operators on $\bar{L}^I$ is given.

Theorem 8 [22] *The mapping* $\mathcal{T}'_\oplus : (L^I)^2 \to L^I$ *defined by, for all* x, y *in* L^I,

$$\mathcal{T}'_\oplus(x, y) = \sup(0_{\mathcal{L}^I}, x \oplus (y \ominus 1_{\mathcal{L}^I})), \tag{9}$$

is a t-norm on $\mathcal{L}^I$ *if and only if* $\oplus$ *satisfies the following conditions:*

$$(\forall a \in L^I)(1_{\mathcal{L}^I} \oplus (a \ominus 1_{\mathcal{L}^I}) = a)$$

and

$$\begin{aligned}
&(\forall (x, y, z) \in (L^I)^3) \\
&\big(((\sup(0_{\mathcal{L}^I}, x \oplus (y \ominus 1_{\mathcal{L}^I})) \oplus (z \ominus 1_{\mathcal{L}^I}))_2 > 0 \text{ and } (x \oplus (y \ominus 1_{\mathcal{L}^I}))_1 < 0\big) \\
&\quad \Longrightarrow (\sup(0_{\mathcal{L}^I}, x \oplus (y \ominus 1_{\mathcal{L}^I})) \oplus (z \ominus 1_{\mathcal{L}^I}))_2 \\
&\qquad = (x \oplus (\sup(0_{\mathcal{L}^I}, y \oplus (z \ominus 1_{\mathcal{L}^I})) \ominus 1_{\mathcal{L}^I}))_2\big).
\end{aligned} \tag{10}$$

Furthermore $\mathcal{T}'_\oplus$ *is a natural extension of* T_L *to* L^I.

Corollary 1 [22] *Assume that* $\oplus$ *satisfies (ADD-5'). The mapping* $\mathcal{T}'_\oplus : (L^I)^2 \to L^I$ *defined by,*[1] *for all* x, y *in* L^I,

$$\mathcal{T}'_\oplus(x, y) = \sup(0_{\mathcal{L}^I}, x \oplus y \ominus 1_{\mathcal{L}^I}),$$

is a t-norm on $\mathcal{L}^I$ *if and only if* $\oplus$ *satisfies* (10). *Furthermore* $\mathcal{T}'_\oplus$ *is a natural extension of* T_L *to* L^I.

---

[1] Since from Theorem 3(iv) it follows that $(x \oplus y) \ominus 1_{\mathcal{L}^I} = x \oplus (y \ominus 1_{\mathcal{L}^I})$, for all x, y in L^I, we will omit the brackets in this formula.

Lemma 1 *Assume that* $\oplus$ *satisfies (ADD-5'). Let* $\mathcal{T}_\oplus : (L^I)^2 \to L^I$ *be the mapping defined by* (7). *Then for all* $x_1 \in [0, 1]$ *and* $y \in L^I$,

$$\mathcal{T}_\oplus([x_1, x_1], y) = [\max(0, x_1 + y_1 - 1), \max(0, x_1 + y_2 - 1)].$$

Proof Using Theorem 3(i) we obtain, for all $x_1 \in [0, 1]$ and $y \in L^I$,

$$\begin{aligned}\mathcal{T}_\oplus([x_1, x_1], y) &= \sup(0_{\mathcal{L}^I}, x \ominus [1 - y_2, 1 - y_1]) \\ &= \sup(0_{\mathcal{L}^I}, [x_1 - (1 - y_1), x_1 - (1 - y_2)]) \\ &= [\max(0, x_1 + y_1 - 1), \max(0, x_1 + y_2 - 1)].\end{aligned}$$

□

Example 3 We give t-norms $\mathcal{T}_\oplus$, $\mathcal{T}_\otimes$ and t-conorms $\mathcal{S}_\oplus$ on $\mathcal{L}^I$ defined using the examples for $\oplus$ and $\ominus$ given in the previous section.

- Let $\oplus$, $\ominus$ and $\otimes$ be the addition, subtraction and multiplication used in the interval calculus, then $\mathcal{T}_\oplus = \mathcal{T}_{T_L, T_L}$, $\mathcal{T}_\otimes = \mathcal{T}_{T_P, T_P}$ and $\mathcal{S}_\oplus = \mathcal{S}_{S_L, S_L}$. Thus the t-norms $\mathcal{T}_\oplus$, $\mathcal{T}_\otimes$ and the t-conorm $\mathcal{S}_\oplus$ obtained using the arithmetic operators from the interval calculus are t-representable.
- Using $\oplus_{\mathcal{L}^I}$, $\ominus_{\mathcal{L}^I}$ and $\otimes_{\mathcal{L}^I}$ we obtain that $\mathcal{T}_{\oplus_{\mathcal{L}^I}} = \mathcal{T}_{T_L}$, $\mathcal{T}_{\otimes_{\mathcal{L}^I}} = \mathcal{T}_{T_P}$ and $\mathcal{S}_{\oplus_{\mathcal{L}^I}} = \mathcal{S}_{S_L}$. Thus the t-norms $\mathcal{T}_{\oplus_{\mathcal{L}^I}}$, $\mathcal{T}_{\otimes_{\mathcal{L}^I}}$ and the t-conorm $\mathcal{S}_{\oplus_{\mathcal{L}^I}}$ are pseudo-t-representable.
- Using $\oplus^t_{\mathcal{L}^I}$, $\ominus^t_{\mathcal{L}^I}$ and $\otimes^t_{\mathcal{L}^I}$ we obtain that $\mathcal{T}_{\oplus^t_{\mathcal{L}^I}} = \mathcal{T}_{T_L,t}$, $\mathcal{T}_{\otimes^t_{\mathcal{L}^I}} = \mathcal{T}_{T_P,t}$ and $\mathcal{S}_{\oplus^t_{\mathcal{L}^I}} = \mathcal{S}_{S_L,t}$.
- Using $\oplus'_{\mathcal{L}^I}$, $\ominus'_{\mathcal{L}^I}$ and $\otimes'_{\mathcal{L}^I}$ we obtain that $\mathcal{T}_{\oplus'_{\mathcal{L}^I}} = \mathcal{T}'_{T_L}$, $\mathcal{T}_{\otimes'_{\mathcal{L}^I}} = \mathcal{T}'_{T_P}$ and $\mathcal{S}_{\oplus'_{\mathcal{L}^I}} = \mathcal{S}'_{S_L}$.

5 Additive Generators on $\mathcal{L}^I$

Definition 5 [2, 4, 6] A mapping $f : [0, 1] \to [0, +\infty]$ satisfying the following conditions:

(ag.1) f is strictly decreasing;
(ag.2) $f(1) = 0$;
(ag.3) f is right-continuous in 0;
(ag.4) $f(x) + f(y) \in \mathrm{rng}(f) \cup [f(0), +\infty]$, for all x, y in $[0, 1]$;

is called an additive generator on $([0, 1], \leq)$.

Definition 6 [4, 6] Let $f : [0, 1] \to [0, +\infty]$ be a strictly decreasing function. The pseudo-inverse $f^{(-1)} : [0, +\infty] \to [0, 1]$ of f is defined by, for all $y \in [0, +\infty]$,

$$f^{(-1)}(y) = \sup(\{0\} \cup \{x \mid x \in [0, 1] \text{ and } f(x) > y\}).$$

We extend these definitions to L^I as follows.

Definition 7 Let $\mathfrak{f} : L^I \to \bar{L}^I_{\infty,+}$ be a strictly decreasing function. The pseudo-inverse $\mathfrak{f}^{(-1)} : \bar{L}^I_{\infty,+} \to L^I$ of $\mathfrak{f}$ is defined by, for all $y \in \bar{L}^I_{\infty,+}$,

$$\mathfrak{f}^{(-1)}(y) = \begin{cases} \sup\{x \mid x \in L^I \text{ and } \mathfrak{f}(x) \gg_{L^I} y\}, & \text{if } y \ll_{L^I} \mathfrak{f}(0_{\mathcal{L}^I}); \\ \sup(\{0_{\mathcal{L}^I}\} \cup \{x \mid x \in L^I \text{ and } (\mathfrak{f}(x))_1 > y_1 \\ \quad \text{and } (\mathfrak{f}(x))_2 \geq (\mathfrak{f}(0_{\mathcal{L}^I}))_2\}), & \text{if } y_2 \geq (\mathfrak{f}(0_{\mathcal{L}^I}))_2; \\ \sup(\{0_{\mathcal{L}^I}\} \cup \{x \mid x \in L^I \text{ and } (\mathfrak{f}(x))_2 > y_2 \\ \quad \text{and } (\mathfrak{f}(x))_1 \geq (\mathfrak{f}(0_{\mathcal{L}^I}))_1\}), & \text{if } y_1 \geq (\mathfrak{f}(0_{\mathcal{L}^I}))_1. \end{cases}$$

Note that if $\mathfrak{f}(0_{\mathcal{L}^I}) \in \bar{D}_{\infty,+}$, then, for all $y \in \bar{L}^I_{\infty,+}$,

$$\mathfrak{f}^{(-1)}(y) = \sup \Phi_y,$$

where

$$\Phi_y = \begin{cases} \{x \mid x \in L^I \text{ and } \mathfrak{f}(x) \gg_{L^I} y\}, & \text{if } y \ll_{L^I} \mathfrak{f}(0_{\mathcal{L}^I}); \\ \{0_{\mathcal{L}^I}\} \cup \{x \mid x \in L^I \text{ and } (\mathfrak{f}(x))_1 > y_1 \\ \quad \text{and } (\mathfrak{f}(x))_2 = (\mathfrak{f}(0_{\mathcal{L}^I}))_2\}, & \text{if } y_2 \geq (\mathfrak{f}(0_{\mathcal{L}^I}))_2. \end{cases}$$

The set $\mathfrak{f}(\Phi_y)$ is depicted in Fig. 2 for two possible values of $y \in \bar{L}^I_{\infty,+}$.

In the following definition we consider continuity w.r.t. the Euclidean metric d^E in $\mathbb{R}^2$ restricted to L^I and $\bar{L}^I_{\infty,+}$. We say that a function $\mathfrak{f} : L^I \to \bar{L}^I_{\infty,+}$ is right-continuous in $a \in L^I$ if

$$(\forall \epsilon > 0)(\exists \delta > 0)(\forall x \in L^I)(d^E(x, a) < \delta \wedge x >_{L^I} a \implies d^E(\mathfrak{f}(x), \mathfrak{f}(a)) < \epsilon).$$

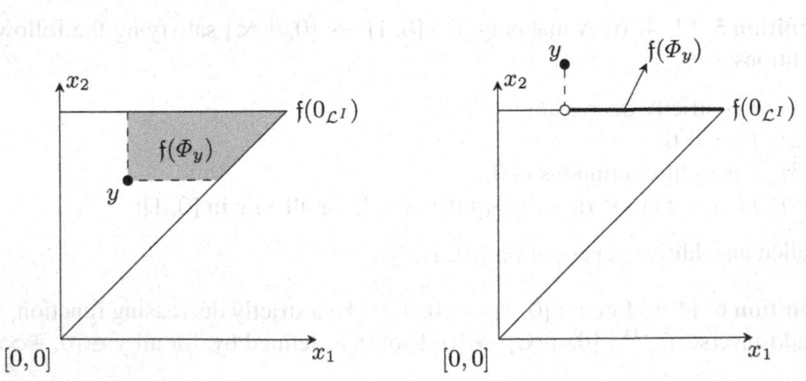

Fig. 2 The largest possible set $\mathfrak{f}(\Phi_y)$ in the case that $y \ll_{L^I} \mathfrak{f}(0_{\mathcal{L}^I})$ (*left*) and in the case that $y_2 \geq (\mathfrak{f}(0_{\mathcal{L}^I}))_2$ (*right*)

Definition 8 A mapping $\mathfrak{f} : L^I \to \bar{L}^I_{\infty,+}$ satisfying the following conditions:

(AG.1) $\mathfrak{f}$ is strictly decreasing;
(AG.2) $\mathfrak{f}(1_{\mathcal{L}^I}) = 0_{\mathcal{L}^I}$;
(AG.3) $\mathfrak{f}$ is right-continuous in $0_{\mathcal{L}^I}$;
(AG.4) $\mathfrak{f}(x) \oplus \mathfrak{f}(y) \in \mathcal{R}(\mathfrak{f})$, for all x, y in L^I, where

$$\begin{aligned}\mathcal{R}(\mathfrak{f}) = \operatorname{rng}(\mathfrak{f}) &\cup \{x \mid x \in \bar{L}^I_{\infty,+} \text{ and } [x_1, (\mathfrak{f}(0_{\mathcal{L}^I}))_2] \in \operatorname{rng}(\mathfrak{f}) \\ &\quad \text{and } x_2 \geq (\mathfrak{f}(0_{\mathcal{L}^I}))_2\} \\ &\cup \{x \mid x \in \bar{L}^I_{\infty,+} \text{ and } [(\mathfrak{f}(0_{\mathcal{L}^I}))_1, x_2] \in \operatorname{rng}(\mathfrak{f}) \\ &\quad \text{and } x_1 \geq (\mathfrak{f}(0_{\mathcal{L}^I}))_1\} \\ &\cup \{x \mid x \in \bar{L}^I_{\infty,+} \text{ and } x \geq_{L^I} \mathfrak{f}(0_{\mathcal{L}^I})\};\end{aligned}$$

(AG.5) $\mathfrak{f}^{(-1)}(\mathfrak{f}(x)) = x$, for all $x \in L^I$;

is called an additive generator on $\mathcal{L}^I$.

If $\mathfrak{f}(0_{\mathcal{L}^I}) \in \bar{D}_{\infty,+}$, then

$$\begin{aligned}\mathcal{R}(\mathfrak{f}) = \operatorname{rng}(\mathfrak{f}) &\cup \{x \mid x \in \bar{L}^I_{\infty,+} \text{ and } [x_1, (\mathfrak{f}(0_{\mathcal{L}^I}))_2] \in \operatorname{rng}(\mathfrak{f}) \text{ and } x_2 \geq (\mathfrak{f}(0_{\mathcal{L}^I}))_2\} \\ &\cup \{x \mid x \in \bar{L}^I_{\infty,+} \text{ and } x \geq_{L^I} \mathfrak{f}(0_{\mathcal{L}^I})\}.\end{aligned}$$

An example of how the set $\mathcal{R}(\mathfrak{f})$ may look like is given in Fig. 3.

In [16] the following three properties are shown. Since their proof does not involve (AG.4), they are also valid for the current definition of additive generator.

Lemma 2 [16] *Let* $\mathfrak{f} : L^I \to \bar{L}^I_{\infty,+}$ *be a mapping satisfying (AG.1), (AG.2), (AG.3) and (AG.5). Then, for all* $x \in L^I$ *such that* $x_1 > 0$*, it holds that* $(\mathfrak{f}(x))_2 < (\mathfrak{f}(0_{\mathcal{L}^I}))_2$ *and* $(\mathfrak{f}(x))_1 < (\mathfrak{f}(0_{\mathcal{L}^I}))_1$.

Lemma 3 [16] *Let* $\mathfrak{f} : L^I \to \bar{L}^I_{\infty,+}$ *be a mapping satisfying (AG.1), (AG.2), (AG.3) and (AG.5). Then* $(\mathfrak{f}([0,1]))_1 = (\mathfrak{f}(0_{\mathcal{L}^I}))_1$ *or* $(\mathfrak{f}([0,1]))_2 = (\mathfrak{f}(0_{\mathcal{L}^I}))_2$.

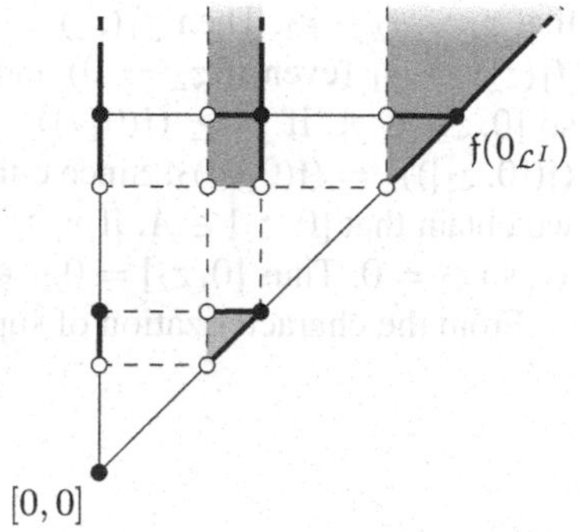

Fig. 3 An example of the set $\mathcal{R}(\mathfrak{f})$ given by the *shaded areas* together with the *thick lines*

Corollary 2 [16] *Let* $\mathfrak{f} : L^I \to \bar{L}^I_{\infty,+}$ *be a mapping satisfying (AG.1), (AG.2), (AG.3), (AG.5) and* $\mathfrak{f}(D) \subseteq \bar{D}_{\infty,+}$. *Then* $(\mathfrak{f}([0,1]))_2 = (\mathfrak{f}(0_{\mathcal{L}^I}))_2$.

Lemma 4 *Let* f_1 *be an additive generator on* $([0,1], \leq)$ *and let* $\mathfrak{f} : L^I \to \bar{L}^I_{\infty,+}$ *be a mapping satisfying (AG.1), (AG.2), (AG.3), (AG.5) and, for all* $x \in L^I$,

$$(\mathfrak{f}(x))_1 = f_1(x_2).$$

Then, for all $y \in \bar{L}^I_{\infty,+}$,

$$(\mathfrak{f}^{(-1)}(y))_2 = f_1^{(-1)}(y_1).$$

Proof Let $y \in \bar{L}^I_{\infty,+}$. Define the set A by,

$$A = \begin{cases} \{x \mid x \in L^I \text{ and } f_1(x_2) > y_1 \text{ and } (\mathfrak{f}(x))_2 > y_2\}, & \text{if } y \ll_{L^I} \mathfrak{f}(0_{\mathcal{L}^I}); \\ \{0_{\mathcal{L}^I}\} \cup \{x \mid x \in L^I \text{ and } f_1(x_2) > y_1 \text{ and } (\mathfrak{f}(x))_2 = (\mathfrak{f}(0_{\mathcal{L}^I}))_2\}, & \\ \quad \text{if } y_2 \geq (\mathfrak{f}(0_{\mathcal{L}^I}))_2; & \\ \{0_{\mathcal{L}^I}\} \cup \{x \mid x \in L^I \text{ and } (\mathfrak{f}(x))_2 > y_2 \text{ and } f_1(x_2) = f_1(0)\}, & \\ \quad \text{if } y_1 \geq f_1(0). & \end{cases}$$

Then $\mathfrak{f}^{(-1)}(y) = \sup A$. Let $a = [(\sup A)_1, \sup A_2]$, where

$$A_2 = \{0\} \cup \{x \mid x \in [0,1] \text{ and } f_1(x) > y_1\};$$

We prove that $(\sup A)_2 = a_2$, then we will have that $(\mathfrak{f}^{(-1)}(y))_2 = a_2 = f_1^{(-1)}(y_1)$. From the characterization of supremum in $\mathcal{L}^I$ it follows that it is sufficient to prove that $x_2 \leq a_2$, for all $x \in A$, and that for all $\epsilon_2 > 0$ there exists $z \in A$ such that $z_2 > a_2 - \epsilon_2$.

If $x \in A$, then $f_1(x_2) > y_1$ or $x_2 = 0$, taking into consideration that $f_1(x_2) = f_1(0) \iff x_2 = 0$, since f_1 is strictly decreasing. So $x_2 \in A_2$. We obtain that $x_2 \leq \sup A_2 = a_2$.

From Lemma 3 and the fact that $(\mathfrak{f}([0,1]))_1 = f_1(1) < f_1(0) = (\mathfrak{f}(0_{\mathcal{L}^I}))_1$, since f_1 is strictly decreasing, it follows that $(\mathfrak{f}([0,1]))_2 = (\mathfrak{f}(0_{\mathcal{L}^I}))_2$.

Let $\epsilon_2 > 0$, then from $\sup A_2 = a_2$ it follows that there exists a $z_2 \in A_2$ such that $z_2 > a_2 - \epsilon_2$. Then $f_1(z_2) > y_1$ or $z_2 = 0$. Let $z_1 = 0$. If $y \ll_{L^I} \mathfrak{f}(0_{\mathcal{L}^I})$, then $f_1(z_2) > y_1$ (even if $z_2 = 0$) and $(\mathfrak{f}([0,z_2]))_2 \geq (\mathfrak{f}([0,1]))_2 = (\mathfrak{f}(0_{\mathcal{L}^I}))_2 > y_2$, so $[0,z_2] \in A$. If $y_2 \geq (\mathfrak{f}(0_{\mathcal{L}^I}))_2$, then $(\mathfrak{f}([0,1]))_2 \leq (\mathfrak{f}([0,z_2]))_2 \leq (\mathfrak{f}(0_{\mathcal{L}^I}))_2$, so $(\mathfrak{f}([0,z_2]))_2 = (\mathfrak{f}(0_{\mathcal{L}^I}))_2$. Since either $f_1(z_2) > y_1$ or $z_2 = 0$ (and so $[0,z_2] = 0_{\mathcal{L}^I}$), we obtain that $[0,z_2] \in A$. If $y_1 \geq f_1(0)$, then $f_1(z_2)$ cannot be strictly greater than y_1, so $z_2 = 0$. Thus $[0,z_2] = 0_{\mathcal{L}^I} \in A$.

From the characterization of supremum in $\mathcal{L}^I$ it now follows that $\sup A = a$. □

Lemma 5 *Let f_2 be an additive generator on* $([0,1],\leq)$ *and let* $\mathfrak{f} : L^I \to \bar{L}^I_{\infty,+}$ *be a mapping satisfying (AG.1), (AG.2), (AG.3), (AG.5),* $\mathfrak{f}(D) \subseteq \bar{D}_{\infty,+}$ *and, for all* $x \in L^I$,

$$(\mathfrak{f}(x))_2 = f_2(x_1).$$

Then, for all $y \in \bar{L}^I_{\infty,+}$,

$$(\mathfrak{f}^{(-1)}(y))_1 = f_2^{(-1)}(y_2).$$

Proof Let $y \in \bar{L}^I_{\infty,+}$. Define the set A by,

$$A = \begin{cases} \{x \mid x \in L^I \text{ and } (\mathfrak{f}(x))_1 > y_1 \text{ and } f_2(x_1) > y_2\}, \text{ if } y \ll_{L^I} \mathfrak{f}(0_{\mathcal{L}^I}); \\ \{0_{\mathcal{L}^I}\} \cup \{x \mid x \in L^I \text{ and } (\mathfrak{f}(x))_1 > y_1 \text{ and } f_2(x_1) = f_2(0)\}, \\ \quad \text{if } y_2 \geq f_2(0). \end{cases}$$

Then $\mathfrak{f}^{(-1)}(y) = \sup A$. Let $a = [\sup A_1, (\sup A)_2]$, where

$$A_1 = \{0\} \cup \{x \mid x \in [0,1] \text{ and } f_2(x) > y_2\};$$

We prove that $(\sup A)_1 = a_1$, then we will have that $(\mathfrak{f}^{(-1)}(y))_1 = a_1 = f_2^{(-1)}(y_2)$. From the characterization of supremum in $\mathcal{L}^I$ it follows that it is sufficient to prove that $x_1 \leq a_1$, for all $x \in A$, and that for all $\epsilon_1 > 0$ there exists $z \in A$ such that $z_1 > a_1 - \epsilon_1$.

If $x \in A$, then $f_2(x_1) > y_2$ or $x_1 = 0$, taking into consideration that $f_2(x_1) = f_2(0) \iff x_1 = 0$, since f_2 is strictly decreasing. So $x_1 \in A_1$. We obtain that $x_1 \leq \sup A_1 = a_1$.

Let $\epsilon_1 > 0$, then from $\sup A_1 = a_1$ it follows that there exists a $z_1 \in A_1$ such that $z_1 > a_1 - \epsilon_1$. Then $f_2(z_1) > y_2$ or $z_1 = 0$. If $y_2 \geq f_2(0)$, then $f_2(z_1) > y_2$ would imply that $f_2(z_1) > f_2(0)$, which is a contradiction. So $z_1 = 0$, and $[z_1, z_1] = 0_{\mathcal{L}^I} \in A$. If $y \ll_{L^I} \mathfrak{f}(0_{\mathcal{L}^I})$, then $f_2(z_1) > y_2$ (even if $z_1 = 0$). Since $\mathfrak{f}(D) \subseteq \bar{D}_{\infty,+}$, $(\mathfrak{f}([z_1, z_1]))_1 = f_2(z_1) > y_2 \geq y_1$. Hence $[z_1, z_1] \in A$.

From the characterization of supremum in $\mathcal{L}^I$ it now follows that $\sup A = a$. □

Theorem 9 *Let f be an additive generator on* $([0,1],\leq)$ *and let* $\mathfrak{f} : L^I \to \bar{L}^I_{\infty,+}$ *the mapping defined by, for all* $x \in L^I$,

$$\mathfrak{f}(x) = [f(x_2), f(x_1)]. \tag{11}$$

Then, for all $y \in \bar{L}^I_{\infty,+}$,

$$\mathfrak{f}^{(-1)}(y) = [f^{(-1)}(y_2), f^{(-1)}(y_1)]. \tag{12}$$

Proof From the fact that f is an additive generator on $([0,1],\leq)$ it follows that the mapping $\mathfrak{f}$ defined by (11) satisfies (AG.1), (AG.2) and (AG.3). Furthermore $\mathfrak{f}(D) \subseteq \bar{D}_{\infty,+}$.

Since, for all $x \in L^I$,

$$\Phi_{\mathfrak{f}(x)} = \begin{cases} \{z \mid z \in L^I \text{ and } f(z_2) > f(x_2) \text{ and } f(z_1) > f(x_1)\}, \\ \quad \text{if } f(x_2) < f(0) \text{ and } f(x_1) < f(0); \\ \{[0,0]\} \cup \{z \mid z \in L^I \text{ and } f(z_2) > f(x_2) \text{ and } f(z_1) = f(0)\}, \\ \quad \text{if } f(x_1) \geq f(0) \end{cases}$$

$$= \begin{cases} ([0, x_1[\times [0, x_2[) \cap L^I, & \text{if } f(x_1) < f(0); \\ \{0\} \times [0, x_2[, & \text{if } f(x_1) = f(0), \end{cases}$$

we obtain that $\mathfrak{f}^{(-1)}(\mathfrak{f}(x)) = \sup \Phi_{\mathfrak{f}(x)} = x$. So (AG.5) holds. From Lemmas 4 and 5 it follows that $\mathfrak{f}^{(-1)}$ is given by (12). □

Lemma 6 *Let $\mathfrak{f} : L^I \to \bar{L}^I_{\infty,+}$ be a continuous mapping satisfying (AG.1), (AG.2), (AG.3), (AG.5) and $\mathfrak{f}(D) \subseteq \bar{D}_{\infty,+}$. Then there exists a continuous additive generator f on $([0, 1], \leq)$ such that $(\mathfrak{f}(x))_1 = f(x_2)$, for all $x \in L^I$.*

Proof Let $x_2 \in\ ]0, 1]$. Since $\mathfrak{f}$ is decreasing, $(\mathfrak{f}([x_2, x_2]))_1 \leq (\mathfrak{f}([0, x_2]))_1$. Assume that $(\mathfrak{f}([x_2, x_2]))_1 < (\mathfrak{f}([0, x_2]))_1$. We first show that $(\mathfrak{f}([y_2, y_2]))_1 \geq (\mathfrak{f}([0, x_2]))_1$, for all $y_2 < x_2$. If this were not the case, then from $\mathfrak{f}(D) \subseteq \bar{D}_{\infty,+}$ it would follow that $[0, x_2] \in \Phi_{\mathfrak{f}([y_2, y_2])}$. Thus $[y_2, y_2] = \mathfrak{f}^{(-1)}(\mathfrak{f}([y_2, y_2])) \geq_{\mathcal{L}^I} [0, x_2]$, so $y_2 \geq x_2$, which is a contradiction.

Define the mapping $f : [0, 1] \to \mathbb{R}$ by $f(z_2) = (\mathfrak{f}([z_2, z_2]))_1$, for all $z_2 \in [0, 1]$. Since $\mathfrak{f}$ is continuous, $f = \text{pr}_1 \circ \mathfrak{f} \circ g$ is continuous, where $g : [0, 1] \to L^I$ is defined by $g(z_2) = [z_2, z_2]$, for all $z_2 \in [0, 1]$. From the above it follows that $f(y_2) \geq (\mathfrak{f}([0, x_2]))_1$, for all $y_2 \in [0, x_2[$. On the other hand, since $\mathfrak{f}$ is strictly decreasing and $\mathfrak{f}(D) \subseteq \bar{D}_{\infty,+}$, f is strictly decreasing, so $f(y_2) \leq f(x_2)$, for all $y_2 \in [x_2, 1]$. Hence $\text{rng}(f) \subseteq [0, f(x_2)] \cup [(\mathfrak{f}([0, x_2]))_1, f(0)]$, taking into consideration that $(\mathfrak{f}([0, x_2]))_1 \leq (\mathfrak{f}(0_{\mathcal{L}^I}))_1 = f(0)$. It follows from the Mean Value Theorem that $a \in \text{rng}(f)$, for any $a \in\ ]f(x_2), (\mathfrak{f}([0, x_2]))_1[$, which is a contradiction. Hence $(\mathfrak{f}([0, x_2]))_1 = (\mathfrak{f}([x_2, x_2]))_1 = f(x_2)$.

From $\mathfrak{f}(1_{\mathcal{L}^I}) = 0_{\mathcal{L}^I}$, it follows that $f(1) = 0$, so from the above it follows that f is a continuous additive generator. □

Corollary 3 *Let $\mathfrak{f} : L^I \to \bar{L}^I_{\infty,+}$ be a continuous mapping satisfying (AG.1), (AG.2), (AG.3), (AG.5) and $\mathfrak{f}(D) \subseteq \bar{D}_{\infty,+}$. Then $\mathfrak{f}([0, 1]) = [0, (\mathfrak{f}(0_{\mathcal{L}^I}))_2]$.*

Proof This follows immediately from Corollary 2, Lemma 6 and (AG.2). □

Lemma 7 *Let $\mathfrak{f} : L^I \to \bar{L}^I_{\infty,+}$ be a continuous mapping satisfying (AG.1), (AG.2), (AG.3), (AG.5) and $\mathfrak{f}(D) \subseteq \bar{D}_{\infty,+}$. Then there exists a continuous additive generator f on $([0, 1], \leq)$ such that $(\mathfrak{f}(x))_2 = f(x_1)$, for all $x \in L^I$.*

Proof From Corollary 3 it follows that $(\mathfrak{f}([x_1, 1]))_1 = 0$, for all $x_1 \in [0, 1]$. Thus, since $\mathfrak{f}$ is strictly decreasing and using Corollary 2, $(\mathfrak{f}([x_1, 1]))_2 < (\mathfrak{f}([0, 1]))_2 = (\mathfrak{f}(0_{\mathcal{L}^I}))_2$, for all $x_1 \in]0, 1]$.

Let $x_1 \in]0, 1[$. Assume that $(\mathfrak{f}([x_1, 1]))_2 < (\mathfrak{f}([x_1, x_1]))_2$. Since for any $y_1 \in]x_1, 1]$, from $(\mathfrak{f}([x_1, 1]))_2 < (\mathfrak{f}([y_1, y_1]))_2$ it would follow that $[y_1, y_1] \in \Phi_{\mathfrak{f}([x_1,1])}$, we obtain similarly as in the proof of Lemma 6 a contradiction. So $(\mathfrak{f}([x_1, 1]))_2 \geq (\mathfrak{f}([y_1, y_1]))_2$, for all $y_1 > x_1$. Define the mapping $f : [0, 1] \to \mathbb{R}$ by $f(z_1) = (\mathfrak{f}([z_1, z_1]))_2$, for all $z_1 \in [0, 1]$. The sequel of the proof is now similar as for Lemma 6. □

Lemma 8 *Let* $\mathfrak{f} : L^I \to \bar{L}^I_{\infty,+}$ *be a continuous mapping satisfying (AG.1), (AG.2), (AG.3), (AG.5) and* $\mathfrak{f}(D) \subseteq \bar{D}_{\infty,+}$. *Then there exists a continuous additive generator* f *on* $([0, 1], \leq)$ *such that, for all* $x \in L^I$,

$$\mathfrak{f}(x) = [f(x_2), f(x_1)].$$

Proof From Lemmas 6 and 7 it follows that there exist continuous additive generators f and f' on $([0, 1], \leq)$ such that $\mathfrak{f}(x) = [f(x_2), f'(x_1)]$, for all $x \in L^I$. Since $\mathfrak{f}(D) \subseteq \bar{D}_{\infty,+}$, we have that $f = f'$. □

Lemma 9 *Let* $\mathfrak{f} : L^I \to \bar{L}^I_{\infty,+}$ *be a continuous mapping satisfying (AG.1), (AG.2), (AG.3), (AG.5) and* $\mathfrak{f}(D) \subseteq \bar{D}_{\infty,+}$. *Then* $\mathcal{R}(\mathfrak{f}) = \bar{L}^I_{\infty,+}$ *and* $\mathfrak{f}$ *satisfies (AG.4).*

Proof By Lemma 8, there exists a continuous additive generator f on $([0, 1], \leq)$ such that $\mathfrak{f}(x) = [f(x_2), f(x_1)]$, for all $x \in L^I$. From the Mean Value Theorem it follows that $[0, f(0)] \subseteq \text{rng}(f)$. Since $\mathfrak{f}(D) \subseteq \bar{D}_{\infty,+}$, we obtain that $\{x \mid x \in \bar{L}^I_{\infty,+}$ and $(x_1, x_2) \in [0, f(0)]^2\} = \{x \mid x \in \bar{L}^I_{\infty,+}$ and $x \leq_{L^I} \mathfrak{f}(0_{\mathcal{L}^I})\} \subseteq \text{rng}(\mathfrak{f})$. Hence $\mathcal{R}(\mathfrak{f}) = \bar{L}^I_{\infty,+}$. It follows immediately that $\mathfrak{f}$ satisfies (AG.4). □

Theorem 10 *A mapping* $\mathfrak{f} : L^I \to \bar{L}^I_{\infty,+}$ *is a continuous additive generator on* $\mathcal{L}^I$ *for which* $\mathfrak{f}(D) \subseteq \bar{D}_{\infty,+}$ *if and only if there exists a continuous additive generator* f *on* $([0, 1], \leq)$ *such that, for all* $x \in L^I$,

$$\mathfrak{f}(x) = [f(x_2), f(x_1)]. \tag{13}$$

Proof From Lemma 8 it follows that if $\mathfrak{f}$ is a continuous additive generator for which $\mathfrak{f}(D) \subseteq \bar{D}_{\infty,+}$, then there exists a continuous additive generator f on $([0, 1], \leq)$ such that (13) holds for all $x \in L^I$.

Let conversely f be a continuous additive generator on $([0, 1], \leq)$ and define the mapping $\mathfrak{f}$ by (13) for all $x \in L^I$. Then clearly $\mathfrak{f}$ is continuous, strictly decreasing, and $\mathfrak{f}(1_{\mathcal{L}^I}) = 0_{\mathcal{L}^I}$. From Theorem 9 it follows that $\mathfrak{f}^{(-1)}(\mathfrak{f}(x)) = [f^{(-1)}(f(x_1)), f^{(-1)}(f(x_2))] = x$, for all $x \in L^I$. Obviously, $\mathfrak{f}(D) \subseteq \bar{D}_{\infty,+}$. From Lemma 9 it follows that $\mathfrak{f}$ satisfies (AG.4), so $\mathfrak{f}$ is an additive generator on $\mathcal{L}^I$. □

Theorem 10 shows that no matter which operator $\oplus$ satisfying (ADD-1)–(ADD-4) is used in (AG.4), a continuous additive generator $\mathfrak{f}$ on $\mathcal{L}^I$ for which $\mathfrak{f}(D) \subseteq \bar{D}_{\infty,+}$ can be represented using an additive generator on $([0, 1], \leq)$.

The following theorem shows that in order to allow additive generators on $\mathcal{L}^I$ to be defined in a componentwise way using an additive generator on the unit interval, the set $\mathcal{R}(\mathfrak{f})$ in (AG.4) must indeed be defined as in (AG.4).

Theorem 11 *Let f be an additive generator on $([0,1],\leq)$. Then the mapping $\mathfrak{f}: L^I \to \bar{L}^I_{\infty,+}$ defined by, for all $x \in L^I$,*

$$\mathfrak{f}(x) = [f(x_2), f(x_1)],$$

is an additive generator on $\mathcal{L}^I$ associated to $\oplus$ if and only if, for all x, y in L^I,

$$\mathfrak{f}(x) \oplus \mathfrak{f}(y) \in (\mathrm{rng}(f) \cup [f(0), +\infty])^2.$$

Proof Since f is strictly decreasing, $\mathfrak{f}$ is strictly decreasing. From $f(1) = 0$ it follows that $\mathfrak{f}([1,1]) = [0,0]$. Since f is right-continuous in 0, $\lim\limits_{x \to 0_{\mathcal{L}^I}} \mathfrak{f}(x) = [\lim\limits_{x \to 0_{\mathcal{L}^I}} f(x_2), \lim\limits_{x \to 0_{\mathcal{L}^I}} f(x_1)] = [\lim\limits_{x_2 \to 0} f(x_2), \lim\limits_{x_1 \to 0} f(x_1)] = [f(0), f(0)] = \mathfrak{f}(0_{\mathcal{L}^I})$, so $\mathfrak{f}$ is right-continuous in $0_{\mathcal{L}^I}$. From Theorem 9 it follows that $\mathfrak{f}^{(-1)}(\mathfrak{f}(x)) = [f^{(-1)}(f(x_1)), f^{(-1)}(f(x_2))] = x$, for all $x \in L^I$.

Finally we check (AG.4). Since f is decreasing, $f(x_2) \leq f(x_1)$, so $\mathfrak{f}(x) \in \bar{L}^I_{\infty,+}$, for all $x \in L^I$. It is easy to see that

$$\mathrm{rng}(\mathfrak{f}) = (\mathrm{rng}(f))^2 \cap \bar{L}^I_{\infty,+},$$
$$\{x_1 \mid x_1 \in [0,+\infty] \text{ and } [x_1, (\mathfrak{f}(0_{\mathcal{L}^I}))^U] \in \mathrm{rng}(\mathfrak{f})\} = \mathrm{rng}(f),$$
$$[(\mathfrak{f}(0_{\mathcal{L}^I}))^U, +\infty] = [f(0), +\infty],$$
$$\{x \mid x \in \bar{L}^I_{\infty,+} \text{ and } x \geq_{L^I} \mathfrak{f}(0_{\mathcal{L}^I})\} = [f(0), +\infty]^2 \cap \bar{L}^I_{\infty,+}.$$

So $\mathcal{R}(\mathfrak{f}) = (\mathrm{rng}(f) \cup [f(0), +\infty])^2 \cap \bar{L}^I_{\infty,+}$. Hence $\mathfrak{f}$ is an additive generator on $\mathcal{L}^I$ if and only if $\mathfrak{f}(x) \oplus \mathfrak{f}(y) \in (\mathrm{rng}(f) \cup [f(0), +\infty])^2$. □

6 Additive Generators and Triangular Norms on $\mathcal{L}^I$

Lemma 10 *Let $\mathfrak{f}$ be an additive generator on $\mathcal{L}^I$ associated to $\oplus$. Then the mapping $\mathcal{T}_{\mathfrak{f}}: (L^I)^2 \to L^I$ defined by, for all x, y in L^I,*

$$\mathcal{T}_{\mathfrak{f}}(x, y) = \mathfrak{f}^{(-1)}(\mathfrak{f}(x) \oplus \mathfrak{f}(y)),$$

is commutative, increasing and $\mathcal{T}_{\mathfrak{f}}(1_{\mathcal{L}^I}, x) = x$, for all $x \in L^I$.

Proof Clearly, since $\oplus$ is commutative, $\mathcal{T}_{\mathfrak{f}}$ is commutative. Since $\mathfrak{f}$ and $\mathfrak{f}^{(-1)}$ are decreasing and $\oplus$ is increasing, $\mathcal{T}_{\mathfrak{f}}$ is increasing. Finally, from (AG.5) it follows that, for all $x \in L^I$, $\mathcal{T}_{\mathfrak{f}}(1_{\mathcal{L}^I}, x) = \mathfrak{f}^{(-1)}(0_{\mathcal{L}^I} \oplus \mathfrak{f}(x)) = \mathfrak{f}^{(-1)}(\mathfrak{f}(x)) = x$. □

Theorem 12 *Let $\mathfrak{f}$ be a continuous additive generator on $\mathcal{L}^I$ associated to $\oplus$ for which $\mathfrak{f}(D) \subseteq \bar{D}_{\infty,+}$. The mapping $\mathcal{T}_{\mathfrak{f}} : (L^I)^2 \to L^I$ defined by, for all x, y in L^I,*

$$\mathcal{T}_{\mathfrak{f}}(x, y) = \mathfrak{f}^{(-1)}(\mathfrak{f}(x) \oplus \mathfrak{f}(y)),$$

is a t-norm on $\mathcal{L}^I$ if and only if $\oplus$ satisfies the following condition:

$$\begin{aligned} &(\forall (x, y, z) \in A^3) \\ &\Big(\big((\inf(\alpha, x \oplus y) \oplus z)_1 < \alpha_1 \text{ and } (x \oplus y)_2 > \alpha_1\big) \\ &\quad \Longrightarrow (\inf(\alpha, x \oplus y) \oplus z)_1 = (x \oplus \inf(\alpha, y \oplus z))_1\Big), \end{aligned} \tag{14}$$

where $\alpha = \mathfrak{f}(0_{\mathcal{L}^I})$ and $A = \{x \mid x \in \bar{L}^I_{\infty,+} \text{ and } x \leq_{L^I} \mathfrak{f}(0_{\mathcal{L}^I})\}$.

Proof Let $\mathfrak{f}$ be a continuous additive generator on $\mathcal{L}^I$ for which $\mathfrak{f}(D) \subseteq \bar{D}_{\infty,+}$. Then from Theorem 10 it follows that there exists a continuous additive generator f on $([0, 1], \leq)$ such that (13) holds. From Theorem 9 it follows that $\mathfrak{f}^{(-1)}(y) = [f^{(-1)}(y_2), f^{(-1)}(y_1)]$, for all $y \in \bar{L}^I_{\infty,+}$.

Let arbitrarily $y \in \bar{L}^I_{\infty,+}$. Then

$$\begin{aligned} \mathfrak{f}(\mathfrak{f}^{(-1)}(y)) &= [f(f^{(-1)}(y_1)), f(f^{(-1)}(y_2))] \\ &= [\min(f(0), y_1), \min(f(0), y_2)] \\ &= \inf(\mathfrak{f}(0_{\mathcal{L}^I}), y). \end{aligned} \tag{15}$$

Define the mapping $\mathcal{T}_{\mathfrak{f}} : (L^I)^2 \to L^I$ by $\mathcal{T}_{\mathfrak{f}}(x, y) = \mathfrak{f}^{(-1)}(\mathfrak{f}(x) \oplus \mathfrak{f}(y))$, for all x, y in L^I. Assume first that $\mathcal{T}_{\mathfrak{f}}$ is a t-norm. From the associativity of $\mathcal{T}_{\mathfrak{f}}$ it follows for all x, y, z in L^I,

$$\mathfrak{f}^{(-1)}\big(\mathfrak{f}(x) \oplus \mathfrak{f}\big(\mathfrak{f}^{(-1)}(\mathfrak{f}(y) \oplus \mathfrak{f}(z))\big)\big) = \mathfrak{f}^{(-1)}\big(\mathfrak{f}\big(\mathfrak{f}^{(-1)}(\mathfrak{f}(x) \oplus \mathfrak{f}(y))\big) \oplus \mathfrak{f}(z)\big). \tag{16}$$

By applying $\mathfrak{f}$ on both sides of this equality and taking into account (15), we obtain

$$\begin{aligned} &\inf\big(\mathfrak{f}(0_{\mathcal{L}^I}), \mathfrak{f}(x) \oplus \inf\big(\mathfrak{f}(0_{\mathcal{L}^I}), \mathfrak{f}(y) \oplus \mathfrak{f}(z)\big)\big) \\ &= \inf\big(\mathfrak{f}(0_{\mathcal{L}^I}), \inf\big(\mathfrak{f}(0_{\mathcal{L}^I}), \mathfrak{f}(x) \oplus \mathfrak{f}(y)\big) \oplus \mathfrak{f}(z)\big). \end{aligned} \tag{17}$$

Since $\mathfrak{f}$ is given by (13) and f is continuous, $\text{rng}(\mathfrak{f}) = A$, where $A = \{x \mid x \in \bar{L}^I_{\infty,+}$ and $x \leq_{L^I} \mathfrak{f}(0_{\mathcal{L}^I})\}$ (see the proof of Lemma 9). Thus, from (17) it follows that for all x, y, z in A,

$$\inf(\alpha, x \oplus \inf(\alpha, y \oplus z)) = \inf(\alpha, \inf(\alpha, x \oplus y) \oplus z), \tag{18}$$

where $\alpha = \mathfrak{f}(0_{\mathcal{L}^I})$.

Assume conversely that (14) holds. From Lemma 10 it follows that $\mathcal{T}_\mathfrak{f}$ is commutative, increasing and $\mathcal{T}_\mathfrak{f}(1_{\mathcal{L}^I}, x) = x$, for all $x \in L^I$. We still have to prove the associativity. Let x, y in L^I such that $x \neq y$, then $x_1 \neq y_1$ or $x_2 \neq y_2$. Since $\mathfrak{f}$ is represented by (13) and f is strictly decreasing, from $x_1 \neq y_1$ it follows that $f(x_1) \neq f(y_1)$, so $\mathfrak{f}(x) \neq \mathfrak{f}(y)$. Similarly, if $x_2 \neq y_2$, then $\mathfrak{f}(x) \neq \mathfrak{f}(y)$. Hence, for all x, y in L^I,

$$\mathfrak{f}(x) = \mathfrak{f}(y) \iff x = y.$$

Thus, if (17) holds, then using (15) we obtain that (16) holds, so $\mathcal{T}_\mathfrak{f}$ is associative. Note that (17) holds as soon as (18) holds. Since $\alpha = \mathfrak{f}(0_{\mathcal{L}^I}) \in \bar{D}$, the proof that (18) holds for all x, y, z in A, is similar to the second part of the proof of Theorem 4. □

Taking into consideration the similarity between the conditions (6) and (14), we consider a condition which is similar to (8) and prove that it is a sufficient condition for $\oplus$ so that a (not necessarily continuous) additive generator associated to $\oplus$ generates a t-norm. First we give a lemma.

Lemma 11 *Let $\mathfrak{f}$ be an additive generator on $\mathcal{L}^I$ associated to $\oplus$. Assume that $\oplus$ satisfies the following condition:*

$$(\forall (x, y) \in \bar{L}^I_+ \times A)$$
$$\big(\big(([x_1, \alpha_2] \oplus y)_1 < \alpha_1 \text{ and } x_2 \in \,]\alpha_2, 2\alpha_2]\big) \implies ([x_1, \alpha_2] \oplus y)_1 = (x \oplus y)_1\big), \tag{19}$$

and

$$(\forall (x, y) \in \bar{L}^I_+ \times A)$$
$$\big(\big(([\alpha_1, x_2] \oplus y)_2 < \alpha_2 \text{ and } x_1 \in \,]\alpha_1, 2\alpha_1]\big) \implies ([\alpha_1, x_2] \oplus y)_2 = (x \oplus y)_2\big), \tag{20}$$

where $\alpha = \mathfrak{f}(0_{\mathcal{L}^I})$ and $A = \{x \mid x \in \bar{L}^I_{\infty,+}$ and $x \leq_{L^I} \mathfrak{f}(0_{\mathcal{L}^I})\}$. Then, for all $x \in L^I$ and $y \in \mathcal{R}(\mathfrak{f})$ such that $y \leq_{L^I} \mathfrak{f}(0_{\mathcal{L}^I}) \oplus \mathfrak{f}(0_{\mathcal{L}^I})$,

$$\mathfrak{f}(x) \oplus \mathfrak{f}(\mathfrak{f}^{(-1)}(y)) \in \mathcal{R}(\mathfrak{f})$$

and

$$\mathfrak{f}^{(-1)}(\mathfrak{f}(x) \oplus \mathfrak{f}(\mathfrak{f}^{(-1)}(y))) = \mathfrak{f}^{(-1)}(\mathfrak{f}(x) \oplus y). \tag{21}$$

Proof Let $x \in L^I$ and $y \in \mathcal{R}(\mathfrak{f})$ such that $y \leq_{L^I} \mathfrak{f}(0_{\mathcal{L}^I}) \oplus \mathfrak{f}(0_{\mathcal{L}^I})$. We have the following cases:

- If $y \in \text{rng}(\mathfrak{f})$, then there exists a $z \in L^I$ such that $\mathfrak{f}(z) = y$, so $\mathfrak{f}(x) \oplus \mathfrak{f}(\mathfrak{f}^{(-1)}(y)) = \mathfrak{f}(x) \oplus \mathfrak{f}(\mathfrak{f}^{(-1)}(\mathfrak{f}(z))) = \mathfrak{f}(x) \oplus \mathfrak{f}(z) = \mathfrak{f}(x) \oplus y \in \mathcal{R}(\mathfrak{f})$, using (AG.4) and (AG.5).
- If $[y_1, (\mathfrak{f}(0_{\mathcal{L}^I}))_2] \in \text{rng}(\mathfrak{f})$ and $y_2 > (\mathfrak{f}(0_{\mathcal{L}^I}))_2$, then there exists a $z \in L^I$ such that $\mathfrak{f}(z) = [y_1, (\mathfrak{f}(0_{\mathcal{L}^I}))_2]$. Since y_2 and $(\mathfrak{f}(z))_2$ are both greater than or equal to $(\mathfrak{f}(0_{\mathcal{L}^I}))_2$, we obtain that $\mathfrak{f}^{(-1)}(y) = \mathfrak{f}^{(-1)}(\mathfrak{f}(z)) = \sup(\{0_{\mathcal{L}^I}\} \cup \{x' \mid x' \in L^I$ and $(\mathfrak{f}(x'))_1 > y_1 = (\mathfrak{f}(z))_1$ and $(\mathfrak{f}(x'))_2 = (\mathfrak{f}(0_{\mathcal{L}^I}))_2\})$. It follows that $\mathfrak{f}(\mathfrak{f}^{(-1)}(y)) = \mathfrak{f}(\mathfrak{f}^{(-1)}(\mathfrak{f}(z))) = \mathfrak{f}(z)$, using (AG.5). Thus

$$\mathfrak{f}(x) \oplus \mathfrak{f}(\mathfrak{f}^{(-1)}(y)) = \mathfrak{f}(x) \oplus \mathfrak{f}(z) \in \mathcal{R}(\mathfrak{f}). \tag{22}$$

  Since $(\mathfrak{f}(z))_2 = (\mathfrak{f}(0_{\mathcal{L}^I}))_2$, we obtain that $(\mathfrak{f}(x) \oplus \mathfrak{f}(z))_2 \geq (\mathfrak{f}(0_{\mathcal{L}^I}))_2$. On the other hand, $(\mathfrak{f}(x) \oplus y)_2 \geq y_2 \geq (\mathfrak{f}(0_{\mathcal{L}^I}))_2$. Note that $\mathfrak{f}(z) = [y_1, \alpha_2]$ and $y_2 \in]\alpha_2, 2\alpha_2]$. If $(\mathfrak{f}(x) \oplus \mathfrak{f}(z))_1 < (\mathfrak{f}(0_{\mathcal{L}^I}))_1 = \alpha_1$, then from (19) it follows that $(\mathfrak{f}(x) \oplus \mathfrak{f}(z))_1 = (\mathfrak{f}(x) \oplus y)_1$. If $(\mathfrak{f}(x) \oplus \mathfrak{f}(z))_1 \geq (\mathfrak{f}(0_{\mathcal{L}^I}))_1$, then, since $y \geq_{L^I} \mathfrak{f}(z)$ and $\oplus$ is increasing, $(\mathfrak{f}(x) \oplus y)_1 \geq (\mathfrak{f}(0_{\mathcal{L}^I}))_1$. It is easy to see that in all cases $\Phi_{\mathfrak{f}(x) \oplus \mathfrak{f}(z)} = \Phi_{\mathfrak{f}(x) \oplus y} = \{0_{\mathcal{L}^I}\} \cup \{x' \mid x' \in L^I$ and $(\mathfrak{f}(x'))_1 > (\mathfrak{f}(x) \oplus y)_1$ and $(\mathfrak{f}(x'))_2 = (\mathfrak{f}(0_{\mathcal{L}^I}))_2\}$. Using the equality in (22), we obtain that (21) holds.
- If $[(\mathfrak{f}(0_{\mathcal{L}^I}))_1, y_2] \in \text{rng}(\mathfrak{f})$ and $y_1 > (\mathfrak{f}(0_{\mathcal{L}^I}))_1$, then it can be similarly proven that $\mathfrak{f}(x) \oplus \mathfrak{f}(\mathfrak{f}^{(-1)}(y)) \in \mathcal{R}(\mathfrak{f})$ and that (21) holds.
- If $y \geq_{L^I} \mathfrak{f}(0_{\mathcal{L}^I})$, then $\mathfrak{f}^{(-1)}(y) = 0_{\mathcal{L}^I}$ and $\mathfrak{f}(x) \oplus \mathfrak{f}(\mathfrak{f}^{(-1)}(y)) = \mathfrak{f}(x) \oplus \mathfrak{f}(0_{\mathcal{L}^I}) \in \{x' \mid x' \in \bar{L}^I_{\infty,+}$ and $x' \geq_{L^I} \mathfrak{f}(0_{\mathcal{L}^I})\} \subseteq \mathcal{R}(\mathfrak{f})$, so $\mathfrak{f}^{(-1)}(\mathfrak{f}(x) \oplus \mathfrak{f}(\mathfrak{f}^{(-1)}(y)) = 0_{\mathcal{L}^I}$. Since $y \geq_{L^I} \mathfrak{f}(0_{\mathcal{L}^I})$, we obtain that $\mathfrak{f}(x) \oplus y \geq_{L^I} \mathfrak{f}(0_{\mathcal{L}^I})$, so $\mathfrak{f}^{(-1)}(\mathfrak{f}(x) \oplus y) = 0_{\mathcal{L}^I} = \mathfrak{f}^{(-1)}(\mathfrak{f}(x) \oplus \mathfrak{f}(\mathfrak{f}^{(-1)}(y)))$. □

Using Lemma 11, the following theorem can be shown.

Theorem 13 *Let $\mathfrak{f}$ be an additive generator on $\mathcal{L}^I$ associated to $\oplus$. If $\oplus$ satisfies* (19) *and* (20), *then the mapping $\mathcal{T}_\mathfrak{f} : (L^I)^2 \to L^I$ defined by, for all x, y in L^I,*

$$\mathcal{T}_\mathfrak{f}(x, y) = \mathfrak{f}^{(-1)}(\mathfrak{f}(x) \oplus \mathfrak{f}(y)),$$

is a t-norm on $\mathcal{L}^I$.

Proof From Lemma 10 it follows that $\mathcal{T}_\mathfrak{f}$ is commutative, increasing and that $\mathcal{T}_\mathfrak{f}(1_{\mathcal{L}^I}, x) = x$, for all $x \in L^I$. We still have to prove the associativity of $\mathcal{T}_\mathfrak{f}$. Let x, y, z in L^I, then $\mathcal{T}_\mathfrak{f}(x, \mathcal{T}_\mathfrak{f}(y, z)) = \mathfrak{f}^{(-1)}\big(\mathfrak{f}(x) \oplus \mathfrak{f}\big(\mathfrak{f}^{(-1)}(\mathfrak{f}(y) \oplus \mathfrak{f}(z))\big)\big)$. From Lemma 11 it follows that

$$\mathfrak{f}(x) \oplus \mathfrak{f}(\mathfrak{f}^{(-1)}(\mathfrak{f}(y) \oplus \mathfrak{f}(z))) \in \mathcal{R}(\mathfrak{f})$$

$$\mathfrak{f}^{(-1)}\big(\mathfrak{f}(x) \oplus \mathfrak{f}\big(\mathfrak{f}^{(-1)}(\mathfrak{f}(y) \oplus \mathfrak{f}(z))\big)\big) = \mathfrak{f}^{(-1)}\big(\mathfrak{f}(x) \oplus (\mathfrak{f}(y) \oplus \mathfrak{f}(z))\big).$$

From the associativity of $\oplus$ it now easily follows that $\mathcal{T}_\mathfrak{f}$ is associative. □

Theorems 10 and 9 show that no matter which operator $\oplus$ is used in (AG.4), a continuous additive generator $\mathfrak{f}$ on $\mathcal{L}^I$ satisfying $\mathfrak{f}(D) \subseteq \bar{D}_{\infty,+}$ is representable and has a representable pseudo-inverse. Therefore it depends on the operator $\oplus$ which classes of t-norms on $\mathcal{L}^I$ can have continuous additive generators that extend additive generators on $([0, 1], \leq)$.

7 Conclusion

In this paper we presented a more general approach to additive generators in interval-valued fuzzy and Atanassov's intuitionistic fuzzy set theory than in [16]. In this paper, instead of choosing one particular set of arithmetic operators, we allow any set of arithmetic operators satisfying certain axioms to be used in the construction of additive generators. We characterized continuous generators which map exact elements to exact elements in terms of generators on the unit interval. We gave a necessary and sufficient condition under which a generator actually generates a t-norm and we showed that the generated t-norm belongs to particular classes of t-norms depending on the arithmetic operators involved in the definition of the generator.

References

1. Schweizer, B., Sklar, A.: Probabilistic Metric Spaces. Elsevier, North-Holland (1983)
2. Fodor, J.C., Roubens, M.: Fuzzy Preference Modelling and Multicriteria Decision Support. Kluwer Academic Publishers, Dordrecht (1994)
3. Hájek, P.: Metamathematics of Fuzzy Logic. Kluwer Academic Publishers, Dordrecht (1998)
4. Klement, E.P., Mesiar, R., Pap, E.: Triangular Norms. Kluwer Academic Publishers, Dordrecht (2000)
5. Faucett, W.M.: Compact semigroups irreducibly connected between two idempotents. Proceedings of the American Mathematical Society **6**, 741–747 (1955)
6. Klement, E.P., Mesiar, R., Pap, E.: Quasi- and pseudo-inverses of monotone functions, and the construction of t-norms. Fuzzy Sets Syst. **104**(1), 3–13 (1999)
7. Ling, C.H.: Representation of associative functions. Publ. Math. Debrecen **12**, 189–212 (1965)
8. Mostert, P.S., Shields, A.L.: On the structure of semigroups on a compact manifold with boundary. Ann. Math. **65**, 117–143 (1957)
9. Gorzałczany, M.B.: A method of inference in approximate reasoning based on interval-valued fuzzy sets. Fuzzy Sets Syst. **21**(1), 1–17 (1987)
10. Sambuc, R.: Fonctions Φ-floues. Application à l'aide au diagnostic en pathologie thyroidienne. Ph.D. thesis, Université de Marseille, France (1975)
11. Atanassov, K.T.: Intuitionistic fuzzy sets (1983) VII ITKR's Session, Sofia (deposed in Central Sci.-Technical Library of Bulg. Acad. of Sci., 1697/84) (in Bulgarian)
12. Atanassov, K.T.: Intuitionistic fuzzy sets. Fuzzy Sets Syst. **20**(1), 87–96 (1986)
13. Atanassov, K.T.: Intuitionistic Fuzzy Sets. Physica-Verlag, Heidelberg (1999)
14. Deschrijver, G., Kerre, E.E.: On the relationship between some extensions of fuzzy set theory. Fuzzy Sets Syst. **133**(2), 227–235 (2003)
15. Goguen, J.A.: L-fuzzy sets. J Math. Anal. Appl. **18**(1), 145–174 (1967)
16. Deschrijver, G.: Additive and multiplicative generators in interval-valued fuzzy set theory. IEEE Trans. Fuzzy Syst. **15**(2), 222–237 (2007)

17. Deschrijver, G.: Arithmetic operators in interval-valued fuzzy set theory. Inf. Sci. **177**(14), 2906–2924 (2007)
18. Deschrijver, G., Vroman, A.: Generalized arithmetic operations in interval-valued fuzzy set theory. J. Intell. Fuzzy Syst. **16**(4), 265–271 (2005)
19. Deschrijver, G., Cornelis, C., Kerre, E.E.: On the representation of intuitionistic fuzzy t-norms and t-conorms. IEEE Trans. Fuzzy Syst. **12**(1), 45–61 (2004)
20. Deschrijver, G., Kerre, E.E.: Classes of intuitionistic fuzzy t-norms satisfying the residuation principle. Int. J. Uncertain. Fuzziness Knowl.-Based Syst. **11**(6), 691–709 (2003)
21. Deschrijver, G., Kerre, E.E.: Implicators based on binary aggregation operators in interval-valued fuzzy set theory. Fuzzy Sets Syst. **153**(2), 229–248 (2005)
22. Deschrijver, G.: Generalized arithmetic operators and their relationship to t-norms in interval-valued fuzzy set theory. Fuzzy Sets Syst. **160**(21), 3080–3102 (2009)
23. Moore, R.E.: Interval Arithmetic. Prentice-Hall, Englewood Cliffs (1966)

On the TOPSIS-Class Methods in the Intuitionistic Fuzzy Environment

Piotr Dworniczak

Abstract TOPSIS is one of the basic methods of multicriteria decision aid. In the paper a classical algorithm TOPSIS and its analogies in the intuitionistic fuzzy environment are presented. The application of intuitionistic fuzzy implication for a new solution of some step in the method is given. The comments about the new conversion of the input decision matrix due to the criteria validity are given. The illustrative example is given.

Keywords Intuitionictic fuzzy sets · TOPSIS · Intuitionistic fuzzy implication

1 Introduction

Multi-criteria Decision Making is, despite of many years of research, a field of presentation of new methods and modifications of the already existing. The intuitionistic fuzzy (for shortly: IF) sets theory is a promising tool in this area. Furthermore, due to the tendency to facilitate the opinion expressed by experts, the linguistic values as evaluation of the variants are being used. Due to the genesis, the linguistic variables are usually vague (fuzzy). For their processing the IF sets theory can also be useful.

The intuitionistic fuzzy sets (IFS) have been introduced by K. Atanassov [2,[1] 3].

Definition 1 (*Atanassov* [2]) The *intuitionistic fuzzy set A* on a universe $U \neq \emptyset$ is understood as

$$A = \{(x, \mu_A(x), \nu_A(x)) : x \in U\},$$

[1]The publication from 1983 was first but IFS became widely known after the publication [3].

P. Dworniczak (✉)
The Great Poland University of Social and Economics, Środa Wielkopolska, Poland
e-mail: p.dworniczak@wwsse.pl

P. Angelov and S. Sotirov (eds.), *Imprecision and Uncertainty in Information Representation and Processing*, Studies in Fuzziness and Soft Computing 332,
DOI 10.1007/978-3-319-26302-1_11

where μ_A and ν_A are the function from U to closed interval [0, 1], and for every $x \in U$ holds $\mu_A(x) + \nu_A(x) \leq 1$.

The ordered pair $<\mu_A(x), \nu_A(x)>$ is called the IF value or IF pair [6].
The value $\pi_A(x) = 1 - \mu_A(x) - \nu_A(x)$ is called a hesitation margin (or a hesitancy degree).
The values $\mu_A(x)$ and $\nu_A(x)$ are, respectively, the degree of membership and degree of non-membership of element x to the set A.

Decision making in the intuitionistic fuzzy environment is an example of an area of research in which, in light of the above comments, the IFSs can be used even in those cases where uncertainty or ignorance exists, concerning the evaluation of options made it difficult or impossible to use other methods for decision support.

Citing Szmidt and Kacprzyk [26] we assume, that IFSs based models may be adequate mainly in the situations when we face human testimonies, opinions, etc. involving answers of three types:

- yes,
- no,
- abstaining, i.e. which can not be classified (because of different reasons, e.g. "I do not know", "I am not sure", "I do not want to answer", "I am not satisfied with any of the options" etc.).

The applications of the IFSs theory for the decision-making appeared a relatively long time after the first publication of Atanassov [10]. The first monograph on this subject was the Szmidt paper [23] where two main approaches for decision support were presented. The first is the intuitionistic generalization of Bellman–Zadeh's approach given by Kacprzyk [18], and the second presented some solution concepts in group decision making with (individual and social) intuitionistic fuzzy preference relations.

One of the directions of further research is the adaptation of the TOPSIS method to the IF environment.

2 The TOPSIS-Class Methods

One of the basic methods of the decision making (decision aid) is the TOPSIS method. It was presented by Hwang and Yoon [17]. The name of method is the acronym derived from *Technique for Order Preference by Similarity to an Ideal Solution*. It is an important method of ordering of the elements in the multidimensional metric space. This method allows to show the best solution (alternative, decision), based on some aggregate measure of the distance of the evaluation of the variants from the ideal- and the anti-ideal solution (positive- and negative-ideal, ideal- and anti-ideal evaluation).

In this paper the review the TOPSIS-like methods in the intuitionistic fuzzy environment is presented. Besides, the idea of simple modification of the method is given. The methods based mainly on the idea of TOPSIS method we will call hereafter *TOPSIS-class* methods.

2.1 The Classical TOPSIS Method

The decision-making problem is to identify the optimal variant or ordering of the set of variants based on the evaluating criterion or criteria. In the multi-criteria decision making problem (MCDM), it is very difficult due to the lack of a unequivocal order in a multidimensional space. The TOPSIS is one of the ways to solve this problem with a finite number of decision variants and a finite number of criteria.

Let $X = \{x_1, x_2, \ldots x_n\}$ be a finite set of the variants (alternatives) and $C_1, C_2, \ldots C_k$ the set of criteria. The classical TOPSIS method is specified in following steps.

1^0 It is given the matrix $\boldsymbol{S} = [s_{ij}]_{n\times k}$, where s_{ij} is the evaluation of the *i*th variant due to the *j*th criterion (or, in other words, the value of the *j*th attribute of the *i*th variant).

On this basis we create the matrix $\boldsymbol{R} = [r_{ij}]_{n\times k}$ of the normalized values, calculated by formula

$$r_{ij} = \frac{s_{ij}}{\sqrt{\sum_{i=1}^{n} s_{ij}^2}}.$$

The $\boldsymbol{S}$ matrix is called the *decision matrix*, where $\boldsymbol{R}$ is called the *normalized decision matrix*.

2^0 We construct the *weighted normalized decision matrix* $\boldsymbol{T} = [t_{ij}]_{n\times k}$, where

$$t_{ij} = w_j \cdot r_{ij}.$$

The value w_j, where $w_j \in [0, 1]$ and $\sum_{j=1}^{k} w_j = 1$, is the weight for the *j*th criterion.

3^0 We determine the ideal solution $x^+ = (t_1^+, t_2^+, \ldots t_k^+)$ and anti-ideal solution $x^- = (t_1^-, t_2^-, \ldots t_k^-)$. For the *benefit criteria* we take

$$t_j^+ = \max_{i=1\ldots n} t_{ij}, t_j^- = \min_{i=1\ldots n} t_{ij}$$

while for *cost criteria*

$$t_j^+ = \min_{i=1\ldots n} t_{ij}, t_j^- = \max_{i=1\ldots n} t_{ij}.$$

4^0 For the ith variant (i = 1, …, n) we compute the distance d_i^+ from the ideal solution and the distance d_i^- from the anti-ideal solution

$$d_i^+ = \sqrt{\sum_{j=1}^{k} (t_{ij} - t_j^+)^2}$$

and

$$d_i^- = \sqrt{\sum_{j=1}^{k} (t_{ij} - t_j^-)^2}.$$

5^0 For the ith variant (i = 1, …, n) we calculate the *index of relative closeness* to the ideal solution as

$$c_i^+ = \frac{d_i^-}{d_i^+ + d_i^-} \in [0, 1].$$

6^0 We order the variants by the rule: the higher is the value c_i^+, the better is the ith variant.

Let us note that the normalization, as given in step 1^0, the weighted normalized decision matrix determined in step 2^0, the ideal and anti-ideal solution in step 3^0, the distance (here—the Euclidean distance) determined in step 4^0 and even the index of relative closeness in 5^0 can be determined in other ways.

2.2 *TOPSIS-Class Methods in the IF Environment—Basic Solutions*

TOPSIS-class methods in the intuitionistic fuzzy environment appeared relatively late. It is possible that the work of Szmidt and Kacprzyk [24, 25, 27] have established finally the standards for determining of the distances, and have opened the way for important methods using a distance or similarity measure of the IFSs. The research on the distance of the IFSs were initiated by Atanassov. In the book [4] Atanassov gives the definition of the Hamming distance and Euclidean distance of IFSs.

Definition 2 (*Atanassov* [4]) For IFSs $A = \{<x, \mu_A(x), \nu_A(x)>: x \in U\}$ and $B = \{<x, \mu_B(x), \nu_B(x)>: x \in U\}$ the Hamming distance is given as

$$H(A,B) = \frac{1}{2}\sum_{x \in U} (|\mu_A(x) - \mu_B(x)| + |\nu_A(x) - \nu_B(x)|),$$

while the Euclidean distance

$$E(A,B) = (\frac{1}{2}\sum_{x \in U}((\mu_A(x) - \mu_B(x))^2 + (\nu_A(x) - \nu_B(x))^2))^{0,5}.$$

Szmidt and Kacprzyk [24] proposed the Hamming distance and the Euclidean distance using also the third parameter—the hesitation degree π.

Definition 3 (*Szmidt and Kacprzyk* [24]) For the IFSs $A = \{<x, \mu_A(x), \nu_A(x)>: x \in U\}$ and $B = \{<x, \mu_B(x), \nu_B(x)>: x \in U\}$, in the finite universe U, the normalized Hamming distance is the number

$$l_{IFS}^1(A,B) = \frac{1}{2n}\sum_{x \in U}(|\mu_A(x) - \mu_B(x)| + |\nu_A(x) - \nu_B(x)| + |\pi_A(x) - \pi_B(x)|),$$

and the normalized Euclidean distance is

$$q_{IFS}^1(A,B) = \frac{1}{2n}\sum_{x \in U}((\mu_A(x) - \mu_B(x))^2 + (\nu_A(x) - \nu_B(x))^2 + (\pi_A(x) - \pi_B(x))^2))^{0,5}.$$

In the subsequent papers [25, 27] Szmidt and Kacprzyk justify the need of taking into account the degree of uncertainty, pointing its important role in the assessing of decisions. The authors emphasize that the degree of uncertainty is an important measure of the lack of information. This lack of information is strictly associated with a risk characterizing almost all decisions, and it is clear that the decision should be assessed otherwise at the changes of the level of risk.

In 2008 Luo and Yu [21] proposed the TOPSIS-class method for the rank of variants using the inclusion degree of the IFSs characterizing the variant in the IFSs characterizing the optimal variant (optimal solution). The method of Luo and Yu addresses the problem by a finite set of variants and a finite set of criteria. This solution indicates one of the basic directions of using of the inclusion for the comprehensive evaluation of variants. The main idea is as follows.

Let $X = \{x_1, x_2, \ldots x_n\}$ be the finite set of variants (alternatives) while $\{C_1, C_2, \ldots, C_k\}$ a finite set of criteria. Each variant x_i is characterized by the IFS

$$X_i = \{<C_1, \mu_{i1}, \nu_{i1}>, <C_2, \mu_{i2}, \nu_{i2}>, \ldots <C_k, \mu_{ik}, \nu_{ik}>\}.$$

The μ_{ij} value is interpreted as the degree to which the x_i satisfies the C_j, while the ν_{ij} value as the degree to which x_i does not satisfy the C_j.

Authors assume that the decision-maker would like to choose a variant, which in the best way satisfies all criteria $C_1, C_2, \ldots, C_{k-1}$ or the C_k criterion.

In the first step, on the base of sets X_i (for $i = 1, \ldots, n$) are determined two IFSs, called the ideal- and the anti-ideal solution. Luo and Yu [21] proposed following formulas.

For $C_1, C_2, \ldots C_{k-1}$ criteria, the IFS

$$IS_1 = \{ <C_1, \mu_{IS1}, \nu_{IS1}>, <C_2, \mu_{IS2}, \nu_{IS2}>, \ldots, <C_{k-1}, \mu_{ISk-1}, \nu_{ISk-1}> \}$$
$$= \{ <C_1, \max_{i=1\ldots n} \mu_{i1}, \min_{i=1\ldots n} \nu_{i1}>, <C_2, \max_{i=1\ldots n} \mu_{i2}, \min_{i=1\ldots n} \nu_{i2}>, \ldots,$$
$$<C_{k-1}, \max_{i=1\ldots n} \mu_{ik-1}, \min_{i=1\ldots n} \nu_{ik-1}> \},$$

is the *ideal solution* of MCDM problem, while

$$AIS_1 = \{ <C_1, \mu_{AIS1}, \nu_{AIS1}>, <C_2, \mu_{AIS2}, \nu_{AIS2}>, \ldots, <C_{k-1}, \mu_{AISk-1}, \nu_{AISk-1}> \}$$
$$= \{ <C_1, \min_{i=1\ldots n} \mu_{i1}, \max_{i=1\ldots n} \nu_{i1}>, <C_2, \min_{i=1\ldots n} \mu_{i2}, \max_{i=1\ldots n} \nu_{i2}>, \ldots,$$
$$<C_{k-1}, \min_{i=1\ldots n} \mu_{ik-1}, \max_{i=1\ldots n} \nu_{ik-1}> \}$$

is the *anti-ideal solution* of the MCDM problem.

For the last criterion C_k the ideal- and anti-ideal solution are, similarly

$$IS_2 = \{ <C_k, \max_{i=1\ldots n} \mu_{ik}, \min_{i=1\ldots n} \nu_{ik}> \},$$

and

$$AIS_2 = \{ <C_k, \min_{i=1\ldots n} \mu_{ik}, \max_{i=1\ldots n} \nu_{ik}> \}.$$

In the next step the inclusion degrees of the ideal solution in all of the variants evaluations, and, similarly, the inclusion degrees of all of the variants evaluations in the anti-ideal solution are defined as

$$D_{INC}(x_i) = \max\{INC_{IFS}(IS_1, X_{i1}), INC_{IFS}(IS_2, X_{i2})\},$$
$$d_{INC}(x_i) = \min\{INC_{IFS}(X_{i1}, AIS_1), INC_{IFS}(X_{i2}, AIS_2)\},$$

where

$$X_{i1} = \{ <C_1, \mu_{i1}, \nu_{i1}>, <C_2, \mu_{i2}, \nu_{i2}>, \ldots <C_{k-1}, \mu_{ik-1}, \nu_{ik-1}> \},$$
$$X_{i2} = \{ <C_k, \mu_{ik}, \nu_{ik}> \},$$

The INC_{IFS} *(A, B)* is a real degree of the inclusion of *A* in *B*.

Luo and Yu in the cited paper give various ways of the determining of the inclusion degree of IF sets based on intuitionistic fuzzy implication or on the cardinality of IF sets.

The value $D_{INC}(x_i)$ as well as $d_{INC}(x_i)$ is the analogy of the distance from classical TOPSIS solution.

In the last step the index p_i (for $i = 1, \ldots, n$) is determined:

$$p_i = \frac{D_{INC}(x_i)}{D_{INC}(x_i) + d_{INC}(x_i)}.$$

The ordering of indexes from largest to smallest determines the ranking of variants from the best to the worst.

The above sets IS_1 (and IS_2) can be called the *max-min ideal solution*. It is the basic form of the ideal solutions considered in the TOPSIS-class methods.

Further papers concerning the TOPSIS-class methods in the IF environment are given by Wang and Wei [28] and Xu [31]. The solution given in the first paper is close to the Luo and Yu solution. Wang und Wei assume also the choice from the finite set of variants $x_1, x_2, \ldots x_n$. Each variant is assigned to k attributes (the variant is evaluated according to k criteria) $C_1, C_2, \ldots C_k$. But, unlike as in Luo and Yu solution, attributes are associated with the weights $w_1, w_2, \ldots w_k$, where $w_j \in [0, 1]$ and $\sum_{j=1}^{k} w_j = 1$. Authors do not specify the method of determining of weights. However, we can suspect that they must be given by the decision maker. Authors assume also that it must be given the matrix $\boldsymbol{R_{IFS}} = [r_{ij}]_{n \times k} = [<\mu_{ij}, \nu_{ij}>]_{n \times k}$ called the *intuitionistic decision matrix*. Each element $<\mu_{ij}, \nu_{ij}>$ of the matrix is the IF value. The μ_{ij} value is interpreted as the degree to which the x_i satisfies the C_j, while the ν_{ij} value as the degree to which x_i does not satisfy the C_j. Similarly, as in the Luo and Yu solution, the ideal solution *IS* and anti-ideal solution *AIS* (called in the paper the positive and the negative ideal) are determined. It is namely

$$\begin{aligned} IS &= \{<C_1, \mu_{IS1}, \nu_{IS1}>, <C_2, \mu_{IS2}, \nu_{IS2}>, \ldots, <C_k, \mu_{ISk}, \nu_{ISk}>\} \\ &= \{<C_1, \max_{i=1\ldots n} \mu_{i1}, \min_{i=1\ldots n} \nu_{i1}>, <C_2, \max_{i=1\ldots n} \mu_{i2}, \min_{i=1\ldots n} \nu_{i2}>, \ldots, \\ &\quad <C_k, \max_{i=1\ldots n} \mu_{ik}, \min_{i=1\ldots n} \nu_{ik}>\}, \end{aligned}$$

and

$$\begin{aligned} AIS &= \{<C_1, \mu_{AIS1}, \nu_{AIS1}>, <C_2, \mu_{AIS2}, \nu_{AIS2}>, \ldots, <C_k, \mu_{AISk}, \nu_{AISk}>\} \\ &= \{<C_1, \min_{i=1\ldots n} \mu_{i1}, \max_{i=1\ldots n} \nu_{i1}>, <C_2, \min_{i=1\ldots n} \mu_{i2}, \max_{i=1\ldots n} \nu_{i2}>, \ldots, \\ &\quad <C_k, \min_{i=1\ldots n} \mu_{ik}, \max_{i=1\ldots n} \nu_{ik}>\}. \end{aligned}$$

Further the authors compute the weighted Hamming distance (not using the hesitation degree)

$$H(r_i, IS) = \frac{1}{2} \sum_{j=1}^{k} w_j \left(\left| \mu_{ij} - \mu_{ISj} \right| + \left| \nu_{ij} - \nu_{ISj} \right| \right)$$

and

$$H(r_i, AIS) = \frac{1}{2}\sum_{j=1}^{k} w_j\left(\left|\mu_{ij} - \mu_{AISj}\right| + \left|\nu_{ij} - \nu_{AISj}\right|\right)$$

of the IFSs r_i (rows of the $\boldsymbol{R_{IFS}}$ matrix), characterizing each variants, from the ideal and the anti-ideal solutions. Finally, they compute indexes c_i of relative closeness of the x_i to the *IS* in the form

$$c_i(r_i, IS) = \frac{H(r_i, AIS)}{H(r_i, IS) + H(r_i, AIS)},$$

for $i = 1, \ldots, n$.

In the conclusion the ranking of variants is obtained, wherein a variant having greater index is considered better.

Similar solution is given by Xu [31]. He proposed to determine the *index of relative similarity*

$$RSim(r_i, IS) = \frac{Sim(r_i, IS)}{Sim(r_i, IS) + Sim(r_i, AIS)},$$

of each variants to the ideal.

The *Sim* is a similarity measure of IFSs. It was introduced in the paper earlier. Xu gives in particular two measures of similarity

$$Sim_l(A, B) = 1 - l_{IFS}^1(A, B)$$

and

$$Sim_q(A, B) = 1 - q_{IFS}^1(A, B).$$

Both measures are based on the distance defined by Szmidt and Kacprzyk.

The larger the $RSim(r_i, IS)$ value is, the better is the variant corresponding to the set r_i.

A solutions similar to the work of Wang und Wei and Xu are presented by Liu [20], Guo and Zhang [15] and Guo et al. [14] but the weights, given by an expert, are used to the transformation of the intuitionistic decision matrix $\boldsymbol{R_{IFS}}$ and not to correct the distance measurement.

A similar paper to that of Wang and Wei is presented also by Li and Zhang [19]. In this paper, weights are computed based on the symmetric *judgment matrix* relating to preferences of attributes (criteria). Elements of the matrix are IF values $<\mu_{ij}, \nu_{ij}>$, where μ_{ij} is a preference degree of the criterion C_i to the criterion C_j, while ν_{ij} is a preference degree of the criterion C_j to the C_i. It is also $\mu_{ij} = \nu_{ji}$, $\nu_{ij} = \mu_{ji}$, $\mu_{ii} = \nu_{ii} = 0{,}5$, $0 \le \mu_{ij}, \nu_{ij} \le 1$, $0 \le \mu_{ij} + \nu_{ij} \le 1$.

Hung and Chen [16] solution is also similar, but the authors, for the computing of weights, used the entropy measures of the IFSs which are the columns of the intuitionistic decision matrix $\boldsymbol{R_{IFS}}$.

In turn, Boran et al. [9] applying the solution, given in the mainstream by Luo and Yu, and Wang and Wei, introduce to the method, besides weights of criteria (given by the experts) also weights of opinion of experts. All the weights are computed on the basis of IF values equivalent to the linguistic terms for rating the importance of criteria and the importance of decision makers, like *very important, important, medium* etc. Variants can be assessed up to 10, while criteria and decision makers up to 5 linguistic terms. Every variant is evaluated by l experts. The mth expert gives the *intuitionistic fuzzy decision matrix* $R_{IFS}^{(m)} = [r_{ij}^{(k)}]_{n \times k} = [<\mu_{ij}, \nu_{ij}>]_{n \times k}$. Basing on the linguistic assessments of decision makers authors compute the weights of the kth decision maker as

$$\lambda_m = \frac{\mu_m + \pi_m \frac{\mu_m}{\mu_m + \nu_m}}{\sum\limits_{m=1}^{l} \left(\mu_m + \pi_m \frac{\mu_m}{\mu_m + \nu_m}\right)}.$$

Taking into account the above weights it is constructed the *aggregated intuitionistic fuzzy decision matrix* $\boldsymbol{R}$ which elements are calculated using the operator IFWA proposed by Xu [30].

The element r_{ij} of the aggregated matrix is the IF value equal to

$$r_{ij} = <\mu_{r_{ij}}, \nu_{r_{ij}}> = <1 - \prod_{m=1}^{l} \left(1 - \mu_{ij}^{(m)}\right)^{\lambda_m}, \quad \prod_{m=1}^{l} \left(\nu_{ij}^{(m)}\right)^{\lambda_m} >.$$

Similarly to the weights λ_m, the weights w_1, w_2 , …, w_l are computed. These weights $w_m = <\mu_{w_m}, \nu_{w_m}>$. correspond to the decision makers. Basing on these values and the aggregated intuitionistic fuzzy decision matrix $\boldsymbol{R}$ authors compute the aggregated weighted intuitionistic fuzzy decision matrix $\mathbf{R}'$ which elements are

$$r'_{ij} = <\mu_{r_{ij}} \cdot \mu_{w_m}, \nu_{r_{ij}} + \nu_{w_m} - \nu_{r_{ij}} \cdot \nu_{w_m}>.$$

The matrix $\mathbf{R}'$ is the basis for next steps. Subsequently are computed, in the classical form, the max-min positive ideal solution and the negative ideal solution, and then the *separation measures* (distances in the Szmidt-Kacprzyk sense) of each variant from positive and negative ideal solutions. In the last step, the index of relative closeness (as in the classical TOPSIS method) is computed and the rank of alternatives is determined.

The method presented above is almost in the same form as presented by Agarwal et al. [1]. The difference lies in the determination of weights assessed by decision makers (experts). Authors proposed namely

$$\lambda_m = \frac{\mu_m + \frac{\mu_m}{\mu_m + \nu_m}}{\sum_{m=1}^{l} (\mu_m + \frac{\mu_m}{\mu_m + \nu_m})},$$

without taking into account the hesitation degree.

For the evaluations of variants or experts importance authors proposed up to 10 different linguistic values with corresponding IF values. As the averaging operator, besides the IFWA operator, is also given the IFWG operator [32].

It means that the element r_{ij} of the aggregated intuitionistic fuzzy decision matrix is the IF value equal to

$$r_{ij} = < \mu_{r_{ij}}, \nu_{r_{ij}} > = < \prod_{m=1}^{l} \left(\mu_{ij}^{(m)}\right)^{\lambda_m}, \quad 1 - \prod_{m=1}^{l} \left(1 - \nu_{ij}^{(m)}\right)^{\lambda_m} >.$$

The subsequent steps are the same as given by Boran et al.

The solution of Boran et al. [9] has been applied, in the economic problems: in the personnel selection [8], in a supply chain management process [29], in the construction safety evaluation [7], in Project and Portfolio Management Information System [13].

A simplest version of TOPSIS in the IF environment has been applied by Zhang and Huang [33] for a supplier selection in information technology service outsourcing.

Maldonado-Macías et al. [22] applied the TOPSIS-class method for ergonomic compatibility evaluation of advanced manufacturing technology. They used mainly the Wang and Wei solution, but the vector of weights of attributes is determined as the eigenvector of the matrix of the pair-wise comparison of attributes corresponding to the maximal eigenvalue of this matrix. The elements of this comparison matrix are determined by decision makers according to the 9 point Saaty importance scale. The distances to the (classical) ideal and anti-ideal solution and the index of relative closeness are computed based on Euclidean distance q_{IFS}^1.

The solutions presented above link directly to the classical method of TOPSIS joint with the IFSs. However, a lot of papers on IFSs use some parts of the TOPSIS. The very developed direction, is currently the use of interval valued IFSs in connection with the TOPSIS-class methods. In this paper these works are omitted.

3 The Intuitionistic Fuzzy Implication as a Tool in the TOPSIS-Class Method

In the intuitionistic fuzzy logic (IFL) the truth-value of variable p is given by ordered pair $<a, b>$, where $a, b, a + b \in [0, 1]$. The numbers a and b are interpreted as the degrees of validity and non-validity of p. We denote the truth-value of p by $V(p)$.

The variable with truth-value *true* (in the classical logic) we denote by $\underline{1}$ and the variable *false* by $\underline{0}$. For this variables holds also $V(\underline{1}) = <1, 0>$ and $V(\underline{0}) = <0, 1>$. Important in the IFL is an intuitionistic fuzzy implication. It is, following [11], the logical connective → fulfilled for any variable p, p_1, p_2, q, q_1, q_2 properties:

(1) if $V(p_1) \preccurlyeq V(p_2)$ then $V(p_1 \rightarrow q) \succcurlyeq V(p_2 \rightarrow q)$,
(2) if $V(q_1) \preccurlyeq V(q_2)$ then $V(p \rightarrow q_1) \preccurlyeq V(p \rightarrow q_2)$,
(3) $V(\underline{0} \rightarrow q) = V(\underline{1})$,
(4) $V(p \rightarrow \underline{1}) = V(\underline{1})$,
(5) $V(\underline{1} \rightarrow \underline{0}) = V(\underline{0})$.

where $\preccurlyeq$ denotes the (typical) order relation on the set of IF values. Namely, for the variable p and q with $V(p) = <a, b>$ and $V(q) = <c, d>$ it holds $V(p) \preccurlyeq V(q)$ if and only if $a \leq c$ and $b \geq d$.

In the literature on the subject, about 150 different IF implication were noticed [5, 6]. Two of the classical are the Kleene-Dienes and the Łukasiewicz implications. They are given by formulas

$$V(p \rightarrow_{KD} q) = <\max\{b, c\}, \min\{a, d\}>,$$
$$V(p \rightarrow_{L} q) = <\min\{1, b + c\}, \max\{0, a + d - 1\}>$$

where p and q are the logical variable with $V(p) = <a, b>$, $V(q) = <c, d>$ and a, b, c, d, $a + b$, $c + d \in [0,1]$.

Each of IF implication can be used as a tool for the aggregation of the validity of criteria / attributes and the assessments of variants according to these criteria.

Let us suppose that each variant x_i, $i = 1, \ldots n$, from a finite set of variants (alternatives) are assigned according to the criteria $C_1, C_2, \ldots C_k$. We assume also that the decision maker gives the *intuitionistic decision matrix* $\boldsymbol{R_{IFS}} = [r_{ij}]_{n \times k} = [<\mu_{ij}, \nu_{ij}>]_{n \times k}$. The values μ_{ij}, ν_{ij} are interpreted as the degrees of validity and non-validity of the judgment *the variant* x_i *satisfies the criterion* C_j.

Let us assume that the criteria are not equally important and each of them is assigned to one of the linguistic assessments, as given in the first column of the Table 1. These assessments must be made by a supervisor, or, in the case of many experts, must be some aggregation of their opinions. Let the linguistic assessment of C_j correspond to the intuitionistic fuzzy value $\text{IFV}_j = <c_j, d_j>$ as in Table 1.

Based on the application of the implication to the assessments $<\mu_{ij}, \nu_{ij}>$, using IFV_j from the Table 1, we will evaluate the *adjusted degrees of validity* of each variants.

The *adjusted degrees of validity* we compute as the truth-value of the expression *if the criterion* C_j *is valid then it is by* x_i *satisfied.*

Therefore we compute the IF value $V(<c_j, d_j> \rightarrow <\mu_{ij}, \nu_{ij}>)$.

The type of IF implication → affects of course the truth-value obtained after its application.

The intuitionistic assessment of the variants according to various criteria, adjusted by degrees of criteria validity, form the matrix $\boldsymbol{T_{IFS}} = [<e_{ij}, f_{ij}>]_{n \times k}$.

Table 1 Linguistic assessments and their intuitionistic counterparts

Linguistic assessments of the criterion C_j	IFV_j
Strongly important	<1.0, 0.0>
Important	<0.8, 0.0>
Rather important	<0.6, 0.0>
Insignificant	<0.0, 0.5>
Almost totally unimportant	<0.0, 0.9>
I do not know, Type 1; I have no opinion, I can not regard this criterion as valid or invalid	<0.0, 0.0>
I do not know, Type 2; some prerequisites suggest that the criterion is important and some prerequisites, on the contrary	<0.5, 0.5>

Source [12]

The *i*th row of the $\boldsymbol{T_{IFS}}$ matrix is the vector of adjusted evaluations of *i*th variant according to all criteria, while the *j*th column is the vector of evaluations of all variants according to *j*th criteron.

Based on elements of $\boldsymbol{T_{IFS}}$ we create, in the usual manner, the IFSs, that are the ideal and anti-ideal solution. Namely

$$\begin{aligned} IS &= \{<C_1, e_{IS1}, f_{IS1}>, <C_2, e_{IS2}, f_{IS2}>, \ldots, <C_k, e_{ISk}, f_{ISk}>\} \\ &= \{<C_1, \max_{i=1\ldots n} e_{i1}, \min_{i=1\ldots n} f_{i1}>, <C_2, \max_{i=1\ldots n} e_{i2}, \min_{i=1\ldots n} f_{i2}>, \ldots, \\ &\quad <C_k, \max_{i=1\ldots n} e_{ik}, \min_{i=1\ldots n} f_{ik}>\}, \\ AIS &= \{<C_1, e_{AIS1}, f_{AIS1}>, <C_2, e_{AIS2}, f_{AIS2}>, \ldots, <C_k, e_{AISk}, f_{AISk}>\} \\ &= \{<C_1, \min_{i=1\ldots n} e_{i1}, \max_{i=1\ldots n} f_{i1}>, <C_2, \min_{i=1\ldots n} e_{i2}, \max_{i=1\ldots n} f_{i2}>, \ldots, \\ &\quad <C_k, \min_{i=1\ldots n} e_{ik}, \max_{i=1\ldots n} f_{ik}>\} \end{aligned}$$

Using the normalized Hamming distance l^1_{IFS}, or the normalized Euclidean distance q^1_{IFS}, for the *i*-th variant, we compute the distance from the ideal solution d_i^+ and the distance from the anti-ideal solution d_i^-. In the last step we obtain the *index of relative closeness* to the ideal solution as given in the classical TOPSIS method

$$c_i^+ = \frac{d_i^-}{d_i^+ + d_i^-} \in [0,1]$$

and also, as a consequence, the ranking of variants, and the best variant with the highest index value.

The new step in this method, mentioned first in [12], is the use of an IF implication for the transformation of input IF values. The use of an IF implication is, in author's opinion, valuable, although it may still be under discussion. An IF implication has the properties (1)–(5). In a simplified wording, the property (1) means that the same input value after adjustment by the criterion with the lower validity should be assessed at least as much as after adjustment by the criterion with the higher validity. We can say that if the criterion is important then the assessment of variants, due to this criterion, should be taken into account, but if the criterion is not important (low degree of importance), then the assessment of variants should not be relevant to the overall assessment of the variant, or, in other words, the *adjusted degree* should be, in this case, considered as high. The property (2) means that the higher input value after adjustment by certain criterion should be assessed at least as much as the lower input value. The property (3) means that, if the criterion had a value of zero, regardless of degree of its fulfillment by the variant, the adjusted value is as high as possible. In practice, the criterion validity of zero is not taken into account. The property (4) means that, regardless of the criterion validity, if the variant fulfills it fully, the revised degree is as high as possible. The property (5) means that, with the highest criterion validity and the lowest degree of its fulfillment, the adjusted degree is as low as possible.

Taking into account that most of the implications are continuous mappings, we can roughly formulate the properties (3)–(5) as follows:

- the low value of importance of criterion implies a practical indifference of adjusted values regardless of the fulfillment of this criterion,
- the very high evaluation of the variant due to certain criterion remains, after adjustment, very high regardless of the assessment of the criterion validity,
- if the criterion is very important and the evaluation of the variant due to them is very low then this very low assessment remains after the adjustment, too.

4 The Numerical Example

Suppose that three alternatives were evaluated according to each of the four criteria by intuitionistic assessment, and these criteria were considered to be either strongly important or rather important or insignificant or not known type 2, respectively (Table 2).

Calculated adjusted degrees are contained in Tables 3 and 4.

The ideal and anti-ideal solution in the both cases are:

$$IS_{K-D} = \{ <C_1, 1.0, 0.0>, <C_2, 0,7,0,1>, <C_3, 0.8, 0.0>, <C_4, 0,7,0,1> \}$$

$$AIS_{K-D} = \{ <C_1, 0.6, 0.4>, <C_2, 0,3,0,6>, <C_3, 0.5, 0.0>, <C_4, 0,5,0,2> \}$$

$$IS_L = \{ <C_1, 1.0, 0.0>, <C_2, 0,7,0,0>, <C_3, 1.0, 0.0>, <C_4, 1,0,0,0> \}$$

$$AIS_L = \{ <C_1, 0.6, 0.4>, <C_2, 0,3,0,3>, <C_3, 0.8, 0.0>, <C_4, 0,7,0,0> \}$$

Table 2 The values of degrees of the criteria met by the variants

Criteria	C_1	C_2	C_3	C_4
Alternatives	Strongly important	Rather important	Insignificant	I do not know, Type 2
x_1	<1.0, 0.0>	<0.7, 0.1>	<0.3, 0.0>	<0.7, 0.2>
x_2	<0.8, 0.0>	<0.3, 0.7>	<0.4, 0.3>	<0.4, 0.1>
x_3	<0.6, 0.4>	<0.6, 0.3>	<0.8, 0.0>	<0.2, 0.1>

Table 3 The assessments, adjusted by using the Kleene-Dienes implication

Criteria	C_1	C_2	C_3	C_4
Alternatives	Strongly important	Rather important	Insignificant	I do not know, Type 2
x_1	<1.0, 0.0>	<0.7, 0.1>	<0.5, 0.0>	<0.7, 0.2>
x_2	<0.8, 0.0>	<0.3, 0.6>	<0.5, 0.0>	<0.5, 0.1>
x_3	<0.6, 0.4>	<0.6, 0.3>	<0.8, 0.0>	<0.5, 0.1>

Table 4 The assessments, adjusted by using the Łukasiewicz implication

Criteria	C_1	C_2	C_3	C_4
Alternatives	Strongly important	Rather important	Insignificant	I do not know, Type 2
x_1	<1.0, 0.0>	<0.7, 0.0>	<0.8, 0.0>	<1.0, 0.0>
x_2	<0.8, 0.0>	<0.3, 0.3>	<0.9, 0.0>	<0.9, 0.0>
x_3	<0.6, 0.4>	<0.6, 0.0>	<1.0, 0.0>	<0.7, 0.0>

Table 5 The Hamming and the Euclidean distances from ideal and anti-ideal solutions

Alternatives	The distance l_{IFS}^1 of alternatives from				The distance q_{IFS}^1 of alternatives from			
	$IS_{K\text{-}D}$	$AIS_{K\text{-}D}$	IS_L	AIS_L	$IS_{K\text{-}D}$	$AIS_{K\text{-}D}$	IS_L	AIS_L
x_1	0.100	0.275	0.050	0.275	0.158	0.320	0.100	0.308
x_2	0.300	0.125	0.200	0.175	0.308	0.180	0.218	0.206
x_3	0.200	0.175	0.200	0.125	0.240	0.218	0.255	0.180

The results presented in Tables 3 and 4 are the basis for calculating of the distances of assessments of alternatives from the ideal and anti-ideal solutions. They are given in the Table 5.

The indexes of relative closeness in both cases are given in the Table 6.

Finally, we obtain the rank (from best to worst) of alternatives. In the case of Kleene-Dienes implication it is x_1, x_3, x_2, and in the case of Łukasiewicz implication x_1, x_2, x_3.

Table 6 The indexes of relative closeness for different distances and implications

Alternatives	The Hamming distance l_{IFS}^1		The Euclidean distance q_{IFS}^1	
	Kleene-Dienes implication	Łukasiewicz implication	Kleene-Dienes implication	Łukasiewicz implication
x_1	0.266	0.154	0.331	0.245
x_2	0.706	0.533	0.631	0.514
x_3	0.533	0.615	0.524	0.586

The numerical example shows, incidentally, that the choice of implications affect the result. Unfortunately, such disadvantage have also other methods of decision aid or inference using the fuzzy or intuitionistic fuzzy implications.

5 Conclusion

Use of IFSs and the IF implications for data processing takes advantage of even partial information about the degrees of fulfilling (and not-fulfilling) criteria by the different variants. In the paper a review of TOPSIS-class methods in the intuitionistic fuzzy environment is given. The new solution of some step in the TOPSIS-class method is proposed. The use of the intuitionistic fuzzy implication are given. The possibility of the applications of the new procedure in the multi-criteria decision making problems with varying degrees of criteria importance is, in the numerical example, presented.

References

1. Agarwal, M., Hanmandlu, M., Biswas, K.K.: Choquet integral versus TOPSIS: an intuitionistic fuzzy approach. In: IEEE International Conference on Fuzzy Systems, pp. 1–8. Hyderabad, India (2013)
2. Atanassov, K.T.: Intuitionistic fuzzy sets. In: VII ITKR's Science Session, Sofia (Deposed in Central Science and Technical Library, Bulgarian Academy of Sciences, Hπ 1697/84, in Bulgarian) (1983)
3. Atanassov, K.T.: Intuitionistic fuzzy sets. Fuzzy Sets Syst. **20**, 87–96 (1986)
4. Atanassov, K.T.: Intuitionistic Fuzzy Sets: Theory and Applications. Springer, Heidelberg (1999)
5. Atanassov, K.T.: On Intuitionistic Fuzzy Sets Theory. Springer, Heidelberg (2012)
6. Atanassov, K., Szmidt, E., Kacprzyk, J.: On intuitionistic fuzzy pairs. Notes Intuit. Fuzzy Sets **19**(3), 1–13 (2013)
7. Bo, Y.: Construction safety evaluation based on intuitionistic fuzzy TOPSIS. CECNet **2011**, 1136–1139 (2011)
8. Boran, F.E., Genç, S., Akay, D.: Personnel selection based on intuitionistic fuzzy sets. Hum. Fact. Ergon. Manuf. Serv. Ind. **21**(5), 493–503 (2011)

9. Boran, F.E., Genç, S., Kurt, M., Akay, D.: A multi-criteria intuitionistic fuzzy group decision making for supplier selection with TOPSIS method. Expert Syst. Appl. **36**(8), 11363–11368 (2009)
10. Bustince, H.: Handling multicriteria fuzzy decision-making problems based on intuitionistic fuzzy sets. Notes Intuit. Fuzzy Sets **1**(1), 42–47 (1995)
11. Czogała, E., Łęski, J.: On equivalence of approximate reasoning results using different interpretations of fuzzy if-then rules. Fuzzy Sets Syst. **117**, 279–296 (2001)
12. Dworniczak, P.: Basic properties of some new class of parametric intuitionistic fuzzy implications. Control Cybern. **40**(3), 793–804 (2011)
13. Gerogiannis, V.C., Fitsilis, P., Kameas, A.D.: Using a combined intuitionistic fuzzy set—TOPSIS method for evaluating project and portfolio management information systems. In: Iliadis, L., Maglogiannis, I., Papadopoulos, H. (eds.) Artifical Inteligence Applications and Innovations, Proceedings of the International Conferences EANN 2011 and AIAI 2011, Part II, pp. 67–81 (2011)
14. Guo, Z.-X., Qi, M., Zhao, X.: A new approach based on intuitionistic fuzzy set for selection of suppliers. In: Proceedings of the ICNC 2010, vol. 7, pp. 3715–3718 (2010)
15. Guo, Z.-X., Zhang, Q.: A new Approach to project risk evaluation based on intuitionistic fuzzy sets. Proceedings of the FSKD 2009, vol. 6, pp. 58–61 (2009)
16. Hung, C.-C., Chen, L.-H.: A fuzzy TOPSIS decision making model with entropy weight under intuitionistic fuzzy environment. In: Proceedings of the IMECS 2009, vol. 1, pp. 13–16 (2009)
17. Hwang, C.-L., Yoon, K.: Multiple Attribute Decision Making—Methods and Applications, A State-of-the-Art Survey. Springer, New York (1981)
18. Kacprzyk, J.: Multistage Fuzzy Control: A Model-Based Approach to Fuzzy Control and Decision Making. Wiley, Chichester (1997)
19. Li, J., Zhang, C.-Y.: a new solution of intuitionistic fuzzy multiple attribute decision-making based on attributes preference. In: Proceedings of the FSKD 2009, vol. 3, pp. 228–232 (2009)
20. Liu, P.-D.: Research on risk evaluation for venture capital based on intuitionistic fuzzy set and TOPSIS. In: Proceedings of the First International Symposium on Data, Privacy, and E-Commerce ISDPE 2007, pp. 415–417 (2007)
21. Luo, Y., Yu, C.: A fuzzy optimization method for multi-criteria decision-making problem based on the inclusion degrees of intuitionistic fuzzy sets. J. Inf. Comput. Sci. **3**(2), 146–152 (2008)
22. Maldonado-Macías, A., Alvarado, A., García, J.L., Balderrama, C.O.: Intuitionistic fuzzy TOPSIS for ergonomic compatibility evaluation of advanced manufacturing technology. Int. J. Adv. Manuf. Technol. **70**(9–12), 2283–2292 (2013)
23. Szmidt, E.: Applications of intuitionistic fuzzy sets in decision making. Working Paper WP-7-2000, Systems Research Institute, Polish Academy of Sciences, Warszawa (2000)
24. Szmidt, E., Kacprzyk, J.: Distances between intuitionistic fuzzy sets. Fuzzy Sets Syst. **114**(3), 505–518 (2000)
25. Szmidt, E., Kacprzyk, J.: a model of case based reasoning using intuitionistic fuzzy sets In: Proceedings of the IEEE International Conference on Fuzzy Systems, pp. 1769–1776 (2006)
26. Szmidt, E., Kacprzyk, J.: Atanassov's intuitionistic fuzzy sets as a promising tool for extended fuzzy decision making models. In: Bustince, H., Herrera, F., Montero, J. (eds.) Fuzzy Sets and Their Extensions: Representation, Aggregation and Models, pp. 335–355. Springer, Berlin (2008)
27. Szmidt, E., Kacprzyk, J.: Dilemmas with distances between intuitionistic fuzzy sets: straightforward approaches may not work. In: Chountas, P., Petrounias, I., Kacprzyk, J. (eds.) Intelligent Techniques and Tools for Novel System Architectures, SCI, vol. 109, pp. 415–430. Springer, Berlin (2008)
28. Wang, H.-J., Wei, G.-W.: An effective supplier selection method with intuitionistic fuzzy information. In: International Conference on Wireless Communications, Networking and Mobile Computing (2008)
29. Wen, L., Zhang, X.-J.: Intuitionistic fuzzy group decision making for supply chain information. In: Collaboration Partner Selection, ISME 2010, vol. 2, pp. 43–46 (2010)

30. Xu, Z.-S.: Intuitionistic fuzzy aggregation operators. IEEE Trans. Fuzzy Syst. **15**(6), 1179–1187 (2007)
31. Xu, Z.-S.: Some similarity measures of intuitionistic fuzzy sets and their applications to multiple attribute decision making. Fuzzy Optim. Decis. Making **6**(2), 109–121 (2007)
32. Xu, Z.-S., Yager, R.R.: Some geometric aggregation operators based on intuitionistic fuzzy sets. Int. J. Gen. Syst. **35**(4), 417–433 (2006)
33. Zhang, Q., Huang, Y.: Intuitionistic fuzzy decision method for supplier selection in information technology service outsourcing. In: Lei, J., Wang, F.L., Deng, H., Miao, D. (eds.) Proceedings of the International Conference, AICI 2012, pp. 432–439 (2012)

Intuitionistic Fuzzy Dependency Framework

Boyan Kolev and Ivaylo Ivanov

Abstract This paper proposes an Intuitionistic Fuzzy Dependency Framework (IFDF) model as a flexible tool for analyzing the cause and effect of events occurring in systems, where causation dependencies might be partial or vague. A core data model with basic operations is introduced. A static approach for dependency analyses is presented using traversals of the dependency graph. A dynamic approach using generalized nets as a simulation tool is also presented for the case of systems with temporal dependencies.

Keywords Intuitionistic fuzzy · Generalized nets · Dependency graph · Causal analysis · Impact analysis

1 Introduction

In a system, where dependencies exist between components, the problem of analyzing the cause of an event or the impact it might have on other components is of major importance. In order to make it possible to perform causation analyses, all dependencies between the possible events that can occur in the system must be preliminarily identified and kept in a directed graph, where each node represents a particular event and each arc represents a dependency of the output node from the input node. Having built the dependency graph, once an event occurs, there exist methods to:

- Determine the effects of this event on causing other events, usually by performing a (breadth first) search in the dependency graph starting from the node, corresponding to the occurred event;

B. Kolev (✉)
Bulgarian Academy of Sciences, Sofia, Bulgaria
e-mail: bkolev@gmail.com

I. Ivanov
SoftConsultGroup Ltd, Sofia, Bulgaria
e-mail: ivaylo.ivanov@softconsultgroup.com

P. Angelov and S. Sotirov (eds.), *Imprecision and Uncertainty in Information Representation and Processing*, Studies in Fuzziness and Soft Computing 332,
DOI 10.1007/978-3-319-26302-1_12

- Determine the possible root causes of this event, usually by performing a (breadth first) search in the dependency graph in reverse direction starting from the node, corresponding to the occurred event.

The problem of performing analytical tasks in a dependency topology has been studied in the context of many applications—software build systems, financial data analytics, problem management in IT infrastructure, medical diagnosing, semantic networks, etc.

In this paper we will introduce the notion of an Intuitionistic Fuzzy Dependency Framework (IFDF), based on the assumption that dependencies between events might be partial or vague. We use intuitionistic fuzzy logic to express dependencies, considering that the proposition "B depends on A" is no longer Boolean, but has degrees of truth and falsity according to the theory of intuitionistic fuzzy sets, proposed by Atanassov in [1] as an extension to the classical fuzzy sets theory. An intuitionistic fuzzy dependency between A and B means that:

- The occurrence of A will cause partial occurrence of B to some extent, expressed by the degree of truth, and
- B has a level of resistance against impacts of A, expressed by the degree of falsity.

In Sect. 2 we will summarize the research that has been made in the field of intuitionistic fuzzy extensions to ITIL's configuration management database. In the following sections we will generalize the proposed concept to fit in a wider range of applications.

In Sect. 3 we will describe the data structures that comprise the IFDF, a semantic view of the information they hold and basic operations that can be performed.

In Sect. 4 we will discuss intuitionistic fuzzy dependencies and will introduce methods for calculating indirect dependencies to estimate the indirect impact that the occurrence of an event might have over distant nodes. We will also introduce methods to investigate possible causes for the occurrence of the event by traversing the dependency graph in reverse direction.

In Sect. 5 we will extend our framework by adding temporal components, thus allowing the expected impact of an event to be estimated as a function of time. In such a framework, dependencies between nodes A and B have an additional attribute, representing the amount of time, after which an occurrence of A will cause occurrence of B. We will propose a methodology for performing dynamic causation analyses using the simulation capabilities of intuitionistic fuzzy generalized nets [2, 3].

2 Related Work

Several studies have been made in the area of performing analyses over intuitionistic fuzzy dependencies in the IT Service Management domain. In [5] the authors propose an Intuitionistic Fuzzy Configuration Management Database

(IFCMDB) model as an extension of the Configuration Management Database (CMDB), part of the IT Infrastructure Library (ITIL) standard [4]. In the terminology of Configuration Management, IT components and the services provided with them are known as Configuration Items (CIs), which can include hardware, software, active and passive network components, servers, documentation, services and all other IT components and the network can be extended to even include IT users, IT staff and business units. The authors of [5] introduce a new type of "dependency relationship" between the CIs that reflects the partial impact that one CI has on another, expressed by means of intuitionistic fuzzy values.

In [5] a methodology for impact analysis is proposed based on calculations of the levels of impact that a failure in one CI can have on other CIs that are indirectly dependent from the failed one.

In [6] a methodology for analyzing the possible candidates for the root cause of a failure is introduced based on traversing the reverse IFCMDB graph and calculating the indirect impacts backwards, starting from the failed node.

In [7] the author maps Service Level Agreement (SLA) concepts to the idea behind component dependencies in an IFCMDB, which makes it possible to provide an adequate service level management of systems with complex multi tier architecture.

The research made so far relies on static analyses of an intuitionistic fuzzy dependency network in a particular domain. However, our current research is motivated by the fact that an intuitionistic fuzzy dependency framework is applicable to a much wider range of domains, therefore needs a generalization. Moreover, the framework can be supplemented by adding temporality to the model, thus making it a robust tool for simulating activities, workloads and problem propagations among components in complex coherent systems.

3 Intuitionistic Fuzzy Dependency Model

Let E be the set of possible events that can happen within the modeled system. An intuitionistic fuzzy dependency D is an intuitionistic fuzzy binary relation over E. The core data structure in the IFDF is an intuitionistic fuzzy directed graph, where each node corresponds to a possible event and an intuitionistic fuzzy arc from node a to node b expresses the dependency of b from a:

$$G = (E, D), \text{ where:}$$
$$D = \{ > (a, b), (\mu_D(a, b), \nu_D(a, b)) < | a \in E, b \in E\}$$

and the functions $\mu_D(a, b)$ and $\nu_D(a, b)$ define the degrees of truth and falsity of the existence of a dependency between the events a and b, meaning "the occurrence of a will cause occurrence of b".

3.1 Indirect Dependencies and Dependency Semantics

The arcs in the core dependency graph represent direct dependencies, i.e. the direct effect that the occurrence of one event has on causing another event. For the purpose of causality analyses indirect dependencies must also be obtained, which expresses the impact that the occurrence of an event will have on distant nodes. The existence of an indirect dependency is determined by the existence of a path in the graph. Indirect dependencies are calculated considering the direct ones and applying a chain of intuitionistic fuzzy logical conjunction and disjunction operations.

An intuitionistic fuzzy dependency from a to b may have different semantics depending on the type of uncertainty it represents, examples of which are listed below:

- The probability of occurrence of b in case of occurrence of a;
- The level of impact that an occurrence of a has on causing occurrence of b;
- Vague information about the dependency.

3.2 Basic Operations

Depending on the semantics behind the intuitionistic fuzziness of the dependencies, an arbitrary combination of variants of existing conjunction and disjunction operations may be used in the computation of indirect dependencies. Thus, the administrator of the system is given the possibility to select the most appropriate calculation method, according to its relevance with the modeled system. Below are listed two possible variants of intuitionistic fuzzy conjunction and disjunction operations (assuming that $a = <\mu_a, \nu_a>$ and $b = <\mu_b, \nu_b>$ are intuitionistic fuzzy propositions with $\mu_a$and ν_a as the degrees of truth and falsity of a and $\mu_b$and ν_b as the degrees of truth and falsity of b respectively):

- Conjunctions:

– $$a\&b = <\min(\mu_a, \mu_b), \max(\nu_a, \nu_b)> \quad (1.1)$$

– $$a.b = <\mu_a\mu_b, \nu_a + \nu_b - \nu_a\nu_b> \quad (1.2)$$

- Disjunctions:

– $$a \vee b = <\max(\mu_a, \mu_b), \min(\nu_a, \nu_b)> \quad (1.3)$$

– $$a + b = <\mu_a + \mu_b - \mu_a\mu_b, \nu_a\nu_b> \quad (1.4)$$

The type of logical operations that can be used within the system is not limited to this list. The implementer can choose the types of operations that are most appropriate for each particular use case. In the rest of the paper we will refer to conjunction and disjunction operations in general, disregarding the chosen variant, and we will denote them respectively with the symbols & and v.

3.3 *Adding Temporality to Dependencies*

The intuitionistic fuzzy dependency model can be extended by adding a temporal component in the dependency measure. Thus, the dependency relation has this form:

$$D = \{ < (a, b), (\mu_D(a, b), \nu_D(a, b)), t > | a \in E, b \in E, t \in R^+ \}$$

In this notation t denotes the amount of time, after which the occurrence of a will cause occurrence of b. Temporal dependency models are subjects to dynamic causation analyses unlike non-temporal models, where the impact as a result of indirect dependency can be computed by traversing the intuitionistic fuzzy dependency graph. Both static and dynamic analytical approaches will be covered in the following two sections.

4 Static Analytical Approach

Static analytical approach is applicable to non-temporal intuitionistic fuzzy dependency models. The objectives of the causation analysis are:

- Forward analysis (impact analysis): Determine the effects of the occurrence of an event on causing other events;
- Backward analysis (root cause analysis): Determine the possible root causes for an occurred event.

Both analytical tasks take into account the direct dependencies and perform computations by traversing the dependency graph, to discover the indirect dependencies in which the starting node is involved.

Let ddep(a, b) denotes the direct dependency of event b from event a:

$$ddep(a, b) = \begin{cases} < \mu_D(a, b), \nu_D(a, b) >, & |(a, b) \in D \\ < 0, 1 >, & |(a, b) \notin D \end{cases} \quad (2)$$

Let idep(c, d) denotes the indirect dependency of event d from event c.

4.1 Forward Analysis

The static forward analysis aims to detect the possible effects from an event by traversing the graph using a breadth-first search algorithm. At each visited node, the algorithm computes the node's dependency from the starting node, taking into account the already computed dependencies at the previous level, using the formula:

$$idep(a,b) = \begin{cases} \bigvee\limits_{(i,b)\in D} idep(a,i)\&ddep(i,b), & |a \neq b \\ <1,0>, & |a = b \end{cases} \tag{3}$$

Finally, each traversed node has a computed dependency from the starting one.

4.2 Backward Analysis

The purpose of the backward analysis is to discover the set of possible root causes for an occurred event. The breadth-first search algorithm now traverses the graph in reverse direction in order to find which events might have caused the occurred one. At each visited node, the algorithm computes the dependency of the starting node from the visited one, taking into account the already computed dependencies at the previous level, using the following formula, analogous to the one for forward analysis:

$$idep(a,b) = \begin{cases} \bigvee\limits_{(a,i)\in} idep(i,b)\&(a,i), & |a \neq b \\ <1,0>, & |a = b \end{cases} \tag{4}$$

4.3 Algebraic Approach

The purpose of the algebraic analysis is to calculate the dependency between two nodes regardless of the context in which the analysis is performed—to identify the effects or to analyze the root cause. To calculate the indirect dependency of b from a, the following procedure is proposed:

1. The graph is traversed in order to find all paths from a to b.
2. For each path, we calculate the partial indirect dependency by applying a chain of logical conjunction operations over all the direct dependencies between nodes along the path.
3. Finally, the indirect impact is the result of applying a logical disjunction operation over all partial impacts, calculated on the previous step.

Another way to perform algebraic analysis on the graph G is to find the transitive closure of D, using the Warshall algorithm proposed in [12]. Then, the indirect dependency of b from a is the membership of the pair (a, b) to the transitive closure of D.

In general, a single indirect dependency may have different values when computed using the three different approaches: forward, backward and algebraic. However, the three methods will produce the same result if the used logical operations have the following properties: the conjunction is distributive over the disjunction and the conjunction and disjunction are idempotent; for example the combination of conjunction (1.1) and disjunction (1.3).

5 Dynamic Analytical Approach

In real life scenarios very often the effects of events are propagated with delays or the dependency of one event from another is a function of time. Dynamic analytical approach is applicable to both temporal and non-temporal intuitionistic fuzzy dependency models. It gives more flexibility to the analytical tasks, by allowing the propagation of the effects of an occurred event to be simulated, taking into account propagation delays, associated to each node. Also, this approach allows more than one event to be involved in the simulation at a time.

In the proposed IFDF we use intuitionistic fuzzy generalized nets [2, 3] as the simulation tool. Whenever an analytical task is assigned to the framework, it performs the following steps:

1. The framework engine automatically transforms the intuitionistic fuzzy dependency graph into an intuitionistic fuzzy generalized net, using the algorithm, defined in [8, 9]. For example, let us consider the following sample dependency graph (the clock symbols denote delayed dependencies):

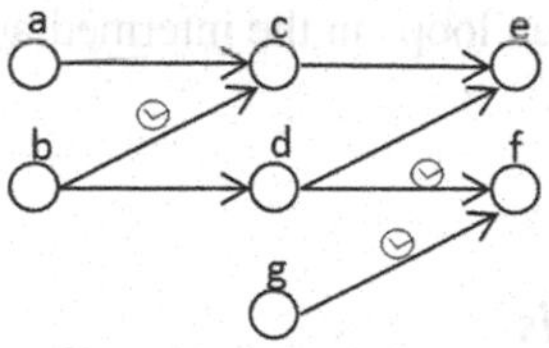

The IFDF engine transforms the graph to the following generalized net:

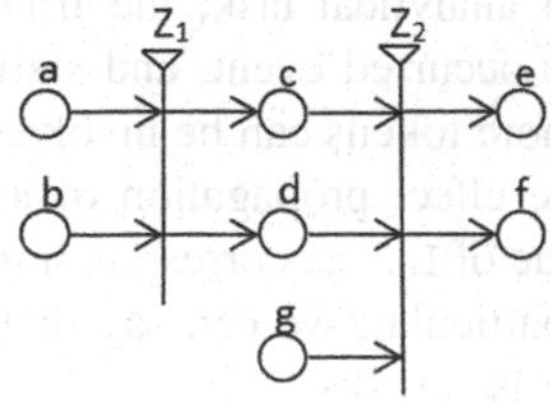

2. Then, IFDF assigns the intuitionistic fuzzy dependencies to the evaluation functions of the corresponding predicates. Each transition is given a type to be activated whenever there is a token in all of the transition's input places.
3. Each token has a characteristic L_{curr} that carries an intuitionistic fuzzy level of impact (L_{curr} consists of two components—the degrees of truth and falsity). Every time a token passes through a transition and enters a place, the following actions are taken:
   a. Before leaving the old place, the token splits, so that it can move to all output places, for which the corresponding predicate function is defined.
   b. Upon entering the new place, the token's L_{curr} obtains a new value, which is the result of applying a logical conjunction operation between L_{curr} and the value of the intuitionistic fuzzy predicate.
   c. If the new place already contains another token or if more than one token enters the same place during the same activation of the transition, all the tokens are merged in one token, whose L_{curr} obtains a value, which is the result of applying a logical disjunction operation over the values of L_{curr} of all merged tokens.
4. Each token has another characteristic T_{curr}, which is the representation of the current moment in time, relative to the time of initialization of the net. T_{curr} is increased at each transition. Each place, which is the right-hand side of a delayed dependency (nodes c and f in our example), is substituted with a single-transition generalized subnet with the following structure:

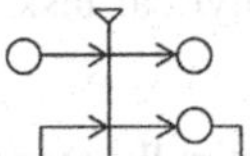

Each token, entering the input place of this transition, is given a numeric characteristic, according to the delay of the corresponding dependency. The transition's predicates and the characteristic function of the intermediate place are adjusted in a way that a token loops in the intermediate place for an amount of time, corresponding to the delay.

5.1 Forward Analysis

When executing a forward analytical task, the IFDF engine puts a token in the place, corresponding to the occurred event, and starts the GN simulation. If more events are observed, then more tokens can be involved in the simulation, so that the simulation can discover the effect propagation of a combination of events. Each token is given an initial value of L_{curr}, corresponding to the level of the occurrence of the observed event. In particular, we can say that if an event definitely occurs, then its level of occurrence is <1, 0>.

Since the calculation of the level of occurrence of a particular event in the chain depends on the level of occurrence of all its preceding events, the corresponding transition must wait until all its input places are populated. To make the task achievable, before starting the simulation, the framework puts tokens in all empty global input places with an initial value of L_{curr}, which in the case of absence of occurrence of the corresponding event is <0, 1>. In case the level of occurrence of a particular event is completely unknown, it is also possible to put a token in the corresponding place with initial L_{curr} = <0, 0>.

After entering each place:

- the token's L_{curr} contains the level of impact that the initially observed events has on it and
- the token's T_{curr} contains the amount of time after the occurrence of the initial event the effect on the corresponding event will be expected.

5.2 Backward Analysis

The backward analytical task is analogous to the forward one, with the following remark: The dependency graph is first reversed, and then transformed into an intuitionistic fuzzy generalized net.

Before starting the simulation, analogously, the IFDF puts tokens in the places, corresponding to the observed events, with initial L_{curr} according to the level of occurrence. The involvement of several initial tokens in the generalized net gives the possibility for the framework to find a common root cause for several observed events. Analogously, the input places, not related to any of the observed events, must be populated with tokens with initial L_{curr} denoting that the event is not observed or information about its state is vague or missing.

After entering each place:

- the token's L_{curr} shows how likely the corresponding event could be the root cause of the observed ones and
- the token's T_{curr} gives an approximation about the latest time of occurrence of the root cause.

6 Conclusions

The proposed framework is a robust tool for simulating and analyzing causality effects in systems, where the occurrences of events and causalities can be partial, probabilistic or vague.

The work was inspired by our previous research on intuitionistic fuzzy extensions of the configuration management database (IFCMDB), part of the ITIL standard [5, 6]. Having defined the IFDF framework, IFCMDB becomes just one of the use cases, applicable within IFDF. In the IFCMDB graph, each node corresponds to an item in the infrastructure and the arcs represent dependencies between items, meaning that the proper work of one item depends on the proper work of another. Mapping IFCMDB graph to the data model of IFDF can be done by retaining the same graph topology, but the nodes in IFDF will represent failure events of the corresponding items and the arcs represent causality of failures of one node to another. However, IFDF goes beyond the analytical capabilities of the current IFCMDB model, because it introduces temporality to the causality links, thus allowing analyzing the possible effects of a failure of an item in the IT infrastructure, considering delays in effect propagation, the way they exist in a real scenario. A typical example of the existence of such delayed causalities appears in a scenario, where hardware is protected from power failures by a UPS—in the event of failure in the power supply, the effect is propagated to the hardware with a certain delay, after the UPS battery is discharged.

More general, the IFDF concept can be applied to a much broader range of domains, where causality dependencies can be partial or vague and the effects of event occurrences are propagated throughout the system either immediately or within a timeframe. As subject to our further research, we will investigate the application of IFDF in several domains, e.g. business process modeling, intuitionistic fuzzy scheduling in business processes, social networks, etc.

References

1. Atanassov, K.: Intuitionistic Fuzzy Sets. Springer-Verlag, Heidelberg (1999)
2. Atanassov, K.: Generalized Nets. World Scientific, Singapore (1991)
3. Atanassov, K.: On Generalized Nets Theory. Prof. Marin Drinov Academic Publishing House, Sofia (2007)
4. Office of Government Commerce (OGC).: Service Delivery Book. The Stationery Office Publishing House, London (2005)
5. Ivanov, I., Kolev, B.: Intuitionistic fuzzy extensions to the information technology infrastructure library configuration management database. Notes Intuitionistic Fuzzy Sets **14** (4), 34–40 (2008)
6. Kolev, B., Ivanov, I.: Fault tree analysis in an intuitionistic fuzzy configuration management database. Notes Intuitionistic Fuzzy Sets **15**(2), 10–17 (2009)
7. Schuetze, R.: Intuitionistic fuzzy component failure impact analysis (IFCFIA)—A gradual method for SLA dependency mapping and bi-polar impact assessment. Notes Intuitionistic Fuzzy Sets **19**(3), 62–72 (2013)
8. Kolev B.: An algorithm for transforming a graph to a generalized net. In: Proceedings of the First International Workshop on Generalized Nets, pp. 26–28, Sofia, 6 July 2000
9. Kolev, B.: An algorithm for transforming an intuitionistic fuzzy graph to an intuitionistic fuzzy generalized net. Notes Intuitionistic Fuzzy Sets **7**(2), 61–62 (2001)
10. Popova B., Atanassov, K.: Opposite generalized nets. I. advances in modelling and analysis. AMSE Press **19**(2), 15–21 (1994)

11. Popova B., K. Atanassov. Opposite generalized nets. II. advances in modelling and analysis. AMSE Press **19**(2), 23–28 (1994)
12. Liu G.: Closures of intuitionistic fuzzy relations. In: Rough Sets and Knowledge Technology, Lecture Notes in Computer Science, vol. 5589, pp. 281–288 (2009)

Consistency and Consensus of Intuitionistic Fuzzy Preference Relations in Group Decision Making

Huchang Liao and Zeshui Xu

Abstract Intuitionistic fuzzy preference relations (IFPRs) have turned out to be a useful structure in expressing the experts' uncertain judgments. In this chapter, we consider a group decision making problem where all the members of the group use the IFPRs to express their preferences over the candidate alternatives. Firstly, we describe such a group decision making problem mathematically in details. Then, different types of definitions for the consistency of an IFPR are reviewed, which can be divided into two sorts, i.e., the additive consistency and the multiplicative consistency. Once all the IFPRs are of acceptable consistency, we then introduce a consensus measure to depict the consensus degree of the experts. A consensus reaching procedure is given to help the experts modify their assessments and then obtain an agreement between the experts as to the choice of a proper decision. A numerical example is given to show the validation and computational process of the consensus reaching procedure.

Keywords Intuitionistic fuzzy preference relation · Consistency · Consensus · Consensus reaching procedure · Group decision making

1 Introduction

Group decision making takes place commonly in many domains of our daily life, including such significant ones as the managerial, financial, engineering, and medical fields. It has gained prominence owing to the complexity of modern-life decision problems. For a group decision making problem, a group of experts are getting together to express their individual opinions over the problem and then yield

H. Liao · Z. Xu (✉)
Business School, Sichuan University, Chengdu 610064, China
e-mail: xuzeshui@263.net

H. Liao
e-mail: liaohuchang@163.com

P. Angelov and S. Sotirov (eds.), *Imprecision and Uncertainty in Information Representation and Processing*, Studies in Fuzziness and Soft Computing 332,
DOI 10.1007/978-3-319-26302-1_13

a final decision which is mutually agreeable. Very often, such group decision making problem involves multiple feasible alternatives, and the objective of the group decision making problem is to select the best alternative(s) from these mutually exclusive alternatives based on the preferences provided by the experts. In many cases, the experts can not determine their preferences in accurate numerical numbers but fuzzy terms [1]. Fuzzy set (FS) was proposed to represent the relationship between a set and an element by membership degrees rather than by crisp membership of classical binary logic. When all the preferences of the experts are determined by fuzzy numbers which denote the relative intensities between each pair of alternatives, a set of fuzzy preference relations can be established [2]. Let $X=\{x_1,x_2,\cdots,x_n\}$ be the set of alternatives under consideration, and $E=\{e_1,e_2,\ldots,e_s\}$ be the set of decision makers, who are invited to evaluate the alternatives. The fuzzy preference relations $B^{(l)}=(b_{ij}^{(l)})_{n\times n}$ $(l=1,2,\cdots s)$ can be generated, where $0\leq b_{ij}^{(l)}\leq 1$ and $b_{ij}^{(l)}+b_{ji}^{(l)}=1$. $b_{ij}^{(l)}$ indicates the degree that the alternative x_i is preferred to x_j. Concretely speaking, the case $b_{ij}^{(l)}=0.5$ indicates that there is indifference between the alternatives x_i and x_j; $b_{ij}^{(l)}>0.5$ indicates that the alternative x_i is preferred to x_j, especially, $b_{ij}^{(l)}=1$ means that the alternative x_i is absolutely preferred to x_j; $b_{ij}^{(l)}<0.5$ indicates that the alternative x_j is preferred to x_i, especially, $b_{ij}^{(l)}=0$ means that the alternative x_j is absolutely preferred to x_i.

Although fuzzy preference relations can be used to represent the fuzzy and uncertain preferences of the experts in the process of group decision making, they still have some flaws due to the limitation of the fuzzy set itself. Since the membership function of a fuzzy set is only single-valued function, it can't be used to express the support and objection evidences simultaneously in many practical situations [3]. If not possessing a precise or sufficient level of knowledge of the problem domain in cognition of things due to the complexity of the socio-economic environment, people usually have some uncertainty in assigning the preference evaluation values to the objects considered, which makes the judgments of cognitive performance exhibit the characteristics of affirmation, negation and hesitation. In 1983, Atanassov [4] proposed the concept of intuitionistic fuzzy set (IFS), which is characterized by a membership function, a non-membership function and a hesitancy function. Such type of fuzzy set extension is essential in representing the imprecision and hesitation of the experts' cognition [5]. Till now it has been applied to many different fields, such as decision making [3, 6], fuzzy logics [7], fuzzy cognitive maps [8], topological space [9], medical diagnosis [10] and pattern recognition [11]. Given the underlying set X of objects, an IFS $\tilde{A}$ is a set of ordered triples, $\tilde{A}=\{(x,\mu_A(x),v_A(x))|x\in X\}$, where μ_A and v_A are the membership and non-membership functions mapping from X into [0, 1] with the condition $0\leq\mu_A+v_A\leq 1$. For each $x\in X$, $\mu_A(x)$ represents the degree of membership of the element x in X to the set $A\subseteq X$, and $v_A(x)$ gives the non-membership degree. The number $\pi_A(x)=1-\mu_A(x)-v_A(x)$ is called the hesitant degree or the intuitionistic index of x to A. The FS do not leave any room for indeterminacy between each membership degree and its negation, but, in the realistic recognition of experts, such "disagreement" and indeterminacy are very common and useful in describing their opinions in decision making. The introduction of this ignorance statement, which is represented as the hesitancy function in an IFS, is the most characteristic of the IFS [12]. In many cases, when the experts are not able to express their preferences accurately or they are unable or unwilling to discriminate explicitly the degree to which alternative is better than others especially at the beginning of evaluation [13], it is suitable to express their preference information in IFS and thus we can get a set of intuitionistic fuzzy preference relations (IFPRs) [14].

As for group decision making with IFPRs, there are several problems raised, the first one of which is how to judge whether the IFPRs are consistent or not. Consistency of IFPRs requires that the preferences given by the experts yield no contradiction. The lack of consistency for IFPRs may lead to inconsistent or incorrect results for a group decision making problem. Thus it has turned out to be a very important research topic in decision making with IFPRs, and many scholars have paid attention to this topic [6, 12, 14–22]. In this chapter, we would give detail review for the different kinds of consistency of IFPRs. As for those IFPRs without consistency, how to repair them is also a problem which needs to be solved. Generally, this can be done by two different kinds of methodologies, which are the automatic methods and the interactive methods [21, 23].

In the next of this chapter, we would focus on another important issue, i.e., the consensus of group decision making with IFPRs. The consistency checking process of IFPRs can be seen as a collection of individual decision making problems and it is easy to be done by extending the methodology of single expert decision making problem. While the consensus of group decision making is much more complicated because of the complexity introduced by the conflicting views of experts and the varying significance of those views in the decision making process [24]. Sometimes, one expert may determine his/her preferences based on his/her perception, but the others may not agree with it unless they are confident about the perception of the former expert. The consensus is very important in group decision making. Although we can yield a decision for a group decision making problem by aggregating all individual IFPRs into an overall IFPR, the result derived by this type of methodologies may be not much reasonable because some experts may not agree with the final result derived by the weighted averaging methodologies. Consensus is viewed as a pathway to a true group decision because it considers concerns and conflicting ideas without hostility and fear [25]. Till now, there is litter research on the consensus of group decision making with IFPRs. In the following of this chapter, we would pay attention to this issue and give some basic definitions.

The rest of this chapter is organized as follows: Sect. 2 describes the group decision making problem mathematically within the context of intuitionistic fuzzy circumstance. Section 3 reviews the different types of consistency for IFPRs, including the additive consistency and multiplicative consistency for IFPRs. The definition of acceptable consistent IFPR is also given in this Section. In Sect. 4, we present the difficulties in reaching consensus in the process of group decision making with IFPRs. Furthermore, we introduce a hard consensus measure to depict the consensus degree of the experts. The consensus reaching procedure is given for helping the experts to reach group agreement. A numerical example is given to validate the procedure in Sect. 5. Section 6 ends the chapter with some concluding remarks.

2 Group Decision Making with Intuitionistic Fuzzy Preference Relations

A group decision making problem with intuitionistic fuzzy preference information can be described as follows: Let $X=\{x_1,x_2,\cdots,x_n\}$ be the set of alternatives under consideration, and $E=\{e_1,e_2,\ldots,e_s\}$ be a set of experts, who are invited to evaluate the alternatives and then provide their preferences through pairwise comparison. The weight vector of the experts $e_l(l=1,2,\ldots,s)$ is $\lambda=(\lambda_1,\lambda_2,\ \ldots,\lambda_s)^T$, where $\lambda_l>0,\ l=1,2,\ldots,s$, and $\sum_{l=1}^{s}\lambda_l=1$, which can be determined subjectively or objectively according to the experts' experience, judgment quality and related knowledge. In general, they can be assigned equal importance if there is no evidence to show significant differences among the decision makers or specific preference on some decision makers [22]. In the existing literature, many techniques have been developed for determining the decision makers' weights (for more information, refer to Refs. [26–28]). In this chapter, we assume that the weights of experts can always be given.

In many cases, if the problem is very complicated or the experts can not be able to give explicit preferences over alternatives because of vague information and incomplete knowledge about the preference degrees between any pair of alternatives, it is suitable to use the IFSs, which express the preference information from three aspects: "preferred", "not preferred", and "indeterminate", to represent their opinions. Motivated by the idea of IFS, Szmit and Kacprzyk [29] firstly proposed the concept of intuitionistic fuzzy preference relation (IFPR). Later, Xu [14] gave the simple and straightforward notion and expression for it.

Definition 1 [14] An intuitionistic fuzzy preference relation (IFPR) on the set $X=\{x_1,x_2,\ldots,x_n\}$ is represented by a matrix $\tilde{R}=\left(\tilde{r}_{ij}\right)_{n\times n}$, where $\tilde{r}_{ij}=<(x_i,x_j),\mu(x_i,x_j),v(x_i,x_j),\ \pi(x_i,x_j)>$ for all $i,j=1,2,\cdots,n$. For convenience, we let $\tilde{r}_{ij}=\left(\mu_{ij},v_{ij},\pi_{ij}\right)$ where μ_{ij} denotes the degree to which the object x_i is preferred to the object x_j, v_{ij} indicates the degree to which the object x_i is not preferred to the object x_j, and $\pi_{ij}=1-\mu_{ij}-v_{ij}$ is interpreted as an indeterminacy degree or a hesitancy degree, with the conditions:

$$\mu_{ij},\ v_{ij}\in[0,1],\mu_{ij}+v_{ij}\leq 1,\mu_{ij}=v_{ji},\mu_{ii}=v_{ii}=0.5,\pi_{ij}=1-\mu_{ij}-v_{ij},\text{ for all } i,j=1,2,\ldots,n \tag{1}$$

Xu [14] also proposed the concept of incomplete IFPR in which some of the preference values are unknown. There are some algorithms to estimate the missing values for the incomplete IFPR [30]. For convenience, in this paper we assume that the experts can provide complete IFPRs.

Suppose that the expert e_l provides his/her preference values for the alternative x_i against the alternative x_j as $\tilde{r}_{ij}^{(l)} = (\mu_{ij}^{(l)}, v_{ij}^{(l)}), (i,j=1,2,\ldots,n, l=1,2,\ldots,s)$ in which $\mu_{ij}^{(l)}$ denotes the degree to which the object x_i is preferred to the object x_j, $v_{ij}^{(l)}$ indicates the degree to which the object x_i is not preferred to the object x_j, and $\pi_{ij}^{(l)} = 1 - \mu_{ij}^{(l)} - v_{ij}^{(l)}$ is interpreted as an indeterminacy degree or a hesitancy degree, subject to $\mu_{ij}^{(l)}, v_{ij}^{(l)} \in [0,1], \mu_{ij}^{(l)} + v_{ij}^{(l)} \leq 1, \mu_{ij}^{(l)} = v_{ij}^{(l)}, \mu_{ii}^{(l)} = v_{ii}^{(l)} = 0.5$, for all $i,j=1,2,\ldots,n, l=1,2,\ldots,s$. The IFPR $\tilde{R}^{(l)} = \left(\tilde{r}_{ij}^{(l)}\right)_{n\times n}$ for the lth expert can be written as:

$$\tilde{R}^{(l)} = \begin{pmatrix} \tilde{r}_{11}^{(l)} & \tilde{r}_{12}^{(l)} & \cdots & \tilde{r}_{1n}^{(l)} \\ \tilde{r}_{21}^{(l)} & \tilde{r}_{22}^{(l)} & \cdots & \tilde{r}_{2n}^{(l)} \\ \vdots & \vdots & \ddots & \vdots \\ \tilde{r}_{n1}^{(l)} & \tilde{r}_{n2}^{(l)} & \cdots & \tilde{r}_{nn}^{(l)} \end{pmatrix} \tag{2}$$

For any a group decision making problem with s decision makers, we can obtain s individual IFPRs $\tilde{R}^{(l)} (l=1,2,\cdots,s)$ with the form of (2).

3 Consistency of Intuitionistic Fuzzy Preference Relations

Consistency is a very important issue for any kinds of preference relations, and the lack of consistency in preference relations may lead to unreasonable conclusions. There are several different forms of definition for the consistency of IFPRs, which mainly involve two sorts: the additive consistency and the multiplicative consistency.

3.1 Additive Consistency

The concept of additive consistency of an IFPR was motivated by the additive transitivity property proposed by Tanino [1] in 1984. It was proposed to represent the relationship among different preferences. A preference relation $R=(r_{ij})_{n\times n}$ is with additive transitivity if it satisfies $(r_{ij}-0.5)+(r_{jk}-0.5)=(r_{ik}-0.5)$ for all $i,j,k=1,2,\cdots,n$. This can be interpreted as the intensity of preference of the alternative x_i over x_k should be equal to the sum of the intensities of preference of x_i over x_j and that of x_j over x_k when $(r_{ij}-0.5)$ is defined as the intensity of preference of x_i over x_j. Let $\omega_i(i=1,2,\cdots,n)$ be the underlying weights of the alternatives and satisfies $\sum_{i=1}^{n}\omega_i=1, \omega_i\in[0,1]$. Then, an additive consistent fuzzy preference relation can be given as [15]: $r_{ij}=0.5(\omega_i+\omega_j-1)$, for all $i,j=1,2,\cdots,n$.

Based on the additive transitivity of a preference relation, different forms of definitions for additive consistency of IFPRs have been proposed.

Xu's additive consistency

For each IFS $\tilde{r}_{ij}=(\mu_{ij},\nu_{ij})$, the condition $\mu_{ij}\le 1-\nu_{ij}(i,j=1,2,\cdots,n)$ always holds. Thus, the IFPR $\tilde{R}=(\tilde{r}_{ij})_{n\times n}$ can be transformed into an interval-valued complementary judgment matrix $\hat{R}=(\hat{r}_{ij})_{n\times n}$ where $\hat{r}_{ij}=(\hat{r}_{ij}^-,\hat{r}_{ij}^+)=[\mu_{ij},1-\nu_{ij}](i,j=1,2,\ldots,n)$, and $\hat{r}_{ij}^-+\hat{r}_{ji}^+=\hat{r}_{ij}^++\hat{r}_{ji}^-=1, \hat{r}_{ij}^+\ge\hat{r}_{ji}^-\ge 0, \hat{r}_{ii}^+\ge\hat{r}_{ii}^-\ge 0.5, i,j=1,2,\ldots,n$. Based on the above transformation, Xu [16] introduced the definition of additive consistent IFPR.

Definition 2 [16] Let $\tilde{R}=(\tilde{r}_{ij})_{n\times n}$ be an IFPR with $\tilde{r}_{ij}=(\mu_{ij},\nu_{ij})(i,j=1,2,\ldots,n)$, if there exists a vector $\omega=(\omega_1,\omega_2,\cdots,\omega_n)^T$, such that

$$\mu_{ij}\le 0.5(\omega_i-\omega_j+1)\le 1-\nu_{ij}, \text{for all } i,j=1,2,\ldots,n \tag{3}$$

where $\omega_i\in[0,1](i=1,2,\ldots,n)$, and $\sum_{i=1}^{n}\omega_i=1$. Then, $\tilde{R}$ is called an additive consistent IFPR.

Gong et al.'s additive consistency

Gong et al. [17]'s definition is also based on the transformation between the IFPR $\tilde{R}=(\tilde{r}_{ij})_{n\times n}$ and its corresponding interval-valued complementary judgment matrix $\hat{R}=(\hat{r}_{ij})_{n\times n}$. As for an interval-valued complementary judgment matrix $\hat{R}=(\hat{r}_{ij})_{n\times n}$, Gong et al. claimed that it is additive consistent if there exists a priority vector $\hat{\omega}=(\hat{\omega}_1,\hat{\omega}_2,\cdots,\hat{\omega}_n)^T=([\omega_1^l,\omega_1^u],[\omega_2^l,\omega_2^u],\cdots,[\omega_n^l,\omega_n^u])^T$, such that $\hat{r}_{ij}=0.5+0.2\log 3$ $\omega_i/-0pt\omega_j=[0.5+0.2\log 3^{\omega_{il}/-0pt\omega_{ju}}, 0.5+0.2\log 3^{\omega_{iu}/-0pt\omega_{jl}}]$ $(i,j=1,2,\ldots,n)$, and the priorities $\hat{\omega}_i$ can be interpreted as the membership degree range of the importance of the alternative x_i. Hence, with the additive consistency condition of interval-valued complementary judgment matrix $\hat{R}=(\hat{r}_{ij})_{n\times n}$, a new form of definition for additive consistent IFPR can be given as follows.

Definition 3 [17] Let $\tilde{R}=(\tilde{r}_{ij})_{n\times n}$ be an IFPR with $\tilde{r}_{ij}=(\mu_{ij},\nu_{ij})(i,j=1,2,\ldots,n)$, if there exists a vector $\hat{\omega}=(\hat{\omega}_1,\hat{\omega}_2,\cdots,\hat{\omega}_n)^T=([\omega_1^l,\omega_1^u],[\omega_2^l,\omega_2^u],\cdots,[\omega_n^l,\omega_n^u])^T$, such that

$$\mu_{ij}=0.5+0.2\log 3^{\omega_{il}/-0pt\omega_{ju}}, \nu_{ij}=0.5+0.2\log 3^{\omega_{jl}/-0pt\omega_{iu}}, \text{for all } i,j=1,2,\ldots,n \tag{4}$$

Then, $\tilde{R}$ is called an additive consistent IFPR.

Wang's additive consistency
According to additive transitivity, Wang [18] introduced a definition of additive consistent IFPR by directly employing the membership and nonmembership degrees of IFSs.

Definition 4 [18] An IFPR $\tilde{R}=(\tilde{r}_{ij})_{n\times n}$ with $\tilde{r}_{ij}=(\mu_{ij},v_{ij})(i,j=1,2,\ldots,n)$ is called additive consistent if it satisfies the following additive transitivity:

$$\mu_{ik}+\mu_{jk}+\mu_{ki}=\mu_{kj}+\mu_{ji}+\mu_{ik}, \text{for all } i,j,k=1,2,\cdots,n \tag{5}$$

Let $\tilde{\omega}=(\tilde{\omega}_1,\tilde{\omega}_2,\cdots,\tilde{\omega}_n)^T=((\omega_1^\mu,\omega_1^v),(\omega_2^\mu,\omega_2^v),\cdots,(\omega_n^\mu,\omega_n^v))^T$ be an underlying intuitionistic fuzzy priority vector of an IFPR $\tilde{R}=(\tilde{r}_{ij})_{n\times n}$, where $\tilde{\omega}_i=(\tilde{\omega}_i^\mu,\tilde{\omega}_i^v)$ $(i=1,2,\cdots,n)$ is an intuitionistic fuzzy value, which satisfies $\tilde{\omega}_i^\mu,\tilde{\omega}_i^v\in[0,1]$ and $\tilde{\omega}_i^\mu+\tilde{\omega}_i^v\leq 1$. $\tilde{\omega}_i^\mu$ and $\tilde{\omega}_i^v$ indicate the membership and non-membership degrees of the alternative x_i as per a fuzzy concept of "importance", respectively. The normalization of $\tilde{\omega}$ can be done via the following definition:

Definition 5 [18] An intuitionistic fuzzy weight vector $\tilde{\omega}=(\tilde{\omega}_1,\tilde{\omega}_2,\cdots,\tilde{\omega}_n)^T$ with $\tilde{\omega}_i=(\omega_i^\mu,\omega_i^v)$, $\omega_i^\mu,\omega_i^v\in[0,1]$ and $\omega_i^\mu+\omega_i^v\leq 1$ for $i=1,2,\cdots,n$ is said to be normalized if it satisfies the following conditions:

$$\sum_{j=1,j\neq i}^{n}\omega_j^\mu\leq\omega_i^v,\ \omega_i^\mu+n-2\geq\sum_{j=1,j\neq i}^{n}\omega_j^v, \text{for all } i=1,2,\cdots,n \tag{6}$$

With the underlying normalized intuitionistic fuzzy priority vector $\tilde{\omega}=(\tilde{\omega}_1,\tilde{\omega}_2,\cdots,\tilde{\omega}_n)^T$, an additive consistent IFPR $\tilde{R}^*=(\tilde{r}_{ij}^*)_{n\times n}$ can be established as:

$$\tilde{r}_{ij}^*=(\mu_{ij},v_{ij})=\begin{cases}(0.5,0.5) & if\ \ i=j\\ (0.5\omega_i^\mu+0.5\omega_j^v,0.5\omega_i^v+0.5\omega_j^\mu) & if\ \ i\neq j\end{cases} \tag{7}$$

where $\omega_i^\mu,\omega_i^v\in[0,1],\omega_i^\mu+\omega_i^v\leq 1$, $\sum_{j=1,j\neq i}^{n}\omega_j^\mu\leq\omega_i^v$, and $\omega_i^\mu+n-2\geq\sum_{j=1,j\neq i}^{n}\omega_j^v$, for all $i=1,2,\cdots,n$.

3.2 Multiplicative Consistency

The additive consistency is, to some extent, inappropriate in modeling consistency due to that its consistency condition is sometimes in conflict with the [0, 1] scale used for providing the preference values [31]. However, the multiplicative consistency does not have this limitation [32]. The main idea of multiplicative consistency is based on the multiplicative transitivity of a preference relation $R=(r_{ij})_{n\times n}$, which is characterized as $r_{ij}/r_{ji}=(r_{ik}/r_{ki})\cdot(r_{kj}/r_{jk})$ for all $i,j,k=1,2,\cdots,n$. This relationship can be interpreted as the ratio of the preference intensity for the alternative x_i to that of x_j should be equal to the multiplication of the ratios of preferences when using an intermediate alternative x_k, in the case where r_{ij}/r_{ji} indicates a ratio of the preference intensity for the alternative x_i to that of x_j. In other words, x_i is r_{ij}/r_{ji} times as good as x_j. Inspired by the multiplicative transitivity and the relationship between the IFPR and its corresponding preference relations, several distinct definitions of multiplicative consistency were proposed for IFPRs.

Xu's multiplicative consistency of IFPR

Based on the transformation relationship between the IFPR $\tilde{R}=(\tilde{r}_{ij})_{n\times n}$ and its corresponding interval complementary judgment matrix $\hat{R}=(\hat{r}_{ij})_{n\times n}$, Xu [16] proposed the definition of multiplicative consistent IFPR.

Definition 6 [24] Let $\tilde{R}=(\tilde{r}_{ij})_{n\times n}$ with $\tilde{r}_{ij}=(\mu_{ij},v_{ij})(i,j=1,2,\ldots,n)$ be an IFPR, if there exists a vector $\omega=(\omega_1,\omega_2,\cdots,\omega_n)^T$, such that

$$\mu_{ij}\leq\frac{\omega_i}{\omega_i+\omega_j}\leq 1-v_{ij},\text{ for all } i=1,2,\ldots,n-1;\ j=i+1,\ldots,n \tag{8}$$

where $\omega_i\geq 0,(i=1,2,\ldots,n)$, $\sum_{i=1}^{n}\omega_i=1$. Then, we call $\tilde{R}$ a multiplicative consistent IFPR.

Gong et al.'s multiplicative consistency of IFPR

Based on the transformation between an IFPR and its corresponding interval-valued fuzzy preference relation, Gong et al. [19] introduced a definition of multiplicative consistent IFPR.

Definition 7 [19] Let $\tilde{R}=(\tilde{r}_{ij})_{n\times n}$ be an IFPR with $\tilde{r}_{ij}=(\mu_{ij},v_{ij})(i,j=1,2,\ldots,n)$, if there exists a vector $\hat{\omega}=(\hat{\omega}_1,\hat{\omega}_2,\cdots,\hat{\omega}_n)^T=([\omega_1^l,\omega_1^u],[\omega_2^l,\omega_2^u],\cdots,[\omega_n^l,\omega_n^u])^T$, such that

$$\mu_{ij}=\frac{\omega_{il}}{\omega_{il}+\omega_{ju}},\ v_{ij}=\frac{\omega_{jl}}{\omega_{jl}+\omega_{iu}},\text{ for all } i,j=1,2,\ldots,n \tag{9}$$

Then, $\tilde{R}$ is called a multiplicative consistent IFPR.

Xu et al.'s multiplicative consistency of IFPR

Xu et al. [30] proposed another definition of multiplicative consistent IFPR, which was based on the membership and non-membership degrees of IFSs, shown as follows:

Definition 8 [30] An IFPR $\tilde{R}=(r_{ij})_{n\times n}$ with $r_{ij}=(\mu_{ij},\nu_{ij})(i,j=1,2,\ldots,n)$ is multiplicative consistent if

$$\mu_{ij}=\begin{cases}0, & (\mu_{ik},\mu_{kj})\in\{(0,1),(1,0)\}\\ \frac{\mu_{ik}\mu_{kj}}{\mu_{ik}\mu_{kj}+(1-\mu_{ik})(1-\mu_{kj})}, & otherwise\end{cases}, \text{for all } i\leq k\leq j, \quad (10)$$

$$\nu_{ij}=\begin{cases}0, & (\nu_{ik},\nu_{kj})\in\{(0,1),(1,0)\}\\ \frac{\nu_{ik}\nu_{kj}}{\nu_{ik}\nu_{kj}+(1-\nu_{ik})(1-\nu_{kj})}, & otherwise\end{cases}, \text{for all } i\leq k\leq j. \quad (11)$$

Liao and Xu's multiplicative consistency of IFPR

Liao and Xu [20] pointed out that the definition of Xu et al. [30] was not reasonable in some cases because with the above consistency conditions, the relationship $\mu_{ij}\cdot\mu_{jk}\cdot\mu_{ki}=\mu_{ik}\cdot\mu_{kj}\cdot\mu_{ji}$ (for all $i,j,k=1,2,\cdots,n$) can not be derived any more. Then, they introduced a general definition of multiplicative consistent IFPR, shown as follows:

Definition 9 [20] An IFPR $\tilde{R}=(\tilde{r}_{ij})_{n\times n}$ with $\tilde{r}_{ij}=(\mu_{ij},\nu_{ij})$ is called multiplicative consistent if the following multiplicative transitivity is satisfied:

$$\mu_{ij}\cdot\mu_{jk}\cdot\mu_{ki}=\nu_{ij}\cdot\nu_{jk}\cdot\nu_{ki}, \text{for all } i,j,k=1,2,\cdots,n. \quad (12)$$

Liao and Xu [20] further clarified that the conditions in Definition 8 satisfy (12), which implies the consistency measured by the conditions given in Definition 8 is a special case of multiplicative consistency defined as Definition 9 for an IFPR. Hence, in general, Definition 8 is not sufficient and suitable to measure the multiplicative consistency of an IFPR.

With the underlying normalized intuitionistic fuzzy priority weight vector $\tilde{\omega}=(\tilde{\omega}_1,\tilde{\omega}_2,\cdots,\tilde{\omega}_n)^T$, a multiplicative consistent IFPR $\tilde{R}^*=(\tilde{r}^*_{ij})_{n\times n}$ can be established as [20]:

$$\tilde{r}^*_{ij}=(\mu_{ij},\nu_{ij})=\begin{cases}(0.5,0.5) & if\ \ i=j\\ \left(\frac{2\omega_i^\mu}{\omega_i^\mu-\omega_i^\nu+\omega_j^\mu-\omega_j^\nu+2},\frac{2\omega_j^\mu}{\omega_i^\mu-\omega_i^\nu+\omega_j^\mu-\omega_j^\nu+2}\right) & if\ \ i\neq j\end{cases} \quad (13)$$

where $\omega_i^\mu,\omega_i^\nu\in[0,1],\omega_i^\mu+\omega_i^\nu\leq 1,\ \sum_{j=1,j\neq i}^{n}\omega_j^\mu\leq\omega_i^\nu$, and $\omega_i^\mu+n-2\geq\sum_{j=1,j\neq i}^{n}\omega_j^\nu$, for all $i=1,2,\cdots,n$.

3.3 Acceptable Consistency of IFPR

Due to the complexity of the problem and the limited knowledge of the experts, the experts often determine some inconsistent IFPR. Perfect consistent IFPR is somehow too strict for the experts to construct especially when the number of objects is very large. Since in practical cases, it is impossible to get the consistent IFPRs, Liao and Xu [22] introduced the concept of acceptable consistent IFPR.

Definition 10 [22] Let $\tilde{R}=(\tilde{r}_{ij})_{n\times n}$ be an IFPR with $\tilde{r}_{ij}=(\mu_{ij},\nu_{ij},\pi_{ij})(i,j=1,2,\cdots,n)$. We call R an acceptable consistent IFPR, if

$$d(\tilde{R},\tilde{R}^{*})\leq\xi, \tag{14}$$

where $d(\tilde{R},\tilde{R}^{*})$ is the distance measure between the given IFPR $\tilde{R}$ and its corresponding underlying consistent IFPR $\tilde{R}^{*}$, which can be calculated by

$$d(\tilde{R},\tilde{R}^{*})=\frac{1}{(n-1)(n-2)}\sum_{1\leq i<j<n}^{n}\left(\left(\left|\mu_{ij}-\mu_{ij}^{*}\right|+\left|\nu_{ij}-\nu_{ij}^{*}\right|+\left|\pi_{ij}-\pi_{ij}^{*}\right|\right),\right. \tag{15}$$

and ξ is the consistency threshold. The corresponding underlying consistent IFPR $\tilde{R}^{*}$ can be yielded by (7) or (13).

As for those IFPRs of inconsistency, there are many procedures to improve the inconsistent IFPRs into acceptable consistent IFPRs (For more details, please refer to [21, 23]).

If all the IFPRs are of acceptable consistency, we can aggregate these IFPRs into an overall IFPR and then derive the ranking of the alternatives. Liao and Xu [22] proposed a simple intuitionistic fuzzy weighted geometric (SIFWG) operator to fuse the IFPRs. For s IFPRs $\tilde{R}^{(l)}=\left(\tilde{r}_{ij}^{(l)}\right)_{n\times n}$ $(l=1,2,\cdots,s)$, their fused IFPR $\bar{R}=(\bar{r}_{ij})_{n\times n}$ with $\bar{r}_{ij}=(\bar{\mu}_{ij},\bar{\nu}_{ij},\bar{\pi}_{ij})$ by the SIFWG operator is also an IFPR, where

$$\bar{\mu}_{ij}=\prod_{l=1}^{s}\left(\mu_{ij}^{(l)}\right)^{\lambda_l},\bar{\nu}_{ij}=\prod_{l=1}^{s}\left(\nu_{ij}^{(l)}\right)^{\lambda_l},\bar{\pi}_{ij}=1-\bar{\mu}_{ij}-\bar{\nu}_{ij},i,j=1,2,\ldots,n. \tag{16}$$

Liao and Xu [22] further proved that if all the individual IFPRs are of acceptable multiplicative consistency, then their fused IFPR by the SIFWG operator is also of acceptable consistency. This is a good property in group decision making with IFPRs because with this property there is no need to check the consistency of the fused IFPR and we can use it to derive the decision making result directly.

4 Consensus of Group Decision Making with IFPRs

4.1 Difficulties in Reaching Consensus

With all the above mentioned different types of consistency and the corresponding inconsistency repairing methods, we can get a set of consistent or acceptable consistent IFPRs. This is the precondition of deriving a reasonable solution for a group decision making problem. The group decision making problem is very complicated owing to the complexity introduced by the conflicting opinions of the experts. As to a group decision making problem with IFPRs, how to find a final solution which is accepted by all the experts is a great challenge. The consensus is very important in any group decision making problems. It can be defined as "*a decision that has been reached when most members of the team agree on a clear option and the few who oppose it think they have had a reasonable opportunity to influence that choice. All team members agree to support the decision.*" [25] Consensus is a pathway to a true group decision and it can guarantee that the final result been supported by all the group members despite their different opinions.

However, to find such a consensus result is very difficult because of some inherent differences in value systems, flexibility of members, etc. Generally, if all experts are wise and rational, they should agree with each other. But, in reality, disagreement among the experts is inevitable. In fact, the disagreement is just the valuation of group decision making.

In the process of group decision making, the target is to find a solution which is accepted by all the experts. Initially, the experts should be with no consensus, and thus they need to communicate with each other and modify their judgments. That is to say, the consensus reaching process should be an iterative procedure and it should be converge finally. In addition, a group decision making problem with too many times of iteration does not make sense because it wastes too many resources and is not worthy to be investigated by the experts.

4.2 Consensus Measures for Group Decision Making with IFPRs

The consensus reaching process refers to how to obtain the maximum degree of consensus or agreement between the set of experts [33]. To do so, we should first know how to measure the consensus degree among the experts. Although there is litter research focused on the consensus of IFPRs [34], we still can found many approaches to model consensus process in group decision making with other preference relations, such as fuzzy preference relation [35], incomplete fuzzy preference relation [37], and linguistic preference relation [36]. These consensus measures involve two parts: hard consensus measure and soft consensus measure. The hard consensus measure uses a number in the interval [0, 1] to represent the consensus degree of the experts, while the soft consensus measure employs a linguistic label such as "most" to describe the truth of a statement such as "most experts agree on almost all the alternatives." As to group decision making with IFPRs, Szmidt and Kacprzyk [34] used an interval-valued measure of distance to represent the consensus of experts. They took the membership degrees and the hesitance degrees as two separate matrices and then used these two matrices to derive the upper bound and lower bound of the interval-valued consensus measure. Their work can be seen as the first attempt to measure the consensus of group decision with IFPRs. However, they did not include any procedures to reaching consensus. In the following, we would define a hard consensus measure of experts whose opinions are represented by IFPRs.

For a set of IFPRs $\tilde{R}^{(l)} = \left(\tilde{r}_{ij}^{(l)}\right)_{n\times n}$ $(l=1,2,\cdots,s)$ given by s independent experts $e_l(l=1,2,\ldots,s)$, since it is known that if all individual IFPRs are of acceptable consistency, their fused IFPRs $\bar{R}=(\bar{r}_{ij})_{n\times n}$ with the SIFWG operator is also of acceptable consistency, then, motivated by the distance measure of two IFPRs given as (15), we can introduce a hard consensus measure of the experts with IFPRs.

Definition 11 For a set of IFPRs $\tilde{R}^{(l)} = \left(\tilde{r}_{ij}^{(l)}\right)_{n\times n}$ $(l=1,2,\cdots,s)$ with $r_{ij}^{(l)} = (\mu_{ij}^{(l)}, v_{ij}^{(l)}, \pi_{ij}^{(l)})(l=1,2,\cdots,s)$ given by s independent experts $e_l(l=1,2,\ldots,s)$, whose weight vector is $\lambda=(\lambda_1,\lambda_2,\cdots,\lambda_s)^T$ with $0\le\lambda_l\le1$, $\sum_{l=1}^{s}\lambda_l=1$, then the consensus of the lth expert is defined as

$$C_l = 1-\frac{1}{(n-1)(n-2)}\sum_{1\le i<j<n}^{n}\left(\left|\mu_{ij}^{(l)}-\bar{\mu}_{ij}\right|+\left|v_{ij}^{(l)}-\bar{v}_{ij}\right|+\left|\pi_{ij}^{(l)}-\bar{\pi}_{ij}\right|\right), \tag{17}$$

where $\bar{R}=(\bar{r}_{ij})_{n\times n}$ with $\bar{r}_{ij}=(\bar{\mu}_{ij},\bar{v}_{ij},\bar{\pi}_{ij}), \bar{\mu}_{ij}=\prod_{l=1}^{s}\left(\mu_{ij}^{(l)}\right)^{\lambda_l}, \bar{v}_{ij}=\prod_{l=1}^{s}\left(v_{ij}^{(l)}\right)^{\lambda_l}, \bar{\pi}_{ij}=1-\bar{\mu}_{ij}-\bar{v}_{ij}$ is the overall IFPR derived by the SIFWG operator.

4.3 *Consensus Reaching Procedure with IFPRs*

With the above consensus measure, the consensus reaching procedure for helping the experts, whose preferences are given as IFPRs, to reach consensus can be given as follows:

- Establish s IFPRs $\tilde{R}^{(l)} = \left(\tilde{r}_{ij}^{(l)}\right)_{n\times n}$ $(l=1,2,\cdots,s)$ for s independent experts $e_l(l=1,2,\ldots,s)$;
- Check the consistency of each IFPR: for those IFPRs of unacceptable consistency, repair them until acceptable;
- Compute the consensus degree of each experts;

- Determine the minimum consensus bound of the experts, τ. For $C_l < \tau$, ask the expert e_l to modify the IFPR. A suggestion for the expert e_l to modify the IFPR is to change the preferences by the following formulas:

$$\mu_{ij}^{(l)'} = \left(\mu_{ij}^{(l)}\right)^{\zeta} \cdot \left(\bar{\mu}_{ij}\right)^{1-\zeta}, \tag{18}$$

$$\nu_{ij}^{(l)'} = \left(\nu_{ij}^{(l)}\right)^{\zeta} \cdot \left(\bar{\nu}_{ij}\right)^{1-\zeta}, \tag{19}$$

$$\pi_{ij}^{(l)'} = 1 - \mu_{ij}^{(l)'} - \nu_{ij}^{(l)'}. \tag{20}$$

- Articulate the decision via aggregating all the IFPRs whose consensus degrees are greater than the threshold τ into an overall IFPR.

5 Numerical Example

The following example concerning the selection of the global suppliers (adapted from [22]) can be used to illustrate the consensus reaching procedure for group decision making with IFPRs.

Example The current globalized market trend identifies the necessity of the establishment of long term business relationship with competitive global suppliers spread around the world. This can lower the total cost of supply chain; lower the inventory of enterprises; enhance information sharing of enterprises; improve the interaction of enterprises and obtain more competitive advantages for enterprises. Thus, how to select different unfamiliar international suppliers according to the broad evaluation is very critical and has a direct impact on the performance of an organization. Suppose a company invites three experts e_1, e_2 and e_3 from different field to evaluate four candidate suppliers x_1, x_2, x_3 and x_4. The weights of the experts are 0.3, 0.4, 0.3, respectively, which is established by the decision making committee according to the experts' expertise and reputation. Global supplier development is a complex problem which includes much qualitative information. In such a case, it is straightforward for the experts to compare the different suppliers in pairs and then construct some preference relations to express their preferences. Since the experts do not have precise information of the global suppliers, it is reasonable for them to use the IFSs to describe their assessments, and then three IFPRs can be established:

$$\tilde{R}^{(1)} = \begin{pmatrix} (0.5,0.5) & (0.5,0.2) & (0.7,0.1) & (0.5,0.3) \\ (0.2,0.5) & (0.5,0.5) & (0.6,0.2) & (0.3,0.6) \\ (0.1,0.7) & (0.2,0.6) & (0.5,0.5) & (0.3,0.6) \\ (0.3,0.5) & (0.6,0.3) & (0.6,0.3) & (0.5,0.5) \end{pmatrix},$$

$$\tilde{R}^{(2)} = \begin{pmatrix} (0.5,0.5) & (0.6,0.1) & (0.8,0.2) & (0.6,0.3) \\ (0.1,0.6) & (0.5,0.5) & (0.5,0.1) & (0.3,0.7) \\ (0.2,0.8) & (0.1,0.5) & (0.5,0.5) & (0.4,0.6) \\ (0.3,0.6) & (0.7,0.3) & (0.6,0.4) & (0.5,0.5) \end{pmatrix},$$

$$\tilde{R}^{(3)} = \begin{pmatrix} (0.5,0.5) & (0.6,0.2) & (0.8,0.1) & (0.7,0.2) \\ (0.2,0.6) & (0.5,0.5) & (0.6,0.1) & (0.2,0.7) \\ (0.1,0.8) & (0.1,0.6) & (0.5,0.5) & (0.2,0.3) \\ (0.2,0.7) & (0.7,0.2) & (0.3,0.2) & (0.5,0.5) \end{pmatrix}.$$

Using the fractional programming models constructed by Liao and Xu [20], the underlying intuitionistic fuzzy weights for these three individual IFPRs are

$$\tilde{\omega}^{(1)} = ((0.3951, 0.4221), (0.1354, 0.8397), (0.0451, 0.8894), (0.2370, 0.6298))^T,$$
$$\tilde{\omega}^{(2)} = ((0.4137, 0.5517), (0.1552, 0.7069), (0.0862, 0.9138), (0.2069, 0.6897))^T.$$
$$\tilde{\omega}^{(3)} = ((0.4686, 0.4143), (0.1406, 0.7891), (0.0586, 0.9414), (0.1538, 0.6700))^T.$$

According to (13), the corresponding multiplicative consistent IFPRs can be generated:

$$\tilde{R}^{(1)*} = \begin{pmatrix} (0.5000,0.5000) & (0.6228,0.2134) & (0.7001,0.0799) & (0.5001,0.3000) \\ (0.2134,0.6228) & (0.5000,0.5000) & (0.5999,0.1998) & (0.2999,0.5250) \\ (0.0799,0.7001) & (0.1998,0.5999) & (0.5000,0.5000) & (0.1182,0.6213) \\ (0.3000,0.5001) & (0.5250,0.2999) & (0.6213,0.1182) & (0.5000,0.5000) \end{pmatrix},$$

$$\tilde{R}^{(2)*} = \begin{pmatrix} (0.5000,0.5000) & (0.6315,0.2369) & (0.7999,0.1667) & (0.5999,0.3000) \\ (0.2369,0.6315) & (0.5000,0.5000) & (0.5001,0.2778) & (0.3215,0.4286) \\ (0.1667,0.7999) & (0.2778,0.5001) & (0.5000,0.5000) & (0.2500,0.6001) \\ (0.3000,0.5999) & (0.4286,0.3215) & (0.6001,0.2500) & (0.5000,0.5000) \end{pmatrix},$$

$$\tilde{R}^{(3)*} = \begin{pmatrix} (0.5000,0.5000) & (0.6667,0.2000) & (0.8000,0.1000) & (0.6093,0.2000) \\ (0.2000,0.6667) & (0.5000,0.5000) & (0.6000,0.2501) & (0.3366,0.3683) \\ (0.1000,0.8000) & (0.2501,0.6000) & (0.5000,0.5000) & (0.1950,0.5118) \\ (0.2000,0.6093) & (0.3683,0.3366) & (0.5118,0.1950) & (0.5000,0.5000) \end{pmatrix}.$$

Thus, via (15), we can obtain $d(\tilde{R}^{(1)}, \tilde{R}^{(1)*}) = 0.1382, d(\tilde{R}^{(2)}, \tilde{R}^{(2)*}) = 0.2569$, and $d(\tilde{R}^{(3)}, \tilde{R}^{(3)*}) = 0.2837$. Suppose $\xi = 0.3$, then all these three individual IFPRs are of acceptable multiplicative consistency. Thus, with the SIFWG operator, the overall IFPR of the group can be aggregated as

$$\bar{R} = \begin{pmatrix} (0.5000, 0.5000) & (0.5681, 0.1516) & (0.7686, 0.1320) & (0.5950, 0.2656) \\ (0.1516, 0.5681) & (0.5000, 0.5000) & (0.5578, 0.1231) & (0.2656, 0.6684) \\ (0.1320, 0.7686) & (0.1231, 0.5578) & (0.5000, 0.5000) & (0.2980, 0.4874) \\ (0.2656, 0.5950) & (0.6684, 0.2656) & (0.4874, 0.2980) & (0.5000, 0.5000) \end{pmatrix}$$

By (16), we can calculate the consensus of each expert, which are $C_1 = 0.8114$, $C_2 = 0.8160$, and $C_3 = 0.8292$, respectively. That is to say, the experts have at least 80 % consensus. If the minimum consensus bound τ of the experts is 0.8, then, we can say that all the experts in this decision making problem reach the group consensus.

On the other hand, if the consensus threshold given by the decision maker is much higher, for instance, $\tau = 0.85$, then, the experts need to modify there assessments to reach a much higher consensus.

Let $\zeta = 0.6$, with (18)–(20), the suggested IFPRs are

$$\tilde{R}^{(1)'} = \begin{pmatrix} (0.5000, 0.5000) & (0.5262, 0.1790) & (0.7267, 0.1117) & (0.5360, 0.2857) \\ (0.1790, 0.5262) & (0.5000, 0.5000) & (0.5827, 0.1647) & (0.2857, 0.6265) \\ (0.117, 0.7267) & (0.1647, 0.5827) & (0.5000, 0.5000) & (0.2992, 0.5521) \\ (0.2857, 0.5360) & (0.6265, 0.2857) & (0.5521, 0.2992) & (0.5000, 0.5000) \end{pmatrix}$$

$$\tilde{R}^{(2)'} = \begin{pmatrix} (0.5000, 0.5000) & (0.5870, 0.1181) & (0.7873, 0.1694) & (0.5980, 0.2857) \\ (0.1181, 0.5870) & (0.5000, 0.5000) & (0.5224, 0.1087) & (0.2857, 0.6872) \\ (0.1694, 0.7873) & (0.1087, 0.5224) & (0.5000, 0.5000) & (0.3556, 0.5521) \\ (0.2857, 0.5980) & (0.6872, 0.2857) & (0.5521, 0.3556) & (0.5000, 0.5000) \end{pmatrix}$$

$$\tilde{R}^{(3)'} = \begin{pmatrix} (0.5000, 0.5000) & (0.5870, 0.1790) & (0.7873, 0.1117) & (0.6559, 0.2240) \\ (0.1790, 0.5870) & (0.5000, 0.5000) & (0.5827, 0.1087) & (0.2240, 0.6872) \\ (0.1117, 0.7873) & (0.1087, 0.5827) & (0.5000, 0.5000) & (0.2346, 0.3643) \\ (0.2240, 0.6559) & (0.6872, 0.2240) & (0.3643, 0.2346) & (0.5000, 0.5000) \end{pmatrix}$$

Since $\tilde{R}^{(l)'}$ $(l = 1, 2, 3)$ is simple geometric aggregated by $\tilde{R}^{(l)*}$ $(l = 1, 2, 3)$ and $\bar{R}$ respectively, according to the theorem of Liao and Xu [17], the new IFPRs $\tilde{R}^{(l)'}$ $(l = 1, 2, 3)$ should be still of acceptable consistency. Then we aggregate them together with the SIFWG operator and yield an overall IFPR:

$$\bar{R}' = \begin{pmatrix} (0.5000, 0.5000) & (0.5681, 0.1516) & (0.7686, 0.1320) & (0.5950, 0.2656) \\ (0.1516, 0.5681) & (0.5000, 0.5000) & (0.5578, 0.1231) & (0.2656, 0.6684) \\ (0.1320, 0.7686) & (0.1231, 0.5578) & (0.5000, 0.5000) & (0.2980, 0.4874) \\ (0.2656, 0.5950) & (0.6684, 0.2656) & (0.4874, 0.2980) & (0.5000, 0.5000) \end{pmatrix},$$

which is the same as $\bar{R}$ (It can be proven that $\bar{R}'$ is always the same as $\bar{R}$ with our aggregation methodology). Then, by (16), the consensus of each expert are calculated as $C_1 = 0.8875$, $C_2 = 0.8936$, and $C_3 = 0.8732$, respectively. Since each consensus degree of the experts is greater than the threshold $\tau = 0.85$, the group reaches the consensus.

6 Conclusion

IFPR is a powerful tool to express the experts' opinions in the process of group decision making. In this chapter, we have discussed the consistency of the IFPRs and the consensus of the experts in group decision making with IFPRs. Firstly, we have described the group decision making problem with IFPRs in details. Then, we have reviewed all the different kinds of definitions for the consistency of IFPRs, which involves two sorts, i.e., the additive consistency and the multiplicative consistency. Based on the consistency, the definition of acceptable consistency can be given. For those IFPRs which are of unacceptable consistency, the consistency repairing process should be employed to modify them. Once all the IFPRs are of acceptable consistency, we have introduced a hard consensus measure to depict the consensus degree of the experts within the group decision making. Furthermore, we have given a simple procedure for aiding the experts to reach group consensus.

Acknowledgment The work was supported in part by the National Natural Science Foundation of China (No. 61273209 and No. 71501135) and the Scientific Research Foundation for Scholars at Sichuan University (No. 1082204112042).

References

1. Liao, H.C., Xu, Z.S., Xia, M.M.: Multiplicative consistency of hesitant fuzzy preference relation and its application in group decision making. Int. J. Inform. Technol. Decis. Making **13**(1), 47–76 (2014)
2. Tanino, T.: Fuzzy preference orderings in group decision making. Fuzzy Sets Syst. **12**, 117–131 (1984)
3. Liao, H.C., Xu, Z.S.: Multi-criteria decision making with intuitionistic fuzzy PROMETHEE. J. Intell. Fuzzy Syst. **27**(4), 1703–1717 (2014)
4. Atanassov, K.: Intuitionistic fuzzy sets. In: Sgurev, V. (ed.) VII ITKR's Session, June, (1983)
5. Atanassov, K.: On intuitionistic fuzzy sets theory. (Springer, Berlin 2012)
6. Xu, Z.S., Liao, H.C.: Intuitionistic fuzzy analytic hierarchy process. IEEE Trans. Fuzzy Syst. **22**(4), 749–761 (2014)

7. Jiang, Y.C., Tang, Y., Wang, J., Tang, S.Q.: Reasoning within intuitionistic fuzzy rough description logics. Inf. Sci. **179**, 2362–2378 (2009)
8. Papageorgiou, E.I., Iakovidis, D.K.: Intuitionistic fuzzy cognitive maps. IEEE Trans. Fuzzy Syst. **21**, 342–354 (2013)
9. Çoker, D.: An introduction to intuitionistic fuzzy topological spaces. Fuzzy Sets Syst. **88**, 81–89 (1997)
10. De, S.K., Biswas, R., Roy, A.R.: An application of intuitionistic fuzzy sets in medical diagnosis. Fuzzy Sets Syst. **117**, 209–213 (2001)
11. Vlachos, I.K., Sergiadis, G.D.: Intuitionistic fuzzy information-applications to pattern recognition. Pattern Recogn. Lett. **28**, 197–206 (2007)
12. Liao, H.C., Xu, Z.S.: Some algorithms for group decision making with intuitionistic fuzzy preference information. Int. J. Uncertain. Fuzz. Knowl.-Based Syst. **22**(4), 505–529 (2014)
13. Herrera-Viedma, E., Chiclana, F., Herrera, F., Alonso, S.: A group decision-making model with incomplete fuzzy preference relations based on additive consistency. IEEE Trans. Syst. Man Cybern. **37**, 176–189 (2007)
14. Xu, Z.S.: Intuitionistic preference relations and their application in group decision making. Inf. Sci. **177**, 2363–2379 (2007)
15. Xu, Z.S.: A survey of preference relations. Int. J. Gen. Syst. **36**, 179–203 (2007)
16. Xu, Z.S.: Approaches to multiple attribute decision making with intuitionistic fuzzy preference information. Syst. Eng. Theory Pract. **27**, 62–71 (2007)
17. Gong, Z.W., Li, L.S., Forrest, J., Zhao, Y.: The optimal priority models of the intuitionistic fuzzy preference relation and their application in selecting industries with higher meteorological sensitivity. Expert Syst. Appl. **38**, 4394–4402 (2011)
18. Wang, Z.J.: Derivation of intuitionistic fuzzy weights based on intuitionistic fuzzy preference relations. Appl. Math. Model. **37**, 6377–6388 (2013)
19. Gong, Z.W., Li, L.S., Zhou, F.X., Yao, T.X.: Goal programming approaches to obtain the priority vectors from the intuitionistic fuzzy preference relations. Comput. Ind. Eng. **57**, 1187–1193 (2009)
20. Liao, H.C., Xu, Z.S.: Priorities of intutionistic fuzzy preference relation based on multiplicative consistency. IEEE Trans. Fuzzy Syst. **22**(6), 1669–1681 (2014)
21. Xu, Z.S., Xia, M.M.: Iterative algorithms for improving consistency of intuitionistic preference relations. J. Oper. Res. Soc. **65**, 708–722 (2014)
22. Liao, H.C., Xu, Z.S.: Consistency of the fused intuitionistic fuzzy preference relation in group intuitionistic fuzzy analytic hierarchy process. Appl. Soft Comput. **35**, 812–826 (2015)
23. Liao, H.C., Xu, Z.S.: Automatic procedures for group decision making with intuitionistic preference relation. J. Intell. Fuzzy Syst. **27**(5), 2341–2353 (2014)
24. Ben-Arieh, D., Chen, Z.F.: Fuzzy group decision making. In: Badiru, A.B. (ed.) Handbook of Industrial and Systems Engineering. CRC Press, Boca Raton (2006)
25. Ness, J., Hoffman, C.: Putting Sense into Consensus: Solving the Puzzle of Making Team Decisions. VISTA Associates, Tacoma (1998)
26. Brock, H.W.: The problem of utility weights in group preference aggregation. Oper. Res. **28**, 176–187 (1980)
27. Ramanathan, R., Ganesh, L.S.: Group preference aggregation methods employed in AHP: an evaluation and an intrinsic process for deriving members' weightages. Eur. J. Oper. Res. **79**, 249–265 (1994)
28. Chen, Z.P., Yang, W.: A new multiple attribute group decision making method in intuitionistic fuzzy setting. Appl. Math. Model. **35**, 4424–4437 (2011)
29. Szmidt, E., Kacprzyk, J.: Group decision making under intuitionistic fuzzy preference relations. In: Proceedings of 7th Information Processing and Management of Uncertainty in Knowledge-Based Systems Conference, Paris, pp. 172–178 (1988)
30. Xu, Z.S., Cai, X.Q., Szmidt, E.: Algorithms for estimating missing elements of incomplete intuitionistic preference relations. Int. J. Intell. Syst. **26**, 787–813 (2011)

31. Chiclana, F., Herrera-Viedma, E., Alonso, S., Marques Pereira, R.A.: Preferences and consistency issues in group decision making. In: Bustince, H., Herrera, F., Montero, J. (eds.) Fuzzy Sets and their Extensions: Representation, Aggregation and Models, vol. 220, pp. 219–237. Springer-Verlag, New York (2008)
32. Chiclana, F., Herrera-Viedma, E., Alonso, S., Herrera, F.: Cardinal consistency of reciprocal preference relations: a characterization of multiplicative transitivity. IEEE Trans. Fuzzy Syst. **17**, 277–291 (2009)
33. Herrera-Viedma, E., Alonso, S., Chiclana, F., Herrera, F.: A consensus model for group decision making with incomplete fuzzy preference relations. IEEE Trans. Fuzzy Syst. **15**, 863–877 (2007)
34. Szmidt, E., Kacprzyk, J.: A consensus-reaching process under intuitionistic fuzzy preference relations. Int. J. Intell. Syt. **18**, 837–852 (2003)
35. Kacprzyk, J., Fedrizzi, M., Nurmi, H.: Group decision making and consensus under fuzzy preferences and fuzzy majority. Fuzzy Sets Syst. **49**, 21–31 (1992)
36. Herrera, F., Herrera-Viedma, E., Verdegay, J.L.: A model of consensus in group decision making under linguistic assessments. Fuzzy Sets Syst. **78**, 73–87 (1996)
37. Xu, Z.S.: Goal programming models for obtaining the priority vector of incomplete fuzzy preference relation. Int. J. Approx. Reason. **36**, 261–270 (2004)

Differential Calculus on *IF* Sets

Alžbeta Michalíková

Abstract This contribution summarizes the theory of differential calculus on *IF* sets. First the definition of the function is given. Then the absolute value and limit of the function are defined and the properties of these functions are studied. By using the limit of the function the derivative of the function is define and Lagrange mean value theorem is proved. Since the main aim of this contribution is to proof the Taylor's theorem the polynomial function and Taylor polynomial are defined. Finally the Taylor's theorem is proved and some examples are given.

1 Introduction

The idea of Fuzzy sets was first time presented in the year 1965 by professor Zadeh [1]. Fuzzy sets were presented like the structures which could be use for computing with the natural language. These structures find their applications in many spheres of the research, e.g. decision making, prediction, classification, image processing, expert systems, etc. Mathematically we can write

$$A = \{\langle x, \mu_A(x)\rangle | x \in \Omega\} \quad \mu_A : \Omega \to [0, 1]$$

where μ_A is called the membership function. The membership function assigns the value which is called the degree of membership function to each element x which belongs to the set A.

Example 1 Let us assume that the average summer temperature in some area is 35 °C. Consider that the summer temperatures achieve the values from the interval $[10, 50]$ and let us think about low temperature in summer in this area. The temperature is sure low if its value is from the interval $[10, 20]$ and it is definitely not low if its value is from the interval $[30, 50]$. Let us connect these two function by the line

A. Michalíková (✉)
Matej Bel University, Banská Bystrica, Slovakia
e-mail: alzbeta.michalikova@umb.sk

P. Angelov and S. Sotirov (eds.), *Imprecision and Uncertainty in Information Representation and Processing*, Studies in Fuzziness and Soft Computing 332,
DOI 10.1007/978-3-319-26302-1_14

on the interval [20, 30] then the fuzzy set *A*—*Low temperature* could be describe by the following membership function

$$\mu_A(x) = \begin{cases} 1, & \text{if } x \in [10, 20] \\ \frac{30-x}{10}, & \text{if } x \in [20, 30] \\ 0, & \text{if } x \in [30, 50] \end{cases}$$

We can see that for each $x \in [10, 50]$ the degree of membership function is from unit interval. For example if $x = 25\,°\text{C}$ than the degree of membership function is $\frac{30-25}{10} = 0.5$.

Now let us determine the fuzzy set *B*—*High temperature*. It could be describe for example by the function

$$\mu_B(x) = \begin{cases} 0, & \text{if } x \in [10, 25] \\ \frac{x-25}{15}, & \text{if } x \in [25, 40] \\ 1, & \text{if } x \in [40, 50] \end{cases}$$

Let us compare the sets A, B. From mathematical point of view the set B could represent the complement of the set A but there is a problem how to describe the function by which we get set B from the set A. Of course in fuzzy set theory there exists the definition of the negations. For example standard negation has the form

$$\mu_{\neg A} = 1 - \mu_A \ .$$

But human thinking doesn't work on this principle. The values of the membership functions of the sets A and B could be independent. We feel that the sum of the membership functions of the sets A and B for one element could be less then or equal to one. And exactly these examples from the real life led to define new structure which name is intuitionistic fuzzy set.

2 Intuitionistic Fuzzy Sets

First publication about Intuitionistic fuzzy sets (*IF* sets in short) was publish in the year 1986 by professor Atanassov [2]. The set $A = \{\langle x, \mu_A(x), \nu_A(x)\rangle | x \in \Omega\}$ is an *IF* set if for each $x \in \Omega$ it holds

$$0 \leq \mu_A(x) + \nu_A(x) \leq 1 \ .$$

Function $\mu_A : \Omega \to [0, 1]$ is called membership function and the function $\nu_A : \Omega \to [0, 1]$ is called nonmembership function.

Since the year 1986 there was a large development of the theory and applications in the area of intuitionistic fuzzy sets therefore we deemed necessary to build theory of differential calculus on these structures.

For any *IF* set we will use following shorter notation $A = (\mu_A, \nu_A)$. Denote by $\mathcal{F}$ the family of all *IF* sets. On $\mathcal{F}$ we shall define two binary operations $\oplus, \odot$ and one unary operation $\neg$

$$A \oplus B = (\min(\mu_A + \mu_B, 1), \max(\nu_A + \nu_B - 1, 0))$$

$$A \odot B = (\max(\mu_A + \mu_B - 1, 0), \min(\nu_A + \nu_B, 1))$$

$$\neg A = (1 - \mu_A, 1 - \nu_A)$$

A partial ordering on $\mathcal{F}$ is given by

$$A \leq B \Longleftrightarrow \mu_A \leq \mu_B, \nu_A \geq \nu_B$$

Consider the set $A = (\mu_A, \nu_A)$, where $\mu_A, \nu_A : \Omega \to \mathbb{R}$. It is not difficult to construct an additive group $\mathcal{G} \supset \mathcal{F}$ with an ordering such that $\mathcal{G}$ is a lattice ordered group, also called ℓ-group, where

$$A + B = (\mu_A + \mu_B, \nu_A + \nu_B - 1)$$

with the neutral element $0 = (0_\Omega, 1_\Omega)$ and

$$A \leq B \Longleftrightarrow \mu_A \leq \mu_B, \nu_A \geq \nu_B \ .$$

Lattice operations are given by

$$A \wedge B = (\mu_A \wedge \mu_B, \nu_A \vee \nu_B)$$

$$A \vee B = (\mu_A \vee \mu_B, \nu_A \wedge \nu_B) \ .$$

Evidently

$$A - B = (\mu_A - \mu_B, \nu_A - \nu_B + 1)$$

and

$$-A = (-\mu_A, 2 - \nu_A) \ .$$

The operations on $\mathcal{F}$ can be derived from operations on $\mathcal{G}$ if we use the unit $u = (1_\Omega, 0_\Omega)$. Then

$$A \oplus B = (A + B) \wedge u$$

$$A \odot B = (A + B - u) \vee 0$$

$$\neg A = u - A \ .$$

In our contribution also the following two operations will be used

$$A.B = (\mu_A.\mu_B, \nu_A + \nu_B - \nu_A.\nu_B)$$

and if $\mu_B \neq 0, \nu_B \neq 1$ then

$$\frac{A}{B} = \left(\frac{\mu_A}{\mu_B}, 1 - \frac{1-\nu_A}{1-\nu_B}\right) .$$

3 Definition of the Function and Its Properties

We will study the functions which are defined on the ℓ-group $\mathcal{G}$. From the previous text it follows that these results could be easily applied also for *IF* sets. The function on the ℓ-group $\mathcal{G}$ is defined by the following way

$$\tilde{f}(X) = (f(\mu_X), 1 - f(1 - \nu_X))$$

where $f : \mathbb{R} \to \mathbb{R}$ and $X \in \mathcal{G}$ [4].

To compare the results of classical analysis with our result we will give some example in the text. For these examples we choose function

$$\tilde{\sin} X = (\sin(\mu_X), 1 - \sin(1 - \nu_X))$$

as a template function.

In the first step we will look better on the domain of the function $\tilde{f}$. Since

$$\tilde{f}(X) = (f(\mu_X), 1 - f(1 - \nu_X))$$

where $f : \mathbb{R} \to \mathbb{R}$ then if the domain of the function $\tilde{f}$ is an interval $[A, B]$ then it must hold

$$A \leq B \Longleftrightarrow \mu_A \leq \mu_B \text{ and } \nu_A \geq \nu_B$$

and therefore

$$[\mu_A, \mu_B] \cup [\nu_B, \nu_A] \subset Domf \ .$$

First special function which we will study is absolute value. Following definition was first mentioned in [6].

Definition 1 Let $A \in \mathcal{G}$. Then absolute value of A is defined by following formula

$$|A| = (|\mu_A|, 1 - |1 - \nu_A|) \ .$$

Lemma 1 *Let $|A|$ be the absolute value of A. Then $|A| = A$ for each $A \geq (0, 1)$ and $|A| = -A$ for each $A < (0, 1)$.*

Proof Let $A \geq (0, 1)$ i.e. $\mu_A \geq 0$ and $\nu_A \leq 1$. Then $1 - \nu_A \geq 0$ and

$$|A| = (|\mu_A|, 1 - |1 - \nu_A|) = (\mu_A, 1 - 1 + \nu_A) = (\mu_A, \nu_A) = A \ .$$

Let $A < (0, 1)$ i.e. $\mu_A < 0$ and $\nu_A > 1$. Then $1 - \nu_A < 0$ and

$$|A| = (|\mu_A|, 1 - |1 - \nu_A|) = (-\mu_A, 1 + 1 - \nu_A) = (-\mu_A, 2 - \nu_A) = -A \ .$$

□

Lemma 2 *For each $A, B \in \mathcal{G}$ it holds*

1. $|A.B| = |A|.|B|$
2. $|A - B| = |B - A|$
3. $|A + B| \leq |A| + |B|$
4. $|A - B| \geq |A| - |B|$
5. $||A| - |B|| \leq |A - B|$

Proof Let $A = (\mu_A, \nu_A), B = (\mu_B, \nu_B)$ then it holds

1.
$$|A|.|B| = \left(|\mu_A|, 1 - |1 - \nu_A|\right) . \left(|\mu_B|, 1 - |1 - \nu_B|\right)$$

$$= (|\mu_A.\mu_B|, 1 - |1 - \nu_A| + 1 - |1 - \nu_B| - (1 - |1 - \nu_A|).(1 - |1 - \nu_B|))$$

$$= \left(|\mu_A.\mu_B|, 1 - |1 - \nu_A - \nu_B + \nu_A.\nu_B|\right) = |A.B| \ .$$

2.
$$|A - B| = \left(|\mu_A - \mu_B|, 1 - |\nu_B - \nu_A|\right) = |B - A| \ .$$

3.
$$|A + B| = \left(|\mu_A + \mu_B|, 1 - |2 - \nu_A - \nu_B|\right)$$

$$|A| + |B| = \left(|\mu_A| + |\mu_B|, 1 - |1 - \nu_A| - |1 - \nu_B|\right) \ .$$

Since
$$|\mu_A + \mu_B| \leq |\mu_A| + |\mu_B|$$

and
$$|2 - \nu_A - \nu_B| = |(1 - \nu_A) + (1 - \nu_B)| \leq |1 - \nu_A| + |1 - \nu_B|$$

then also
$$|A + B| \leq |A| + |B| \ .$$

4. Similarly

$$|A - B| = \left(|\mu_A - \mu_B|, 1 - |1 - \nu_A + 1 - \nu_B|\right)$$

$$|A| - |B| = \left(|\mu_A| - |\mu_B|, 1 - |1 - \nu_A| + |1 - \nu_B|\right) \ .$$

It holds

$$|\mu_A - \mu_B| \geq |\mu_A| - |\mu_B|$$

and also

$$|(1 - \nu_A) - (1 - \nu_B)| \geq |1 - \nu_A| - |1 - \nu_B|$$

therefore

$$|A - B| \geq |A| - |B| \ .$$

5.

$$||A| - |B|| = \left(||\mu_A| - |\mu_B||, 1 - |1 - (1 - |1 - \nu_A| + |1 - \nu_B|)|\right) \ .$$

It holds

$$||\mu_A| - |\mu_B|| \leq |\mu_A - \mu_B|$$

and

$$||1 - \nu_A| - |1 - \nu_B|| \leq |(1 - \nu_A) - (1 - \nu_B)|$$

therefore

$$||A| - |B|| \leq |A - B| \ .$$

□

Lemma 3 *Let $A, B \in \mathcal{G}$ and let $\tilde{\delta} = (\delta, 1 - \delta)$. Then*

$$|A - B| < \tilde{\delta} \Longleftrightarrow A - \tilde{\delta} < B < A + \tilde{\delta} \ .$$

Proof The inequality $|A - B| < \tilde{\delta}$ holds if and only if

$$(|\mu_A - \mu_B|, 1 - |\nu_A - \nu_B|) < (\delta, 1 - \delta)$$

and it holds when

$$|\mu_A - \mu_B| < \delta \text{ and } |\nu_A - \nu_B| < \delta$$

and that is

$$\mu_A - \delta < \mu_B < \mu_A + \delta \text{ and } \nu_A - \delta < \nu_B < \nu_A + \delta \ .$$

On the other hand $A - \tilde{\delta} < B$ if and only if

$$(\mu_A - \delta, \nu_A + \delta) < (\mu_B, \nu_B)$$

and that is

$$\mu_A - \delta < \mu_B \text{ and } \nu_A + \delta > \nu_B \ .$$

Similarly $B < A + \tilde{\delta}$ if and only if

$$(\mu_B, \nu_B) < (\mu_A + \delta, \nu_A - \delta)$$

and that is

$$\mu_B < \mu_A + \delta \text{ and } \nu_B > \nu_A - \delta \ .$$

Therefore it holds

$$\mu_A - \delta < \mu_B < \mu_A + \delta \text{ and } \nu_A - \delta < \nu_B < \nu_A + \delta \ .$$

□

Definition 2 Let $A_0, A, \tilde{\delta} = (\delta, 1 - \delta)$ be from the ℓ-group $\mathcal{G}$. A point A is in the $\tilde{\delta}$-neighborhood of a point A_0 if it holds

$$|A - A_0| < \tilde{\delta} \ .$$

We will use the definition of the neighborhood of the point a lot of times therefore it is useful denote it by $\tilde{\mathcal{U}}(A_0)$. Then the notation $A \in \tilde{\mathcal{U}}(A_0)$ means that

$$A \in (A_0 - \tilde{\delta}, A_0 + \tilde{\delta}) \ .$$

Lemma 4 *Let $A_0, A, \tilde{\delta} = (\delta, 1 - \delta)$ be the elements from the ℓ-group $\mathcal{G}$. The element $A = (\mu_A, \nu_A)$ belongs to $\tilde{\mathcal{U}}(A_0)$ if and only if*

$$\mu_A \in (\mu_{A_0} - \delta, \mu_{A_0} + \delta)$$

and at the same time

$$\nu_A \in (\nu_{A_0} - \delta, \nu_{A_0} + \delta) \ .$$

Proof Since

$$\tilde{\mathcal{U}}(A_0) = (A_0 - \tilde{\delta}, A_0 + \tilde{\delta})$$

then

$$A_0 - \tilde{\delta} = (\mu_{A_0} - \delta, 1 + \nu_{A_0} - (1 - \delta)) = (\mu_{A_0} - \delta, \nu_{A_0} + \delta)$$

and on the other hand

$$A_0 + \tilde{\delta} = (\mu_{A_0} + \delta, 1 + \nu_{A_0} - (1 + \delta)) = (\mu_{A_0} + \delta, \nu_{A_0} - \delta) \ .$$

Therefore

$$A = (\mu_A, \nu_A) \in \tilde{\mathcal{U}}(A_0) \Longleftrightarrow A_0 - \tilde{\delta} < A < A_0 + \tilde{\delta} \ .$$

From the previous inequality it follows

$$\mu_{A_0} - \delta < \mu_A < \mu_{A_0} + \delta$$

$$\nu_{A_0} + \delta > \nu_A > \nu_{A_0} - \delta \ .$$

Therefore

$$\mu_A \in (\mu_{A_0} - \delta, \mu_{A_0} + \delta)$$

and

$$\nu_A \in (\nu_{A_0} - \delta, \nu_{A_0} + \delta) \ .$$

□

Definition 3 Function $\tilde{f}$ is bounded on the interval $[A, B]$ if there exist such $H \in \mathcal{G}$ that for each $X \in [A, B]$ it holds

$$|\tilde{f}(X)| \leq H \ .$$

Now we are ready to define the limit of the function on ℓ-group $\mathcal{G}$.

Definition 4 Denote $\tilde{\varepsilon} = (\varepsilon, 1 - \varepsilon)$ and $\tilde{\delta} = (\delta, 1 - \delta)$. Let $\tilde{f}$ be a function defined on the ℓ-group $\mathcal{G}$ and let $X_0, X, L, \tilde{\varepsilon}, \tilde{\delta}$ be from $\mathcal{G}$. For a function $\tilde{f}$ of a variable X defined on a $\tilde{\delta}$-neighborhood of a point X_0 except possibly the point X_0 itself, for each $\tilde{\varepsilon} > (0, 1)$ there exists $\tilde{\delta} > (0, 1)$ such that $|\tilde{f}(X) - L| < \tilde{\varepsilon}$ holds whenever $(0, 1) < |X - X_0| < \tilde{\delta}$. Then we say that the function $\tilde{f}(X)$ tends to the limit L as X approaches X_0 and we write

$$\lim_{X \to X_0} \tilde{f}(X) = L \ .$$

Lemma 5 *Let* $\lim_{X \to X_0} \tilde{f}(X) = L$. *Then for the point* X_0 *there exist such* $\tilde{\delta} > (0, 1)$ *that the function* $\tilde{f}$ *is bounded on the set* $(0, 1) < |X - X_0| < \tilde{\delta}$.

Proof Since $\lim_{X \to X_0} \tilde{f}(X) = L$ then for $\tilde{\varepsilon} = (1, 0)$ there exist such $\tilde{\delta} > (0, 1)$ that for each X

$$(0, 1) < |X - X_0| < \tilde{\delta}$$

it holds

$$|\tilde{f}(X) - L| < (1, 0)$$

or else (Lemma 3)

$$L - (1, 0) < \tilde{f}(X) < L + (1, 0) \ .$$

Therefore $\tilde{f}$ is the bounded function on

$$(0,1) < |X - X_0| < \tilde{\delta} \ .$$

□

Theorem 1 *Let* $\lim_{X \to X_0} \tilde{f}(X) = L$ *and* $\lim_{X \to X_0} \tilde{g}(X) = K$*. Then there are satisfied the following properties*

1. $\lim_{X \to X_0} \tilde{f}(|X|) = |L|$
2. $\lim_{X \to X_0} (\tilde{f}(X) + \tilde{g}(X)) = L + K$
3. $\lim_{X \to X_0} (\tilde{f}(X).\tilde{g}(X)) = L.K$
4. If $\mu_K \neq 0$ *and* $\nu_K \neq 1$ *then* $\lim_{X \to X_0} \frac{\tilde{f}(X)}{\tilde{g}(X)} = \frac{L}{K}$

Proof 1. Since $\lim_{X \to X_0} \tilde{f}(X) = L$ then for each $\tilde{\varepsilon} > (0,1)$ there exists such $\tilde{\delta} > (0,1)$ that for each $X \in \mathcal{G}$, $(0,1) < |X - X_0| < \tilde{\delta}$ holds $|\tilde{f}(X) - L| < \tilde{\varepsilon}$. However

$$||\tilde{f}(X)| - |L|| \leq |\tilde{f}(X) - L| < \tilde{\varepsilon}$$

and therefore

$$\lim_{X \to X_0} \tilde{f}(|X|) = |L| \ .$$

2. Since $\lim_{X \to X_0} \tilde{f}(X) = L$ then for each $\tilde{\varepsilon} > (0,1)$ there exists such $\tilde{\delta_1} > (0,1)$ that for each $X \in \mathcal{G}$, $(0,1) < |X - X_0| < \tilde{\delta_1}$ it holds

$$|\tilde{f}(X) - L| < \frac{\tilde{\varepsilon}}{2} \ .$$

And since $\lim_{X \to X_0} \tilde{g}(X) = K$ then for each $\tilde{\varepsilon} > (0,1)$ there exists such $\tilde{\delta_2} > (0,1)$ that for each $X \in \mathcal{G}$, $(0,1) < |X - X_0| < \tilde{\delta_2}$ it holds

$$|\tilde{g}(X) - L| < \frac{\tilde{\varepsilon}}{2} \ .$$

Put

$$\tilde{\delta} = \tilde{\delta_1} \wedge \tilde{\delta_2} = (\delta_1 \wedge \delta_2, (1 - \delta_1) \vee (1 - \delta_2))$$

and consider such X that it holds $(0,1) < |X - X_0| < \tilde{\delta}$. Then from Lemma 2 it holds

$$|\tilde{f}(X) + \tilde{g}(X) - (L + K)| \leq |\tilde{f}(X) - L| + |\tilde{g}(X) - K| < \frac{\tilde{\varepsilon}}{2} + \frac{\tilde{\varepsilon}}{2} = \tilde{\varepsilon} \ .$$

3. Since $\lim_{X \to X_0} \tilde{g}(X) = K$, then from Lemma 5 follows that there exist such $\tilde{\delta}_1 > (0,1)$ and such $M \in \mathcal{G}$, $M > (0,1)$ that for each $X \in \mathcal{G}$, $(0,1) < |X - X_0| < \tilde{\delta}_1$ it holds

$$|\tilde{g}(X)| \leq M \ .$$

Put $N = M \vee |L|$ then

$$|\tilde{f}(X).\tilde{g}(X) - L.K| = |\tilde{f}(X).\tilde{g}(X) - L.\tilde{g}(X) + L.\tilde{g}(X) - L.K|$$

$$= |(\tilde{f}(X) - L).\tilde{g}(X) + L.(\tilde{g}(X) - K)| \leq |(\tilde{f}(X) - L)|.|\tilde{g}(X)| + |L|.|(\tilde{g}(X) - K)|$$

$$\leq N.(|(\tilde{f}(X) - L)| + |(\tilde{g}(X) - K)|) \ .$$

Put

$$\tilde{\varepsilon}' = \frac{\tilde{\varepsilon}}{2N}$$

and find such $\tilde{\delta}_2 > (0,1)$, $\tilde{\delta}_3 > (0,1)$ such for each $X \in \mathcal{G}$, $(0,1) < |X - X_0| < \tilde{\delta}_2$ it holds

$$|\tilde{f}(X) - L| < \tilde{\varepsilon}'$$

and for $X \in \mathcal{G}$, $(0,1) < |X - X_0| < \tilde{\delta}_3$ it holds

$$|\tilde{g}(X) - K| < \tilde{\varepsilon}' \ .$$

Then for $\tilde{\delta} = \tilde{\delta}_1 \wedge \tilde{\delta}_2 \wedge \tilde{\delta}_3$ it holds

$$N.(|(\tilde{f}(X) - L)| + |(\tilde{g}(X) - K)|) < N.(\tilde{\varepsilon}' + \tilde{\varepsilon}') = \tilde{\varepsilon} \ .$$

4. For each $L, K \in \mathcal{G}$ where $\mu_K \neq 0$ and $\nu_K \neq 1$ it holds

$$\frac{L}{K} = L.\frac{(0,1)}{K}$$

therefore it is suffice to prove that it holds

$$\lim_{X \to X_0} \frac{(1,0)}{\tilde{g}(X)} = \frac{(1,0)}{K} \ .$$

From the point 1. of this Theorem it follows

$$\lim_{X \to X_0} \tilde{g}(|X|) = |K| > \left|\frac{K}{2}\right| > (0,1) \ .$$

Therefore there exists such $\tilde{\delta_1}$ that for each $X \in \mathcal{G}$, $(0,1) < |X - X_0| < \tilde{\delta_1}$ it holds

$$|\tilde{g}(X)| > \frac{K}{2}$$

and

$$\left|\frac{(1,0)}{\tilde{g}(X)} - \frac{(1,0)}{K}\right| = \left|\frac{K - \tilde{g}(X)}{\tilde{g}(X).K}\right| = \frac{|K - \tilde{g}(X)|}{|\tilde{g}(X)|.|K|} < \frac{|\tilde{g}(X) - K|}{\left|\frac{K}{2}\right|.|K|} = \frac{|\tilde{g}(X) - K|}{\left|\frac{K^2}{2}\right|} .$$

Put

$$\tilde{\varepsilon}' = \frac{K^2.\tilde{\varepsilon}}{2} .$$

Since $\tilde{\varepsilon} > (0,1)$ then also $\tilde{\varepsilon}' > (0,1)$ and there exist such $\tilde{\delta_2}$ that for each $X \in \mathcal{G}$, $(0,1) < |X - X_0| < \tilde{\delta_2}$ it holds

$$|\tilde{g}(X) - K| < \tilde{\varepsilon}' .$$

Then for $\tilde{\delta} = \tilde{\delta_1} \wedge \tilde{\delta_2}$ it holds

$$\frac{|\tilde{g}(X) - K|}{\left|\frac{K^2}{2}\right|} < \frac{\tilde{\varepsilon}'}{\frac{K^2}{2}} = \tilde{\varepsilon}$$

and

$$\lim_{X \to X_0} \frac{(1,0)}{\tilde{g}(X)} = \frac{(1,0)}{K} .$$

From these results we get

$$\lim_{X \to X_0} \frac{\tilde{f}(X)}{\tilde{g}(X)} = \lim_{X \to X_0} \left(\tilde{f}(X).\frac{(1,0)}{\tilde{g}(X)}\right)$$

$$= \lim_{X \to X_0} \tilde{f}(X). \lim_{X \to X_0} \frac{(1,0)}{\tilde{g}(X)} = L.\frac{(1,0)}{K} = \frac{L}{K} .$$

Theorem 2 *Let* $X = (\mu_X, \nu_X)$, $X_0 = (\mu_{X_0}, \nu_{X_0})$, $L = (\mu_L, \nu_L)$ *be the elements from* ℓ*-group* $\mathcal{G}$ *and* $\tilde{f}(X) = (f(\mu_X), 1 - f(1 - \nu_X))$. *Then*

$$\lim_{X \to X_0} \tilde{f}(X) = L$$

if and only if

$$\lim_{\mu_X \to \mu_{X_0}} f(\mu_X) = \mu_L$$

and at the same time

$$\lim_{\nu_X \to \nu_{X_0}} f(1-\nu_X) = 1-\nu_L \ .$$

Proof Let $\lim_{X \to X_0} \tilde{f}(X) = L$. Then it holds $(0,1) < |X - X_0| < \tilde{\delta}$ if and only if $0 < |\mu_X - \mu_{X_0}| < \delta$ and $0 < |\nu_X - \nu_{X_0}| < \delta$. Similarly $|\tilde{f}(X) - L| < \tilde{\varepsilon}$ if and only if $|f(\mu_X) - \mu_L| < \varepsilon$ and $|f(1-\nu_X) - (1-\nu_L)| < \varepsilon$. Therefore

$$\lim_{\mu_X \to \mu_{X_0}} f(\mu_X) = \mu_L$$

and

$$\lim_{\nu_X \to \nu_{X_0}} f(1-\nu_X) = 1-\nu_L \ .$$

□

Example 2 In the classical calculus it holds $\lim_{x \to 0} \frac{\sin x}{x} = 1$. Let $X \in \mathcal{G}$ then function $\tilde{sin}X$ is defined by the following way $\tilde{sin}X = (sin\mu_X, 1 - sin(1-\nu_X))$. Therefore we could show that for each $X \in \mathcal{G}$ it holds

$$\lim_{X \to (0,1)} \frac{\tilde{sin}X}{X} = (1,0) \ .$$

Proof We have

$$\frac{\tilde{sin}X}{X} = \left(\frac{\sin \mu_X}{\mu_X}, 1 - \frac{\sin(1-\nu_X)}{1-\nu_X} \right) \ .$$

Of course

$$\lim_{\mu_X \to 0} \frac{\sin \mu_X}{\mu_X} = 1 \ .$$

Since

$$\lim_{1-\nu_X \to 0} \frac{\sin(1-\nu_X)}{1-\nu_X} = \lim_{\nu_X - 1 \to 0} \frac{\sin(1-\nu_X)}{1-\nu_X} = \lim_{\nu_X \to 1} \frac{\sin(1-\nu_X)}{1-\nu_X} = 1$$

then

$$\lim_{\nu_X \to 1} 1 - \frac{\sin(1-\nu_X)}{1-\nu_X} = 1 - 1 = 0 \ .$$

Therefore

$$\lim_{X \to (0,1)} \frac{\tilde{sin}X}{X} = (1,0) \ .$$

□

Definition 5 The function $\tilde{f}$ is continuous at the point X_0 if it holds

$$\lim_{X \to X_0} \tilde{f}(X) = f(X_0) \ .$$

Function $\tilde{f}$ is continuous if it is continuous at each point of its domain.

Remark 1 Function $\tilde{f} = (f(\mu_X), 1 - f(1 - \nu_X))$ is continuous on the interval $[A, B]$ if and only if f is continuous on the interval $[\mu_A, \mu_B]$ and at the same time on the interval $[\nu_B, \nu_A]$.

Theorem 3 *If function $\tilde{f}$ is continuous on an interval $[A, B]$ then $\tilde{f}$ is also bounded on interval the $[A, B]$.*

Proof We need to show that there exist such $K, L \in \mathbb{R}$ that it holds

$$|\tilde{f}(X)| = (|f(\mu_X)|, 1 - |f(1 - \nu_X)|) \leq (K, 1 - L) \ .$$

Let take some fix point $X_0 \in [A, B]$. Let $X = (\mu_X, \nu_X) \in \tilde{U}(X_0)$ and let $\tilde{f}$ be continuous on an interval $[A, B]$. Then from the previous definitions it holds

$$\tilde{f}(X) = (f(\mu_X), 1 - f(1 - \nu_X)) \in \tilde{V}(\tilde{f}(X_0)) \ .$$

Therefore

$$f(\mu_X) \in (f(\mu_{X_0}) - \varepsilon, f(\mu_{X_0}) + \varepsilon)$$

but

$$(f(\mu_{X_0}) - \varepsilon, f(\mu_{X_0}) + \varepsilon) = |f(\mu_{X_0}) - \varepsilon| = K_0$$

and therefore

$$|f(\mu_X)| \leq K_0 \ .$$

Similarly

$$f(1 - \nu_X) \in (f(1 - \nu_{X_0}) - \varepsilon, f(1 - \nu_{X_0}) + \varepsilon)$$

but

$$(f(1 - \nu_{X_0}) - \varepsilon, f(1 - \nu_{X_0}) + \varepsilon) = |f(1 - \nu_{X_0}) - \varepsilon| = L_0 \ .$$

Therefore

$$|f(1 - \nu_X)| \leq L_0 \ .$$

Then

$$|\tilde{f}(X)| = (|f(\mu_X)|, 1 - |1 - (1 - f(1 - \nu_X))|)$$

$$= (|f(\mu_X)|, 1 - |f(1 - \nu_X)|) \leq (K_0, 1 - L_0) \ .$$

This relations hold for each $X_0 \in [A, B]$. Therefore we could take such $X_1, X_2, \ldots, X_k$ that it hold

$$\bigcup_{i=1}^{k} (\mu_{X_i} - \delta, \mu_{X_i} + \delta) \supset [\mu_A, \mu_B]$$

and

$$\bigcup_{i=1}^{k} (\nu_{X_i} - \delta, \nu_{X_i} + \delta) \supset [\nu_B, \nu_A] \ .$$

Then for any $X \in [A, B]$ it holds

$$|f(\mu_X)| \leq \max\{K_{X_1}, K_{X_2}, ..., K_{X_k}\} = K$$

and similarly

$$|f(1 - \nu_X)| \leq \max\{L_{X_1}, L_{X_2}, ..., L_{X_k}\} = L$$

where $K_{X_i}, L_{X_i}, i = 1, 2, ..., k$ are the values appertaining to X_i. Therefore there exist such $K, L \in \mathbb{R}$ that for each $X \in [A, B]$ it hold

$$|\tilde{f}(X)| = (|f(\mu_X)|, 1 - |f(1 - \nu_X)|) \leq (K, 1 - L)$$

i.e. $\tilde{f}$ is bounded on interval $[A, B]$. □

4 Derivative of the Function

Like in the classical calculus we can also define the derivative of the function with the help of the limit of the function [7].

Definition 6 Let $X_0, H, \tilde{\delta} = (\delta, 1 - \delta)$ be from $\mathcal{G}$ and let $\tilde{f}$ be the function defined for such H that it hold $(0, 1) < |H - X_0| < \tilde{\delta}$. Let

$$\lim_{H \to (0,1)} \frac{\tilde{f}(X_0 + H) - \tilde{f}(X_0)}{H}$$

exist. Then this limit is the derivative of the function $\tilde{f}$ at the point X_0 and we can denote it as $\tilde{f}'(X_0)$.

Remark 2 If we denote $X = X_0 + H$ then of course $H = X - X_0$ and since $H \to (0, 1)$ then $X \to X_0$. Then the definition of the derivative of the function can have also following form

$$\tilde{f}'(X_0) = \lim_{X \to X_0} \frac{\tilde{f}(X) - \tilde{f}(X_0)}{X - X_0} \ .$$

Theorem 4 *Let the function $\tilde{f}$ has the derivative at the point X_0. Then the function $\tilde{f}$ is continuous at the point X_0.*

Proof First we will prove that

$$\tilde{f}(X) - \tilde{f}(X_0) = \frac{\tilde{f}(X) - \tilde{f}(X_0)}{X - X_0}(X - X_0) .$$

$$\frac{\tilde{f}(X) - \tilde{f}(X_0)}{X - X_0}(X - X_0)$$

$$= \left(\frac{f(\mu_X) - f(\mu_{X_0})}{\mu_X - \mu_{X_0}}, \frac{f(1 - \nu_X) - f(1 - \nu_{X_0}) - \nu_X + \nu_{X_0}}{-\nu_X + \nu_{X_0}}\right).(\mu_X - \mu_{X_0}, 1 + \nu_X - \nu_{X_0})$$

$$= (f(\mu_X) - f(\mu_{X_0}), 1 + f(1 - \nu_X) - f(1 - \nu_{X_0})) = \tilde{f}(X) - \tilde{f}(X_0) .$$

Then

$$\tilde{f}(X) = \tilde{f}(X_0) + \frac{\tilde{f}(X) - \tilde{f}(X_0)}{X - X_0}(X - X_0)$$

and therefore

$$\lim_{X \to X_0} \tilde{f}(X) = \tilde{f}(X_0) + \lim_{X \to X_0} \frac{\tilde{f}(X) - \tilde{f}(X_0)}{X - X_0} . \lim_{X \to X_0}(X - X_0)$$

$$= \tilde{f}(X_0) + \tilde{f}'(X_0).(0, 1) = \tilde{f}(X_0) .$$

□

Theorem 5 *Let the functions $\tilde{f}, \tilde{g}$ have the derivatives $\tilde{f}', \tilde{g}'$ at the point X_0. Then there also exist the derivatives of the functions $\tilde{f} + \tilde{g}, \tilde{f} - \tilde{g}, \tilde{f}.\tilde{g}$ and $\frac{\tilde{f}}{\tilde{g}}$ at the point X_0 and it holds*

1. $(\tilde{f} + \tilde{g})'(X_0) = \tilde{f}'(X_0) + \tilde{g}'(X_0)$
2. $(\tilde{f} - \tilde{g})'(X_0) = \tilde{f}'(X_0) - \tilde{g}'(X_0)$
3. $(\tilde{f}.\tilde{g})'(X_0) = \tilde{f}'(X_0).\tilde{g}(X_0) + \tilde{f}(X_0).\tilde{g}'(X_0)$
4. *If* $g(\mu_X) \neq 0$ *and* $g(1 - \nu_X) \neq 1$ *then* $\left(\frac{\tilde{f}}{\tilde{g}}\right)'(X_0) = \frac{\tilde{f}'(X_0).\tilde{g}(X_0) - \tilde{f}(X_0).\tilde{g}'(X_0)}{\tilde{g}^2(X_0)}$.

Proof From the definition of the derivative and from the properties of the limit of functions it holds

1.

$$(\tilde{f} + \tilde{g})'(X_0) = \lim_{X \to X_0} \frac{(\tilde{f}(X) + \tilde{g}(X)) - (\tilde{f}(X_0) + \tilde{g}(X_0))}{X - X_0}$$

$$= \lim_{X \to X_0} \frac{\tilde{f}(X) - \tilde{f}(X_0)}{X - X_0} + \lim_{X \to X_0} \frac{\tilde{g}(X) - \tilde{g}(X_0)}{X - X_0} = \tilde{f}'(X_0) + \tilde{g}'(X_0) .$$

2. The proof of the second property is similar like the proof of the point 1.
3.

$$(\tilde{f}.\tilde{g})'(X_0) = \lim_{X \to X_0} \frac{\tilde{f}(X).\tilde{g}(X) - \tilde{f}(X_0).\tilde{g}(X_0)}{X - X_0}$$

$$= \lim_{X \to X_0} \frac{\tilde{f}(X).\tilde{g}(X) - \tilde{f}(X_0).\tilde{g}(X) + \tilde{f}(X_0).\tilde{g}(X) - \tilde{f}(X_0).\tilde{g}(X_0)}{X - X_0}$$

$$= \lim_{X \to X_0} \frac{\tilde{f}(X) - \tilde{f}(X_0)}{X - X_0}.\tilde{g}(X_0) + \lim_{X \to X_0} \frac{\tilde{g}(X) - \tilde{g}(X_0)}{X - X_0}.\tilde{f}(X_0)$$

$$= \tilde{f}'(X_0).\tilde{g}(X_0) + \tilde{f}(X_0).\tilde{g}'(X_0) \ .$$

4. Let $g(\mu_X) \neq 0$ and $g(1 - \nu_X) \neq 1$ then

$$\left(\frac{\tilde{f}}{\tilde{g}}\right)'(X_0) = \lim_{X \to X_0} \frac{\frac{\tilde{f}(X)}{\tilde{g}(X)} - \frac{\tilde{f}(X_0)}{\tilde{g}(X_0)}}{X - X_0} = \lim_{X \to X_0} \frac{\tilde{f}(X).\tilde{g}(X_0) - \tilde{f}(X_0).\tilde{g}(X)}{\tilde{g}(X).\tilde{g}(X_0).(X - X_0)}$$

$$= \lim_{X \to X_0} \frac{\frac{\tilde{f}(X) - \tilde{f}(X_0)}{X - X_0}.\tilde{g}(X) - \frac{\tilde{g}(X) - \tilde{g}(X_0)}{X - X_0}.\tilde{f}(X_0)}{\tilde{g}(X).\tilde{g}(X_0)} = \frac{\tilde{f}'(X_0).\tilde{g}(X_0) - \tilde{f}(X_0).\tilde{g}'(X_0)}{\tilde{g}^2(X_0)} \ .$$

□

Remark 3 Let for each $X \in [A, B]$ there exist $\tilde{f}'(X)$ then it holds

$$\tilde{f}'(X) = (f'(\mu_X), 1 - f'(1 - \nu_X)) \ .$$

5 Lagrange Mean Value Theorem

Our main aim is to proof the Taylor's theorem. Of course before we do this we need to prove some more theorems. Lagrange mean value theorem is one of them. The results which are mentioned in this section were published in the paper [8].

Theorem 6 *Let $\tilde{f}$ be continuous on $[A, B]$ and differentiable on (A, B). Then there exists $C \in (A, B)$ such that*

$$\tilde{f}(B) - \tilde{f}(A) = \tilde{f}'(C)(B - A) \ .$$

Proof By the definition

$$\tilde{f}(B) - \tilde{f}(A) = (f(\mu_B), 1 - f(1 - \nu_B)) - (f(\mu_A), 1 - f(1 - \nu_A))$$

$$= (f(\mu_B) - f(\mu_A), f(1 - \nu_A) - f(1 - \nu_B) + 1) \ .$$

In the classical calculus for each function $f : \mathbb{R} \to \mathbb{R}$ which is continuous on $[a, b]$ and differentiable on (a, b) there exists such $c \in (\mu_A, \mu_B)$ that it holds

$$f(b) - f(a) = f'(c)(b - a) .$$

Since these assumptions hold we could put $a = \mu_A$, $b = \mu_B$ and $c = \mu_C$. Then $\mu_C \in (\mu_A, \mu_B)$ and

$$f(\mu_B) - f(\mu_A) = f'(\mu_C)(\mu_B - \mu_A) .$$

Similarly there exists such $\nu_C \in (\nu_B, \nu_A)$ that $1 - \nu_A \leq 1 - \nu_C \leq 1 - \nu_B$ and

$$f(1 - \nu_A) - f(1 - \nu_B) = f'(1 - \nu_C)(1 - \nu_A - (1 - \nu_B))$$

$$= f'(1 - \nu_C)(\nu_B - \nu_A) .$$

Define $C = (\mu_C, \nu_C)$. Then $\mu_A \leq \mu_C \leq \mu_B, \nu_A \geq \nu_C \geq \nu_B$, hence $A \leq C \leq B$. Moreover

$$\tilde{f}'(C) = (f'(\mu_C), 1 - f'(1 - \nu_C)) .$$

Therefore

$$\tilde{f}(B) - \tilde{f}(A) = (f(\mu_B) - f(\mu_A), f(1 - \nu_A) - f(1 - \nu_B) + 1)$$

$$= (f'(\mu_C)(\mu_B - \mu_A), f'(1 - \nu_C)(\nu_B - \nu_A) + 1) .$$

On the other hand

$$\tilde{f}'(C)(B - A) = (f'(\mu_C), 1 - f'(1 - \nu_C))(\mu_B - \mu_A, \nu_B - \nu_A + 1)$$

$$= (f'(\mu_C)(\mu_B - \mu_A), 1 - (1 - (1 - f'(1 - \nu_C)))(1 - (\nu_B - \nu_A + 1)))$$

$$= (f'(\mu_C)(\mu_B - \mu_A), 1 - f'(1 - \nu_C)(\nu_A - \nu_B))$$

$$= (f'(\mu_C)(\mu_B - \mu_A), f'(1 - \nu_C)(\nu_B - \nu_A) + 1) = \tilde{f}(B) - \tilde{f}(A) .$$

□

6 Polynomial Functions

In this section the definition of the polynomial function is given. In previous text we prove that it holds

$$\tilde{f}'(X) = \big(f'(\mu_X), 1 - f'(1 - \nu_X)\big) .$$

It is easy to prove (see [9]) that for the nth derivative it holds

$$\tilde{f}^{(n)}(X) = \left(f^{(n)}(\mu_X), 1 - f^{(n)}(1 - \nu_X)\right) \ .$$

Definition 7 Let $n \in \mathbb{N}$, let $X=(\mu_X, \nu_X)$ be the variable and $A_i \in \mathbb{R}^2$, $\mathrm{A_i} = (\mu_{\mathrm{A_i}}, \nu_{\mathrm{A_i}})$, $i = 0, 1, 2, \dots, n$ be the constants, $A_n \neq (0, 1)$. Then

$$\tilde{p}_n(X) = A_0 + A_1 X + A_2 X^2 + \cdots + A_n X^n$$

is called the polynomial function.

Theorem 7 *Let $n \in \mathbb{N}$, let $X = (\mu_X, \nu_X)$ be the variable, $A_i \in \mathbb{R}^2$, $\mathrm{A_i} = (\mu_{\mathrm{A_i}}, \nu_{\mathrm{A_i}})$, $i = 0, 1, 2, \dots, n$ be the constants and $\tilde{p}_n$ be a polynomial function. Then it holds*

$$\tilde{p}_n((\mu_X, \nu_X)) = (\mu_{A_0} + \mu_{A_1}\mu_X + \mu_{A_2}\mu_X^2 + \cdots + \mu_{A_n}\mu_X^n,$$

$$\nu_{A_0} + (\nu_{A_1} - 1)(1 - \nu_X) + (\nu_{A_2} - 1)(1 - \nu_X)^2 + \cdots + (\nu_{A_n} - 1)(1 - \nu_X)^n)$$

Proof From the definitions it follows

$$X^n = ((\mu_X)^n, 1 - (1 - \nu_X)^n) \ .$$

For clarification of the next text we will use shorter notation

$$X^n = (\mu_X^n, 1 - (1 - \nu_X)^n) \ .$$

Then

$$AX^n = (\mu_A \mu_X^n, \nu_A + (1 - (1 - \nu_X)^n) - \nu_A(1 - (1 - \nu_X)^n)) \ .$$

The second component could be modified by the following way

$$\nu_A + (1 - (1 - \nu_X)^n) - \nu_A(1 - (1 - \nu_X)^n) - 1 + 1$$

$$= \nu_A(1 - (1 - (1 - \nu_X)^n)) - 1(1 - (1 - (1 - \nu_X)^n)) + 1$$

$$= (\nu_A - 1)(1 - \nu_X)^n + 1 \ .$$

Therefore

$$AX^n = (\mu_A \mu_X^n, (\nu_A - 1)(1 - \nu_X)^n + 1) \ .$$

Then we obtain

$$A_0 = A_0X^0 = (\mu_{A_0}\mu_X^0, (\nu_{A_0} - 1)(1 - \nu_X)^0 + 1) = (\mu_{A_0}, \nu_{A_0}) = B_0$$
$$A_1X^1 = (\mu_{A_1}\mu_X^1, (\nu_{A_1} - 1)(1 - \nu_X)^1 + 1) = B_1$$
$$A_2X^2 = (\mu_{A_2}\mu_X^2, (\nu_{A_2} - 1)(1 - \nu_X)^2 + 1) = B_2$$
$$\vdots$$
$$A_nX^n = (\mu_{A_n}\mu_X^n, (\nu_{A_n} - 1)(1 - \nu_X)^n + 1) = B_n$$

Therefore

$$\tilde{p}_n(X) = \sum_{i=0}^{n} B_i = \left(\sum_{i=0}^{n} \mu_{A_i}\mu_X^i, \sum_{i=0}^{n} (((\nu_{A_i} - 1)(1 - \nu_X)^i + 1) - 1) + 1 \right)$$

$$= \left(\sum_{i=0}^{n} \mu_{A_i}\mu_X^i, \sum_{l=0}^{n} (((\nu_{A_i} - 1)(1 - \nu_X)^i) + 1) \right)$$

$$= \Big(\mu_{A_0} + \mu_{A_1}\mu_X + \mu_{A_2}\mu_X^2 + \cdots + \mu_{A_n}\mu_X^n,$$

$$\nu_{A_0} + (\nu_{A_1} - 1)(1 - \nu_X) + (\nu_{A_2} - 1)(1 - \nu_X)^2 + \cdots + (\nu_{A_n} - 1)(1 - \nu_X)^n\Big). \quad \square$$

Remark 4 We could use also the approach that polynomial function is a special type of a function. Then

$$\tilde{p}_n((\mu_X, \nu_X)) = (p_n(\mu_X), 1 - p_n(1 - \nu_x))$$

where

$$p_n(\mu_X) = \mu_{A_0} + \mu_{A_1}\mu_X + \mu_{A_2}\mu_X^2 + \cdots + \mu_{A_n}\mu_X^n,$$

and

$$1 - p_n(1 - \nu_X) = 1 - [(1 - \nu_{A_0}) + (1 - \nu_{A_1})(1 - \nu_X)$$

$$+(1 - \nu_{A_2})(1 - \nu_X)^2 + \cdots + (1 - \nu_{A_n})(1 - \nu_X)^n]$$

$$= \nu_{A_0} + (\nu_{A_1} - 1)(1 - \nu_X) + (\nu_{A_2} - 1)(1 - \nu_X)^2 + \cdots + (\nu_{A_n} - 1)(1 - \nu_X)^n$$

7 Taylor Polynomial and Taylor's Theorem

In the previous section there was proved that Lagrange mean value theorem could be established also on the ℓ-group $\mathcal{G}$. Since this theorem holds we could answer the question if it possible to prove Taylor's theorem on the ℓ-group $\mathcal{G}$. Let us start with the definition of the nth-order Taylor polynomial.

Definition 8 Let $n \in \mathbb{N}$, $X = (\mu_X, \nu_X)$ be the variable and $X_0 = (\mu_{X_0}, \nu_{X_0})$, $X_0 \in \mathbb{R}^2$ be the fixed point. Let $\tilde{f}(X) = (f(\mu_X), 1 - f(1 - \nu_X))$ be a function defined on ℓ-group $\mathcal{G}$. Let the derivatives $\tilde{f}^{(i)}(X) = (f^{(i)}(\mu_X), 1 - f^{(i)}(1 - \nu_X))$ exist for $i = 1, 2, \ldots, n$. Then the nth-order Taylor polynomial at the point X_0 has the following form

$$\tilde{T}_n(X) = \tilde{f}(X_0) + \frac{\tilde{f}^{(1)}(X_0)}{1!}(X - X_0) + \frac{\tilde{f}^{(2)}(X_0)}{2!}(X - X_0)^2 + \cdots + \frac{\tilde{f}^{(n)}(X_0)}{n!}(X - X_0)^n .$$

Theorem 8 *Let the assumptions of the previous definition hold. Function $\tilde{T}_n$ is the nth-order Taylor polynomial at the point $X_0 = (\mu_{X_0}, \nu_{X_0})$ if and only if for any $X = (\mu_X, \nu_X)$ it holds*

$$\tilde{T}_n((\mu_X, \nu_X)) = \Big(f(\mu_{X_0}) + \frac{f^{(1)}(\mu_{X_0})}{1!}(\mu_X - \mu_{X_0}) + \cdots + \frac{f^{(n)}(\mu_{X_0})}{n!}(\mu_X - \mu_{X_0})^n,$$

$$1 - (f(1 - \nu_{X_0}) + \frac{f^{(1)}(1 - \nu_{X_0})}{1!}(\nu_{X_0} - \nu_X) + \cdots + \frac{f^{(n)}(1 - \nu_{X_0})}{n!}(\nu_{X_0} - \nu_X)^n) \Big) .$$

Proof Since

$$\begin{aligned} X^n &= (\mu_X^n, 1 - (1 - \nu_X)^n) \\ X - X_0 &= (\mu_X - \mu_{X_0}, \nu_X - \nu_{X_0} + 1) \\ \tilde{f}^{(n)}(X_0) &= (f^{(n)}(\mu_{X_0}), 1 - f^{(n)}(1 - \nu_{X_0})) \\ cX &= (c\mu_X, 1 - c(1 - \nu_X)) \end{aligned}$$

where c is any real number. Then

$$\begin{aligned} (X - X_0)^n &= ((\mu_X - \mu_{X_0})^n, 1 - (1 - (\nu_X - \nu_{X_0} + 1))^n) \\ &= ((\mu_X - \mu_{X_0})^n, 1 - (\nu_{X_0} - \nu_X)^n) \end{aligned}$$

and

$$\tilde{f}^{(n)}(X_0)(X - X_0)^n = (f^{(n)}(\mu_{X_0})(\mu_X - \mu_{X_0})^n,$$

$$1 - f^{(n)}(1 - \nu_{X_0}) + 1 - (\nu_{X_0} - \nu_X)^n - ((1 - f^{(n)}(1 - \nu_{X_0}))(1 - (\nu_{X_0} - \nu_X)^n))) .$$

After the modification of the second part we obtain

$$1 - f^{(n)}(1 - \nu_{X_0}) + 1 - (\nu_{X_0} - \nu_X)^n - ((1 - f^{(n)}(1 - \nu_{X_0}))(1 - (\nu_{X_0} - \nu_X)^n)))$$

$$= 1 - f^{(n)}(1 - \nu_{X_0}) + 1 - (\nu_{X_0} - \nu_X)^n$$

$$-1 + (\nu_{X_0} - \nu_X)^n + f^{(n)}(1 - \nu_{X_0}) - f^{(n)}(1 - \nu_{X_0})(\nu_{X_0} - \nu_X)^n$$

$$= 1 - f^{(n)}(1 - \nu_{X_0})(\nu_{X_0} - \nu_X)^n \ .$$

Therefore

$$\tilde{f}^{(n)}(X_0)(X - X_0)^n = (f^{(n)}(\mu_{X_0})(\mu_X - \mu_{X_0})^n, 1 - f^{(n)}(1 - \nu_{X_0})(\nu_{X_0} - \nu_X)^n) \ .$$

Finally for any $c \in \mathbb{R}$ it holds

$$c\tilde{f}^{(n)}(X_0)(X - X_0)^n = (cf^{(n)}(\mu_{X_0})(\mu_X - \mu_{X_0})^n,$$

$$1 - c(1 - (1 - f^{(n)}(1 - \nu_{X_0})(\nu_{X_0} - \nu_X)^n)))$$

hence

$$c\tilde{f}^{(n)}(X_0)(X - X_0)^n = (cf^{(n)}(\mu_{X_0})(\mu_X - \mu_{X_0})^n, 1 - c(f^{(n)}(1 - \nu_{X_0})(\nu_{X_0} - \nu_X)^n)) \ .$$

Put $c_i = \frac{1}{i!}$ for $i = 1, 2, \ldots, n$ then

$$\frac{\tilde{f}^{(i)}(X_0)}{i!}(X - X_0)^i = \left(\frac{f^{(i)}(\mu_{X_0})}{i!}(\mu_X - \mu_{X_0})^i, 1 - \frac{f^{(i)}(1 - \nu_{X_0})}{i!}(\nu_{X_0} - \nu_X)^i\right)$$

is the ith member of the Taylor polynomial. After the summation of all members of Taylor polynomial we get

$$\tilde{T}_n((\mu_X, \nu_X))$$

$$= \left(\sum_{i=0}^{n} \frac{f^{(i)}(\mu_{X_0})}{i!}(\mu_X - \mu_{X_0})^i, \sum_{i=0}^{n}\left(1 - \frac{f^{(i)}(1 - \nu_{X_0})}{i!}(\nu_{X_0} - \nu_X)^i - 1\right) + 1\right)$$

$$= \left(\sum_{i=0}^{n} \frac{f^{(i)}(\mu_{X_0})}{i!}(\mu_X - \mu_{X_0})^i, \sum_{i=0}^{n}\left(-\frac{f^{(i)}(1 - \nu_{X_0})}{i!}(\nu_{X_0} - \nu_X)^i\right) + 1\right)$$

$$= \left(f(\mu_{X_0}) + \frac{f^{(1)}(\mu_{X_0})}{1!}(\mu_X - \mu_{X_0}) + \cdots + \frac{f^{(n)}(\mu_{X_0})}{n!}(\mu_X - \mu_{X_0})^n,\right.$$

$$\left. 1 - \left(f(1 - \nu_{X_0}) + \frac{f^{(1)}(1 - \nu_{X_0})}{1!}(\nu_{X_0} - \nu_X) + \cdots + \frac{f^{(n)}(1 - \nu_{X_0})}{n!}(\nu_{X_0} - \nu_X)^n\right)\right) . \qquad \square$$

Theorem 9 *Let $\tilde{f}$ be a function that has continuous derivatives $\tilde{f}^{(i)}$, $i = 0, 1, 2, \ldots, n$ defined on interval $[X_0, X]$, let there exists derivative $\tilde{f}^{(n+1)}$ on interval (X_0, X) and $\tilde{T}_n$ be the nth-order Taylor polynomial appertaining to $\tilde{f}$ in the point X_0. Then there exist such function $\tilde{R}_n$ and such $C = (\mu_C, \nu_C)$, $C \in (X_0, X)$ that it holds*

$$\tilde{f}(X) = \tilde{T}_n(X) + \tilde{R}_n(X) .$$

The function $\tilde{R}_n$ is usually called remainder and it could have following form

$$\tilde{R}_n(X) = \frac{\tilde{f}^{(n+1)}(C)}{(n+1)!}(X - X_0)^{n+1}$$

(Lagrange's form).

Proof In the first step we will specify the remainder

$$\tilde{R}_n(X) = \frac{\tilde{f}^{(n+1)}(C)}{(n+1)!}(X - X_0)^{n+1}$$

by using membership and nonmembership functions.

$$\tilde{R}_n((\mu_X, \nu_X)) = \left(\frac{f^{(n+1)}(\mu_C)}{(n+1)!}(\mu_X - \mu_{X_0})^{n+1}, 1 - \frac{f^{(n+1)}(1 - \nu_C)}{(n+1)!}(\nu_{X_0} - \nu_X)^{n+1}\right) .$$

It is also important not to forget that if $C \in (X_0, X)$ then it holds

$$\mu_{X_0} \leq \mu_C \leq \mu_X$$

and

$$\nu_{X_0} \geq \nu_C \geq \nu_X .$$

Then from equality

$$\tilde{f}(X) = \tilde{T}_n(X) + \tilde{R}_n(X)$$

for membership function it follows

$$f(\mu_X) = f(\mu_{X_0}) + \frac{f^{(1)}(\mu_{X_0})}{1!}(\mu_X - \mu_{X_0}) + \cdots +$$

$$+\frac{f^{(n)}(\mu_{X_0})}{n!}(\mu_X-\mu_{X_0})^n+\frac{f^{(n+1)}(\mu_C)}{(n+1)!}(\mu_X-\mu_{X_0})^{n+1}$$

and for nonmembership function it follows

$$f(1-\nu_X)=f(1-\nu_{X_0})+\frac{f^{(1)}(1-\nu_{X_0})}{1!}(\nu_{X_0}-\nu_X)+\cdots+$$

$$+\frac{f^{(n)}(1-\nu_{X_0})}{n!}(\nu_{X_0}-\nu_X)^n+\frac{f^{(n+1)}(1-\nu_C)}{(n+1)!}(\nu_{X_0}-\nu_X)^{n+1}\ .$$

Since μ_{X_0}, μ_X, μ_C are the real numbers $\mu_{X_0}\le\mu_C\le\mu_X$ and in addition Lagrange's theorem holds then for membership function we get the formula which is equal with Taylor's theorem in classical calculus. Therefore also the proof of this part has the same steps as in real numbers calculus.

On the other hand ν_{X_0}, ν_X, ν_C are also real numbers and it holds $\nu_{X_0}\ge\nu_C\ge\nu_X$. This inequality is the same as $1-\nu_{X_0}\le 1-\nu_C\le 1-\nu_X$. Denote $1-\nu_{X_0}=y_0$, $1-\nu_X=y$ and $1-\nu_C=c$. Then

$$\nu_{X_0}-\nu_X=1-\nu_X-(1-\nu_{X_0})=y-y_0\ .$$

After substitution the values and rewriting the formula for nonmembership function we get

$$f(y)=f(y_0)+\frac{f^{(1)}(y_0)}{1!}(y-y_0)+\cdots+\frac{f^{(n)}(y_0)}{n!}(y-y_0)^n+\frac{f^{(n+1)}(c)}{(n+1)!}(y-y_0)^{n+1}$$

what is again the formula equal with formula in classical calculus. □

Remark 5 From the proof of previous theorem it is easy to see that for remainder holds

$$\lim_{n\to\infty}\tilde{R}_n(X)=(0,1)\ .$$

Therefore we could define Taylor series for the elements from ℓ-group $\mathcal{G}$.

Definition 9 Let $n\in\mathbb{N}$, $X=(\mu_X,\nu_X)$ be the variable and $X_0=(\mu_{X_0},\nu_{X_0})$, $X_0\in\mathbb{R}^2$ be the fixed point. Let $\tilde{f}(X)=(f(\mu_X), 1-f(1-\nu_X))$ be a function defined on ℓ-group $\mathcal{G}$. Let $\tilde{f}$ be infinitely differentiable at the point X_0. Then the Taylor series at the point X_0 has the following form

$$\tilde{T}(X)=\tilde{f}(X_0)+\frac{\tilde{f}^{(1)}(X_0)}{1!}(X-X_0)+\frac{\tilde{f}^{(2)}(X_0)}{2!}(X-X_0)^2+\cdots+\frac{\tilde{f}^{(n)}(X_0)}{n!}(X-X_0)^n\ldots\ .$$

Since

$$\lim_{n\to\infty}\tilde{R}_n(X)=(0,1)$$

then it holds

$$\tilde{f}(X) \approx \tilde{T}(X) .$$

Example 3 Since Definition 9 holds then each function $\tilde{f}$ can be replaced by corresponding Taylor series. For example let $\tilde{f}(X) = \tilde{\sin}(X) = (\sin(\mu_X), 1 - \sin(1 - \nu_X))$ and $X_0 = (0, 1)$. Then $1 - \nu_{X_0} = 1 - 1 = 0$ and therefore

$$\begin{aligned} f(\mu_{X_0}) &= f(1 - \nu_{X_0}) &= \sin 0 = 0 \\ f^{(1)}(\mu_{X_0}) &= f^{(1)}(1 - \nu_{X_0}) &= \cos 0 = 1 \\ f^{(2)}(\mu_{X_0}) &= f^{(2)}(1 - \nu_{X_0}) &= -\sin 0 = 0 \\ f^{(3)}(\mu_{X_0}) &= f^{(3)}(1 - \nu_{X_0}) &= -\cos 0 = -1 \\ &\vdots \end{aligned}$$

Then

$$\tilde{\sin}(X) \approx \tilde{T}((\mu_X, \nu_X))$$

$$= \left(\frac{1}{1!}\mu_X - \frac{1}{3!}\mu_X^3 + \cdots + \frac{(-1)^n}{(2n+1)!}\mu_X^{2n+1} \ldots, \right.$$

$$\left. 1 - \left(\frac{1}{1!}(1 - \nu_X) - \frac{1}{3!}(1 - \nu_X)^3 + \cdots + \frac{(-1)^n}{(2n+1)!}(1 - \nu_X)^{2n+1} \ldots \right)\right) .$$

8 Conclusion

In this contribution the basic structures of differential calculus for ℓ-groups $\mathcal{G}$ were defined. Since we showed that each operation defined on *IF* set can be derived from the operation defined on ℓ-group $\mathcal{G}$ we could use these results also for *IF* sets.

Acknowledgments The support of the grant VEGA 1/0120/14 is kindly announced.

References

1. Zadeh, L.A.: Fuzzy sets. In: Information and Control, vol. 8, pp. 338–353 (1965)
2. Atanassov, K.: Intuitionistic fuzzy sets. Fuzzy Sets Syst. **20**(1), 87–96 (1986)
3. Atanassov, K.: Intuitionistic Fuzzy Sets. Springer, Berlin (1999)
4. Hollá, I., Riečan, B.: Elementary function on *IF* sets. In: Advances in Fuzzy Stes, Intuitionistic Fuzzy Sets, Generaliyed Nets and Related Topics: Foundations. vol. I, pp. 193–201. Academic Publishing House EXIT, Warszawa (2008)

5. Michalíková, A.: Some notes about boundaries on IF sets. In: New Trends in Fuzzy Sets, Intuitionistic Fuzzy Sets, Generalized nets and Related Topics: Foundations. Systems Research Institute, Polish Academy of Sciences, vol I, pp. 105–113. Poland (2013)
6. Michalíková, A: Absolute value limit of the function defined on IF sets. In: Proceedings of the sixteenth International Conference on Intuitionistic Fuzzy Sets. vol. 18, issu. 3, pp. 8–15. Sofia, Bulgaria (2012)
7. Michalíková, A.: The differential calculus on IF sets. In: FUZZ-IEEE 2009. International Conference on Fuzzy Systems. Proccedings [CD-ROM], pp. 1393–1395. Jeju Island, Korea (2009)
8. Riečan, B.: On Lagrange mean value theorem for functions on Atanassov *IF* sets. In: Proceedings of the Eighth International Workshop on Intuitionistic Fuzzy Sets, vol. 18, issu. 4, pp. 8–11. Banská Bystrica, Slovakia, 9 Oct 2012
9. Michalíková, A.: Taylor's theorem for functions defined on Atanassov IF-sets. In: Notes on Intuitionistic Fuzzy Sets. Academic Publishing House, Sofia, Bulgaria. vol. 19, issu. 3, pp. 34–41 (2013)

Recognizing Imbalanced Classes by an Intuitionistic Fuzzy Classifier

Eulalia Szmidt, Janusz Kacprzyk and Marta Kukier

Abstract The recognition of imbalanced classes is not an easy task for classifiers. Imbalanced classes are classes that are considerably smaller than other classes but not necessarily small ones. Most often smaller classes are more interesting from the user's point of view but more difficult to be derived by a classifier. In this paper, which is a continuation of our previous works, we discuss a classifier using some inherent features of Atanassov's intuitionistic fuzzy sets (A-IFSs, for short) making them a good tool for recognizing imbalanced classes. We illustrate our considerations on benchmark examples paying attention to the behavior of the classifier proposed (several measures in addition to the most popular accuracy are examined). We use a simple cross validation method (with 10 experiments). Results are compared with those obtained by a fuzzy classifier known as a good one from the literature. We also consider a problem of granulation (a symmetric or asymmetric granulation, and a number of the intervals used) and its influence on the results.

Keywords Classification · Imbalanced classes · Intuitionistic fuzzy sets · Intuitionistic fuzzy classifier · Granulation

1 Introduction

The construction of a good classifier for imbalanced classes is a difficult task. An imbalanced class need not be a small class—it may be a class with lots of elements but still far less that other class. Usually, a two-category problem (Duda [16]),

E. Szmidt (✉) · J. Kacprzyk · M. Kukier
Systems Research Institute, Polish Academy of Sciences,
Ul. Newelska 6, 01–447 Warsaw, Poland
e-mail: szmidt@ibspan.waw.pl

J. Kacprzyk
e-mail: kacprzyk@ibspan.waw.pl

E. Szmidt · J. Kacprzyk · M. Kukier
Warsaw School of Information Technology, Ul. Newelska 6,
01–447 Warsaw, Poland

P. Angelov and S. Sotirov (eds.), *Imprecision and Uncertainty in Information Representation and Processing*, Studies in Fuzziness and Soft Computing 332,
DOI 10.1007/978-3-319-26302-1_15

positive/negative, called also the *legal/illegal* classification problem with a relatively small class is considered. The construction of a classifier for such classes is both an interesting theoretical challenge and a problem often met in different types of real tasks like, e.g., medical diagnosis, medical monitoring, fraud detection, bioinformatics, text categorization (cf. Fawcett and Provost [17], Japkowicz [19], Kubat et al. [20], Lewis and Catlett [21], Mladenic and Grobelnik [22], He and Garcia [18]). To solve the imbalance classes problems usually up-sampling and down-sampling are used but both methods interfere with the structure of the data, and in the case of overlapping classes even the artificially obtained balance does not solve the problem (some data points may appear as valid examples in both classes).

This paper is a continuation of our previous works (cf. Szmidt and Kukier [37–39]) on intuitionistic fuzzy approach to the problem of classification of imbalanced and overlapping classes. We consider a two–class classification problem (*legal*—a relatively small class, and *illegal*—a much bigger class).

The concept of a classifier using the A-IFSs has its roots in the fuzzy set approach proposed by Baldwin et al. [11]. In that approach classes are represented by fuzzy sets generated from relative frequency distributions representing data points used as examples of the classes [11]. In the process of generating fuzzy sets a mass assignment based approach is adopted (Baldwin et al. [8, 11]). For the model obtained (fuzzy sets describing the classes), using the chosen classification rule, a testing phase is performed to evaluate the performance of the proposed method.

In the case of the intuitionistic fuzzy classifier we perform the same steps as in the case of the above mentioned fuzzy classifier. The main difference is in the use of the A-IFSs for the representation of classes, and in taking advantage of the use of the A-IFSs to obtain a classifier which better recognizes relatively small classes.

The crucial step of the method is the representation of classes by the A-IFSs (first, in the training phase). The A-IFSs are generated from the relative frequency distributions representing the data considered according to the procedure given by Szmidt and Baldwin [27]. Trying to recognize the smaller class as good as possible, we use information about the hesitation margins making it possible to improve the results of data classification in the (second) testing phase.

The results obtained in the testing phase are examined using confusion matrices making it possible to explore the behavior of the classifier in a broader sense, not only in the sense of the widely used error/accuracy evaluation. We have used a simple cross validation method (with 10 experiments). the results obtained are compared with those obtained by a fuzzy classifier. Several benchmark data sets are used, exemplified by the "Glass", and "Wine" (cf. [44]).

We have also taken into account other measures of classifier errors, namely, the geometric mean, and the so called F-*value* (Sect. 3.1). The last two measures were used to assess the influence of the parameters used in one of the important steps while constructing the classifier, namely, granulation. We compared results for the symmetric and asymmetric granulation, and for an increasing number of intervals. The influence of granulation on the results has been verified using some data sets, notably "Glass", and "Wine", "Heart" and "Breast Cancer" (cf. [44]).

2 A Brief Introduction to the A-IFSs

One of the possible generalizations of a fuzzy set in X (Zadeh [42]) given by

$$A' = \{< x, \mu_{A'}(x) > | x \in X\} \tag{1}$$

where $\mu_{A'}(x) \in [0, 1]$ is the membership function of the fuzzy set A', is the A-IFS (Atanassov [1, 3, 4]) A is given by

$$A = \{< x, \mu_A(x), \nu_A(x) > | x \in X\} \tag{2}$$

where: $\mu_A : X \to [0, 1]$ and $\nu_A : X \to [0, 1]$ such that

$$0 \leq \mu_A(x) + \nu_A(x) \leq 1 \tag{3}$$

and $\mu_A(x)$, $\nu_A(x) \in [0, 1]$ denote a degree of membership and a degree of non-membership of $x \subset A$, respectively. (An approach to the assigning memberships and non-memberships for A-IFSs from data is proposed by Szmidt and Baldwin [28]).

Obviously, each fuzzy set may be represented by the following A-IFS: $A = \{< x, \mu_{A'}(x), 1 - \mu_{A'}(x) > | x \in X\}$.

An additional concept for each A-IFS in X, that is not only an obvious result of (2) and (3) but which is also relevant for applications, will be called (Atanassov [3]) a *hesitation margin* of $x \in A$, written

$$\pi_A(x) = 1 - \mu_A(x) - \nu_A(x) \tag{4}$$

which expresses a lack of knowledge of whether x belongs to A or not (cf. Atanassov [3]). It is obvious that $0 \leq \pi_A(x) \leq 1$, for each $x \in X$.

The hesitation margin turns out to be important while considering the distances (Szmidt and Kacprzyk [29, 30, 34], entropy (Szmidt and Kacprzyk [31, 35]), similarity (Szmidt and Kacprzyk [36]) for the A-IFSs, etc. i.e., the measures that play a crucial role in virtually all information processing tasks (Szmidt [24]).

The hesitation margin turns out to be relevant for applications—in image processing (cf. Bustince et al. [12, 13]) and the classification of imbalanced and overlapping classes (cf. Szmidt and Kukier [37–39]), group decision making (e.g., [5]), negotiations, voting and other situations (cf. Szmidt and Kacprzyk papers).

In our further considerations we will use the $D_\alpha(A)$ operator (Atanassov [3]) with $\alpha \in [0, 1]$:

$$D_\alpha(A) = \{\langle x, \ \mu_A(x) + \alpha\pi_A(x), \ \nu_A(x) + (1 - \alpha)\pi_A(x)\rangle \, | x \in X\} \tag{5}$$

Operator $D_\alpha(A)$ makes it possible to better "see" imbalanced classes (information about the hesitation margins is most important here).

3 An Intuitionistic Fuzzy Classifier

The main idea of the intuitionistic fuzzy classifier in the sense discussed here has been proposed by Szmidt and Kukier [37–39]. Here we present in detail the consecutive steps of the algorithm.

While constructing a classifier, the data are to be divided into two subsets—a training subset (used in a learning phase), and a testing subset (used in a testing phase).

The first, training (called also learning) phase, using a training subset, generates decision rules: IF <condition> THEN <conclusion>, where <condition> means aggregated membership of each attribute of an instance considered to an interval chosen.

The algorithm

1. Training phase (input: learning data, output: decision rules)
   - granulation (discretization of the universes of the attributes),
   - construction of histograms for the discretized universes of the attributes,
   - representation of classes via the intuitionistic fuzzy sets with characteristics derived on the basis of the histograms,
   - validation, tuning (adjusting) of the classifier parameters.
2. Testing phase (input: a new instance to classify, output: assignment of instance to a class)
   - finding a granule for each new instance,
   - applying decision rules to classify a new instance to a class.

Finally, it is necessary to measure the quality of the classifier.

3.1 Classifier Error Measures

The most often used measure of a classifier error is called an *accuracy* being the percentage of instances that are correctly classified. Another measure called an *error* is the percentage of incorrectly classified instances (unseen data). Unfortunately, for the imbalanced classes or for not equal misclassification costs, neither the accuracy nor the error are sufficient.

Confusion Matrix

The confusion matrix (Table 1) is often used to evaluate a two–class classifier. One class consists of so called legal instances, i.e., instances we are especially interested in, another class consists of so called illegal instances, i.e., instances from other class or classes considered together (multicategory classification problems can be reduced to the two-category cases—see Duda [16]).

Table 1 The confusion matrix

	Tested legal	Tested illegal
Actual legal	a	b
Actual illegal	c	d

The meaning of the symbols in Table 1 is
a—the number of correctly classified legal instances,
b—the number of incorrectly classified legal instances,
c—the number of incorrectly classified illegal instances,
d—the number of correctly classified illegal instances,
As a result, the most often used measures for the evaluation of a classifier are:

$$Acc = \frac{legalls\ \ and\ \ illegals\ \ correctly\ \ classified}{total} = \frac{a+d}{a+b+c+d} \tag{6}$$

$$TPR = \frac{legalls\ \ correctly\ \ classified}{total\ \ legalls} = \frac{a}{a+b} \tag{7}$$

$$FPR = \frac{illegals\ \ incorrectly\ \ classified}{total\ \ illegals} = \frac{c}{c+d} \tag{8}$$

The geometric mean (Kubat et al. [20]) is another, often used measure of error:

$$GM = \sqrt{TPR * PPV} \tag{9}$$

where $PPV = \frac{legalls\ \ correctly\ \ classified}{total\ \ legalls} = \frac{a}{a+c}$. Both TPR and PPV are "treated" in the same way (neither is more important) in GM.

If one of the TPR and PPV is most important from the point of view of evaluation, another measure, a so called F-*value* can be used:

$$FV = \frac{(1+\beta^2)TPR * PPV}{\beta^2 PPV * TPR} \tag{10}$$

To better recognize relatively small classes, parameter β should be greater than 1.

In the following the main problems to solve while constructing an intuitionistic fuzzy classifier are presented.

3.2 Granulation

One of the problems to solve for the classification tasks with continuous attributes is granulation (discretization), i.e., the partitioning of universes of the attributes. The idea of replacing a continuous domain with a discrete one has been extended

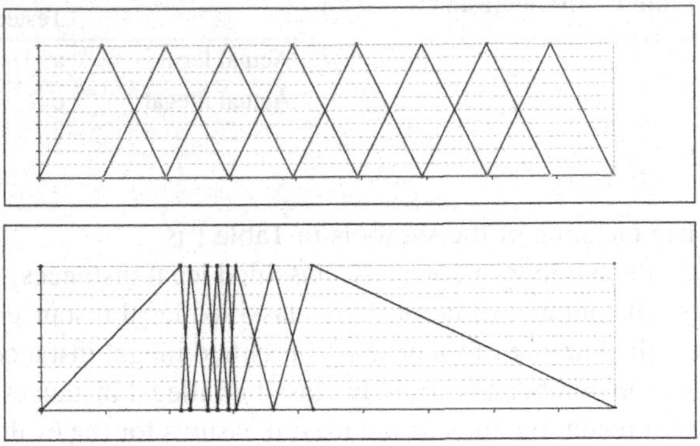

Fig. 1 Example of a symmetric granulation (*upper* part of the figure) and an asymmetric granulation (*bottom* part of the figure)

to fuzzy sets by Ruspini [23]. It is possible to apply the symmetric or asymmetric granulation. In the symmetric granulation (example presented at the upper part of Fig. 1) an attribute universe is partitioned using evenly spaced triangular fuzzy sets (a symmetric fuzzy partition). The asymmetric fuzzy partition (example presented at the bottom part of Fig. 1) is performed with unevenly spaced triangular fuzzy sets which imply each partition to contain an equal number of data instances [23]. For both kinds of granulation the partitioning of triangular fuzzy sets are such that for any attribute value the sum of memberships of the partitioning fuzzy sets is 1.

Another problem, in addition to choosing the sort of granulation, is pointing out the number of used intervals. This problem is discussed in Sect. 4.

3.3 The Core of the Intuitionistic Fuzzy Classifier

In the case of an intuitionistic fuzzy classifier, data is presented via the A-IFSs. After expressing data by relative frequency distributions, the algorithm presented by Szmidt and Baldwin [25–27] is applied to describe smaller, so called *legal*, and bigger, so called *illegal*, classes in the space of all the attributes. As a result, a data instance e is described as an intuitionistic fuzzy element (all three terms are taken into account: the membership value μ, non-membership value ν, and hesitation margin π), i.e.,

$$e : (\mu_e, \nu_e, \pi_e) \tag{11}$$

To enhance the possibility of a proper classification of instances belonging to a smaller (*legal*) class, while training the intuitionistic fuzzy classifier, hesitation margins which assign the (width of) intervals where the unknown values of memberships lie, are used. Namely, the $D_\alpha(A)$ operator (5) is applied making it possible to see as well as possible the elements of the class we are interested in. To be more precise, the values of the hesitation margins are divided so as to "see" better the smaller class—each instance e (11) was expressed as

$$e : (\mu_e + \alpha\pi_e, \nu_e + (1 - \alpha)\pi_e) \quad (12)$$

where $\alpha \in (0.5, 1)$ is a parameter. To guarantee the best behavior of the intuitionistic fuzzy classifier, the parameter α is chosen separately for each attribute, and next, the results are aggregated (see Szmidt and Kukier [37–39]).

It is worth emphasizing that:

- for $\alpha = 0.5$ we obtain a fuzzy classifier;
- the case $\alpha = 1$ does not produce the best results.

The above model is built for each attribute separately, and then the obtained results are aggregated (cf. Szmidt and Kukier [37–39]).

The following aggregation operation is used

$$Agg_1 : Agg_1^{CLASS}(e) = \sum_{k=1}^{n} w_k \mu_{CLASS}^k(e) \quad (13)$$

where e—an examined instance from a database,
$w_k = \frac{n_k}{\sum_{k=1}^{n} n_k}$ for $k = 1, \ldots, n$ is a set of weights for each attribute: n_k is the number of correctly classified training data by the kth attribute.

Knowing the Agg_1 aggregation operator, the classification of an instance examined is done by

$$D_1(e) = arg\max[Agg_1^{CLASS}(e), CLASS \in \{legal, illegal\}] \quad (14)$$

4 Results Obtained

It is worth emphasizing that a whole array of the measures assessing how good is a classifier for recognizing imbalanced classes results from the fact that it is not possible to attain the best behavior of a classifier with respect to all interesting criteria. In the case of imbalanced classes the most important seems to be to attain as good as possible a value of *TPR* (i.e., seeing as good as possible a smaller class) and at the same time not loosing much from the *accuracy* of a classifier. The goal can be achieved by a proper assignment of the parameters of the intuitionistic fuzzy classifier which can operate with two types of granulation, different number of intervals, different values of the parameter α, types of aggregation, etc. We will examine the influence of the parameters chosen on the results obtained by the intuitionistic fuzzy classifier.

First, the results obtained by a fuzzy classifier (for parameter $\alpha = 0.5$ and intuitionistic fuzzy classifier are compared.

Table 2 Results obtained by the intuitionistic fuzzy classifier and the fuzzy classifier: "Glass", $\alpha = 0.7$, a symmetric granulation

No class		Acc FS	Acc IFS	TPR FS	TPR IFS	FPR FS	FPR IFS
1	Average	71.5	60.3	0.25	0.94	0.05	0.57
	Standard deviation	2.8	3.2	0.13	0.06	0.04	0.05
2	Average	73.0	54.0	0.46	0.91	0.11	0.67
	Standard deviation	2.6	3.2	0.14	0.07	0.07	0.07
3	Average	89.4	44.4	0.06	0.84	0.03	0.59
	Standard deviation	2.6	4.1	0.09	0.12	0.04	0.05
5	Average	94.0	92.4	0.56	0.74	0.03	0.06
	Standard deviation	2.2	3.3	0.2	0.15	0.02	0.04
6	Average	96.2	94.3	0.48	0.64	0.01	0.04
	Standard deviation	1.5	2.6	0.22	0.22	0.01	0.03
7	Average	94.7	92.5	0.8	0.86	0.03	0.07
	Standard deviation	1.9	1.5	0.12	0.1	0.02	0.02

Table 3 Results obtained by the intuitionistic fuzzy classifier and the fuzzy classifier: "Glass", $\alpha = 0.7$, an asymmetric granulation

No class		Acc FS	Acc IFS	TPR FS	TPR IFS	FPR FS	FPR IFS
1	Average	79.4	76.3	0.56	0.9	0.09	0.31
	Standard deviation	3.3	3.3	0.12	0.04	0.05	0.05
2	Average	74.8	60.9	0.48	0.85	0.48	0.85
	Standard deviation	3.6	4.2	0.11	0.09	0.11	0.09
3	Average	90.8	84.9	0	0.21	0.01	0.09
	Standard deviation	1.0	3.7	0	0.14	0.01	0.04
5	Average	93.3	93.3	0.09	0.17	0.01	0.01
	Standard deviation	1.0	1.2	0.09	0.12	0.01	0.01
6	Average	95.6	97.0	0.12	0.42	0	0
	Standard deviation	1.2	1.1	0.18	0.23	0	0
7	Average	92.7	94.7	0.44	0.68	0.01	0.02
	Standard deviation	1.4	1.7	0.14	0.17	0.01	0.01

In Tables 2–3 there is a comparison of the results obtained by both classifiers with respect to several evaluating measures for the database "Glass" (cf. [44]). It is the database with 214 instances, 7 classes (4th class is empty), 10 attributes. We use the simple cross validation method (with 10 experiments). For each experiment the mean of the measures examined, and their standard deviation are calculated.

Table 2 presents results for the database "Glass", with a symmetric granulation, and parameter $\alpha = 0.7$. For a symmetric granulation applied to classes 1–3, the intuitionistic fuzzy classifier has a lower *accuracy AccIFS* than the corresponding fuzzy classifier *AccFS*. However, the values of *TPRIFS* are considerably better than the counterpart values of *TPRFS*. Better values of *TPRIFS*, i.e., better recognition of relatively smaller class by the intuitionistic fuzzy classifier, accompany worse (bigger) values of *FPRIFS*. The results mean that the intuitionistic fuzzy classifier with a symmetric granulation better recognizes relatively small classes at the expense of the general accuracy and a worse recognition of other classes.

In Table 3 there are results for an asymmetric granulation with the same $\alpha = 0.7$. Accuracy (6) of the fuzzy classifier *AccFS* is better than that of the intuitionistic fuzzy classifier *AccIFS* for classes 1–3, is the same for both classifiers for class 5, and is better for the intuitionistic fuzzy classifier for classes 6–7. The same time, for all the cases considered, the *TPRIFS* is better than *TPRFS* which means that the intuitionistic fuzzy classifier "sees" better the smaller class we are interested in. We can observe that the improvement of *TPR* for the intuitionistic fuzzy classifier is at the expense of bigger values of *FPR* for classes 1–3. However, in the case of classes 5–7 we obtain both a better accuracy and a better *TPR* for the intuitionistic classifier whereas *FPR* remains practically the same.

More results illustrating the influence of the discretization applied (symmetrical or asymmetrical) on *accuracy* and *TPR* for other classes chosen from the same repository (cf. [44]) are given in Figs. 2 and 3. We use 10 intervals, (in the literature usually ca. 10 intervals are used for discretization), assuming $\alpha = 0.5$.

Accuracy (Fig. 2) is a little better for the asymmetric discretization. The asymmetric discretization gives in general better values of *TPR*, i.e., results in a better recognition of an interesting class (Fig. 3).

It is interesting to analyze the distribution of the membership and non-membership values in the successive intervals for both types of discretization. Typical shapes of the symmetric, and the asymmetric granulations are presented in Figs. 4 and 5, respectively (for the data base "Wine", second attribute, class 2). We can see that in the case of the symmetric granulation, usually the maximal values of the memberships and non-memberships are close (e.g., Fig. 4) whereas for the asymmetric granulation the maximal values are more distant (e.g., Fig. 5). As a result the effects of classification may be worse for the symmetric discretization as a more substantial concentration of the largest membership values and non-membership values may cause that small differences in the values decide to which class an instance is clas-

Fig. 2 *Accuracy* for a symmetric and asymmetric granulation, 10 intervals

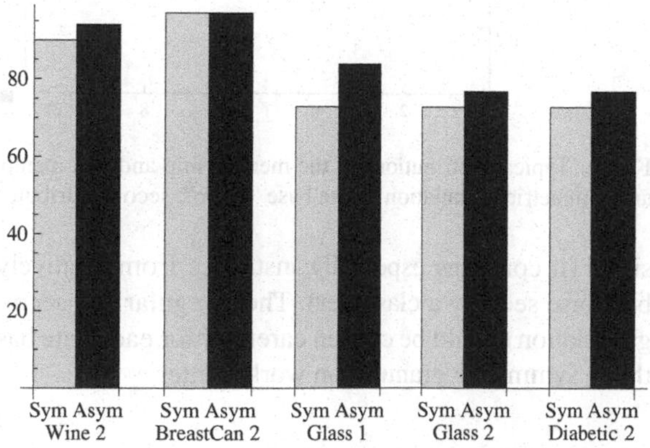

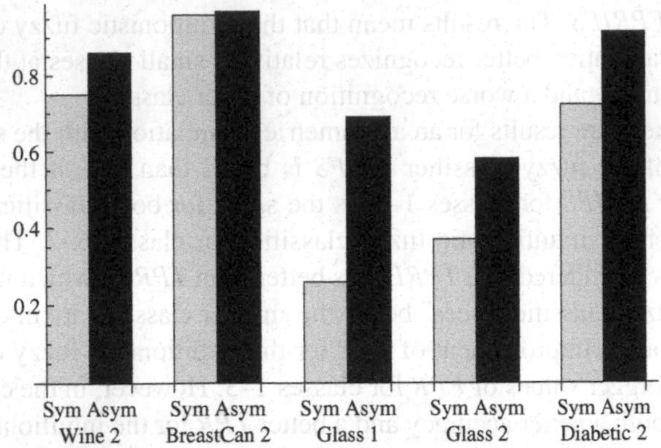

Fig. 3 *TPR* for a symmetric and asymmetric granulation, 10 intervals

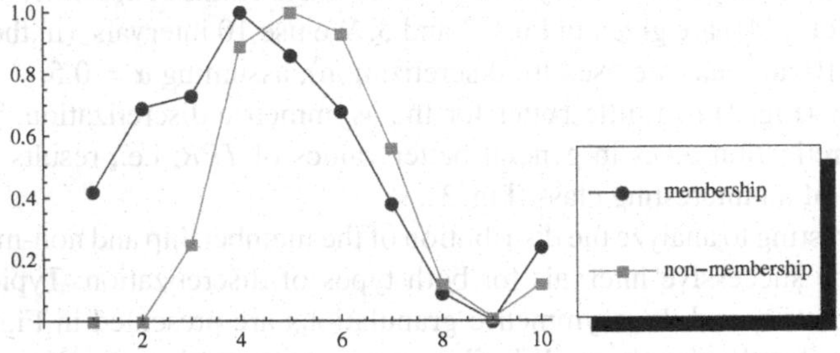

Fig. 4 Typical distributions of the membership and non-membership values in 10 intervals with a symmetric granulation—data base "Wine", second attribute, class 2

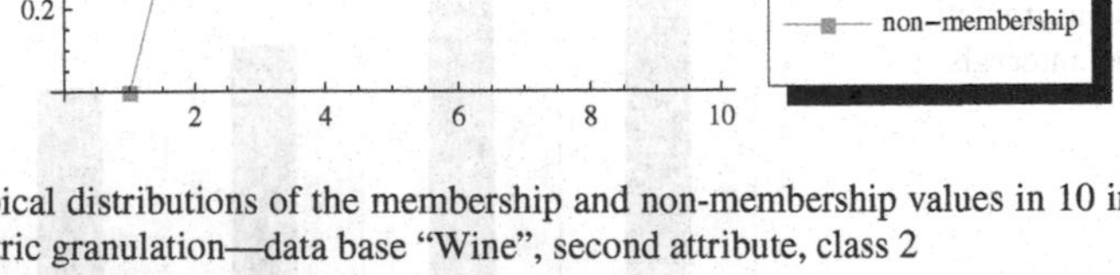

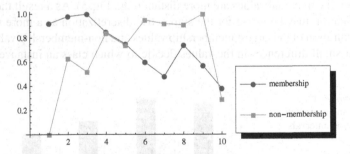

Fig. 5 Typical distributions of the membership and non-membership values in 10 intervals with an asymmetric granulation—data base "Wine", second attribute, class 2

sified (it concerns especially instances from relatively smaller classes which may be worse seen by a classifier). These regularities seem most general but the type of granulation should be chosen carefully for each data base as it happens for some data that a symmetric granulation works better.

4.1 Number of Intervals

The results discussed are obtained for the discretization with the same number of intervals (in each case 10 intervals). But it seems worthwhile to verify the influence of the number of intervals on the results obtained by the intuitionistic fuzzy classifier. In the following we compare several measures characterizing a classifier as a function of the number of the intervals used.

Figures 6 and 7 present the *accuracy* obtained for the symmetric and asymmetric granulation, respectively. The *accuracy* increases with the number of intervals so having in mind the *accuracy* only, we could come to a conclusion that the more intervals the better.

The values of *TPR* as a function of the number of intervals used for the symmetric and asymmetric granulation are in Figs. 8 and 9, respectively. The effects are different from those obtained for *accuracy*—we may observe that the more intervals the smaller values of *TPR* which means that relatively smaller classes are worse recognized. In general, the *accuracy* increases with the number of the intervals (as shown above) but at the expense of a worse recognition of relatively smaller classes (bigger classes are better seen).

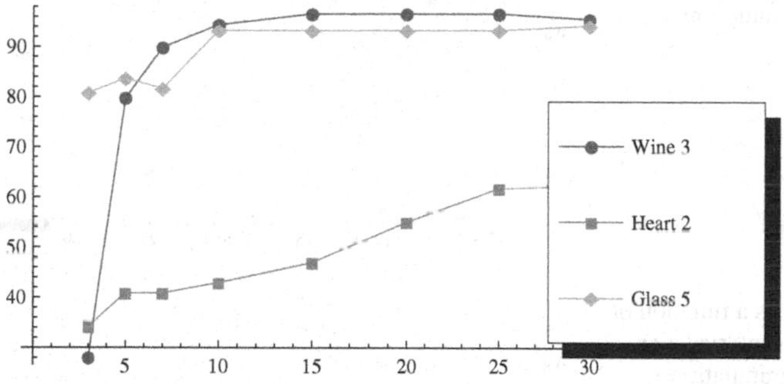

Fig. 6 *Accuracy* as a function of the number of intervals; a symmetric granulation

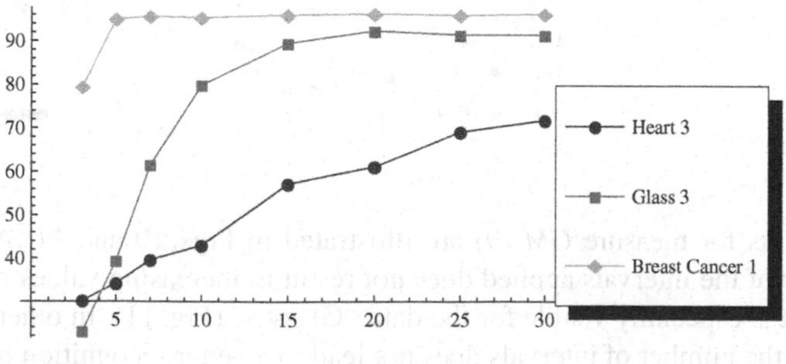

Fig. 7 *Accuracy* as a function of the number of intervals; an asymmetric granulation

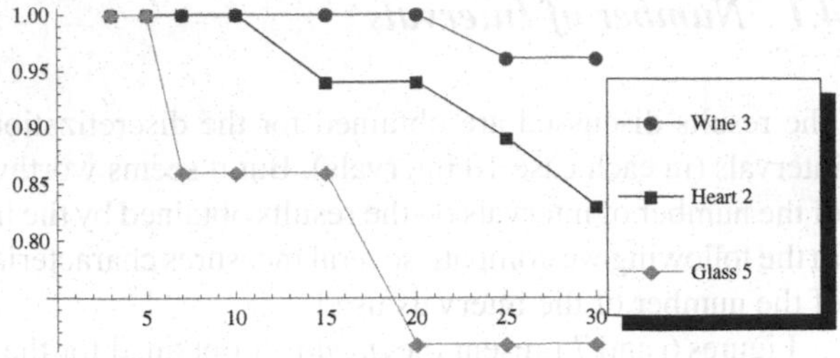

Fig. 8 *TPR* as a function of the number of intervals; a symmetric granulation

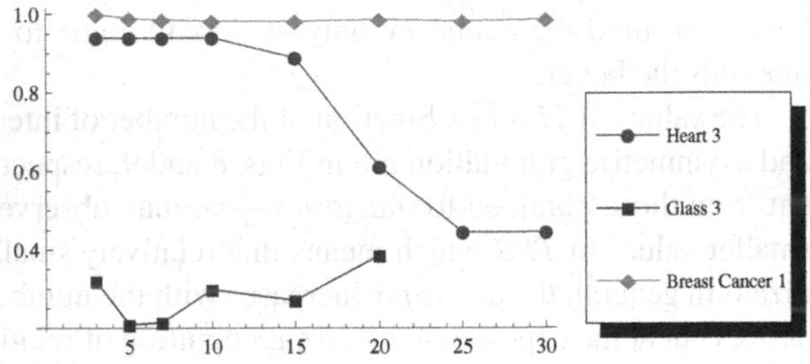

Fig. 9 *TPR* as a function of the number of intervals; an asymmetric granulation

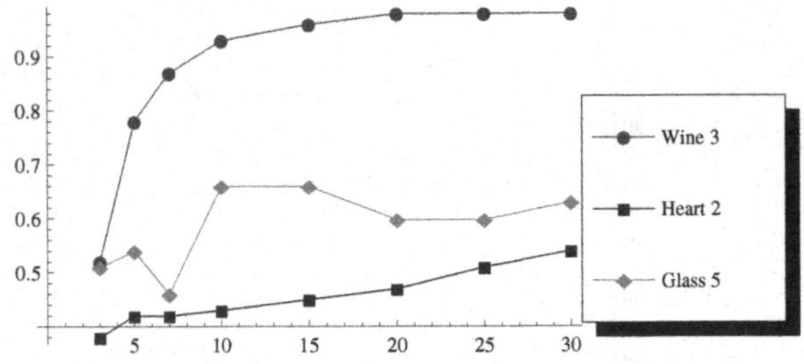

Fig. 10 *GM* as a function of the number of intervals; a symmetric granulation

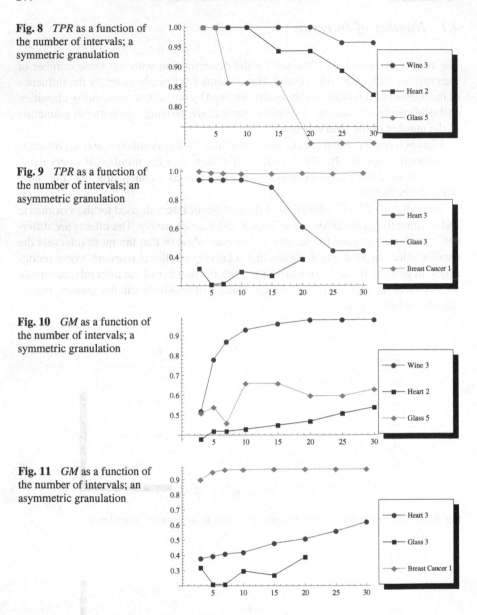

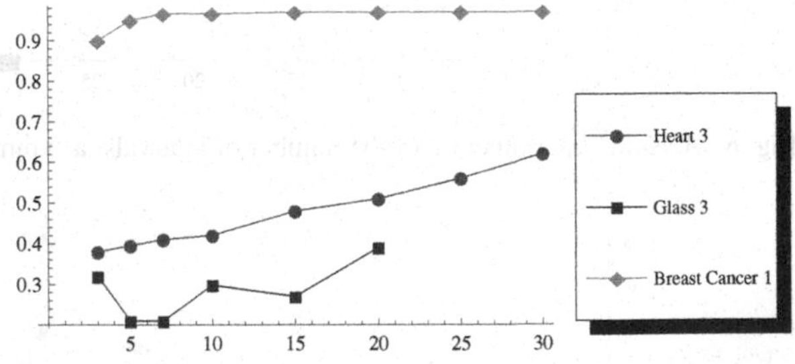

Fig. 11 *GM* as a function of the number of intervals; an asymmetric granulation

The results for measure *GM* (9) are illustrated in Figs. 10 and 11. An increasing number of the intervals applied does not result in increasing values of the measure *GM*. It is especially visible for the data "Glass 3" (Fig. 11). In other words, an increase of the number of intervals does not lead to a better recognition of a smaller class.

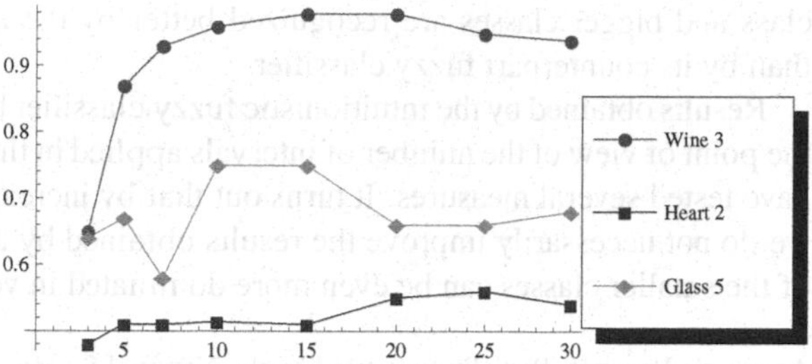

Fig. 12 *FV* as a function of the number of intervals; a symmetric granulation

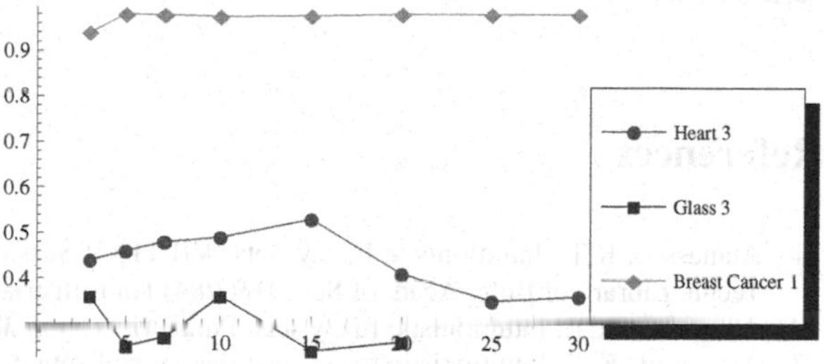

Fig. 13 *FV* as a function of the number of intervals; an asymmetric granulation

The same result is also obtained for measure *FV* (10); $\beta = 2$ which means that a relatively smaller class is slightly preferred (Figs. 10 and 11). To be more precise, when the number of the intervals increases, we can see increasing values of *FV* but after obtaining a maximal value, they start to decrease. Having in mind the results for *TPR*, we come to conclusion, that with an increasing number of intervals a smaller class is worse visible (Figs. 12 and 13).

In general and in conclusion, an increase of the number of intervals during the granulation is not the best practice while constructing a classifier for recognizing imbalanced classes. It results from the fact that while using more intervals during the granulation, instances from relatively smaller classes are even more substantially dominated in a separate interval (worse "visible"). On the other hand, as each data base, and each class in a data base is specific, the only solution is a careful process of assigning the number of the intervals while constructing the classifier considered.

5 Conclusions

An effective although simple intuitionistic fuzzy classifier has been tested on some imbalanced data. Some experiments confirm that the intuitionistic fuzzy classifier fulfills our main demand, i.e., recognizes better relatively smaller classes. The comparison of the results was done with a fuzzy classifier known from literature to be good for recognizing imbalanced classes.

However, the price is a lower accuracy of recognizing all instances because bigger classes might be seen worse. But it is not a rule as sometimes both relatively smaller class and bigger classes are recognized better by the intuitionistic fuzzy classifier than by its counterpart fuzzy classifier.

Results obtained by the intuitionistic fuzzy classifier have been also assessed from the point of view of the number of intervals applied in the process of granulation. We have tested several measures. It turns out that by increasing the number of intervals we do not necessarily improve the results obtained by the classifier as the elements of the smaller classes can be even more dominated in very narrow intervals.

Acknowledgments Partially supported by the National Science Centre under Grant UMO-2012/05/B/ST6/03068.

References

1. Atanassov, K.T.: Intuitionistic Fuzzy Sets. VII ITKR Session. Sofia (Deposed in Centr. Sci. Techn. Library of Bulg. Acad. of Sci., 1697/84) (in Bulgarian) (1983)
2. Atanassov, K.T.: Intuitionistic fuzzy sets. Fuzzy Sets Syst. **20**, 87–96 (1986)
3. Atanassov, K.T.: Intuitionistic fuzzy sets: theory and applications. Springer, Berlin (1999)
4. Atanassov, K.T.: On Intuitionistic Fuzzy Sets Theory. Springer, Berlin (2012)
5. Atanassova, V.: Strategies for Decision Making in the Conditions of Intuitionistic Fuzziness. In: International Conference 8th Fuzzy Days, Dortmund, Germany, pp. 263–269 (2004)
6. Baldwin, J.F.: Combining evidences for evidential reasoning. Int. J. Intell. Syst. **6**, 569–616 (1991)
7. Baldwin, J.F.: The management of fuzzy and probabilistic uncertainties for knowledge based systems. In: Shapir, S.A. (ed.) Encyclopaedia of AI, 2nd edn, pp. 528–537. Wiley (1992)
8. Baldwin, J.F., Martin, T.P., Pilsworth, B.W.: FRIL–Fuzzy and Evidential Reasoning in Artificial Intelligence. Wiley, New York (1995)
9. Baldwin, J.F., Lawry, J., Martin, T.P.: A Mass Assignment Theory of the Probability of Fuzzy Events. ITRC Report 229, University of Bristol (1995)
10. Baldwin, J.F., Coyne, M.R., Martin, T.P.: Intelligent reasoning using general knowledge to update specific information: a database approach. J. Intell. Inf. Syst. **4**, 281–304 (1995b)
11. Baldwin, J.F., Lawry, J., Martin, T.P.: The application of generalized fuzzy rules to machine learning and automated knowledge discovery. Int. J. Uncertain. Fuzzyness Knowl-Based Syst. **6**(5), 459–487 (1998)
12. Bustince, H., Mohedano, V., Barrenechea, E., Pagola, M.: Image thresholding using intuitionistic fuzzy sets. In: Atanassov, K., Kacprzyk, J., Krawczak, M., Szmidt, E. (Eds.) Issues in the Representation and Processing of Uncertain and Imprecise Information. Fuzzy Sets, Intuitionistic Fuzzy Sets, Generalized Nets, and Related Topics. EXIT, Warsaw (2005)
13. Bustince, H., Mohedano, V., Barrenechea, E., Pagola, M.: An algorithm for calculating the threshold of an image representing uncertainty through A-IFSs. In: IPMU'2006, pp. 2383–2390 (2006)
14. Dubois, D., Prade, H.: Unfair coins and necessity measures: towards a possibilistic interpretation of histograms. Fuzzy Sets Syst. **10**(1983), 15–20 (1983)
15. Dubois, D., Prade, H.: The three semantics of fuzzy sets. Fuzzy Sets Syst. **90**, 141–150 (1997)
16. Duda, R.O., Hart, P., Stork, D.: Pattern Classification. Wiley, Chichester (2000)
17. Fawcett, T., Provost, F.: Adaptive fraud detection. Data Min. Knowl. Discov. **3**(1), 291–316 (1997)
18. He, H., Garcia, E.A.: Learning from imbalanced data. IEEE Trans. Knowl. Data Eng. **21**(3), 1263–1284 (2009)
19. Japkowicz, N.: Class Imbalances: Are We Focusing on the Right Issue?. ICML, Washington (2003)

20. Kubat, M., Holte, R., Matwin, S.: Machine learning for the detection of oil spills in satellite radar images. Mach. Learn. **30**, 195–215 (1998)
21. Lewis, D., Catlett, J.: Heterogeneous uncertainty sampling for supervised learning. In: Proceedings of the 11th Conference on Machine Learning, pp. 148–156 (1994)
22. Mladenic, D., Grobelnik, M.: Feature selection for unbalanced class distribution and Naive Bayes. In: Proceedings of the 16th International Conference on Machine Learning, pp. 258–267 (1999)
23. Ruspini, E.H.: A new approach to clustering. Inf. Control **15**, 22–32 (1969)
24. Szmidt, E.: Distances and Similarities in Intuitionistic Fuzzy Sets. Springer, Berlin (2014)
25. Szmidt, E., Baldwin, J.: New similarity measure for intuitionistic fuzzy set theory and mass assignment theory. Notes IFSs **9**(3), 60–76 (2003)
26. Szmidt, E., Baldwin, J.: Entropy for intuitionistic fuzzy set theory and mass assignment theory. Notes IFSs **10**(3), 15–28 (2004)
27. Szmidt, E., Baldwin, J.: Assigning the parameters for intuitionistic fuzzy sets. Notes IFSs **11**(6), 1–12 (2005)
28. Szmidt, E., Baldwin, J.: Intuitionistic fuzzy set functions, mass assignment theory, possibility theory and histograms. IEEE World Congr. Comput. Intell. **2006**, 237–243 (2006)
29. Szmidt, E., Kacprzyk, J.: On measuring distances between intuitionistic fuzzy sets. Notes IFS **3**(4), 1–13 (1997)
30. Szmidt, E., Kacprzyk, J.: Distances between intuitionistic fuzzy sets. Fuzzy Sets Syst. **114**(3), 505–518 (2000)
31. Szmidt, E., Kacprzyk, J.: Entropy for intuitionistic fuzzy sets. Fuzzy Sets Syst. **118**(3), 467–477 (2001)
32. Szmidt, E., Kacprzyk, J.: An intuitionistic fuzzy set base approach to intelligent data analysis (an application to medical diagnosis). In: Abraham, A., Jain, L., Kacprzyk, J. (Eds.) Recent Advances in Intelligent Paradigms and Applications, pp. 57–70. Springer, Berlin (2002)
33. Szmidt, E., Kacprzyk, J.: A new concept of similarity measure for intuitionistic fuzzy sets and its use in group decision making. In: Torra, V., Narukawa, Y., Miyamoto, S. (Eds.) Modelling Decisions for AI. LNAI 3558, pp. 272–282. Springer, Berlin (2005)
34. Szmidt, E., Kacprzyk, J.: Distances between intuitionistic fuzzy sets: straight forward approaches may not work. In: IEEE IS'06, pp. 716–721 (2006)
35. Szmidt, E., Kacprzyk, J.: Some problems with entropy measures for the Atanassov intuitionistic fuzzy sets. In: Applications of Fuzzy Sets Theory. LNAI 4578, pp. 291–297. Springer (2007)
36. Szmidt, E., Kacprzyk, J.: A new similarity measure for intuitionistic fuzzy sets: straight forward approaches may not work. In: 2007 IEEE Conference on Fuzzy Systems, pp. 481–486 (2007)
37. Szmidt, E., Kukier, M.: Classification of imbalanced and overlapping classes using intuitionistic fuzzy sets. In: IEEE IS'06, London, pp. 722–727 (2006)
38. Szmidt, E., Kukier, M.: A new approach to classification of imbalanced classes via Atanassov's intuitionistic fuzzy sets. In: Wang, H.-F. (Ed.) Intelligent Data Analysis : Developing New Methodologies Through Pattern Discovery and Recovery. Idea Group, pp. 85–101 (2008)
39. Szmidt, E., Kukier, M.: Atanassov's intuitionistic fuzzy sets in classification of imbalanced and overlapping classes. In: Chountas, P., Petrounias, I., Kacprzyk, J. (Eds.) Intelligent Techniques and Tools for Novel System Architectures. Seria: Studies in Computational Intelligence, pp. 455–471. Springer, Berlin (2008)
40. Yager, R.R.: Level sets for membership evaluation of fuzzy subsets. Technical Report RRY-79-14, Iona Colledge, New York (1979)
41. Yamada, K.: Probability–possibility transformation based on evidence theory. In: Proceedings of the IFSA–NAFIPS'2001, pp. 70–75 (2001)
42. Zadeh, L.A.: Fuzzy sets. Inf. Control **8**, 338–353 (1965)
43. Zadeh, L.A.: Fuzzy sets as the basis for a theory of possibility. Fuzzy Sets Syst. **1**, 3–28 (1978)
44. http://archive.ics.uci.edu/ml/machine-learning-databases

IT Business Service-Level-Management—An Intuitionistic Fuzzy Approach

Roland Schuetze

Abstract Bridging from IT-centric service levels, written in IT technical terms, to business-oriented service achievement is a hot topic in today's service research. The proposed 'IFSFIA' methodology will help for Service Level Agreements (SLAs) to relate metrics for business application's into measurable parameters for technical services that can be defined and reported against a SLA and monitored under Service Level Management. It allows assessing the complex dependency and impact relationships of low-level backend components to the quality of the frontend service. This work defines dependency couplings in a practical and feasible manner in order to satisfy aspects of the distributed nature of SLAs in a multi-tier-architectural environment. The concept starts from the idea of naturally approaching impact relationships by separately envisaging positive and negative aspects with the notion of bipolarity. Performing a multi-level impact analysis by means of intuitionistic fuzzy-mathematical models it unveils business insights into how service accounts as a whole can improve quality and allows pro-actively tracking measures of backend components to gather the overall SLA quality status of a business service.

Keywords Service level management · SLA · Business impact · Intuitionistic fuzzy sets

This concept was first published at ICIFS conference, Nov. 2013, Sofia Bulgaria [1], the following version includes several conceptual extensions.

R. Schuetze (✉)
Department of Informatics, University of Fribourg, Fribourg, Switzerland
e-mail: roland.schuetze@unifr.ch

P. Angelov and S. Sotirov (eds.), *Imprecision and Uncertainty in Information Representation and Processing*, Studies in Fuzziness and Soft Computing 332,
DOI 10.1007/978-3-319-26302-1_16

1 The Complexity of Multi-Layered Service Level Requirements

In an increasingly service-oriented world, "best effort" service delivery is not good enough. But how does the business know whether it is getting an adequate service? Service level requirements are set to ensure that the business goals underlying IT services are met. The Service Level Agreements (SLAs) incorporate the expectations and the obligations about the properties of a service. The most significant part of a SLA is the range of the duties of a service. The SLA objectives are mostly the concerns that are associated with the Quality of a Service (QoS). To guarantee business-focused SLAs results in optimization problem solving across multiple domains (e.g. networking, computer systems, and software engineering). The landscape of today's IT service providers is inherently integrated. It consists of all kinds of elements, namely networks, servers, storage, and software stacks. The fulfilment of any higher-level objective requires proper enforcements on multiple resources at several levels.

The challenge with such enterprise SLAs is translating metrics for business applications into measurable parameters for technical services that can be defined and reported against an SLA and monitored under Service Level Management (SLM). Service compositions, translation and mappings lies therefore in the core of SLA management, in that it correlates metrics and parameters within and across layers [2]. For example, in order to guarantee certain bounds on the response times for ERP-type, it involves the ERP software, the application and database servers, the network configuration, and more [3]. When knowing the relation and dependency of this backend service to the end-user service (or composite service), service administrators can then pro-actively track and verify these dependencies by periodically polling the measures of individual services and gathering the overall quality status of the end-user service. This will allow administrators responsible for the functioning of a service to monitor its quality based on the measurements typically already done for the infrastructure components.

2 SLA Dependency Mapping

2.1 The Concept of Key Quality and Performance Indicators

Open Group [4] defined a concept of key quality- and performance indicators (KQI/PI). Service Level Specification parameters can be one of two types: Key Quality Indicators (KQIs) and (most technical) Service Performance Indicators (PIs). At the highest level, a KQI or group of KQIs are required to monitor the quality of the business service offered to the end-user. These KQIs will often form part of the contractual SLA, whereas the monitoring instrumentation is established for the lower level components to ensure the fulfillment of the service quality objectives (Fig. 1).

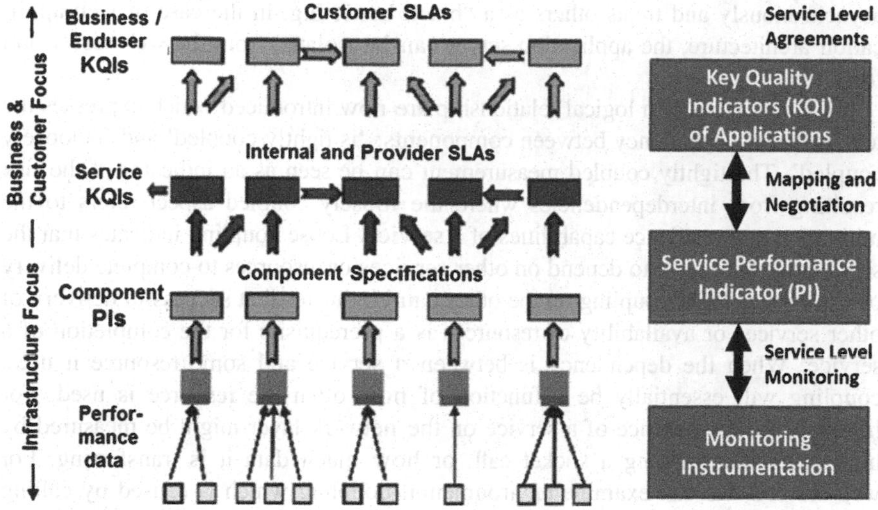

Fig. 1 KQI, PI & SLA relationship [5]

The KQI is derived from a number of sources, including performance metrics of the service or underlying support services with PIs. Different PIs may be assembled to calculate a particular KQI. The mapping between the PI and KQI may be simple or complex, empirical or formal. The automated process of translating and correlating high-level requirements and policies for all kinds down to infrastructure level creates a set of related PIs, which is termed now a KQI/PI hierarchy. While the association relationship only relates adjacent sets of KQIs/PIs, the hierarchy establishes associations across the whole stack in a distributed multi-tier architecture. In the following a Coupling *C* association is defined, which can be constructed in a practical and feasible manner in order to satisfy aspects of the different types of component interdependencies.

2.2 Dependence Coupling as Measurement

Dependence Coupling is a measure that we propose to capture how dependent the component or service is on other services or resources for its delivery. The goal is to build components that do not have tight dependencies on each other, so that if one service component were to die (fail), sleep (not respond) or remain busy (slow to respond) for some reason, the other components in the system are built to still continue to work. Loose coupling describes an approach where integration interfaces are developed with minimum assumptions between the sending/receiving parties, thus reducing the risk that failure in one module will affect others. Loose coupling isolates the components of an application so that each component interacts asynchronously and treats others as a "black box". E.g. in the case of web application architecture, the application server can be isolated from the web server and from the database.

Two new types of a logical relationship are now introduced which expresses the level of inter-dependency between components: 'is tightly coupled' and 'is loosely coupled'. The tightly coupled measurement can be seen as an indicator of the risk resulting from interdependencies where the loosely coupled aspect refers to the mitigation and resilience capabilities of a service. Loose coupling indicates that the service does not have to depend on other services or resources to complete delivery of its service. Tight coupling on the other hand indicates that successful delivery of other services or availability of resources is a prerequisite for the completion of a service. When the dependency is between a service and some resource it uses, coupling will essentially be a function of how often the resource is used. For instance, the dependence of a service on the network layer might be measured by how often it is making a socket call, or how much data it is transferring. For web-services we can examine environmental coupling which is caused by calling and being called. Traditional components are more tightly and statically integrated and measurements are related mostly to procedural programming languages e.g. proposed by Dhama [5] or Fenton and Melton [6]. More advanced are object-oriented coupling measures [7] and further several metrics are proposed to evaluate the coupling level real-time by runtime monitoring, introduced as dynamic coupling metrics [8].

2.3 Application Dependency Discovery Management (ADDM)

Application discovery is the process of automatically analyzing artefacts of a software application and physical elements that constitute a network (e.g., servers, firewalls, etc.). ADDM products [9] deliver a powerful enabler that minimize IT organizations expend on the information assimilation function and can also provide a basis for further higher level, logical dependency assessments. According to [10] these tool assert networks mainly based on three different approaches: middleware or instrumenting applications; analyzing program configuration files or analyzing application traffic. ADDM products deliver a point-in-time view of the "truth" and unveil dependencies, but do not measure a granular truth value of an impact two service components may have on each other. Dependency graphs created by an automated discovery tool can be leveraged as a great starting point for advanced methods to calculate granular degrees of dependence.

An inductive approach can also be chosen by calculating couplings between servers or services based on historical data collected from the actual server network. As opposite a deductive method would be applicable, where dependencies are not calculated based on data the system produces, but rather the system itself, for

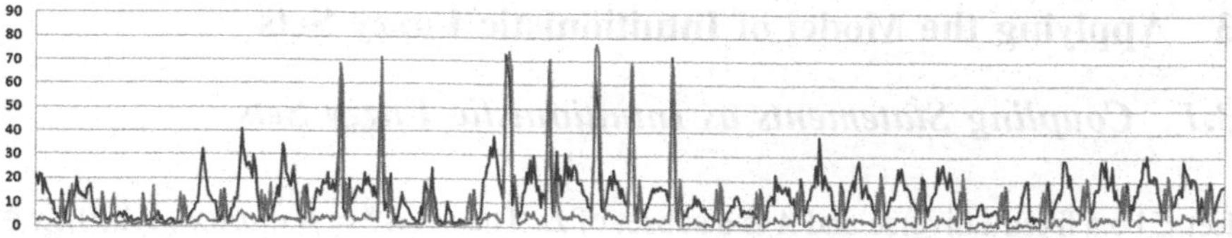

Fig. 2 Inductive coupling assessment between database and application performance

example plans system architects make or comparisons to other systems, which have a similar layout (Fig. 2).

For inductive coupling measurements statistical methods can be applied or an expert can determine coupling effects based on the given data-series and his experience.

2.4 *Bi-Polar Coupling Aspects*

A key principle of the following proposed impact assessment method is the idea of naturally envisaging positive and negative instances of the dependency relation and simultaneous consideration by pulling both strengths together. For a complex IT system the risk are the dependencies through interactions, the controversy mitigation ability are the built-in system resilience capabilities. The simultaneous and free play of contrary forces, dependence and resilience together will define the overall system behavior and the expected impact to the business. Considering and judging positive and negative aspects isolated will not lead to reliable assessments. This leads to the question whether traditional impact analysis methods can be applied for such integrated model. In general the ITIL v3 methods already cover both aspects [11]. Fault Tree Analysis (FTA), like the word fault tree indicates, work in the "failure space" and looks at system failure combinations. So the FTA method covers the aspect of negative risk of interdependencies and negative impacts on failure. On the other side, the ITIL Component Failure Impact Analysis (CFIA) approach [12] is assessing on the mitigation, restoration and resilience capabilities, which represents the positive aspect of independence.

There are several scenarios how an incident may interfere indirectly with other components which is mainly resulting out of the combination of the contrary forces. IT systems try to implement strategies that the resilience capabilities of each component should pro-actively limit the inference and impact of the incident to related components or the business services. In praxis impacts are complex which constitutes uncertainty. They involve a multitude of effects that cannot be easily assessed and may involve complex causalities, non-linear relationships as well as interactions between effects [13]. This may render it difficult to determine exactly what may happen.

3 Applying the Model of Intuitionistic Fuzzy Sets

3.1 *Coupling Statements as Intuitionistic Fuzzy Sets*

Let E be a fixed universe and A is a subset of E. The set $A^* = \{(x, \mu A(x), \nu A(x)) | x \in E\}$ where $0 \leq \mu A(x) + \nu A\,(x) \leq 1$ is called Intuitionistic Fuzzy Set (IFS) [14]. Every element has a degree of membership (validity, etc.) $\mu A(x)$: $E \rightarrow [0,1]$and a degree of non-membership (non-validity, etc.) $\nu A(x)$: $E \rightarrow [0,1]$. Intuitionistic Fuzzy Sets have only loosely related membership and non-membership values unlike classical (Zadeh) [16] fuzzy sets. An IFS is a generalization of the classical fuzzy set which defines another degree of freedom into the set description, the independent judgment of validity and non-validity. This two-sided view, including the possibility to represent formally also a third aspect of imperfect knowledge could be used to describe many real-world problems in a more adequate way—by independent rating of both, positive and negative aspects—for each variable in the model. For each IFS A in E, $\pi(x) = 1 - \mu A(x) - \nu A(x)$ is called the intuitionistic index of x in A which represents the third aspect, the degree of uncertainty, indeterminacy, limited knowledge etc. In the following approach let now a be the intuitionistic fuzzy logical statement of tightly coupling and b of loosely coupling with estimations respectively $< \mu a, \nu a >$ and $< \mu b, \nu b >$. The tightly coupling a degree of truth is $< \mu a >$ and the degree of falsity $< \nu a >$. The same assessment is done for loosely coupling b where $< \mu b, \nu b >$ represent the degrees of truth and falsity. This maps service quality impacts to the idea behind intuitionistic fuzzy service dependencies, where the level of tightly coupling between service components corresponds to the intuitionistic fuzzy degrees of truth and falsity of the dependency impact and the loosely coupling index assesses the resilience capabilities of a service.

3.2 *Defining the Fuzzy Intuitionistic Direct Coupling Between Components*

The validities (membership degrees) for tightly and loosely couplings are independently estimated by separate approaches, for 'tightly' using the described inter-modular coupling metrics and for 'loosely' applying assessed intrinsic component resilience capabilities. In praxis dependencies are naturally expressed by positive forms (membership) only, which is also the way human assessments work. Thus, the proposed method does only require the experts to judge on the validity of the tightly and loosely coupling and to specify a level of certainty of these statements (Fig. 3).

The vagueness is expressed in linguistic terms and mapped into a crisp number with regard to the applied complement function, omitting that $\lambda >= 0$ (Sugeno) or $w <= 1$ (Yager). The non-validity is then automatically set by the fuzzy complement function (Fig. 4).

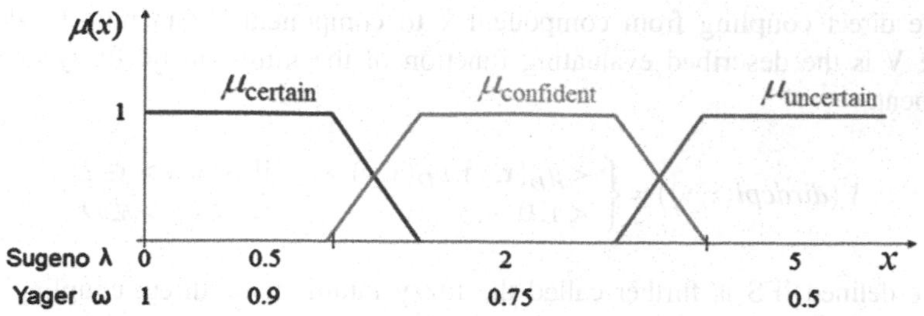

Fig. 3 Certainty mappings to define Sugeno and Yager complements

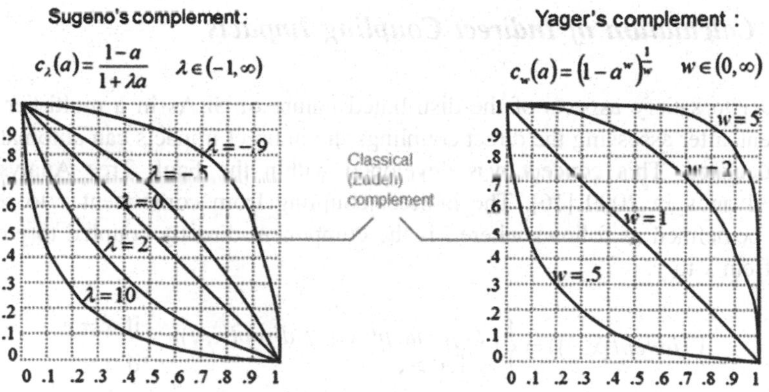

Fig. 4 Sugeno and Yager fuzzy complements

To define now the direct Coupling C association between two components the intuitionistic fuzzy logical statements of tightly coupling and loosely coupling are pulled together in a single IFS. Several operations over IFS are possible. As tightly and loosely couplings have contrary effects a meaningful operation for building the combined IFS C is for instance $A@\neg B$ by adding membership 'tightly' with non-membership 'loosely' and vice versa divided by 2. The combined degrees are further referred as μ_D and ν_D for direct coupling index and are called the intuitionistic fuzzy probabilistic direct impact between two related components.

$$\mu_{combined}(x) = \frac{\mu_A(x) + \nu_B(x)}{2} \text{ and } \nu_{combined}(x) = \frac{\nu_A(x) + \mu_B(x)}{2} \quad (1)$$

It implements the idea that although the coupling effects and component resilience are independent, only the simultaneous consideration of both strengths together defines the impact. This implies a beforehand a normalization of the positive and negative effects (even there are independent measurements used) for getting comparable weights, which is a key challenge to get accurate results applying the proposed method.

The direct coupling from component x to component y can now be defined where V is the described evaluating function of the intuitionistic fuzzy coupling statement.

$$V(dirdcpl(x,y)) = \begin{cases} <\mu_D(x,y), \nu_D(x,y)>, & \text{if } <x,y> \in D \\ <1,0>, & \text{if } <x,y> \notin D \end{cases} \tag{2}$$

The defined IFS is further called the fuzzy intuitionistic direct coupling index between the two components x and y.

3.3 Calculation of Indirect Coupling Impacts

In order to satisfy aspects of the distributed nature of SLAs in a multi-tier environment, after assessing the direct couplings the indirect impacts can automatically be calculated. This concept was developed within the Fault Tree Analysis by Kolev/Ivanov in 2009 [16]. The indirect coupling from component x to service y can be defined as follows where i is the component directly coupled to y on the path from x to y.

$$V(indcpl(x,y)) = \begin{cases} \bigvee_{i,y \in D} indcpl(x,i) \wedge dircpl(i,y), & \text{if } x \neq y \\ <1,0>, & \text{if } x = y \end{cases} \tag{3}$$

Within the KQI/KPI hierarchy model the methodology for calculating the indirect coupling follows the forward dependency direction (Forward Coupling Calculation FCC). In case of an incident this means starting from the failed node in the hierarchy and traversing through its direct or indirect dependants to the business service. Vice versa a root cause analysis is a top down approach and requires the reverse task to be solved, i.e. "To which components is the business application B coupled to (depends on)" The second method implies the definition of methodology for calculating indirect impacts starting from the dependant and traversing through its impact arcs in the reverse direction. We refer to this method as Reverse Coupling Calculation (RCC).

$$V(indcpl(x,y)) = \begin{cases} \bigvee_{x,i \in D} dircpl(x,i) \wedge indcpl(i,y), & \text{if } x \neq y \\ <1,0>, & \text{if } x = y \end{cases} \tag{4}$$

The possibility of both, a classical, probabilistic interpretation of the logical operations conjunction (∧) and disjunction (∨) is a key concept in the indirect impact calculations. The partial impact between the component PI and business KPI is now expressed by means of intuitionistic fuzzy values carrying probabilistic information. These IFS operations are proposed for classical, moderate, worst and best case impact analysis [16]:

Worst Case $$\begin{aligned} V(p \wedge q) &= \langle \min(\mu(p), \mu(q)), \max(\nu(p), \nu(q)) \rangle \\ V(a \vee b) &= \langle \mu(a), \mu(b) - \mu(a).\mu(b), \nu(a).\nu(b) \rangle \end{aligned} \tag{5}$$

Moderate Case $$\begin{aligned} V(p \wedge q) &= \langle \mu(p).\mu(q), \nu(p) + \nu(q) - \nu(p).\nu(q) \rangle \\ V(a \vee b) &= \langle \mu(a) + \mu(b) - \mu(a).\mu(b), \nu(a).\nu(b) \rangle \end{aligned} \tag{6}$$

Best Case $$\begin{aligned} V(p \wedge q) &= \langle \mu(p).\mu(q), \nu(p) + \nu(q) - \nu(p).\nu(q) \rangle \\ V(a \vee b) &= \langle \max(\mu(a), \mu(b)), \min(\nu(a), \nu(b)) \rangle \end{aligned} \tag{7}$$

Fuzzy Classical $$\begin{aligned} V(p \wedge q) &= \langle \min(\mu(p), \mu(q)), \max(\nu(p), \nu(q)) \rangle \\ V(a \vee b) &= \langle \max(\mu(a), \mu(b)), \min(\nu(a), \nu(b)) \rangle \end{aligned} \tag{8}$$

Depending on which operations are applied, classical or probabilistic, the results will be greater or smaller. The indirect intuitionistic fuzzy dependencies between components may have different kinds of semantics (functional and probabilistic) depending on the type of information they represent. Combinations of classical and probabilistic applications of the logical operations can as result be interpreted either as a probabilistic indirect dependency between component PI and the business KQI (means the probability that a KQI breaches the SLA in case the component PI fails) or an ordinary indirect fuzzy dependency (means that the KQI is partially out of specification or degraded in functioning in case the component PI fails).

4 Intuitionistic Fuzzy Service Failure Impact Analysis (IFSFIA)

A complete methodical assessment approach, which is practically usable in data-centre environments, includes several sequential steps to be processed. It starts from automated exploring the details of the managed resources and backend components, the grouping of components to impacted frontend services and the enrichment in several tasks and calculation steps up to the gradual business impact assessments, including monetary cost-of-failure information and business objectives. The overall frame for incorporating all data is the CFIA grid (described in step 3). This matrix can be freely extended with different kind of variables showing failure modes, reliability parameters, financial data, operational capabilities and techniques and extends the pure system view to include also the processes, tools and people (e.g. helpdesk) that are necessary for functioning of a distributed information system.

4.1 IFSFIA Structured Step-by-Step Approach

Step 1: Auto-Discovery by ADDM Tools
All infrastructure component items and technical dependencies of a defined scope will be auto-discovered using ADDM (Application Dependency Discovery Management) tools. This provides trust that the discovered information is real by automatically discovering interdependencies among applications and underlying systems and minimize IT organizations expend on the complex information assimilation. The discovered components with corresponding relations can be extracted by commercial ADDM tools in a structured data format e.g. XML for further automated processing. For the later use cases IBM's Tivoli Application Dependency Discovery Manager (TADDM) is chosen as auto-discovery solution that provides in depth automated application dependency mapping and configuration auditing [21].

Step 2: Defining the Business Service
The in-scope discovered component items are grouped to form the business applications, as the top level in the component hierarchy is the business service. A business service is the way to group the different kinds of IT resources into a logical group which acts together as one unit to provide the service. Business services can contain any number of the lower-level resources. This grouping step creates implicitly the fault tree to the business service by chaining all directly and indirectly linked components. In case an incident occurs, a list of possible components which may be the root cause of the incident can now be identified.

Step 3: Creating the CFIA Grid
After auto-discovering of the in-scope infrastructure components, there relationships and the configurations, the next step is to create a grid with components on one axis and the IT services which have a dependency on the component. This matrix is called CFIA (Component Failure Impact Analysis). This enables the identification of critical components (that could cause the failure of multiple IT services) and fragile IT services (that have multiple single points of failure). A basic CFIA will target a specific section of the infrastructure; just looking at simple binary choices (e.g. if we lose component x, will a service stop working? More advanced CFIAs can be expanded to include a number of variables, such as likelihood of failure, repair and recovery time, recovery procedures, organizational assignments and integration into wider service management processes and also can also consider and evaluate for different component failure modes. So within the IFSFIA method in the matrix all data is added which is relevant for the loosely coupling assessment including the business recovery time objectives. The grid is complemented with the evaluated degrees for loosely and tightly coupling. The tightly coupling index is defined as inter-modular coupling metric, which calculate the coupling between each pair of directly related components. For loosely coupling an intrinsic coupling metric is chosen as this refers to the individual components' resilience capabilities. The CFIA will also verbally indicate the assessed level of certainty.

Step 4: Define the Fuzzy Intuitionistic Direct Impact
As next step for the two independent loosely-and tightly coupling indexes a combined representation into an integrated Intuitionistic Fuzzy Set (IFS) is created. This requires the two coupling indexes A and B to be normalized and combined by IFS operations (we may choose the basic IFS operation A@¬B). The result of step 4 is the fuzzy intuitionistic direct coupling impact between two components. The direct coupling IFS can be now added to the CFIA grid (Fig. 5).

Step 5: Calculating the Fuzzy Intuitionistic Indirect Couplings
Based on the direct couplings, described as inter-modular IFS, the indirect impacts can be calculated. By involving different probabilistic variants of the logical operations when calculating the indirect impacts, the strength of the impact transferred throughout the distributed and multi-tiered system can be modelled. For impact analysis the Forward Coupling Calculation (FCC) is applied which follows the forward dependency direction from the component where the incident occurs and traversing through its direct or indirect dependants. In the KQI/KPI Hierarchy a forward looking coupling calculation means a bottom-up direction. Vice versa a root cause analysis is a top down approach and requires the reverse task to be solved, i.e. "to which components is the business application coupled to (depend on)" as Reverse Coupling Calculation (RCC).

In the following example using the forward (FCC) approach for impact assessments in case a component C_2 fails to the business service B_O:

$indcpl(C_2,B_O) = (dircpl(C_2,C_3) \vee (dircpl(C_2,C_4) \wedge dircpl(C_4,C_3))) \wedge dircpl(C_3, B_O)$. *Using* classical operations the $indcpl_{classic}(C_2,B_O) = (0.60,0.30)$, moderate impact $indcpl_{moderate}(C_2,B_O) = (0.43,0.43)$, worst case impact $indcpl_{worst}(C_2, B_O) = (0.60,0.30)$ and best case impact assessment $indcpl_{best}(C_2,B_O) = (0.36,0.51)$.

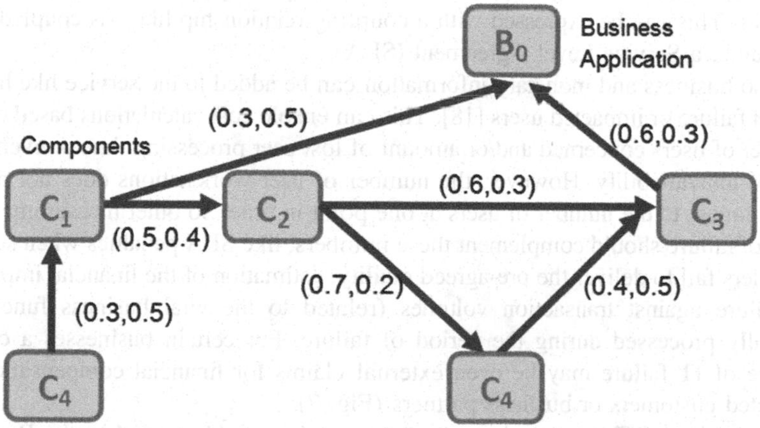

Fig. 5 Directed graph with direct couplings as IFS

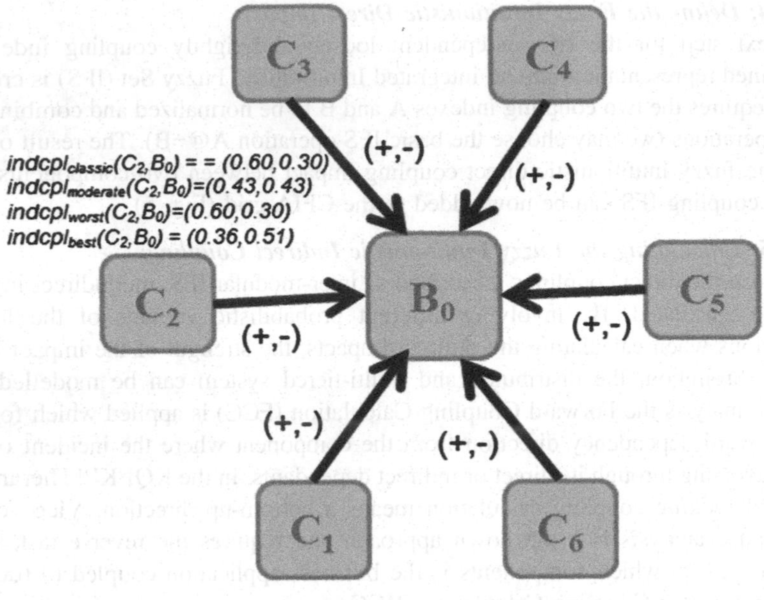

Fig. 6 One-Level dependency map as star schema

The result of step 5 is the fuzzy intuitionistic coupling index of each component to the business service represented as indirect coupling IFS (Fig. 6).

Step 6 (Optional): Extending the Business View

The IFSFIA may be optional extended with additional logical dependencies and business impact information. For operation of IT systems we need to know also about dependencies to e.g. IT users and roles, supporting processes or maintenance services. This can be expressed with a coupling relationship like—is coupled to: a procedure, a Service Level Agreement (SLA).

Also business and monetary information can be added to the service like hourly cost of failure or impacted users [18]. This can enable cost calculations based on the number of users concerned and/or amount of lost user processing time or even total cost of unavailability. However, the number of user workstations does not necessarily equate to the number of users at one point in time. So other measurements of costs of failure should complement these numbers, like SLA penalties when service providers fail to deliver the pre-agreed quality, estimation of the financial impact of IT failure against transaction volumes (related to the vital business functions) normally processed during the period of failure. For certain businesses a consequence of IT failure may be even external claims for financial compensation by impacted customers or business partners (Fig. 7).

The created CFIA matrix is expanded to include fields related to the Business Value and the Cost of Failure of a Service. These fields can simply show the hourly failure cost to the business or can map the number of users supported by each business service.

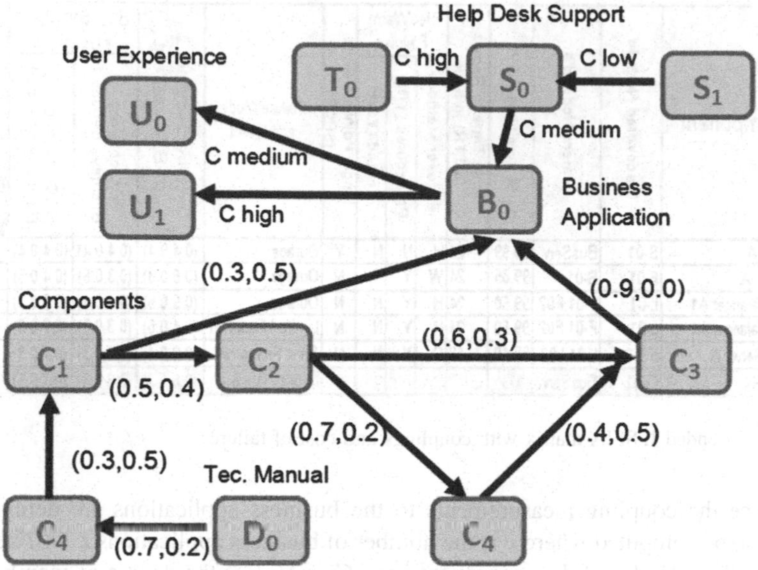

Fig. 7 Extended directed graph with couplings including it enabled services

Step 7: Performing Business Impact and Root Cause Analysis

A high tightly coupling index indicates a higher risk to the affected business service, which means this infrastructure component, is vital to business. A high loosely coupling index for a component indicates a strong resilience capability which allows smaller buffer overhead in the individual component's capacity planning and sizing.

The IFSFIA can be used in two principal ways, bottom-up as impact assessment or top-down for root cause analysis.

7(a) Business Impact Analysis (BIA)

Business Impact Analysis identifies vital business functions and their dependencies. These dependencies may include suppliers, business processes, IT Services etc. BIA defines as an output the requirements which include recovery time objectives and minimum Service Level Targets for each IT Service. The impact analysis using the IFSFIA can answer the question "Which are the indirect dependant business services of a particular component x and to which level are they tightly or loosely coupled?" starting from the low-level infrastructure component in the dependency hierarchy and traversing through its direct or indirect dependants to the business application services. The same BIA estimate used during operation to assess the business impact of incidents, can also be used to justify IT Infrastructure improvements by quantifying the total cost to the organization of an IT Service failure(s). These costs can then be used to support a business case for additional IT Infrastructure investment and provide an objective 'cost versus benefit' assessment.

Component	Discovered Node Id	Parent Node Id's	% Availibility QPI	Mean Time to Repair (MTTR) in hours	Hot/Warm Failover			Single Point of Failure (SPOF)	Failure Mode and Effect	Direct Impact (IFS) on parent	Bus. Service 1 # Users 700		total cost of unavailability per hour
					Failover Method	Procedures (Y/N)	Tested (Y/N)				FCC coupling to Business Service 1	RCC coupling from Business Service 1	
Switch A	S-01	BusServ	99.99	24	H	N	N	Y	Outage	(0.4,0.4)	(0.4,0.4)	(0.4,0.4)	4000
Firewall A	F-01	S-01	99.85	24	W	Y	N	N	Outage	(0.6,0.3)	(0.3,0.5)	(0.4,0.5)	3000
Load-Balancer A1	L-01	F-01,F02	99.50	24	H	Y	N	N	Outage	(0.5,0.5)	(0.4,0.4)	(0.6,0.3)	4000
Load-Balancer A1	L-01	F-01,F02	99.50	24	H	Y	N	N	Limited Function	(0.4,0.5)	(0.3,0.5)	(0.2,0.6)	3000
HTTP Server A	HS-01	L-01,L02	99.10	48	C	Y	N	N	Slow Response	(0.8,0.1)	(0.6,0.3)	(0.7,0.3)	6000
Helpdesk	Org1	BusServ							Quality Issue	(0.6,0.3)	(0.6,0.3)	(0.7,0.3)	6000

Fig. 8 Extended IFSFIA matrix with couplings and cost of failure

Since the coupling measurements to the business applications are defined the cost can be computed where *n* is the number of business applications *i, CCI* denotes the hourly cost of a of the component item *Ci*, *μA(x)i* is the degree of membership of tightly coupling of the component up to the business application *i* and *Ci* denotes the hourly cost of a failure of the business application *i*.

$$C_{CI} = \sum_{i=1..n}^{n} \mu_A(x)_i * C_i \tag{9}$$

The calculated total cost of failure per component can then be added as column to the IFCIA grid which allows assessing at one glance the monetary impact (Fig. 8).

In praxis business impact is hard to measure, as it could have several consequences, from financial impact to fuzzy aspects like feeling of dissatisfaction if IT service problems occur. Measurements on business impact of a failure are hard to quantify in monetary value, like "user productivity loss" or "lost business cost" etc.

7(b) Root Cause Analysis (RCA)
A root cause analysis (RCA) is a top down approach and requires the reverse task then the impact analysis to be solved, i.e. "To which components is the business application B coupled to (depends on)". The IFSFIA analysis procedure takes into account direct and indirect impacts of other components over the failed components. The result of the analysis is an intuitionistic fuzzy distribution of components giving an ordered set of possible root causes. Having the IFSFIA grid created, we simply can sort for the highest level of IFS coupling to get an order for the probability of possible root causes. The infrastructure component with the highest coupling is most likely and should therefore first being considered for causing the impact on a higher business service [17].

RCA implies the calculating of indirect impacts starting from the top and traversing through its impact arcs in the reverse direction. For RCA the Reverse

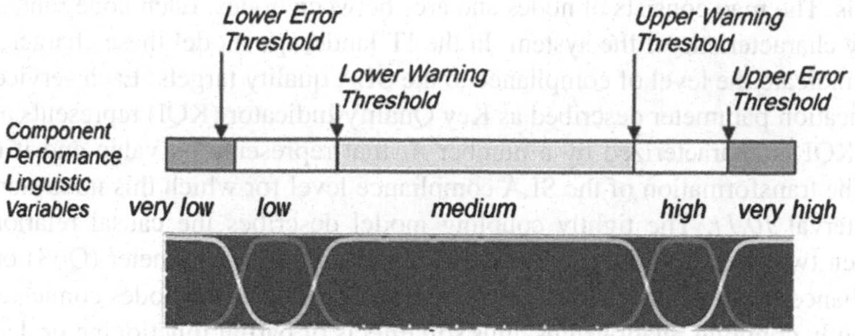

Fig. 9 Mapping of thresholds into linguistic variables

Coupling Calculation (RCC) index in the IFSFIA grid is chosen which may differ from the Forward Coupling Calculation (FCC) index which is applied for bottom-up impact calculations.

Step 8 (Optional): Applying Intuitionistic Fuzzy Reasoning
As final step the IFSFIA allows the application for two-sided (intuitionistic) fuzzy reasoning by combining both aspects including the vagueness of the fact into inference rules and logics. Thresholds can be used as natural limits to assign fuzzy linguistic variables to performance values (Fig. 9).

Using two-sided fuzzy logic, the complex system behaviour can be closely analysed by considering both contrary coupling aspects simultaneously. Two-sided fuzzy if-then rules can consider different interpretations of fuzzy implications, by applying bi-polar operations and interpretations. Once we have determined the fuzzy rules to define the performance measures, we can create linguistic rules for the service that will help to predict the impact to the front-stage service quality (QoS).

E.g.: *If {"Component Service" is (tightly coupled > 0.5) and (loosely coupled < 0.4) to "Business Service" and ("Component Service Performance" is LOW or "Component Service Reliability" is LOW)} then "Business Service" performance is LOW.*

4.2 Impact Analysis for Gradual Failure Modes

To reduce the complexity compliance for technical performance parameters will in praxis mostly measured bi-modal (either they operate correctly or they fail). This model can now be extended for granular failure impacts or service degradation effects and the consideration of several parallel incidents which causes the total impact. Thus a forecast can be given on the effect of e.g. 80 % SLA achievement or 60 % compliance to performance specifications. The direct coupling dependencies can be visualized within a directed graph representing the direct fuzzy intuitionistic impacts. The map consists of nodes and arcs between nodes. Each node represents a quality characteristic of the system. In the IT landscape model these characteristics could indicate the level of compliance to the SLA quality targets. Each service level specification parameter described as Key Quality Indicator (KQI) represents a node. Each KQI is characterized by a number *Ai* that represents its value and it results from the transformation of the SLA compliance level for which this node stands, in the interval *[0,1]*. The tightly coupling model describes the causal relationships between two nodes. A decrease in the value of a quality parameter (QoS) or SLA compliance level would yield a corresponding decrease at the nodes connected to it via tightly coupling relationships, thus soft effects of partial functioning or degraded SLA compliance between IT components can be directly modeled by the same approach. This concept is briefly derived from the mathematical model of cognitive maps. In 1986 Bart Kosko [19] introduced the notion of fuzziness to cognitive maps and created the theory of Fuzzy Cognitive Maps (FCMs). A Fuzzy cognitive map is a cognitive map within which the relations between the elements (e.g. components, IT resources) can be used to compute the "strength of impact" of these elements. FCMs are used in a much wider range of applications [20] which all have to deal with creating and using models of impacts in complex processes and systems. In the IT landscape scenario FCMs can be used to describe mutual dependencies between infrastructure and higher level IT components. The activation level of a quality parameter indicates in this extended model the level of SLA compliance The model of the classical FCM is now leveraged to compute the value of each quality parameter that influenced by the values of the coupled quality indicator with the appropriate weights and by its previous value.

$$A_i = f\Big(\sum_{\substack{j=1 \\ j \neq i}}^{n} A_j W_{ji}\Big) + A_i^{old} \tag{10}$$

So the value *Ai* for each quality indicator *KQIi* can be calculated by the rule where Ai is the activation level of quality parameter *KQIi* at time *t + 1, Aj* is the activation level of quality parameter *KQIj* at time *t, Ai* old is the activation level of quality parameter *KQIi* at time *t*, and *Wji* is the weight of the dependence coupling between *KQIj* and *KQIi*, and *f* is a threshold function. The weights of the dependencies between the *KQIi* and *KQIj* could be positive *(Wji > 0)* which means that an increase in the value of *KQIi* leads to the increase of the value of *KQIj*, and a decrease in the value of KQIi leads to the decrease of the value of *KQIj*. In case of a negative causality *(Wji < 0)* which means that an increase in the value of *KQIi* leads to the decrease of the value of *KQIj* and vice versa (Fig. 10).

By adding also the activation levels of the KQIs, each KQI is characterized by a number *Ai* that represents its value and it results from the transformation of the SLA compliance level for which this KPI stands, in the interval *[0,1]*.

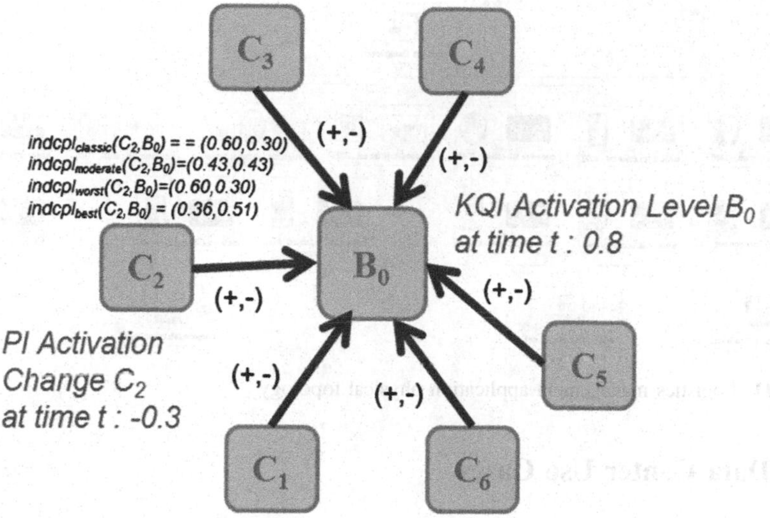

Fig. 10 Couplings related to KQI activation levels

As example: Using the Forward Coupling Calculation (FCC) method (applicable for Impact Analysis) of *indcpl(C_2,B_O)* depicted in the example graph shows the indirect coupling dependency of the Business Application B_O on the Component C_2.

- $indcpl_{classic}(C_2,B_O) = (0.60,0.30)$
- $indcpl_{moderate}(C_2,B_O) = (0.43,0.43)$
- $indcpl_{worst}(C_2,B_O) = (0.60,0.30)$
- $indcpl_{best}(C_2,B_O) = (0.36,0.51)$

Now the calculation of the KQI Activation Level for B_O at time $t + 1$ can be done as follows using an activation level of $KQI_T\ B_O = 0.8$ at point in time t.

- $KQI_{T+1}\ B_{O\ classic} = (0.8 - 0.3 * 0.6) = 0.62$
- $KQI_{T+1}\ B_{O\ moderate} = (0.8 - 0.3 * 0.43) = 0.671$
- $KQI_{T+1}\ B_{O\ worst} = (0.8 - 0.3 * 0.6) = 0.62$
- $KQI_{T+1}\ B_{O\ best} = (0.8 - 0.3 * 0.36) = 0.692$

In case the performance indicator C_2 decreases of 0.3, an impact between a decrease 0.108 and 0.18 to the quality indicator KQI B_O is estimated. This simple approach can be helpful where it is required to consider how several smaller improvements at different infrastructure components (e.g. improvements in performance or throughput) in total will impact a business service performance parameter KQI. All impacts will be pulled together so all single impacts are aggregated to the total effect on the business.

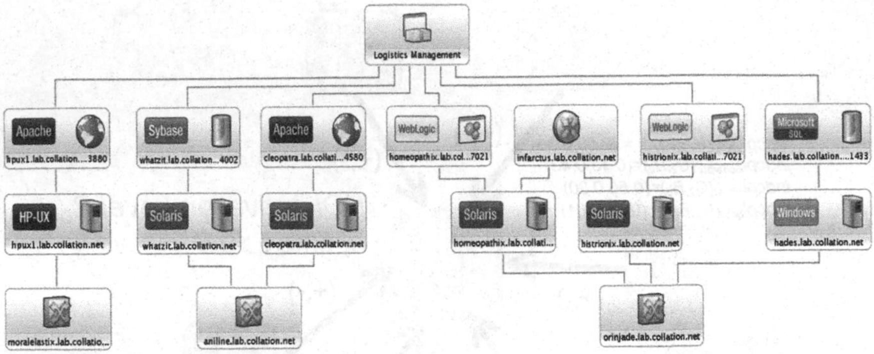

Fig. 11 Logistics management application physical topology

5 Data Center Use Cases

Several real world datacenter use cases have been developed for the IFSFIA framework [21]. These comprise Business Impact Analysis, Root Cause Analysis, Advanced Service Level Monitoring and Capacity Optimization which have been developed as use cases for the business application "Logistics Management" (Fig. 11).

5.1 Performing the IFSFIA Analysis

In the use case the IBM Tivoli Application Dependency Discovery Manager (TADDM) component affinity report extracts all related components which have a dependency (IP dependency, transactional dependency or configuration dependency) on those components which are directly related to the in-scope business services. It creates a table of all servers within the specified scope that are sources of relationships, and the connections from those servers to other server and middleware applications [17] (Fig. 12).

The Intuitionistic Coupling Index is now determined with regard to an appropriate formula (e.g. Dhama's metric) or alternatively assessed by the experts via inductive monitoring of relevant performance indicators. The tightly coupling index is defined as inter-modular coupling metric, which calculate the coupling between each pair of directly related components. For loosely coupling an intrinsic coupling metric is chosen as this refers to the individual components' resilience capabilities. Both index are normalized and pulled together into a single IFS, the fuzzy intuitionistic direct impact (Fig. 13).

The methodology for calculating the indirect coupling follows the forward dependency direction. Following it the indirect dependants of the failed component x are determined, starting from the node x in the dependency graph and traversing

cleopatra.lab .collation.net	10									
		AppServer	Apache	10.10.10.30	3580	ComputerSystem	hpux2.lab.collation.net	10.10.50.2		IP Dependency
		AppServer	Apache	10.10.10.30	3580	AppServer	IBM WebSphere Application Server	10.10.10.72	9811	Transactional Dependency
		AppServer	Apache	10.10.10.30	3580	AppServer	IBM WebSphere Application Server	10.10.10.72	2810	Transactional Dependency
		AppServer	Apache	10.10.10.30	4580	AppServer	WebLogic	10.10.31.21	7021	Transactional Dependency

Fig. 12 TADDM Server affinity report

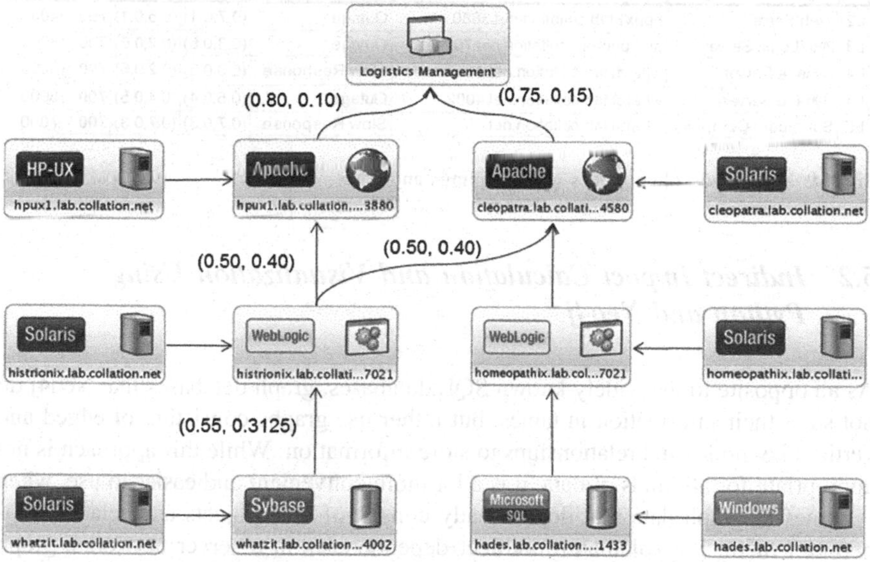

Fig. 13 Logistics management intuitionistic application dependency map

through its direct or indirect dependants. Different types of impact analysis involve the usage of classical or probabilistic variants of the logical operations conjunction and disjunction in calculation of indirect impacts. Depending on which combination of operations will be used, the indirect impacts may be greater or smaller. Within a grid all data relevant for the loosely coupling assessment is shown including the business repair time targets and estimated cost of failures. In the following IFSFIA grid two attitudes are expressed leading to an optimistic (best case) or moderate (mediate case) assessment of the impact caused by an incident situation (Fig. 14).

The result of the IFSFIA analysis is an intuitionistic sorted fuzzy distribution of the components, providing an ordered set by the probability of incident root causes. It can be now a guide for discovering roots for SLA violations and to justify IT investments.

Extended IFCFIA Grid with indirect couplings and cost of failure				Logistics Management Application cost of failure 10.000 per hour # Users 700 RTO 2 hours RPO 4 hours			
Hierarchy Level	Discovered Component Configuration Item (CI)	Discovered Component Id	Failure Mode and Effect	FCC coupling Moderate Risk	FCC coupling Best Case	total End Users impacted	total cost of failure per hour
L2	Web Server	hpux1.lab.collation.net:3880	Outage	(0.7,0.1)	(0.5,0.1)	700	4000
L3	WebLogic Server	histronix.lab.collation.net:7021	Outage	(0.3,0.5)	(0.2,0.6)	700	3000
L4	Sybase Server	whatzit.lab.collation.net:4002	Slow Response	(0.3,0.5)	(0.2,0.6)	700	3000
L4	Sybase Server	whatzit.lab.collation.net:4002	Outage	(0.5,0.4)	(0.4,0.5)	700	5000
L5	Sun Sparc Computer	whatzit.lab.collation.net	Slow Response	(0.7,0.3)	(0.7,0.3)	700	7000

Fig. 14 Monitored failure modes with couplings and costs

5.2 *Indirect Impact Calculation and Visualization Using Python and Neo4j*

As an opposite to the widely known SQL databases, graph databases like Neo4j do not store their information in tables, but rather use graphs consisting of edged and vertices i.e. nodes and relationships to store information. While this approach is not appropriate for all kinds of data, it is a lot more convenient and easier to use, when it comes to graph data that does already consist of data objects and relationships between them. For calculating indirect dependencies in a server networks, graph databases suit perfectly well, since the given data is already in shape of a connected network and actions like path-finding, which are required for the impact calculations, are already implemented in the used graph database Neo4j.

The following image shows the discovered servers of the Logistics Management application including the fuzzy intuitionistic direct impact loaded into the Neo4j database (Fig. 15).

Being able to calculate the indirect dependency index for the discovered network, the impact of any component to any other can be expressed as fuzzy intuitionistic indirect impact by either getting the direct coupling for adjacent servers or calculating the indirect coupling based on the chosen IFS operations. To present the results to the user, the Neo4j browser is used, where a temporary graph is inserted into the database, which forms a star showing the chosen service in the center and all other components connected to it with the calculated indirect coupling levels (Fig. 16).

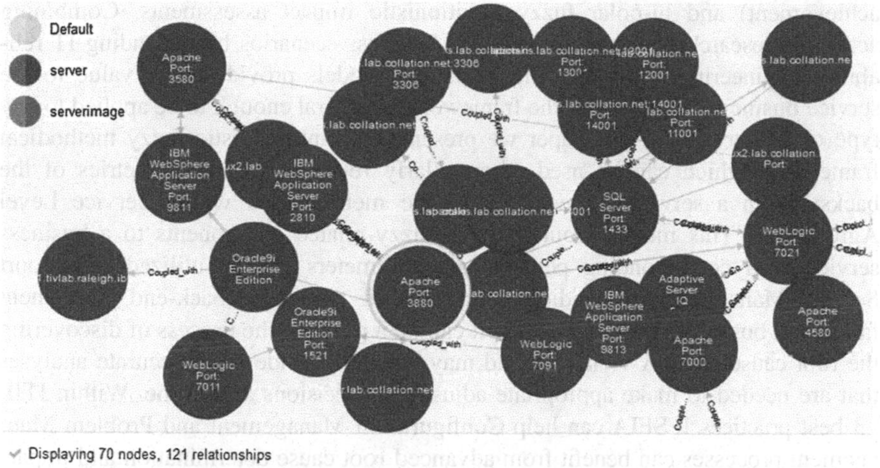

Fig. 15 Loaded components and direct dependencies into Neo4j

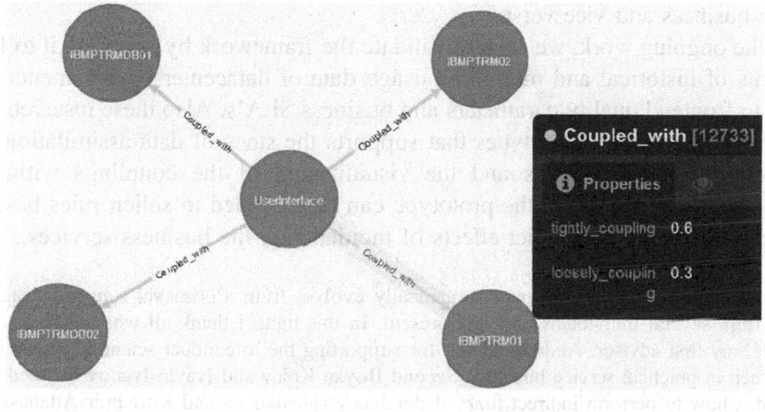

Fig. 16 Star representation of indirect dependencies

6 Conclusions and Ongoing Work

Managing the quality of virtualized, distributed and multi-tiered services is a key challenge in today's service management. Traditional approaches are measured bi-modal (means either operate correctly or fail) and concentrate on local technical IT performance measurements rather than with business-oriented service achievement. There are some more advanced approaches [22], including proposed models of QoS ontologies [23] or works that are based on Fuzzy Rules [3] e.g. Performance Relation Rules and Artificial Intelligence. The novelty of our approach lies in an integrated step-wise methodology, including automated information assimilation, support of gradual failures or service degradations (e.g. predicting a percentual SLA achievement) and bi-polar fuzzy intuitionistic impact assessments. Combining academic research with practice oriented business scenarios by expanding IT reliability engineering with fuzzy mathematical models provides high value to the service business, especially as the framework is general enough to be applied to any type of IT service. In this paper we presented an intuitionistic fuzzy methodical framework, which can be used to granularly relate performance metrics of the backstage in a service orchestration to the metrics used within Service Level Agreements. This model about a set of fuzzy-related components to a business service with corresponding performance parameters can be utilized to support Service Management to predict on impacts of monitored back-end component failures to business services. Further, it can be a guide in the process of discovering the root cause of SLA violations and may help to provide more accurate analyses that are needed to make appropriate adjustment decisions at runtime. Within ITIL v3 best practices IFSFIA can help Configuration Management and Problem Management processes can benefit from advanced root cause determination and impact assessments. The proposed IFSFIA framework enables transformation of availability and performance data into knowledge about the real-time status of business services that allows understanding and communicating the true impact of incidents on the business and vice versa.

In the ongoing work, we seek to validate the framework by applying it to larger amounts of historical and monitored usage data of datacenter environments compared to frontend quality parameters and business SLA's. Also these research ideas are implemented with prototypes that supports the steps of data assimilations, the indirect impact calculations and the visualization of the couplings within the dependency graph. Further the prototype can be extended to solicit rules based on the derived impacts to predict effects of incidents on the business services.

Acknowledgment A scientific paper organically evolves from a primeval soup of ideas originating from several individuals past and present. In this light, I thank all who made this paper possible, my first advisor, Andreas Meier, for supporting me to conduct scientific research after years deep in practical service business. Second Boyan Kolev and Ivaylo Ivanov for feeding the great idea how to perform indirect fuzzy dependency calculations and Krassimir Attanassov for inventing the principal concept and mathematical foundation about intuitionistic fuzzy sets.

References

1. Schütze, R.: ICIFS Conference Nov. 13 Sofia, Bulgaria—Intuitionistic Component Failure Impact Analysis "Notes on IFS", **19**(3), 62–72 (2013)
2. Li, H.: Challenges in SLA Translation—SLA@SOI European Commission Seventh Framework Programme (2007–2013) SAP Research (2009)
3. Joshi, K.P., Joshi, A. Yesha, Y.: Managing the quality of virtualized services. In: Proceedings of the SRII Service Research Conference (2011)
4. Tele Forum: SLA Management Handbook—Vol. 4: Enterprise Perspective ISBN: 1-931624-51-8
5. Dhama, H.: Quantitative models. J. Syst. Softw. **29**, 65–74 (1995)

6. Fenton, N.E., Melton, A.: Deriving software measures. J. Syst. Softw. **12**, 177–187 (1990)
7. Alghamdi, J.S.: Measuring software coupling. In Proceedings of the 6th WSEAS Conference on Software Engineering (2007)
8. Quynh, T.: Dynamic Coupling Metrics for Service–Oriented Software 2009. University of Science, Hanoi Vietnam (2009)
9. Craig, J.: EMA Radar for application discovery and dependency mapping (ADDM): Q4 2010 summary and vendor profile (2010)
10. Chen, X., Zhang, M., Mao, Z.M., Bahl, P.: Automating Network Application Dependency Discovery: Experiences, Limitations, and New Solutions (2008)
11. Van Haren: Continual Service Improvement Based On ITIL V3 ISBN 9087532407 (2008)
12. Three reliability engineering techniques, IBM Systems Journal, Vol 47, No 4, 2008
13. Yang, H.Y.: Measuring indirect coupling. Doctor Thesis, University of Auckland, New Zealand (2010)
14. Atanassov, K.: On Intuitionistic Fuzzy Sets Theory (Studies in Fuzziness and Soft Computing), 1st edn. Springer, Berlin (1999 edn: 2010)
15. Zadeh, Lofti, Computing, Soft, Logic, Fuzzy: IEEE Softw. **11**(6), 48–56 (1994)
16. Kolev, I.: Fault Tree Analysis in an Intuitionistic Fuzzy Configuration Management Database 2009. In: International Conference on Intuitionistic Fuzzy Sets, Sofia Bulgaria (2009)
17. Jacob, B., Adhia, B., Badr, K., Huang, Q.C., Lawrence, C.S., Marino, M., Unglaub-Lloyd, P.: IBM Tivoli Application Dependency Discovery Manager: Capabilities and Best Practices, IBM Redbook, Poughkeepsie (2008)
18. Kieninger, A., Berghoff, F., Fromm, H., Satzger, G.: Simulation-based quantification of business impacts caused by service incidents. In: Proceeding of the Fourth International Conference on Exploring Service Science (Portugal), LNBIP vol. 143, pp. 170–185. Springer (2013)
19. Kosko, B.: Fuzzy cognitive maps. Int. J. Man Mach. Stud. **24**, 65–75 (1986)
20. Stylios, C., Georgopoulos, V., Groumpos, P.: The Use of Fuzzy Cognitive Maps in Modeling Systems. Department of Electrical and Computer Engineering, University of Patras (1997)
21. Schütze, R.: IFCFIA, a gradual method for SLA dependency mapping and bi-polar impact assessment. Technical report 07/2013, Information Systems, University of Fribourg, Switzerland, (2013)
22. Robak, P.: Fuzzy modeling of QoS. J. Comput. Sci. **16**, 69–79 (2008)
23. Sora, T.: Translating User Preferences into Fuzzy Rules. Politehnica University, Romania (2009)

Part III
Generalized Nets

Modifications of the Algorithms for Transition Functioning in GNs, GNCP, IFGNCP1 and IFGNCP3 When Merging of Tokens is Permitted

Velin Andonov and Nora Angelova

Abstract Modifications of the algorithms for transition functioning in Generalized Nets (GNs), Generalized Nets with Characteristics of the Places (GNCP), Intuitionistic Fuzzy Generalized Nets with Characteristics of the Places of type 1 (IFGNCP1) and type 3 (IFGNCP3) are proposed. We consider GNs for which the merging of tokens is permitted. In the standard algorithm for transition functioning no tokens can be transferred to output place of a given transition if it has reached its capacity. The algorithms described in this paper allow the transfer of a token to full output place if it can merge with some of the tokens there. Two versions of the characteristic function Ψ in IFGNCP1 and IFGNCP3 are discussed.

1 Introduction

Generalized Nets (GNs) [5, 6] are extensions of Petri Nets [9]. They are defined in a way that is principally different from the ways of defining the other types of Petri nets. Each part of the net which looks like the one shown on Fig. 1, is called *transition* (more precisely graphic representation of transition).

Formally, every transition is described by a seven-tuple:

$$Z = \langle L', L'', t_1, t_2, r, M, \square \rangle,$$

where:

(**a**) L' and L'' are finite, non-empty sets of places (the transition's input and output places, respectively); for the transition in Fig. 1 these are

V. Andonov (✉) · N. Angelova
Institute of Biophysics and Biomedical, Engineering Bulgarian Academy of Sciences, Acad. G. Bonchev Str., Bl. 105, 1113 Sofia, Bulgaria
e-mail: velin_andonov@yahoo.com

N. Angelova
e-mail: metida.su@gmail.com

P. Angelov and S. Sotirov (eds.), *Imprecision and Uncertainty in Information Representation and Processing*, Studies in Fuzziness and Soft Computing 332,
DOI 10.1007/978-3-319-26302-1_17

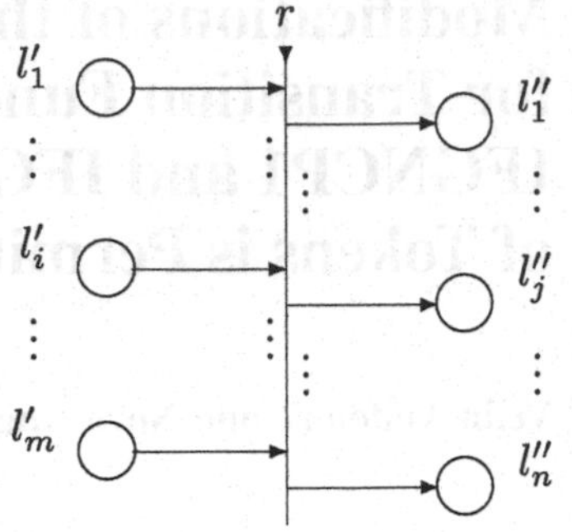

Fig. 1 Transition

$$L' = \{l'_1, l'_2, \dots, l'_m\}$$

and

$$L'' = \{l''_1, l''_2, \dots, l''_n\};$$

(b) t_1 is the current time-moment of the transition's firing;
(c) t_2 is the current value of the duration of its active state;
(d) r is the transition's *condition* determining which tokens will transfer from the transition's inputs to its outputs. Parameter r has the form of an IM:

$$r = \begin{array}{c|c} & l''_1 \ \dots \ l''_j \ \dots \ l''_n \\ \hline l'_1 & \\ \vdots & r_{i,j} \\ l'_i & (r_{i,j} \ - \ \text{predicate}) \\ \vdots & (1 \le i \le m, 1 \le j \le n) \\ l'_m & \end{array};$$

where $r_{i,j}$ is the predicate which expresses the condition for transfer from the ith input place to the jth output place. When $r_{i,j}$ has truth-value "*true*", then a token from the ith input place can be transferred to the jth output place; otherwise, this is impossible;
(e) M is an IM of the capacities of transition's arcs:

$$M = \begin{array}{c|c} & l''_1 \ \dots \ l''_j \ \dots \ l''_n \\ \hline l'_1 & \\ \vdots & m_{i,j} \\ l'_i & (m_{i,j} \ge 0 - \ \text{natural number or } \infty) \\ \vdots & (1 \le i \le m, 1 \le j \le n) \\ l'_m & \end{array};$$

(f) $\square$ is called transition type and it is an object having a form similar to a Boolean expression. It may contain as variables the symbols that serve as labels for transition's input places, and it is an expression constructed of variables and the Boolean connectives $\wedge$ and $\vee$ determining the following conditions:

$\wedge(l_{i_1}, l_{i_2}, \dots, l_{i_u})$—every place $l_{i_1}, l_{i_2}, \dots, l_{i_u}$ must contain at least one token,

$\vee(l_{i_1}, l_{i_2}, \dots, l_{i_u})$—there must be at least one token in the set of places $l_{i_1}, l_{i_2}, \dots, l_{i_u}$, where $\{l_{i_1}, l_{i_2}, \dots, l_{i_u}\} \subset L'$.

When the value of a type (calculated as a Boolean expression) is "*true*", the transition can become active, otherwise it cannot.

The ordered four-tuple

$$E = \langle\langle A, \pi_A, \pi_L, c, f, \theta_1, \theta_2\rangle, \langle K, \pi_K, \theta_K\rangle, \langle T, t^0, t^*\rangle, \langle X, \Phi, b\rangle\rangle$$

is called a *Generalized Net* if:

(a) A is a set of transitions (see above);

(b) π_A is a function giving the priorities of the transitions, i.e., $\pi_A : A \rightarrow \mathcal{N}$;

(c) π_L is a function giving the priorities of the places, i.e., $\pi_L : L \rightarrow \mathcal{N}$, where

$$L = pr_1 A \cup pr_2 A$$

and obviously, L is the set of all GN-places;

(d) c is a function giving the capacities of the places, i.e., $c : L \rightarrow \mathcal{N}$;

(e) f is a function that calculates the truth values of the predicates of the transition's conditions;

(f) θ_1 is a function giving the next time-moment, for which a given transition Z can be activated, i.e., $\theta_1(t) = t'$, where $pr_3 Z = t, t' \in [T, T + t^*]$ and $t \leq t'$; the value of this function is calculated at the moment when the transition terminates its functioning;

(g) θ_2 is a function giving the duration of the active state of a given transition Z, i.e., $\theta_2(t) = t'$, where $pr_4 Z = t \in [T, T + t^*]$ and $t' \geq 0$; the value of this function is calculated at the moment when the transition starts functioning;

(h) K is the set of the GN's tokens. In some cases, it is convenient to consider this set in the form

$$K = \bigcup_{l \in Q^I} K_l,$$

where K_l is the set of tokens which enter the net from place l, and Q^I is the set of all input places of the net;

(i) π_K is a function giving the priorities of the tokens, i.e., $\pi_K : K \rightarrow \mathcal{N}$;

(j) θ_K is a function giving the time-moment when a given token can enter the net, i.e., $\theta_K(\alpha) = t$, where $\alpha \in K$ and $t \in [T, T + t^*]$;

(k) T is the time-moment when the GN starts functioning; this moment is determined with respect to a fixed (global) time-scale;
(l) t^0 is an elementary time-step, related to the fixed (global) time-scale;
(m) t^* is the duration of the GN functioning;
(n) X is a function which assigns initial characteristics to every token when it enters input place of the net;
(o) Φ is a characteristic function that assigns new characteristics to every token when it makes a transfer from an input to an output place of a given transition;
(p) b is a function giving the maximum number of characteristics a given token can receive, i.e., $b : K \to N$.

For the algorithms of transition and GN functioning the reader can refer to [6, 8].

Generalized Nets with Characteristics of the Places (GNCP) is one of the most recent extensions of GNs. It is introduced in [1] and again there it is proved that Σ_{CP}—the class of all GNCP—is conservative extension of the class Σ of the ordinary GNs. The connection between GNCP and the Intuitionistic Fuzzy Generalized Nets of type 1 (IFGN1) and type 2 (IFGN2) is studied in [2]. It is proved that $\Sigma_{CP} \equiv \Sigma_{IFGN1}$ and $\Sigma_{CP} \equiv \Sigma_{IFGN2}$. In [10] some possible applications of GNCP are discussed. In particular, GNCP can be used for evaluation of the work of the places based on the characteristics of the tokens and also to simplify the graphical representation of the net. Two new extensions are defined in [3]. The first one—Intuitionistic Fuzzy Generalized Nets with Characteristics of the places of type 1 (IFGNCP1)—combines the characteristics of GNCP and IFGN1. The second—Intuitionistic Fuzzy Generalized Nets with Characteristics of the Places of type 3 (IFGNCP3)—combines the characteristics of GNCP and IFGN3. It is proved that $\Sigma_{IFGNCP1} \equiv \Sigma$ and $\Sigma_{IFGNCP3} \equiv \Sigma$.

In GNCP the places can obtain characteristics at the end of a time step if tokens have have entered the place, i.e. if something related to the place has happened. Assigning characteristics to the places is a convenient way to keep the data which is relevant to the places. Formally, GNCP E is the ordered four-tuple

$$E = \langle \langle A, \pi_A, \pi_L, c, f, \theta_1, \theta_2 \rangle, \langle K, \pi_K, \theta_K \rangle, \langle T, t^0, t^* \rangle, \langle X, Y, \Phi, \Psi, b \rangle \rangle .$$

All other components with the exception of Y and Ψ have the same meaning as in the definition of the standard GN. Here Y is a function which assigns initial characteristics to some of the places. Initial characteristics can be assigned also to places in which no tokens enter in the initial time moment. Ψ is a function which assigns characteristics to some of the output places of the transitions when tokens enter them. Not all places of the net can receive characteristics. We consider that for every place l that does not receive characteristics during the functioning of the net we have $\Psi_l = \}\}\emptyset$".

The algorithm for transition functioning in GNCP is proposed in [4]. Algorithms for transition functioning in IFGNCP1 and IFGNCP3 are proposed in [3]. We shall shortly discuss these algorithms and propose a modification related to the characteristic function Ψ. Here, we also propose algorithms for transition functioning in GNCP, IFGNCP1 and IFGNCP3 when merging of tokens is permitted. The idea is when an output place of a given transition has reached its capacity but the current token in some input place can merge with some of the tokens in that same output place the transfer will happen when the other conditions allow it.

IFGNCP1 have the same components as GNCP but their meaning is slightly different. In analogy to the Intuitionistic Fuzzy Generalized Nets of type 1 (IFGN1), in IFGNCP1 the function f evaluates the truth value of the predicates in the form of Intuitionistic Fuzzy Pairs (IFP) (see [7]). To every predicate $r_{i,j}$ the function f assigns the ordered couple $\langle \mu_{i,j}, \nu_{i,j} \rangle$ where $\mu_{i,j}$ is the degree of truth of the predicate $r_{i,j}$ and $\nu_{i,j}$ is the degree of falsity. They satisfy the conditions $\mu_{i,j}, \nu_{i,j} \in [0, 1]$ and $\mu_{i,j} + \nu_{i,j} \leq 1$. The number $\pi_{i,j} = 1 - \mu_{i,j} - \nu_{i,j}$ is the degree of uncertainty. The characteristic function Φ which in the ordinary GNs assigns characteristics to the tokens now adds to these characteristics the degrees of truth and falsity of the predicate corresponding to the input and output places. In this way when the token finishes its transfer in the net we can determine the degrees of validity and non-validity of the transfer. In [3] the characteristic function Ψ assigns characteristic to the output places of a given transition for every token that has entered them in the form *"value of Ψ for the respective place, $\langle \mu_{i,j}, \nu_{i,j} \rangle$"*. In the algorithm for transition functioning which we present here, the way the characteristics are assigned to the places is changed to be analogous to the algorithm of transition functioning in GNCP.

IFGNCP3 are defined in analogy to the Intuitionistic Fuzzy Generalized Nets of type 3 (IFGN3). IFGN3 have the same components as IFGN1 but now the characteristic function Φ assigns characteristics to the tokens in the form

$$x^{\alpha}_{cu} = \langle \overline{x}^{\alpha}_{cu}, \mu(r_{i,j}), \nu(r_{i,j}), \mu(x^{\alpha}_{cu}), \nu(x^{\alpha}_{cu}) \rangle$$

where $\overline{x}^{\alpha}_{cu}$ is the standard characteristic of the token in the sense of GNs, $\mu(r_{i,j})$ and $\nu(r_{i,j})$ are the degrees of truth and falsity of the corresponding predicate and $\mu(x^{\alpha}_{cu})$ and $\nu(x^{\alpha}_{cu})$ are estimations in intuitionistic fuzzy sense of the characteristics of the tokens. The pair $\langle \mu(x^{\alpha}_{cu}), \nu(x^{\alpha}_{cu}) \rangle$ represents the validity and non-validity of the characteristics and through them the model determines its status. IFGNCP3 has the same components as IFGNCP1 but the characteristic functions Φ and Ψ have different meaning. Φ assigns characteristics to the tokens as in the case of IFGN3. The tokens obtain characteristics only when the estimations of the characteristics satisfy one of the following conditions:

C1 $\mu(x^{\alpha}_{cu}) = 1, \nu(x^{\alpha}_{cu}) = 0$ (the case of ordinary GN) ;

C2 $\mu(x^{\alpha}_{cu}) > \frac{1}{2}$ $(> \nu(x^{\alpha}_{cu}))$;

C3 $\mu(x^{\alpha}_{cu}) \geq \frac{1}{2}$ $(\geq \nu(x^{\alpha}_{cu}))$;

C4 $\mu(x^{\alpha}_{cu}) > \nu(x^{\alpha}_{cu})$;

C5 $\mu(x^{\alpha}_{cu}) \geq \nu(x^{\alpha}_{cu})$;

C6 $\mu(x^{\alpha}_{cu}) > 0$;

C7 $\nu(x^{\alpha}_{cu}) < 1$.

Ψ assigns characteristics to the places in the form

$$\psi^{l}_{cu} = \langle \overline{\psi}^{l}_{cu}, \mu(r_{i,j}), \nu(r_{i,j}), \mu(\psi^{l}_{cu}), \nu(\psi^{l}_{cu}) \rangle ,$$

where $\mu(\psi^{l}_{c}u), \nu(\psi^{l}_{cu}) \in [0, 1]$ and $\mu(\psi^{l}_{cu}) + \nu(\psi^{l}_{cu}) \leq 1$. ψ^{l}_{cu} is the standard characteristic of the place in the sense of GNCP. The places obtain characteristics only when the estimations of the characteristics satisfy one of the following conditions:

C1* $\mu(\psi^{l}_{cu}) = 1, \nu(\psi^{l}_{cu}) = 0$ (the case of ordinary GN) ;

C2* $\mu(\psi^{l}_{cu}) > \frac{1}{2} (> \nu(\psi^{l}_{cu}))$;

C3* $\mu(\psi^{l}_{cu}) \geq \frac{1}{2} (\geq \nu(\psi^{l}_{cu}))$;

C4* $\mu(\psi^{l}_{cu}) > \nu(\psi^{l}_{cu})$;

C5* $\mu(\psi^{l}_{cu}) \geq \nu(\psi^{l}_{cu})$;

C6* $\mu(\psi^{l}_{cu}) > 0$;

C7* $\nu(\psi^{l}_{cu}) < 1$.

2 Algorithm for Transition Functioning in GNs When Merging of Tokens is Permitted

We consider that the tokens with which an arbitrary token α can merge are part of its initial characteristic. For instance, $x^{\alpha'}_{0} = \}\}\langle\{\beta_1, \ldots, \beta_k\}, x^{\alpha}_{0}\rangle''$, where $\{\beta_1, \ldots, \beta_k\}$ is the set of tokens that can merge with α and x^{α}_{0} is the standard initial characteristic of the token. We shall denote the modified algorithm for transition functioning in GNs when merging of tokens is allowed by **Algorithm A**.

Algorithm A.

(A01) The input places are sorted by priority in descending order.

(A02) Sort the tokens from the input places by their priorities. The tokens from a given input place are divided into two groups. The first one contains those tokens that can be transferred to the transition's outputs, the second contains the rest. In the beggining, the second group is empty.

(A03) An empty index matrix R*, which corresponds to the index matrix of the predicates of the given transition R, is generated. We put values "0" (corresponding to value "false") of all of the elements of this index matrix which:

(a) are placed in a row, corresponding to an empty input place, i.e. the input place doesn't contain tokens that can be transferred to the transition's outputs.
(b) are placed in (i, j)th position, for which $m_{i,j} = 0$, i.e. the current capacity of the arc between ith input and jth output place is zero.

(A04) The sorted places are passed sequentially by their priority, starting with the place having the highest priority, which has at least one token and through which no transfer has occurred on the current time-step. For the highest priority token (from the first list) we determine whether it can split or not. **If it can split**, then the algorithm proceeds to **Step (A05)**, else **if it can't split**, then the algorithm proceeds to **Step (A06)**.

(A05)

(a) Get a predicate corresponding to the relevant row of the index matrix R which is not checked on the current time-step. If there isn't a predicate, then the algorithm proceeds to Step (e). In the opposite case it proceeds to Step (b).
(b) If **the output place is full**, then the algorithm proceeds to step (c). In the opposite case it proceeds to Step (d).
(c) If **the selected token can be merged** with other token in the full output place, then the algorithm proceeds to step (d). In the opposite case we put value "0" (corresponding to value "false") of the element of R* and the algorithm returns to Step (a).
(d) The predicate corresponding to the relevant row of the index matrix R* is checked. If the calculated value is truth, then we put value "1" of the element of R*. In the opposite case we put value "0". The algorithm proceeds to Step (a).
(e) **If the values are only "0"**, then **Step (A05)** stops and the algorithm proceeds to **Step (A07)**.
If the row contains at least one "1", then the corresponding token splits as many times as necessary. These tokens are moved to the output places. The values of the characteristic function for the corresponding output places are assigned as a next token characteristic. Step **(A05)** stops and the algorithm proceeds to **Step (A08)**

(A06)

(a) The predicates corresponding to the relevant row of the index matrix R which is not check on the current time-step are passed sequentially. If there isn't a predicate, then the algorithm proceeds to **Step (A07)**. In the opposite case it proceeds to Step (b).
(b) If **the output place is full**, then the algorithm proceeds to step (c). In the opposite case it proceeds to Step (d).

(c) If **the selected token can be merged** with other token in the full output place, then the algorithm proceeds to step (d). In the opposite case we put value "0" (corresponding to value "false") of the element of R^* and the algorithm return to step (a).
(d) The predicate corresponding to the relevant row of the index matrix R^* is checked. If the calculated value is truth, then the algorithm proceeds to step (e). In the opposite case we put value"0" and the algorithm returns to Step (a).
(e) The token is moved to the output places. The values of the characteristic function for the corresponding output place are assigned as a next token characteristic. **Step (A06)** stops and the algorithm proceeds to **Step (A08)**.

(A07) If one token cannot pass through a given transition on this time interval, it is moved to the second group of tokens of the corresponding input place.

(A08) The current number of tokens for all places in which a token has entered and has not merged with any of the tokens there increment with 1.

(A09) The current number of tokens for all places from which a token has left decrement with 1. If the number of tokens for a given input place has reached 0, the elements of the corresponding row of the index matrix R^* are made "0".

(A10) The capacities of all arcs through which a token has passed decrement with 1.

(A11) If there are still input places with lower priority from which the tokens have not been transferred, the algorithm proceeds to **Step (A04)**. In the opposite case it proceeds to **Step (A12)**.

(A12) The current model time t is increased with t^o.

(A13) Is the current time moment equal to or greater than $t_1 + t_2$?

(A14) If the answer to the question **Step (A13)** is "no", then return to **Step (A04)**, otherwise "Termination of the transition functioning".

3 Algorithm for Transition Functioning in GNCP When Merging of Tokens is Permitted

We denote the modified algorithm for transition functioning in GNCP when merging of tokens is allowed by **Algorithm A'**.

Algorithm A'.

(A'01) The input and output places are ordered by their priorities.

(A'02) For every input place two lists are compounded. One with all tokens in the place ordered by their priorities and an empty list.

(A'03) An empty index matrix *R* which corresponds to the index matrix of the predicates *r* is generated. A value "0" (corresponding to truth-value "false") is assigned to all elements of *R* which:

– are in a row corresponding to empty input place;
– are placed in a position (i,j) for which the current capacity of the arc between the ith input and jth output place is 0.

(A'04) The places are passed sequentially by order of their priorities starting with the place with the highest priority for which transfer has not occurred on the current time step and which has at least one token. For the token with highest priority from the first group the capacities of the output places are checked. If the current capacity of the output place is 0 and the current token cannot merge with any of the tokens in the output place, value "0" is assigned to the corresponding element of R.

(A'05) For the token with the highest priority in the current place it is determined if it can split or not. The predicates in the row corresponding to the current input place are checked. If the token cannot split the checking of the predicates stops with the first predicate whose truth value is not "0". If the token can split, the truth values of all predicates in the row for which the elements of R are not equal to 0 are evaluated. Value 0 is assigned to the elements in the current row of R corresponding to predicates whose truth-value is "*false*". Value "1" is assigned to the elements in the current row of R corresponding to predicates whose truth-value is "*true*".

(A'06) Depending on the execution of the operator for permission or prohibition of tokens' splitting, the token from **(A'05)** is transferred either to all permitted output places or to the only output place which corresponds to the element in the row with value "1". If a token cannot be transferred at the current time step, it is moved to the second group of the corresponding input place. The tokens which have been transferred are moved into the second group of the output places. The tokens which have entered the input place after the activation of the transition are moved to the second group too.

(A'07) If transfer of the token is possible, the number of tokens in the input place is decreased by 1. If there are no tokens in the input place, the elements in the corresponding row of R are assigned value "0".

(A'08) The current number of tokens in every output place is increased with 1 for each token that has entered the place at the current time step and has not merged there with other token. The current number of tokens in the output places in which token has entered and merged with other token remains the same.

(A'09) The capacities of all arcs through which a token has passed decrement with 1. If the current capacity of an arc has reached 0, value "0" is assigned to the element from the index matrix R that corresponds to this arc.

(A'10) The values of the characteristic function Φ for the corresponding output places (one or more) in which tokens have entered according to **(A'05)** are calculated. These values are assigned to the tokens.

(A'11) If there are more input places (with lower priority) at the current time step from which tokens can be transferred, the algorithm proceeds to **(A'04)**, otherwise it proceeds to **(A'12)**.

(A'12) The values of the characteristic function Ψ for all output places to which tokens have been transferred are calculated. These values are assigned to the places.

(A'13) The current model time t is increased with t^0.

(A'14) Is the current time moment equal or greater than $t^1 + t^2$? If the answer to the question is "no", go to **(A'04)**. Otherwise, go to step **(A'15)**.

(A'15) Termination of the transition's functioning.

The novelty of the proposed algorithm can be seen in steps **(A'03)**, **(A'04)**, **(A'08)** and **(A'12)**. In comparison to the algorithms for transition functioning proposed in [4, 6, 8] value "0" is assigned only to the elements of R corresponding to empty input places and arcs with current capacity 0. In the standard algorithm this is also done for all elements in a column corresponding to full output place. Step **(A'04)** is where the capacities of the output places are checked. The proposed modification allows the transfer of tokens to output places which has reached their capacity if they can merge with some of the tokens in the output place. If splitting of tokens is not allowed, the evaluation of the predicates stops with the first predicate whose truth-value allows the transfer. Step **(A'08)** is also new as the current number of tokens in the output places is increased only when the entering token does not merge with any of the tokens in the place.

4 Algorithm for Transition Functioning in IFGNCP1 and IFGNCP3 When Merging of Tokens is Permitted

In the algorithm for transition functioning in IFGNCP1 as suggested in [3], the characteristic function of the places Ψ is evaluated for every token that has entered output place of the transition. Here we modify this algorithm so that the function Ψ is evaluated after all input places of the transition have been passed. The other modification of the algorithm is with regard to the merging of tokens. A token can be transferred from input to output place even when the output place has reached its capacity if it can merge with some of the tokens in the output place.

Algorithm A'

(A'01) The input and output places are ordered by their priorities.

(A'02) For every input place two lists are compounded. One with all tokens in the place ordered by their priorities and an empty list.

(A'03) An empty index matrix R which corresponds to the index matrix of the predicates r is generated. A value "$\langle 0, 1\rangle$" is assigned to all elements of R which:

– are in a row corresponding to empty input place;
– are placed in a position (i, j) for which the current capacity of the arc between the ith input and jth output place is 0.

(A'04) The places are passed sequentially by order of their priorities starting with the place with the highest priority for which transfer has not occurred on the current time step and which has at least one token. For the token with highest priority from the first group the capacities of the output places are checked. If the current capacity of the output place is 0 and the current token cannot merge with any of the tokens in the output place, value "$\langle 0, 1\rangle$" is assigned to the corresponding element of R.

(A'05) For the token with the highest priority in the current place it is determined if it can split or not. The predicates in the row corresponding to the current input place are checked. If the token cannot split the checking of the predicates stops with the first predicate whose truth value is not $\langle 0, 1\rangle$. If the token can split, the truth values of all predicates in the row for which the elements of R are not equal to "$\langle 0, 1\rangle$" are evaluated.

(A'06) Depending on the execution of the operator for permission or prohibition of tokens' splitting, the token from **(A'05)** is transferred either to all permitted output places or to the place with the highest priority. The transfer depends on one of the following conditions:

C1 $\mu(r_{i,j}) = 1, \nu(r_{i,j}) = 0$ (the case of ordinary GN) ;

C2 $\mu(r_{i,j}) > \frac{1}{2}$ $(> \nu(r_{i,j}))$

C3 $\mu(r_{i,j}) \geq \frac{1}{2}$ $(\geq \nu(r_{i,j}))$;

C4 $\mu(r_{i,j}) > \nu(r_{i,j})$,

C5 $\mu(r_{i,j}) \geq \nu(r_{i,j})$;

C6 $\mu(r_{i,j}) > 0$;

C7 $\nu(r_{i,j}) < 1$, i.e. at least $\pi(r_{i,j}) > 0$, where $\pi(r_{i,j}) = 1 - \mu(r_{i,j}) - \nu(r_{i,j})$ is the degree of uncertainty (indeterminacy) and $f(r_{i,j}) = \langle \mu(r_{i,j}), \nu(r_{i,j})\rangle$.

The condition for transfer of the tokens which will be used is determined for every transition before the firing of the net. If a token cannot be transferred at the current time step, it is moved to the second group of the corresponding input place. The tokens which have been transferred are moved into the second group of the output places. The tokens which have entered the input place after the activation of the transition are moved to the second group too.

(A'07) If transfer of the token is possible, the current number of tokens in the input place is decreased by 1. If there are no tokens in the input place, the elements in the corresponding row of R are assigned value "$\langle 0, 1\rangle$".

(A'08) The current number of tokens in every output place is increased with 1 for each token that has entered the place at the current time step and has not merged there with other token. The current number of tokens in the output places in which token has entered and merged with other token remains the same.

(A'09) The capacities of all arcs through which a token has passed decrement with 1. If the current capacity of an arc has reached 0, value "$\langle 0, 1\rangle$" is assigned to the element from the index matrix R that corresponds to this arc.

(A'10) The values of the characteristic function Φ for the output places (one or more) in which tokens have entered according to **(A'06)** are calculated. The token obtains the next characteristic in the form:

$$\text{"value of } \Phi \text{ for the current token, } \langle \mu_{i,j}, \nu_{i,j} \rangle\text{"}$$

(A'11) If there are still tokens in the input places that can be transferred and there are arcs with non-zero capacities, then the algorithm proceeds to Step **(A'04)** otherwise it proceeds to Step **(A'12)**.

(A'12) The values of the characteristic function Ψ for all output places to which tokens have been transferred are calculated. These values are assigned to the places.

(A'13) The current model time t is increased with t^0.

(A'14) Is the current time moment equal to or greater than $t_1 + t_2$? If the answer to the question is no, return to Step **(A'04)**, otherwise go to Step **(A'15)**.

(A'15) Termination of the transition's functioning.

As in the algorithm for transition functioning in GNCP, the novelty of the proposed algorithm can be seen in steps **(A'03)**, **(A'04)**, **(A'08)** and **(A'12)**. In comparison to the algorithm for transition functioning proposed in [3] value "$\langle 0, 1 \rangle$" is assigned only to the elements of R corresponding to empty input places and arcs with current capacity 0. In the standard algorithm this is also done for all elements in a column corresponding to full output place. The proposed modification allows the transfer of tokens to output places which have reached their capacity if they can merge with some of the tokens in the output place. If splitting of tokens is not allowed, the evaluation of the predicates stops with the first predicate whose truth-value allows the transfer. Step **(A'08)** is also new as the current number of tokens in the output places is increased only when the entering token does not merge with any of the tokens in the place.

The other novelty is with regard to the characteristics of the places which are evaluated in step **(A'12)**. In the algorithm proposed in [3] the function Ψ assigns characteristics to the output places after each transfer of token. Here this is done after all input places have been passed. Therefore, we have two possibilities of preserving the IFPs corresponding to the predicates. The first is to define Ψ so that it assigns to the output places characteristic in the form:

$$\text{"}\overline{\Psi}^{l_j}_{cu}, \langle \alpha_1, \mu(r_{i_1,j}), \nu(r_{i_1,j}) \rangle, \ldots, \langle \alpha_k, \mu(r_{i_k,j}), \nu(r_{i_k,j}) \rangle\text{"},$$

where $\alpha_1, \ldots, \alpha_k$ are the tokens that have entered the output place l_j and the IFPs $\langle \mu(r_{i_1,j}), \nu(r_{i_1,j}) \rangle, \ldots, \mu(r_{i_k,j}), \nu(r_{i_k,j})$ are the truth-values of the predicates.

The other possibility is to assign only one IFP to the output places:

$$\text{"}\overline{\Psi}^{l_j}_{cu}, \langle \mu_{l_j}, \nu_{l_j} \rangle\text{"},$$

where

$$\mu_{l_j} = \frac{\sum_{t=1}^{k} \mu(r_{i_t,j})}{k},$$

$$\nu_{l_j} = \frac{\sum_{t=1}^{k} \nu(r_{i_t,j})}{k}.$$

The algorithm for transition functioning in IFGNCP3 when merging of tokens is permitted is the same as the above proposed algorithm for IFGNCP1. Again Ψ is evaluated after all input places have been passed while in the algorithm suggested in [3] this is done after each transfer of token. As in IFGNCP1, we can preserve the truth-values in two ways. The first is to assign to the output place l_j together with the standard characteristic of the place $\overline{\psi}^{l_j}$ a list of all tokens and the IFPs corresponding to the predicates in the form:

$$\text{“}\overline{\psi}^{l_j}_{cu}, \langle \alpha_1, \mu(r_{i_1,j}), \nu(r_{i_1,j})\rangle, \ldots, \langle \alpha_k, \mu(r_{i_k,j}), \nu(r_{i_k,j})\rangle, \langle \mu(\psi^{l_j}_{cu}), \nu(\psi^{l_j}_{cu})\rangle\text{”},$$

where $\alpha_1, \ldots, \alpha_k$ are the tokens that have entered the output place l_j, the IFPs $\langle \mu(r_{i_1,j}), \nu(r_{i_1,j})\rangle, \ldots, \mu(r_{i_k,j}), \nu(r_{i_k,j})$ are the truth-values of the predicates corresponding to each transfer and the IFP $\langle \mu(\psi^{l_j}_{cu}), \nu(\psi^{l_j}_{cu})\rangle$ is evaluation of the characteristic of the place. The second way is to assign only one IFP taking the average of the truth-values of the predicates:

$$\text{“}\overline{\psi}^{l_j}_{cu}, \langle \mu_{l_j}, \nu_{l_j}\rangle, \langle \mu(\psi^{l_j}_{cu}), \nu(\psi^{l_j}_{cu})\rangle\text{”}.$$

5 Conclusion

The proposed modifications of the algorithms for transition functioning in the standard GNs, GNCP, IFGNCP1 and IFGNCP3 take into account the possibility of some tokens to merge after the transfer from input to output place. Up to now, in the algorithms for transition functioning in GNs this has not been considered. The new algorithms would allow more tokens to be transferred during one time step. The same result can be achieved by increasing the capacity of some output places in which tokens can merge. However, when a token merges with other token in output place the current number of tokens in the place remains the same and thus increasing the capacity to allow such transfer is artificial way to treat the problem. Also, greater capacity may allow the transfer of other tokens which cannot merge with any of the tokens in the output place. In future we shall discuss analogous algorithms for other GN extensions.

References

1. Andonov, V., Atanassov, K.: Generalized nets with characteristics of the places. C. R. Acad. Bulg. Sci. **66**(12), 1673–1680 (2013)
2. Andonov, V.: Connection between generalized nets with characteristics of the places and intuitionistic fuzzy generalized nets of type 1 and type 2. Notes Intuitionistic Fuzzy Sets **19**(2), 77–88 (2013)
3. Andonov, V.: Intuitionistic fuzzy generalized nets with characteristics of the places of type 1 and type 3. Notes Intuitionistic Fuzzy Sets **19**(3), 99–110 (2013)
4. Andonov, V.: Reduced generalized nets with characteristics of the places. ITA 2013—ITHEA ISS Joint International Events on Informatics, Winter Session, Sofia, Bulgaria, 18–19 December 2013 (in press)
5. Atanassov, K.: Generalized Nets. World Scientific, Singapore (1991)
6. Atanassov, K.: On Generalized Nets Theory. Prof. M. Drinov Academic Publishing House, Sofia (2007)
7. Atanassov, K., Szmidt, E., Kacprzyk, J.: On intuitionistic fuzzy pairs. Notes Intuitionistic Fuzzy Sets **19**(3), 1–13 (2013)
8. Dimitrov, D.: Optimized algorithm for tokens transfer in generalized nets. In: Recent Advances in Fuzzy Sets, Intuitionistic Fuzzy Sets, Generalized Nets and Related Topics, vol. 1, pp. 63–68. SRI PAS, Warsaw (2010)
9. Murata, T.: Petri nets: properties, analysis and applications. Proc. IEEE **77**(4), 541–580 (1989)
10. Ribagin, S., Andonov, V., Chakarov, V.: Possible applications of generalized nets with characteristics of the places. A medical example. Proceedings of 14th Interational Workshop on Generalized Nets, Burgas, pp. 56–64 (2013)

Generalized Net Model for Monitoring the Degree of Disability in Patients With Multiple Sclerosis Based on Neurophysiologic Criteria

Lyudmila Todorova, Valentina Ignatova, Stefan Hadjitodorov and Peter Vassilev

Abstract Multiple sclerosis (MS) is a chronic disease of the central nervous system, which affects most often the young age. The disease has a social importance, as it leads to disability in active age. The course of the disease is individual and requires an adequate approach in selecting the appropriate treatment strategy. This requires periodic monitoring of the patients for early registration of the disease progression. Most commonly used tests are clinical investigation, brain MRI and other paraclinical methods, including neurophysiological assessment (usually evoked potentials). Evoked potential (EP) is a reliable method for quantifying the severity of damage to the white matter in patients with MS. The aim of this study is to develop a model for Generalized NET (GN)-registration of the direction of the course of disease based on neurophysiological evaluation of multimodal evoked potentials. Also, to make a comparison of the results obtained by the EP with the degree of disability as measured by the scale of Kurtzke (Expanded disability Status Scale-EDSS). We have followed up 48 patients with clinically definite MS over a period of 1 year. The patients were tested both clinically and neurophysiologically at Clinic of Neurology in MHAT- NHH, Sofia. The three main modalities evoked potentials were applied: visual evoked potentials with reversive pattern (VEPRP); Brainstem auditory evoked potentials (BAEP); Somatosensory evoked potentials (SSEP), taken during stimulation of median nerve. As a result it is established that the abnormalities of EP correlate significantly with the clinical findings. Based on the obtained results is

L. Todorova · S. Hadjitodorov · P. Vassilev (✉)
Institute of Biophysics and Biomedical Engineering, Bulgarian Academy of Sciences, Acad. G. Bonchev Str., Bl. 105, 1113 Sofia, Bulgaria
e-mail: peter.vassilev@gmail.com

L. Todorova
e-mail: lpt@biomed.bas.bg

S. Hadjitodorov
e-mail: sthadj@bas.bg

V. Ignatova
Clinic of Neurology, Multiprofile Hospital for Active Treatment, National Heart Hospital, Konyovitsa Str. 65, 1309 Sofia, Bulgaria
e-mail: valyaig@abv.bg

P. Angelov and S. Sotirov (eds.), *Imprecision and Uncertainty in Information Representation and Processing*, Studies in Fuzziness and Soft Computing 332,
DOI 10.1007/978-3-319-26302-1_18

developed a GN-model generating candidate predictive rules for the progression of the illness. In the end it has been found that abnormalities of EP significantly correlated with clinical findings. Based on the obtained results a GN model was developed. The model has a high sensitivity (SEN), specificity (SPE), PPV (Positive predictive value) and NPV (Negative predictive value) for the disease progression.

1 Introduction

Multiple sclerosis (MS) is an autoimmune, inflammatory, chronic and demyelinating degenerative disease of the central nervous system affecting young age [1]. The largest number of the patients is between 30 and 40 years of age. The onset of the disease was recorded in childhood in approximately 10 % of the patients. Susceptibility of the female to the disease is twice as much. Most affected are Caucasian subjects [2, 3]. The highest incidence of MS is set at the Orkney Islands north of Scotland-250/100, 000. A similar incidence of the disease is found in North America, Canada and New Zealand [4]. The incidence of MS is lowest in Japan 6/100 000 and in the other parts of Asia, sub-Saharan Africa and the Middle East. MS is a disease with social significance. The disease is chronic, with a duration of 10–15 years depending on the clinical form [5]. It was found a genetically predisposition, which in combination with certain environmental factors, leads to cascade of immune responses and disruption of the blood-brain barrier (BBB). As a result an inflammatory demyelination of the white matter in the CNS occurs due to activation of T-lymphocytes and macrophages. Impaired conduction of nerve impulses in the CNS, leading to clinical manifestation of the neurological deficit, appears [5]. Four different pathogenic mechanism of the disease are known, one of which is typical for each of the patients. This mechanism can be changed in time, which perhaps associated with different severity of disease [1, 6].

The disease is manifested clinically with multiple neurological dysfunctions from different systems (e.g., visual and sensory disturbances, weakness at limbs, gait disturbance, bowel and bladder disorders), followed by recovery or reinforcing of the disability.

According to the time profile of the disease, four major clinical forms of MS are distinguished.

(1) **Relapsing-remitting form**—affects approximately 65–85 % of the patients with MS. It is characterized with well identifiable attacks, expressed with neurological symptoms, which resolve for a period of few weeks or could lead to slow increase of disability. There is no progression between the attacks.
(2) **Secondary-progressive form**—when gradual accumulation of irreversible disability with or without separate attacks it occurs. After 6–10 years of the MS onset 4–70 % of patients with RRMS are passing into SPMS.
(3) **Primary-progressive form**—at about 10 % of the patients. When the symptoms of the disease got worse constantly and slowly from the beginning

without distinct attacks, although there may be observed also plates and temporary improvement. Usually this form is related to early development of severe disability

(4) **Progressive-remitting form**—affects 5 % of the patients with MS. It is characterized by combination of attacks and progression in the early periods of the disease. There are pronounced attacks with or without full recovery. A gradual progression between relapses occurs. This form has a poor prognosis [5, 7].

Evoked potentials are routinely used in patients with MS [8–11]. They are included in supporting revised diagnostic criteria of MacDonald revision since 2005 [7, 12, 13]. EP are a reliable method for assessing the integrity of afferent and efferent pathways and to quantify the severity of damage of the white matter in MS [14]. It has been found that abnormalities of EP significantly correlates with clinical findings, while the majority of MRI-lesions are not associated with symptoms of the disease. Transverse and longitudinal studies have demonstrated that changes in the EP in MS are more closely associated with disability than MRI lesions [15]. This assumes that the EP could be useful in monitoring evolution of the disease and to serve as surrogate endpoints in clinical studies.

While MRI provides information on the spatial distribution of lesions at white matter, neurophysiological tests reflect their impact on the functions of the nervous system [16]. Neurodegeneration correlates with progression of MS and sensitive marker for its evaluation are still in process of searching.

Most often used EP modalities are: visual, somatosensory and brainstem auditory Abnormal visual evoked potentials with reversal pattern (VEPRP) typical for MS, are characterized with delayed latencies and less often- with altered waveforms [17, 18]. Brainstem auditory evoked potentials (BAEP) give information about the brainstem functioning. In MS, they are usually characterized with prolonged interpeak latencies (IPL) and/or abnormal amplitude ratios (AR) [19]. Somatosensory evoked potentials (SSEP) provide information about the presence of sensory damage as well as the topic of damage along the medial lemnisc system [20]. The values latencies in themselves do not carry information about the function of brain circuitry, but longitudinal changes in latencies during patient's follow up shows deterioration of neurological dysfunction.

Importance of EP in monitoring the MS patients over the course of the disease is still a an object of investigation. Need to further investigate the role of multimodal EP in follow up the MS subjects is determined by insufficient data from previous studies.

2 Materials and Methods

For the purposes of this work 48 patients with clinically definite MS were tested twice at Clinic of Neurology at MHAT-NHH-Sofia. Demographic data concerning our sample are given on the following table:

Examined twice with multimodal EP and EDSS	48	
Mean age	39.4 ± 9.6	
Male patients	17	35 %
Female patients	31	65 %
EDSS ≤ 3.5	18	37 %
EDSS > 3.5	30	63 %

The following 3 modalities EP were followed up:

✓ Visual evoked potentials with reverse pattern (VEPRP)
✓ Brainstem auditory evoked potentials (BAEP)
✓ Somatosensory EP, elicited by stimulation of mediane nerve (SSEP)

The reason for the this choice is the frequent involvement of the three sensory modalities in MS and possibility for an objective neurophysiological monitoring of their changes in the course of the disease. The main indicators that were taken into account in the analysis of the EP in this work are:

✓ latency, i.e. time from the stimulus to the wave response registration
✓ amplitudes, measuring the distance from peak to peak (especially informative in SSEP)
✓ configuration violations.

2.1 *Visual Evoked Potentials (VEP)*

VEP were performed with two-channel installation by applying reversive pattern in order to improve the the objectivity the results. Checkerboard pattern is used with frequency reversion 1 Hz, low-pass filter—1 Hz, high-pass filter—100 Hz, a band filter 50 Hz. Administration of sequential monocular foveolar (15') and peripheral (60') retinal stimulation for objectifying the function of the visual pathways starting from foveolar and peripheral retinal neurons. The epoch of analysis is 300 ms. One hundred averaging in each assay were done. The following parameters were recognized: Delayed latencies of N75 and P100 (L N75, L P100), Configuration violations, Interocular and interhemisphere asymmetry (IO A, IH A). On the next figure are given the graphical interpretation of VEP—examinations a healthy control and a patient with pathologically changed result (Fig. 1).

2.2 *Brainstem Auditory Evoked Potentials (BAEP)*

In BAEP stimulation monaural square click stimulus was used which lasts 100 μs. The frequency of stimulation was 10 Hz, the intensity −90 dB nHL and masking noise to the contralateral ear of 40 dB was applied. Polarity of stimulus:

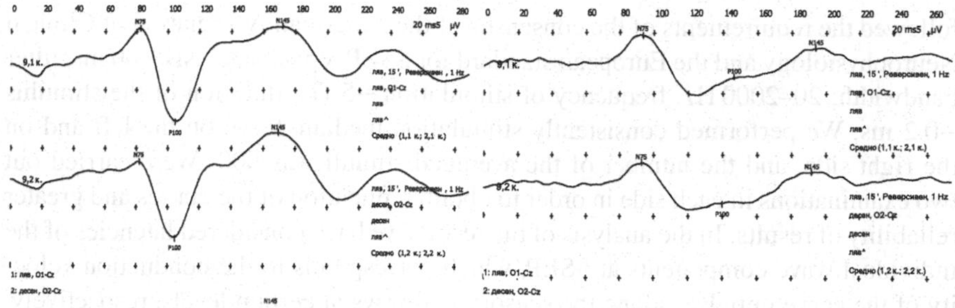

Fig. 1 On the *left* the healthy control result (44 aged woman); on the *right* the pathologically changed result (51 aged woman)

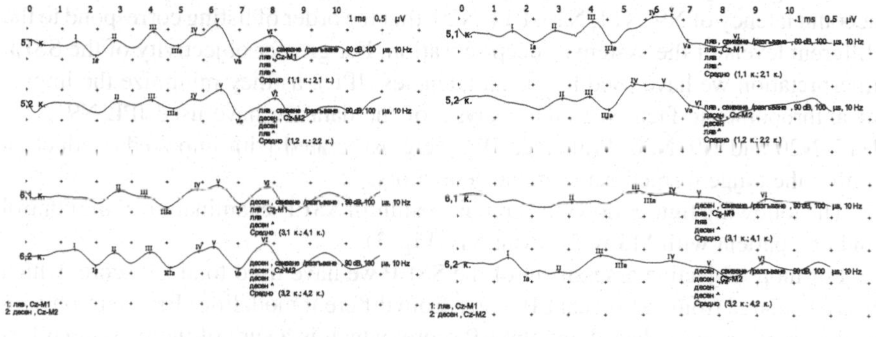

Fig. 2 On the *left* the healthy control result (47 aged woman); on the *right* the pathologically changed result (45 aged woman)

rarefaction-condensation. Band filter was used within the range of 100–2000 Hz. Epoch analysis is 10 ms. In each stimulation 2000 averaging were carried out. Two-channel apparatus was used. The following parameters were analyzed: interpeak latencies (IPL), amplitude ratios (AR), also mono-and binaural impairment was taken into account during interpretation of the results. Hearing impairment was excluded in all of the patients. The montage which we used is approved by the Recommended Standard for short latency EP of American Society of Clinical Neurophysiology (2009). On the following figure are given graphic results of BAEP-tests of healthy control and of a patient with pathologically changed result (Fig. 2).

2.3 *Somatosensory Evoked Potentials at Stimulation of Median Nerve*

The functional state of the pathways of deep sensibility in the tested sample with MS was monitored by examining SSEP, elicited from sequential bilateral stimulation of mediane nerve. For this purpose, 4-channel montage was used which followed the requirements of the consensus of the American Association of Clinical Neurophysiology and the European standard for SSEP: epoch analysis −60 ms, filter bandwidth: 20–2000 Hz, frequency of stimulation −5 Hz, duration of the stimulus −0.2 ms. We performed consistently stimulation median nerve on the left and on the right side, and the number of the averaged stimuli was 600. Were carried out two examinations in each side in order to obtain replication of the curves and greater reliability of results. In the analysis of the results we have considered latencies of the individual wave components at SSEP, which corresponds to the conduction velocity of the nerve impulses along the sensory pathways at certain levels, respectively: peripheral nerve structures, cervical myelon (posterior lemniscus and/or rear posterior horns for surface sensitivity), lemniscus medialis along the brainstem, thalamus, the primary somatosensory representation of the cortex. Were were taken into account latency of N9, N11, N13, P14, N20, that the order of listing correspond to the different levels of the system of deep sensation. For greater objectivity of the SSEP interpretation we have used interpeak latencies (IPL), as they minimize the impact of anthropometric factors. In the analysis of the data we have used IPL N9–N13, N13–N20 and N9–N20. Prolonged IPL were associated with impaired conduction within the range formed between the generators.

The following figure shows the graphic results of SSEP-examination of the control and of a patient with MS with deviations (Fig. 3).

For more objective assessment of the SSEP we have used total EP-score, which was calculated from individual EP scores from different modalities EP. Neurophysiological assessment is based on total EP score, which is a sum of the individual EP-scores of the three types of modalities.

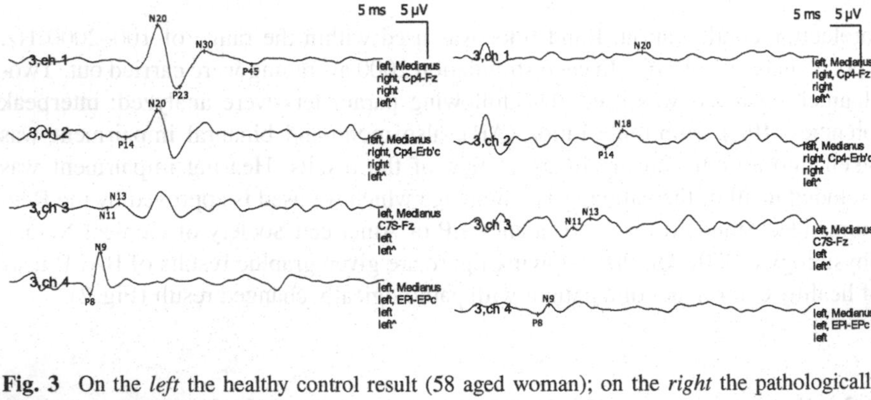

Fig. 3 On the *left* the healthy control result (58 aged woman); on the *right* the pathologically changed result (38 aged man)

	VEP-score	BAEP-score	SSEP-score
0	Normal result	Normal result	Normal result
1	Configuration abnormalities/ Interocular asymmetry/ interhemispheral asymmetry but preserved conduction time	At least one abnormal amplitude ratio (AR) at unilateral monoaural stimulation	Prolongation of 1 IPL at unilateral stimulation of median nerve
2	Prolonged latencies up to 20 ms	Prolonged IPL at monoaural unilateral stimulation and/or bilaterally impaired	Prolongation of ≥2 IPL and/or bilateral stimulation
3	Prolonged latencies within 20–40 ms	Both bilaterally impaired IPL at monoaural bilateral stimulation	Prolongation of 3 IPL at unilateral median nerve stimulation and abnormality of 1 IPL at contralateral stimulation, or lack of identification of 1 wave
4	Prolonged latencies >40 ms	Severe abnormalities and difficult identification of waves in up to 2 leads	Prolongation of both 3 IPL at unilateral median nerve stimulation and ≥2 IPL at contralateral stimulation, or impossible identification of 2 waves
5	Severe abnormalities and difficult identification of the waves	Severe abnormalities and impossible identification of waves bilaterally in all the leads	Severe abnormalities and impossible identification of waves bilaterally

3 Generalized Net Model

The Generalized nets (GN) were introduced by Atanassov as an extension of the Petri nets see [21, 22]. We have used the previously established values for the Visual Evoked Potentials (see [23]). For the SSEP, the most indicative in the preliminary investigation proved to be the number of lesions and the interpeak latencies (IPL). The IPL N9–N13, N13–N20 and N9–N20 were investigated in the model. The BAEP were investigated mainly for the IPL III-V cross simulation and I/III amplitude ratio.

We have adopted the approach taken in [23]. In fact the rule established there for the VEPs is used here unchanged. The setup of the Generalized net is the following: the patients, represented by α-tokens, are placed at L_4. The candidate rules represented by β-tokens are placed in places L_1, L_2 and L_3 and are initially passed to L_9 where they form a γ-token with characteristic "list of current rules; counter for each of the current rules". Places L_6, L_7, L_8 are used to supply the current rules to the next transition and to split the α-tokens representing the patients into three for estimation in the next transition. The next transition corresponds to the satisfaction of the rules and separating the patients in three possible classes (EDSS ≤ 3.5), (EDSS > 3.5) and failed to be classified, and also also assigning a score to all used rules (L_{19}), and sending feedback for the following rules to be created (Fig. 4).

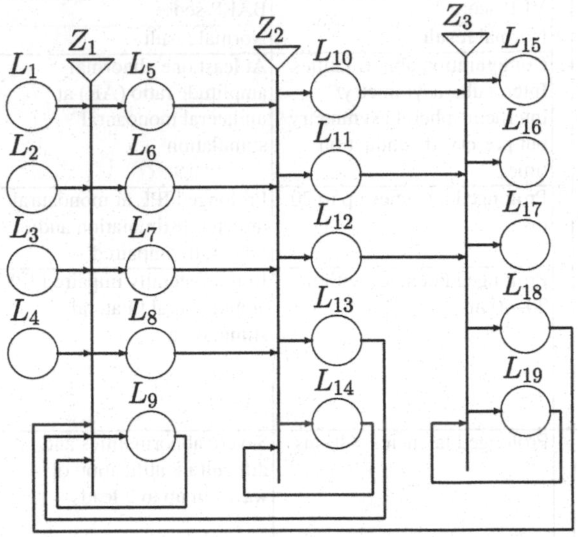

Fig. 4 The generalized net model

The rules are of the kind: $\wedge p * \vee \neg p * \wedge q *$ with $p \in P$, $q \in Q$ and p* being conjunctions of predicates from P and q* being conjunctions of predicates from Q for the first class (EDSS $\leq$ 3.5), while for the second they are from the type $\neg p * \wedge \neg q *$, where P is the set of predicates of the kind "Parameter value less or equal to a threshold value" and Q is the set of predicates of the kind "Parameter value greater or equal to a threshold value".

For every parameter we consider the predicates p (or q) for 120 possible threshold values (i.e. $p \leq ((k + 1)/2)$*MEAN(PARAM_VALUE)/60, with $k = 0, 1 \ldots, 119$).

The generated rules were scored based on the values of calculated sensitivity and specificity.

The model has three transitions, nineteen places and three types of tokens—α, β and γ.

Below is a formal description of the transitions of the net.

$$Z_1 = \langle \{L_1, L_2, L_3, L_4, L_9, L_{13}, L_{18}\}, \{L_5, L_6, L_7, L_8, L_9\},$$

	L_5	L_6	L_7	L_8	L_9
L_1	*true*	*false*	*false*	*false*	*true*
L_2	*false*	*true*	*false*	*false*	*true*
L_3	*false*	*false*	*true*	*false*	*true*
L_4	*false*	*false*	*false*	*false*	*true*
L_9	*false*	*false*	*false*	$W_{9,8}$	*true*
L_{13}	*false*	*false*	*false*	*false*	*true*
L_{14}	*false*	*false*	*false*	*false*	*true*
L_{18}	*false*	*false*	*false*	*false*	*true*

$\rangle$,

where $W_{9,8}$ = "New candidate rules are needed and not all rules have been tried."

$$Z_2 = \langle \{L_5, L_6, L_7, L_8, L_{14}\}, \{L_{10}, L_{11}, L_{12}, L_{13}, L_{14}\},$$

	L_{10}	L_{11}	L_{12}	L_{13}	L_{14}
L_5	$W_{5,10}$	*false*	*false*	*false*	*false*
L_6	*false*	$W_{6,11}$	*false*	*true*	*false*
L_7	*false*	*false*	$W_{7,12}$	*true*	*false*
L_8	$W_{8,10}$	$W_{8,11}$	*false*	*true*	*true*
L_{14}	$W_{14,10}$	$W_{14,11}$	$W_{14,12}$	$W_{14,13}$	*true*

$\rangle$,

where
$W_{5,10}$ = "there is no β token in place L_{10}"
$W_{6,11}$ = "there is no β token in place L_{11}"
$W_{7,12}$ = "there is no β token in place L_{12}"
$W_{8,10}$ = "there is no α token in place $L_{10} \vee$ the current β-token needs to be replaced"
$W_{8,11}$ = "there is no α token in place $L_{11} \vee$ the current β token needs to be replaced"
$W_{8,12}$ = "there is no α token in place $L_{12} \vee$ the current β-token needs to be replaced"
$W_{14,10}$ = "the current β-token in place L_{10} has to be replaced"
$W_{14,11}$ = "the current β-token in place L_{11} has to be replaced"
$W_{14,12}$ = "the current β-token in place L_{12} has to be replaced"
$W_{14,13}$ = "new α-token is required"

$$Z_3 = \langle \{L_{10}, L_{11}, L_{12}, L_{13}, L_{14}, L_{19}\}, \{L_{15}, L_{16}, L_{17}, L_{18}, L_{19}\},$$

	L_{15}	L_{16}	L_{17}	L_{18}	L_{19}
L_{10}	$W_{10,15}$	$W_{10,16}$	$W_{10,17}$	*false*	*true*
L_{11}	$W_{11,15}$	$W_{11,16}$	$W_{11,17}$	*false*	*true*
L_{12}	$W_{12,15}$	$W_{12,16}$	$W_{12,17}$	*false*	*true*
L_{19}	*false*	*false*	*false*	$W_{19,18}$	*true*

$\rangle$,

where
$W_{10,15}$ = "the current α token satisfies the current BAEP rule for belonging to the class (EDDS ≤ 3.5)"
$W_{10,16}$ = "the current α token satisfies the current BAEP rule for belonging to the class (EDDS > 3.5)"
$W_{10,17}$ = "$\neg W_{10,15} \wedge W_{10,16}$"
$W_{11,15}$ = "the current α token satisfies the current SSEP rule for belonging to the class (EDDS ≤ 3.5)"
$W_{11,16}$ = "the current α token satisfies the current SSEP rule for belonging to the class (EDDS > 3.5)"
$W_{11,17}$ = "$\neg W_{11,15} \wedge W_{11,16}$"
$W_{12,15}$ = "the current α token satisfies the current VEP rule for belonging to the class (EDDS ≤ 3.5)"
$W_{12,16}$ = "the current α token satisfies the current VEP rule for belonging to the class

(EDDS > 3.5)"
$W_{12,17}$ = "$\neg W_{12,15} \wedge W_{12,16}$"
$W_{19,18}$ = "the generation of new rules is required (i.e. the number of identified by the current rules α-tokens is divisible by 48.)"

The derived rules from the work of the net (all rules that fail to separate the patients in two classes were discarded) were the following:

For the SSEP:

EDSS> 3.5 if p = (Number of spinal lesions >3) ∧ (IPL N13–N20 > 5.7 ms) ∨ ((Number of spinal lesions ≤3) ∧ (IPL N9–N20 > 13.5 ms))
EDSS≤ 3.5 if $\neg p$.

For the BAEP:

EDSS> 3.5 if p = (IPI III-V under cross stimulation > 2.18 ms) ∨ ((IPI III-V cross stimulation ≤ 2.18 ms) ∧ (I/III amplitude ratio > 6.75))
EDSS≤ 3.5 if $\neg p$.

For the VEP:

EDSS≤ 3.5 if q = ((Foveolar stimulation L N75 ≤ 86.56) ∧ (Foveolar stimulation I 100 ≤ 7.25)) ∨ ((Foveolar stimulation L N75 > 86.56) ∧ (Peripheral stimulation IH N75 ≤ 1.9) ∧ (Foveolar stimulation IH 100 ≤ 3.65) ∧ (Foveolar stimulation IH N75 ≤ 3.65)) ∨ ((Foveolar stimulation L N75 > 86.56) ∧ (Peripheral stimulation IH N75 > 1.9) ∧ (Foveolar stimulation I N75 ≤ 0.275))
EDSS< 3.5 if $\neg q$.

4 Results

The results from both clinical and neurophysiological examination from the baseline and follow up study were compared. The dynamics of the grade of disability is presented at the following figure:

The figure visualizes a linear relationship between the values of the total EP score and EDSS-score in both studies. It is evident that the higher the neurophysiological damage is associated with higher degree of disability. In the follow up study, the tangent-line trend is higher, which corresponds to largest values of the monitored scores, i.e. with deterioration of the patients (Figs. 5 and 6).

Following the suggestion of the GN model, if there are more than 3 demyelinating lesions in the cervical myelon (visualized by MRI study) and when IPL N13–N20 is greater than 5.7 ms, practically all patients from our sample are with high degree of disability, i.e. they belong to class 1 EDSS (EDSS > 3.5). At presence of MRI lesions in the cervical myelomas, while IPL N13–N20 is <5.7 ms, the classification of patients according to the severity of the disability is not reliable. At presence of less than three spinal lesions in MRI study whilst IPL N9–N20 is > 13.4 ms, almost all patients in our sample have a high degree of disabled, i.e. they belong to class 1

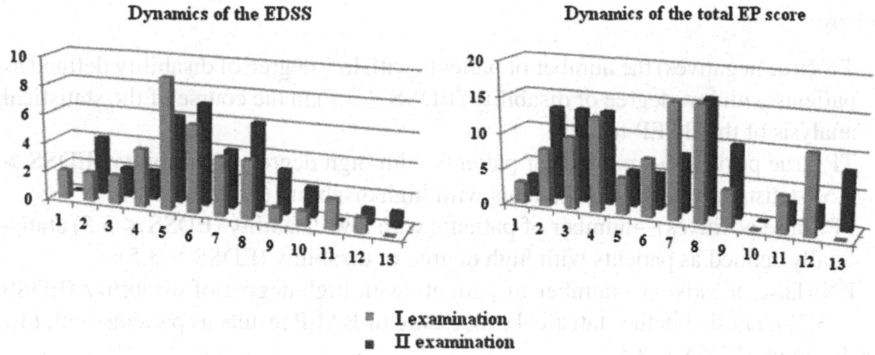

Fig. 5 EDSS versus EP score

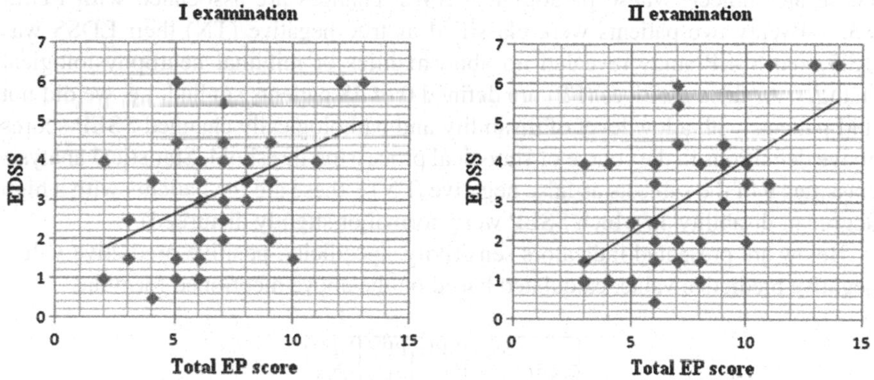

Fig. 6 EDSS versus EP score

EDSS (EDSS > 3.5). If there are less than 3 spinal lesions, also IPL N9–N20 is ≤13.4 ms, the majority of patients are with low degree of disability according to EDSS—class 0 (EDSS ≤ 3.5). The sensitivity (SEN), specificity (SPE), positive predictive value (PPV) and negative predictive value (NPV) of the elected rule were calculated. The calculation is made according to following formulas

$$SEN = \frac{TP}{TP + FN};$$

$$SPE = \frac{TN}{TN + FP};$$

$$PPV = \frac{TP}{TP + FP};$$

$$NPV = \frac{TN}{TN + FN},$$

where

TN (true negatives) the number of patients with low degree of disability defined as patients with low degree of disability (EDSS ≤ 3.5) in the course of the statistical analysis of the BAEP-results;
TP (true positives)—number of patients with high degree of disability (EDSS > 3.5) statistically defined as patients with high disability (EDSS > 3.5);
FP (false positives)—number of patients with low disability (EDSS ≤ 3.5) statistically defined as patients with high degree of disability (EDSS > 3.5);
FN (false negatives)—number of patients with high degree of disability (EDSS > 3.5) identified in the statistical processing of BAEP results as patients with low disability (EDSS ≤ 3.5).

Statistical analysis defined as true positive (TP) 17 patients from our sample. These are subjects whose pathological SSEP changes are associated with EDSS >3.5. Twenty two patients were classified as true negative (TN) their EDSS was ≤3.5, and SSEP study revealed no abnormalities or minimal neurophysiological changes. As false positives (FP) are defined 0 of 48 patients studied, i.e. we did not find patients with a low level of disability and pathologically changed SSEP-scores above definitions in the course of statistical processing threshold. Statistical analysis classified 9 of the subjects as false negative (FN)—these are the patients with a high degree of disability, in which SSEP were normal or slightly disturbed).

Below are presented indicators sensitivity, specificity, positive predictive value, negative predictive value, calculated based on the above mentioned formulas

	SEN	SPE	PPV	NPV
SSEP	61 %	70 %	71 %	67 %

Statistical analysis of the results from the examined patients identified as leading neurophysiological criteria for prediction the severity of disability the following indicators: IPL III-V under cross-stimulation and AR I / III under ipsilateral stimulation. As “Class 0” are categorized the patients with a low degree of disability, i.e. with EDSS ≤ 3.5 and as a “Class 1”—patients with a high degree of disability, i.e. with EDSS > 3.5. In cases where the IPL III-V under cross stimulation is higher than 2.18 ms, 80 % of the patients have a high degree of disability. Where IPL III-V under cross stimulation is ≤2.18 ms, the majority of cases have a low degree of disability (60 %). When IPL III-V under cross stimulation is ≤2.18 ms and AR I/III under ipsilateral stimulation is >6.75, 100 % of the patients express high degree of disability (EDSS > 3.5). In cases when IPL III-V under cross stimulation is <2.18 ms, whilst AR I/III under ipsilateral stimulation is ≤6.75, 72 % of the examined revealed low degree of disability, i.e. their EDSS was less or equal to 3.5.

Below are presented indicators sensitivity, specificity, positive predictive value, negative predictive value, calculated based on the above mentioned formulas

	SEN	SPE	PPV	NPV
BAEP	70 %	81 %	77 %	75 %

We have developed a classification based on N75 latency under foveolar stimulation (respectively less or more than 86.6 ms) combined with other neurophysiological VEPRP-abnormalities can be predictive indicator to the degree of invalidity in the patients, respectively, low level (EDSS ≤ 3.5) and high (EDSS > 3.5). In the analysis of our results the greatest responsiveness of the latency of the early wave component N75 as a marker identifying the severity of the disease has been established, which was confirmed by other authors. Cuypers et al. found that N75 latency is more sensitive marker than P100 latency, when classifying the severity of MS, despite its great influence of age, gender and stimulus pattern [24]. The chart shows that if N75 latency under foveolar stimulation is >86.6 ms and interocular asymmetry of N75 (IOA N75) in peripheral stimulation is >1.9 ms, the patient probably has a disability over 3.5 points on the scale of Kurtzke. If N75 latency under foveolar stimulation is <86.6 ms , and interhemisphere asymmetry of P100 (IHA P100) at foveolar stimulation is <1.9 ms, the probability of predicting disability is reduced. If N75 latency at foveolar stimulation is ≤86.6 ms and interocular asymmetry of P100 (IOA P100) is >7.25 ms, most likely the patient belongs to Class 1, i.e. EDSS score is >3.5 If N75 latency at foveolar stimulation is ≤86.6 ms, while interocular asymmetry of P100 (IOA P100) is ≤7.25 ms, most likely the patient has a low degree of disability (EDSS score ≤ 3.5).

The following table shows the values for the VEP rule:

	SEN	SPE	PPV	NPV
VEPRP	100 %	76 %	83 %	100 %

5 Discussion

For the predictive value of pathologically changed VEPRP concerning the degree of disability in patients with MS it is possible to use lower threshold margins, e.g.:

✓ patients with low grade of disability (EDSS < 3)
✓ patients with high grade of disability (EDSS ≥ 3)

The threshold set in relation to severity of disability assessed by EDSS may be chosen lower in VEPRP, as SSEP have changed earliest from all other modalities EP and often they register deterioration before reaching of significant clinical disability.

Parameters sensitivity and specificity of VEP calculated in the present work are close to the results obtained by other authors. Grover et al. set sensitivity 84.2 % and specificity 90 % VEP-study, using size to the boxes 60' [25]. Ko calculated sensitivity to 100 % VEP [26]. Balnyte et al. determined 90.5 % sensitivity and 82.5 % specificity of VEP [27]. Gnezditskiy and Korepina calculated 72 % sensitivity and 100 % specificity of VEP [28]. Our study confirms the conclusion of most researchers that

VEP are extremely sensitive method for assessing the degree of disability in MS and for prediction the course of the disease.

Regarding the sensitivity of the BAEP, the results in the literature are controversial. Burina et al. [29] revealed 95 % sensitivity of BAEP in terms of detection of brainstem lesions in a study of 60 patients. Berger and Blum [30] registered 46 % sensitivity of BAEP. The divergences are probably due to differences in methodology as well as various parameters for the evaluation; the majority of researchers prefer IPL measurement as a decisive criterion for evaluation. In our study the higher percentage of BAEP abnormalities includes mostly damaged AR and less frequent prolonged IPL. We assume that abnormal ARs are an early marker of brainstem abnormality and this may be a result from gray matter damage.

All of the above suggest using VEPs as leading method in the assessment of the progression of the disease with the other two EP used as an auxiliary tool to diminish the error in the estimate. Further investigation in this direction is currently in progress.

6 Conclusion

Safety, non invasiveness, relatively low cost of the study and the possibility of continuous monitoring make this neuropsychological test appropriate in assessing progression of the disease and the degree of involvement of different sensory systems. The study is informative in follow up the effect of immunomodulatory therapy, also the recovery after attack. The utilized Generalized net models allows for fast, cost efficient way of electing candidate rules to be reviewed and evaluated by the doctors in the course of their practice. If such rules prove to be reliable and have predictive nature, they can be used to alleviate some of the problems the patients experience.

References

1. Milanov, I.: Multiple sclerosis and demyelinating diseases, p. 242. Medicina i Fiskultura, Sofia (2010)
2. Pugliattia, M., Sotgiu, S., Rosati, G.: The worldwide prevalence of multiple sclerosis. Clin. Neurol. Neurosurg. **104**, 182–191 (2002)
3. Pugliattia, M., Rosatia, G., Cartonc, H., et al.: The epidemiology of multiple sclerosis in Europe. Eur. J. Neurol. **13**, 700722 (2006)
4. Taylor, B., Pearson, J., Clarke, G., et al.: MS prevalence in New Zealand, an ethnically and latitudinally diverse country. Mult. Scler. **16**, 1422–1431 (2010)
5. Milanov, I.: National consensus for diagnosis and treatment of multiple sclerosis. Mov. Disord. (Bulgaria) **9**(2), 40 (2012)
6. Lucchinetti, C., Bruck, W., Parisi, J., et al.: Heterogeneity of multiple sclerosis lesions: implications for the pathogenesis of demyelination. Ann. Neurol. **47**, 707–717 (2000)
7. McDonald, W.I., Compston, A, Edan, G., Goodkin, D.: Recommended diagnostic criteria for multiple sclerosis: guidelines from the international panel on the diagnosis of multiple sclerosis. Ann. Neurol. **50**(1) (2010)

8. Capra, R., Mattioli, F., Vignolo, L.A., et al.: Lesion detection in ms patients with and without clinical brainstem disorders: magnetic resonance imaging and brainstem auditory evoked potentials compared. Eur. Neurol. **29**, 317–322 (1989)
9. Carter, J., Stevens, J.: Somatosensory evoked potentials. In: Jasper, D., Devon, R. (eds.) Clinical Neurophysiology, 3rd edn, pp. 18-257–268. Oxford University Press, New York (2009)
10. Carter, J.: Brainstem auditory evoked potentials in central disorders. In: Jasper, D., Devon, R. (eds.) Clinical Neurophysiology, 3rd edn, pp. 19-281–286. Oxford University Press, New York (2009)
11. Carter, J.: Visual evoked potentials. In: Jasper, D., Devon, R. (eds.) Clinical Neurophysiology, 3rd edn, pp. 22- 311–318. Oxford University Press, New York (2009)
12. Epstein, C.: Visual evoked potentials. In: Daly, D., Pedley, T. (eds.) Current Practice of Clinical Electroencephalography, 2nd edn, pp. 593–623. Raven Press, Ltd, New York (1990)
13. Frohman, E., Costello, F., Stüve, O., et al.: Modeling axonal degeneration within the anterior visual system. Arch. Neurol. **65**(1), 26–35 (2008)
14. Haarmeier, T., Dichgans, J.: Die Wertigkeit der evozierten Potenziale in der Diagnostik der multiplen Sclerose. Evozierte Potenziale, 433–445 (2005)
15. Comi, G., Leocani, L., Medaglini, S., Locatelli, T., Martinelli, V., Santuccio, G., et al.: Measuring evoked responses in multiple sclerosis. Mult. Scler. **5**, 263–267 (1999)
16. Rösler, K., Hess, C.: Conduction studies in multiple sclerosis. In: Kesselring, J., Comi, G., Thompson, A. (eds.) Multiple Sclerosis: Recovery of Function and Neurorehabilitation, 1: 1 10. Cambridge University Press (2010)
17. Gilbert, M., Sergott, R.: New directions in optic neuritis and multiple sclerosis. Curr. Neurol. Neurosci. Rep. **7**, 259–264 (2007)
18. McDonald, I., Compston, A., Edan, G., et al.: Recommended diagnostic criteria for multiple sclerosis: guidelines from the international panel on the diagnosis of multiple sclerosis. Ann. Neurol. **50**, 121–127 (2001)
19. Guideline 9C.: Guideline on short latency auditory evoked potentials. Am. Clin. Neurophysiol. Soc. (2008)
20. Movassat, M., Piri, N., AhmadAbadi, M.: Visual evoked potential study in multiple sclerosis disease. Iran. J. Ophthalmol. **21**(4), 37–44 (2009)
21. Atanassov, K.: Generalized Nets. World Scientific, Singapore (1991)
22. Atanassov, K.: On Generalized Nets Theory. Prof. M. Drinov Academic Publishing House, Sofia (2007)
23. Todorova L., Vassilev, P., Ignatova, V.: A generalized net model for assessment of the degree of disability in patients with multiple sclerosis based on the abnormalities of visual evoked potentials. Issues in Intuitionistic Fuzzy Sets and Generalized Nets, vol 10, pp. 173–182 (2013)
24. Cuypers, M., Dickson, K., Pinckers, A., Thijssen, J., Hommes, O.: Discriminative power of visual evoked potential characteristics in multiple sclerosis. Doc. Ophthalmol. **90**, 247–257 (1995)
25. Grover, L., Hood, D., Ghadiali, Q.: A comparison of mulrifocal and conventional visual evoked potential techniques in patients with optic neuritis/multiple sclerosis. Doc. Ophthalmol. **117**(2), 121–128 (2008)
26. Ko, K.: The role of evoked potential and MR imaging in assessing multiple sclerosis: a comparative study. Singapore Med. J. **51**(9), 716 (2010)
27. Balnyte, R., Uloziene, I., Rastenyte, D., et al.: Diagnostic value of conventional visual evoked potentials applied to patients with multiple sclerosis. Medicina (Kaunas) **47**(5), 263–9 (2011)
28. Gnezditskiy, V.V., Korepina, O.S.: Atlas Po Vyzvannym Potentsialam Mozga, p. 532. PressSto, Ivanovo (2011)
29. Burina, A., Sinanovic, O., Smajlovich, D., et al.: Some aspects of balance disorder in patients with multiple sclerosis. Bosn. J. Basic Med. Sci. **8**(1), 80–85 (2008)
30. Berger, J.R., Blum, A.S.: Somatosensory evoked potentials. The Clinical Neurophysiology Primer (2007)

Generalized Net Models of Basic Genetic Algorithm Operators

Tania Pencheva, Olympia Roeva and Anthony Shannon

Abstract Generalized nets (GN) are applied here to describe some basic operators of genetic algorithms, namely *selection*, *crossover* and *mutation* and different functions for *selection* (*roulette wheel selection method* and *stochastic universal sampling*), different *crossover* techniques (*one-point crossover, two-point crossover*, and "*cut and splice*" *technique*), as well as *mutation* operator (*mutation operator of the Breeder genetic algorithm*). The resulting GN models can be considered as separate modules, but they can also be accumulated into a single GN model to describe a whole genetic algorithm.

1 Introduction

Genetic algorithms (GA) originated from the studies of cellular automata, conducted by John Holland and his colleagues at the University of Michigan. Holland's book [12], published in 1975, is generally acknowledged as the beginning of the research of GA. The GA is a model of machine learning which derives its behavior from a metaphor of the processes of evolution in nature [11]. The basic techniques of GA are designed to simulate the processes in natural systems which are necessary for evolution, especially those which follow the principles first laid down by Charles Darwin of "survival of the fittest". GA are based on an analogy with the

T. Pencheva (✉) · O. Roeva
Institute of Biophysics and Biomedical Engineering, Bulgarian Academy of Sciences,
105 Acad. G. Bonchev Str., 1113 Sofia, Bulgaria
e-mail: tania.pencheva@biomed.bas.bg

O. Roeva
e-mail: olympia@biomed.bas.bg

A. Shannon
Faculty of Engineering & IT, University of Technology, Sydney, NSW 2007, Australia
e-mail: Anthony.Shannon@uts.edu.au

P. Angelov and S. Sotirov (eds.), *Imprecision and Uncertainty in Information Representation and Processing*, Studies in Fuzziness and Soft Computing 332,
DOI 10.1007/978-3-319-26302-1_19

genetic structure and behaviour of chromosomes within a population of individuals using the following foundations [9, 14]:

- individuals in a population compete for resources and mates;
- those individuals most successful in each "competition" will produce more offspring than those individuals that perform poorly;
- genes from "good" individuals propagate throughout the population so that two good parents will sometimes produce offspring that are better than either parent;
- thus each successive generation will become more suited to their environment.

The GA maintains a population of chromosomes (solutions) with associated fitness values. Parents are selected to mate, on the basis of their fitness, producing offspring via a reproductive plan. Consequently highly fit solutions are given more opportunities to reproduce, so that offspring inherit characteristics from each parent. After the random generation of an initial population, the algorithm evolves through operators:

- *selection* which equates to survival of the fittest;
- *crossover* which represents mating between individuals;
- *mutation* which introduces random modifications.

The key features of the genetic operators are briefly listed below [14]:

Selection operator

- key idea: give a preference to better individuals, allowing them to pass on their genes to the next generation;
- the goodness of each individual depends on its fitness;
- fitness may be determined by an objective function or by a subjective judgement.

Crossover operator

- prime distinguished factor of GA from other optimization techniques;
- two individuals are chosen from the population using the *selection* operator;
- a *crossover* site along the bit strings is randomly chosen;
- the values of the two strings are exchanged up to this point;
- the two new offspring created from this mating are put into the next generation of the population;
- by recombining portions of good individuals, this process is likely to create even better individuals.

Mutation operator

- with some low probability, a portion of the new individuals will have some of their bits flipped;
- its purpose is to maintain diversity within the population and inhibit premature convergence;
- *mutation* alone induces a random walk through the search space.

The effects of the basic genetic operators could be summarized as follows [14]:

- using *selection* alone will tend to fill the population with copies of the best individual from the population;
- using *selection* and *crossover* operators will tend to cause the algorithms to converge on a good but sub-optimal solution;
- using *mutation* alone induces a random walk through the search space;
- using *selection* and *mutation* creates a parallel, noise-tolerant, hill-climbing algorithm.

Based on the operators' key features and advantages, recently GA have established themselves as an important component from the field of artificial intelligence. GA are quite popular and are applied in many domains, such as industrial design, scheduling, network design, routing, time series prediction, database mining, control systems, artificial life systems, as well as in many fields of science [1, 6, 10, 16, 17, 26, 27, 32].

On the other hand, up until now the approach of Generalized nets (GN) has mainly been used as a tool for the description of parallel processes in several areas —economics, transport, medicine, computer technologies, etc. [2, 4, 15]. That is why the idea of using GN for the description of GA has appeared. A few GN models regarding GA performance have been developed [3, 23–25, 28–31]. For instance, the GN model in [28] describes the GA search procedure. The model simultaneously evaluates several fitness functions, ranks the individuals according to their fitness and provides the opportunity to choose the best fitness function for a specific problem domain. In [3] a GN model for GA learning has also been proposed.

The aim of this current research is to present some GN models of basic GA operators—*selection*, *crossover* and *mutation*. Some of the most used techniques and functions are considered, namely *roulette wheel selection*, *stochastic universal sampling*; *one-*, *two-point crossover*, as well as "*cut and splice*" techniques; and *mutation* operator of the Breeder genetic algorithm [8, 22].

The simple genetic algorithm (SGA), described by Goldberg [11], is used here to illustrate the basic components of a GA. A pseudo-code outline of the SGA is shown in Fig. 1. The population at time t is represented by the time-dependent variable *Pop*, with the initial population of random estimates being *Pop*(0). The GA maintains a population of individuals, $Pop(t) = y_1^t, \ldots, y_n^t$ for generation t. Each individual represents a potential solution to the problem and is implemented as some data structure U. Each solution is evaluated to give some measure of its "fitness". The fitness of an individual is assigned proportionally to the value of the objective function of the individuals. Then, a new population (generation $t + 1$) is formed by selecting more fit individuals (selected step). Some members of the new population undergo transformations by means of "genetic" operators to form a new solution.

Following the outline of a GA (Fig. 1), further major elements of a GA (genetic operators of *selection*, *crossover* and *mutation*) will be described.

```
procedure GA
begin
        t = 0;
        initialize Pop(t);
        evaluate Pop(t);
        while not finished do
        begin
                t = t + 1;
                select Pop(t) from Pop(t - 1);
                reproduce pairs in Pop(t);
                mutate Pop(t);
                evaluate Pop(t);
        end
end
```

Fig. 1 A simple genetic algorithm

2 Basic Genetic Algorithm Operators

2.1 Selection Operator

The *selection* of individuals to produce successive generations plays an extremely important role in a GA. A probabilistic *selection* is performed based upon the individual's fitness such that the better individuals have an increased chance of being selected. An individual in the population can be selected more than once with all individuals in the population having a chance of being selected to reproduce into the next generation. There are several schemes for the *selection* process: *roulette wheel selection* and its extensions, *scaling techniques*, *tournament*, *elitist models*, *ranking methods*, *stochastic universal sampling*, and so on [5, 11, 13, 19].

A common *selection* approach assigns a probability of *selection*, P_j, to each individual j based on its fitness value. A series of N random numbers is generated and compared against the cumulative probability of the population:

$$C_i = \sum_{j=1}^{i} P_j. \tag{1}$$

The appropriate individual, i, is selected and copied into the new population if $C_{i-1} < U(0, 1) \leq C_i$. Various methods exist to assign probabilities to individuals: roulette wheel, linear ranking and geometric ranking.

2.1.1 Roulette Wheel Selection

Roulette wheel, developed by Holland [12], is the first *selection* method. The probability, P_i, for each individual is defined by [13]:

$$P[\text{Individual } i \text{ is chosen}] = \frac{F_i}{\sum_{j=1}^{k} F_j} \tag{2}$$

where F_i represents the fitness of individual *i*, and *k* is the population size. The use of *roulette wheel selection* limits the GA to maximization since the evaluation function must map the solutions to a fully ordered set of values on $\Re^+$. Extensions, such as windowing and scaling, have been proposed to allow for minimization and negativity.

Ranking methods only require the evaluation function to map the solutions to a partially ordered set, thus allowing for minimization and negativity. Ranking methods assign P_i based on the rank of solution *i* when all solutions are sorted. Normalized geometric ranking, [13] defines P_i for each individual by:

$$P[\text{Selecting the } i\text{th individual}] = q'(1-q)^{r-1}, \tag{3}$$

where *q* is the probability of selecting the best individual, and *r* is the rank of the individual, where 1 is the best; and

$$q' = \frac{q}{1-(1-q)^k}. \tag{4}$$

Tournament selection, like ranking methods, only requires the evaluation function to map solutions to a partially ordered set, however, it does not assign probabilities. *Tournament selection* works by selecting *j* individuals randomly, with replacement, from the population, and inserts the best of the *j* into the new population. This procedure is repeated until *N* individuals have been selected.

2.1.2 Stochastic Universal Sampling

Stochastic universal sampling (*SUS*) developed by Baker [5] is a single-phase sampling algorithm with minimum spread and zero bias. Instead of the single *selection* pointer employed in *roulette wheel* methods, *SUS* uses *N* equally spaced pointers, where *N* is the number of selections required. The population is shuffled randomly and a single random number *pointer1* in the range [0, 1/*N*] is generated. The *N* individuals are then chosen by generating the *N* pointers, starting with *pointer1* and spaced by 1/*N*, and selecting the individuals whose fitness spans the positions of the pointers. If *et*(*i*) is the expected number of trials of individual *i*, $\lfloor et(i) \rfloor$ is the floor of *et*(*i*) and $\lceil et(i) \rceil$ is the ceiling, then an individual is thus guaranteed to be selected a minimum of times $\lfloor et(i) \rfloor$ and no more than $\lceil et(i) \rceil$, thus achieving minimum spread. In addition, as individuals are selected entirely on their positions in the population, *SUS* has zero bias. For these reasons, *SUS* has become one of the most widely used *selection* algorithms in current GA.

Fig. 2 One-point crossover

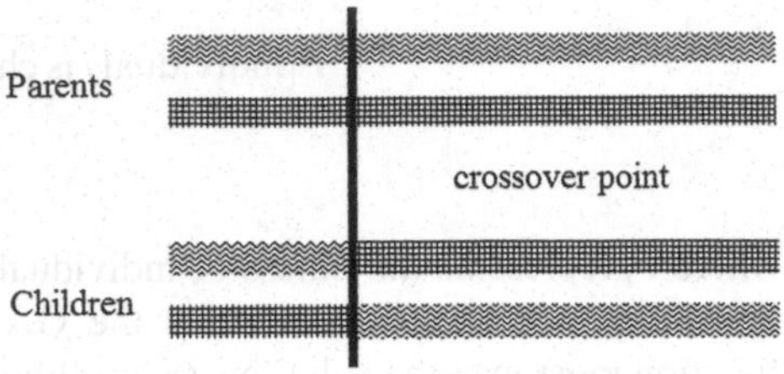

Fig. 3 Two-point crossover

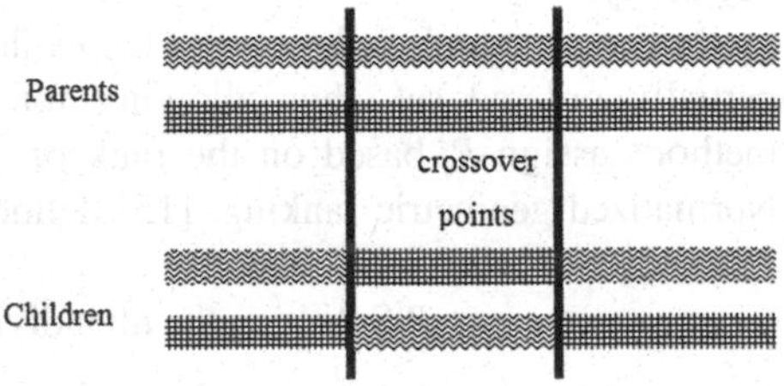

2.2 *Crossover Techniques*

In GA, *crossover* is a genetic operator used to vary the programming of a chromosome (or chromosomes) from one generation to the next. It is analogous to reproduction and biological *crossover*, upon which GA are based. Many *crossover* techniques exist for organisms which use different data structures to store themselves. Here the focus is on the *one-point*, *two-point crossover* as well as the technique of "*cut and splice*".

2.2.1 One-Point Crossover

A single *crossover* point on both parents' organism strings is selected. All data beyond that point in either organism string is swapped between the two parent organisms. The resulting organisms are the children (Fig. 2) [11].

2.2.2 Two-Point Crossover

Two-point crossover calls for two points to be selected on the parent organism strings. Everything between the two points is swapped between the parent organisms, producing two child organisms (Fig. 3) [11].

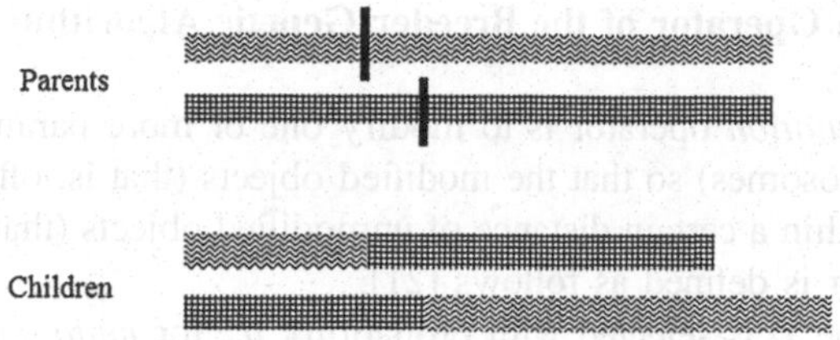

Fig. 4 "Cut and splice" technique

2.2.3 "Cut and Splice" Technique

Another *crossover* variant, the "*cut and splice*" technique, results in a change in the length of the children strings. The reason for this difference is that each parent string has a separate choice of *crossover* point (Fig. 4) [11].

2.3 *Mutation Operator*

Mutation is a genetic operator that alters one ore more gene values in a chromosome from its initial state. This can result in entirely new gene values being added to the gene pool. With these new gene values, the GA may be able to arrive at a better solution than previously possible. *Mutation* is an important part of the genetic search as it helps to prevent the population from stagnating at any local optima. *Mutation* occurs during evolution according to a user-definable *mutation* probability. This probability should usually be set fairly low. If it is set too high, the search will turn into a primitive random search.

There are many ways to perform *mutation*. Some examples of *mutation* according to the type of encoding are: binary encoding—*binary mutation* (*bit inversion*); permutation encoding—*order changing mutation*; value encoding—adding a small number (for real value encoding); tree encoding—changing operator; number—selected nodes are changed. If different genome types have been considered, the following *mutation* types are suitable: *bit string mutation*, *flip bit*, *boundary*, *non-uniform*, *uniform*, *Gaussian*.

In this paper, a *mutation* operator of the Breeder genetic algorithm (BGA) [8, 22], based on the science of breeding, is considered. The BGA is inspired by artificial *selection* as performed by human breeders. But *mutation* and *crossover* are based on mathematical search techniques. The BGA *mutation* scheme is able to optimize many multimodal functions [20–22].

2.3.1 Mutation Operator of the Breeder Genetic Algorithm

The goal of a *mutation* operator is to modify one or more parameters of z_i (individuals or chromosomes) so that the modified objects (that is, offspring) appear in the landscape within a certain distance of unmodified objects (that is, parents). The *mutation* operator is defined as follows [21].

A chromosome x_i is selected with probability p_m for *mutation*. The BGA normally uses $p_m = 1/n$. At least one variable will be mutated. A value out of an interval [$-range_i$, $range_i$] is added to the selected variable. $range_i$ defines the *mutation range*. It is normally set to $a \cdot$ *search-interval*$_i$, where a is a constant. *search-interval*$_i$ (*upper-bound, lower-bound*) is the domain of definition of variable x_i.

The new value z_i (*new chromosome*) is computed according to

$$z_{i\,(NewChrom)} = x_{i\,(OldChrom)} \pm range_i \cdot \delta, \tag{5}$$

where

$$range = 0.5(upper - bound - lower - bound). \tag{6}$$

The (+) or (−) sign is chosen with a probability 0.5 (see Eqs. (5) and (6)). δ is computed from a distribution which prefers small values. This is realized as follows

$$\delta = \sum_{i=0}^{accur-1} \alpha_i 2^{-i}, \quad \alpha_i \in [0, 1], \tag{7}$$

where

$$\alpha_i = rand(accur, 1) < 1/accur. \tag{8}$$

The standard BGA *mutation* operator is able to generate any point in the hypercube with center x defined by $x_i \pm range_i$. But it tests much more often in the neighborhood of x. In Eq. (7), "*accur*" (precision of *mutation* steps) is a parameter originally related to the machine precision, that is, the numbers of bits used to represent a real variable in the machine we are working with; traditionally the values of 8 and 16 were used.

Before *mutation* the value of $\alpha_i = 0$ is set. Then each α_i is mutated to 1 with probability $p_\delta = 1/accur$. Only $\alpha_i = 1$ contributes to the sum. On the average there will be just one α_i with value 1, say α_j. Then δ is given by

$$\delta = 2^{-j} \tag{9}$$

The *mutation* operator is similar in spirit to that used by the parallel GA [21], but the BGA operator is much easier to understand. Furthermore, it is independent of the location in phenotype space.

3 GN Models of Basic Genetic Algorithm Operators

3.1 GN Model of Selection Operator

3.1.1 Roulette Wheel Selection

The widely used Matlab Toolbox for GA contains two functions for the *selection* function, namely *roulette wheel selection* (also known as *stochastic sampling with replacement* (SSR)), and *stochastic universal sampling*. Figure 5 presents the Matlab code of the *rws.m* function of the GA Toolbox of Matlab [7, 18].

A GN model, described the *roulette wheel selection* (*RWS*), as described in the function *rws.m*, is presented in Fig. 6.

The token α enters GN in place l_1 with an initial characteristic "pool of possible parents". The token α is split into new tokens β and γ, which take on corresponding characteristics "identify the population size (*Nind*) and assign fitness values of the individuals in the population (*FitnV*)" in place l_2 and "number of individuals to be

```
% RWS.m - Roulette Wheel Selection
%
%   Syntax:
%          NewChrIx = rws(FitnV, Nsel)
%
%   This function selects a given number of individuals Nsel
%   from a population. FitnV is a column vector containing
%   the fitness values of the individuals in the population.
%
%   The function returns another column vector containing the
%   indexes of the new generation of chromosomes relative to
%   the original population matrix, shuffled. The new
%   population, ready for mating, can be obtained by
%   calculating OldChrom(NewChrIx, :).

% Author: Carlos Fonseca,  Updated: Andrew Chipperfield
% Date: 04/10/93,          Date: 27-Jan-94

function NewChrIx = rws(FitnV,Nsel);

% Identify the population size (Nind)
  [Nind,ans] = size(FitnV);

% Perform Stochastic Sampling with Replacement
  cumfit  = cumsum(FitnV);
  trials = cumfit(Nind)  .* rand(Nsel, 1);
  Mf = cumfit(:, ones(1, Nsel));
  Mt = trials(:, ones(1, Nind))';
  [NewChrIx, ans] = find(Mt < Mf & ...
          [ zeros(1, Nsel); Mf(1:Nind-1, :) ] <= Mt);
```

Fig. 5 Matlab function *rws.m*

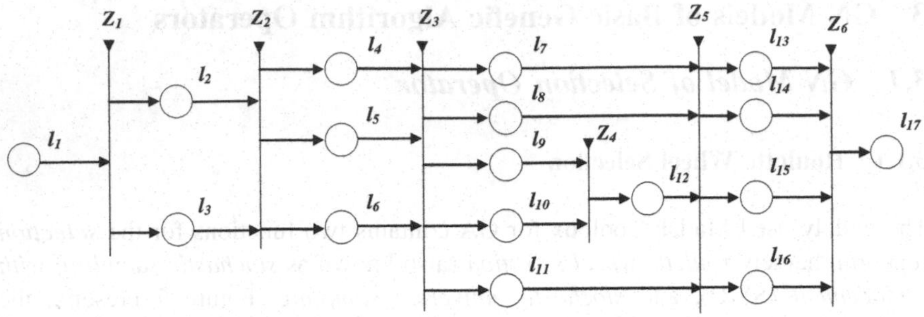

Fig. 6 GN model of roulette wheel selection

selected (*Nsel*)" in place l_3. The form of the first transition of the GN model is as follows:

$$Z_1 = \langle\{l_1\}, \{l_2, l_3\}, r_1, \wedge(l_1)\rangle$$

$r_1 =$

	l_2	l_3
l_1	*true*	*true*

The token β is split into two new tokens δ and ε, which gain corresponding characteristics "identify the population size (*Nind*)" in place l_4 and "calculation of the function *cumfit* = *cumsum*(*FitnV*)" in place l_5, according to Fig. 5. The token γ retains its characteristic "number of individuals to be selected (*Nsel*)" in place l_6. The form of the second transition of the GN model is as follows:

$$Z_2 = \langle\{l_2, l_3\}, \{l_4, l_5, l_6\}, r_2, \wedge(l_2, l_3)\rangle,$$

$r_2 =$

	l_4	l_5	l_6
l_2	*true*	*true*	*false*
l_3	*false*	*false*	*true*

Further, the token δ keeps its characteristic "identify the population size (*Nind*)" in place l_7.

The token ε is split into two new tokens μ and η, which obtain corresponding characteristics "calculation of the function M_f = *cumfit*(:, *ones*(1, *Nsel*))" in place l_8 and "calculation of the function *cumfit*(*Nind*)", in place l_9, according to Fig. 5. The token γ is also split into new tokens θ and π, which are given the corresponding characteristics "calculation of the function *rand*(*Nsel*, 1)" in place l_{10} and keeps characteristic "number of individuals to be selected (*Nsel*)" in place l_{11}. The form of the third transition of the GN model is as follows:

$$Z_3 = <\{l_4, l_5, l_6\}, \{l_7, l_8, l_9, l_{10}, l_{11}\}, r_3, \wedge(l_4, l_5, l_6)>$$

$r_3 =$

	l_7	l_8	l_9	l_{10}	l_{11}
l_4	*true*	*false*	*true*	*false*	*false*
l_5	*false*	*true*	*true*	*false*	*false*
l_6	*false*	*true*	*false*	*true*	*true*

The tokens η and θ are combined in a new token λ in place l_{12} with a characteristic "calculation of the function *trials* = *cumfit(Nind)*. * *rand(Nsel*, 1)", according to Fig. 5. The form of the fourth transition of the GN model is as follows:

$$Z_4 = <\{l_9, l_{10}\}, \{l_{12}\}, r_4, \wedge(l_9, l_{10})>$$

$r_4 =$

	l_{12}
l_9	*true*
l_{10}	*true*

In the next step, the token λ obtains a characteristic "calculation of the function M_t = *trials*(:, *ones*(1, *Nind*))′" in place l_{15}, according to Fig. 5, while the remaining tokens δ, μ and π keep their characteristics as described above, correspondingly in places l_{13}, l_{14} and l_{16}. The form of the fifth transition of the GN model is as follows:

$$Z_5 = <\{l_7, l_8, l_{11}, l_{12}\}, \{l_{13}, l_{14}, l_{15}, l_{16}\}, r_5, \wedge(l_7, l_8, l_{11}, l_{12})>$$

$r_5 =$

	l_{13}	l_{14}	l_{15}	l_{16}
l_7	*true*	*false*	*true*	*false*
l_8	*false*	*true*	*false*	*false*
l_{11}	*false*	*false*	*false*	*true*
l_{12}	*false*	*false*	*true*	*false*

At the final step, all tokens δ, μ, λ and π are combined in a new token σ, which according to Fig. 5 has a characteristic in place l_{17} "calculation of the function

[*NewChrIx*, *ans*] = *find*(M_t < M_f & [*zeros*(1, *Nsel*); M_f(1: *Nind* – 1, :)] <= M_t)". The form of the sixth transition of the GN model is then:

$$Z_6 = \langle \{l_{13}, l_{14}, l_{15}, l_{16}\}, \{l_{17}\}, r_6, \wedge(l_{13}, l_{14}, l_{15}, l_{16})\rangle$$

$r_6 =$

	l_{17}
l_{13}	*true*
l_{14}	*true*
l_{15}	*true*
l_{16}	*true*

In the place l_{17} the new chromosome is created and the *selection* function, performing *roulette wheel selection*, is completely fulfilled.

3.1.2 Stochastic Universal Sampling

Another implemented *selection* function in Matlab Toolbox for GA is *stochastic universal sampling*. Figure 7 lists the Matlab code of the *sus.m* function of the GA Toolbox of Matlab [7, 18].

The GN model, which should describe the *SUS*, could be obtained based on the GN model for *RWS*, with very slight additions. The GN model of *SUS* is presented in Fig. 8.

The GN model of the *SUS* function is identical to the GN model of the *RWS* function up to the fourth transition. In contrast to the GN model of the *RWS* function, there is one additional input. The token π from place l_{11} together with η and θ, as in *RWS*, are combined in a new token λ. This token takes on a characteristic "calculation of the function

$$trials = cumfit(Nind)/Nsel*(rand + (0{:}Nsel - 1)')$$

" in place l_{12}, according to Fig. 7. The token θ passes through the transition, obtaining a new characteristic "calculation of the function *sort*(*rand*(*Nsel*, 1))" in place l_{13}, while the token π passes through the transition, keeping its characteristic "number of individuals to be selected (*Nsel*)". Then the form of the fourth transition of the GN model is as follows:

$$Z_4 = \langle \{l_9, l_{10}, l_{11}\}, \{l_{12}, l_{13}, l_{14}\}, r_4, \wedge(l_9, l_{10}, l_{11})\rangle$$

$r_4 =$

	l_{12}	l_{13}	l_{14}
l_9	*true*	*false*	*false*
l_{10}	*true*	*true*	*false*
l_{11}	*true*	*false*	*true*

```
% SUS.M      (Stochastic Universal Sampling)
%
% This function performs selection with
% STOCHASTIC UNIVERSAL SAMPLING.
%
% Syntax:  NewChrIx = sus(FitnV, Nsel)
%
% Input parameters:
%  FitnV    - Column vector containing the fitness values of
%             the individuals in the population
%  Nsel     - number of individuals to be selected
%
% Output parameters:
%  NewChrIx   - column vector containing the indexes of the
%               selected individuals relative to the original
%               population, shuffled. The new population, ready
%               for mating, can be obtained by calculating
%               OldChrom(NewChrIx,:).

% Author:     Hartmut Pohlheim (Carlos Fonseca)
% History:    12.12.93     file created
%             22.02.94     clean up, comments

function NewChrIx = sus(FitnV,Nsel);

% Identify the population size (Nind)
   [Nind,ans] = size(FitnV);

% Perform stochastic universal sampling
  cumfit = cumsum(FitnV);
  trials = cumfit(Nind) / Nsel * (rand + (0:Nsel-1)');
  Mf = cumfit(:, ones(1, Nsel));
  Mt = trials(:, ones(1, Nind))';
  [NewChrIx, ans] = find(Mt < Mf & [zeros(1, Nsel); ...
  Mf(1:Nind-1, :)] <= Mt);


% Shuffle new population
   [ans, shuf] = sort(rand(Nsel, 1));
   NewChrIx = NewChrIx(shuf);

% End of function
```

Fig. 7 Matlab function *sus.m*

The form of the fifth transition of the GN model is as follows:

$$Z_5 = <\{l_7, l_8, l_{12}, l_{13}, l_{14}\}, \{l_{15}, l_{16}, l_{17}, l_{18}, l_{19}\}, r_5, \wedge(l_7, l_8, l_{12}, l_{13}, l_{14})>$$

$r_5 =$

	l_{15}	l_{16}	l_{17}	l_{18}	l_{19}
l_7	*true*	*false*	*true*	*false*	*false*
l_8	*false*	*true*	*false*	*false*	*false*
l_{12}	*false*	*false*	*true*	*false*	*false*
l_{13}	*false*	*false*	*false*	*true*	*false*
l_{14}	*false*	*true*	*false*	*false*	*true*

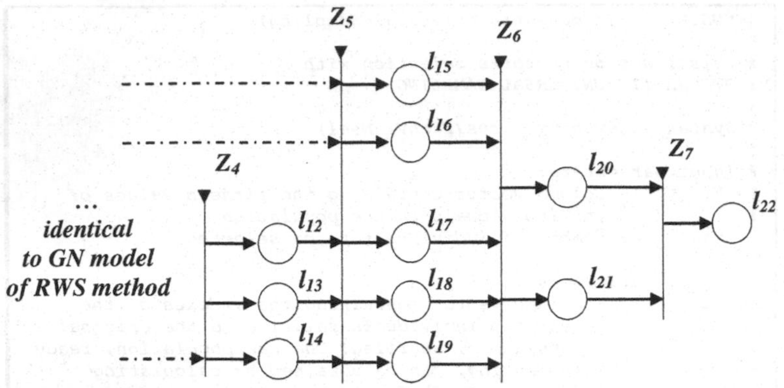

Fig. 8 GN model of stochastic universal sampling

After this transition the tokens characteristics are as follows:

- δ in place l_{15} keeps its characteristic "identify the population size (*Nind*)";
- μ in place l_{16} keeps its characteristic "calculation of the function $M_f = cumfit(:, ones(1, Nsel))$";
- λ in place l_{17} obtains a new characteristic "calculation of the function $M_t = trials(:, ones(1, Nind))'$ ";
- θ in place l_{18} keeps its characteristic "calculation of the function *sort*(*rand*(*Nsel*, 1))";
- π in place l_{19} keeps its characteristic "number of individuals to be selected (*Nsel*)".

At the next step the tokens δ, μ, λ and π are combined in a new token σ, which in place l_{20}, according to Fig. 7, gain a characteristic "calculation of the function $[NewChrIx, ans] = find(M_t < M_f \ \& \ [zeros(1, Nsel); M_f(1: Nind - 1, :)] <= M_t)$". The token θ keeps its characteristic in places l_{21}. Then the form of the sixth transition of the GN model is as follows:

$$Z_6 = <\{l_{15}, l_{16}, l_{17}, l_{18}, l_{19}\}, \{l_{20}, l_{21}\}, r_6, \wedge(l_{15}, l_{16}, l_{17}, l_{18}, l_{19})>$$

$r_6 =$

	l_{20}	l_{21}
l_{15}	*true*	*false*
l_{16}	*true*	*false*
l_{17}	*true*	*false*
l_{18}	*false*	*true*
l_{19}	*true*	*false*

At the final step, the tokens σ and θ are combined in a new token χ with a characteristic "*shuffle new population*" in place l_{22}, according to Fig. 7. The form of the seventh transition of the GN model is as follows:

$$Z_7 = <\{l_{20}, l_{21}\}, \{l_{22}\}, r_7, \wedge(l_{20}, l_{21})>$$

$r_7 =$

	l_{22}
l_{20}	*true*
l_{21}	*true*

In the place l_{22} the new chromosome is created and the *selection* function, performing *stochastic universal sampling*, is completely fulfilled.

3.2 GN Model of Crossover Operator

GN models of three of the most common used techniques of *crossover* (*one-*, *two-point crossover*, as well as "*cut and splice*") have been developed [25]. Here they are combined in a "generalized" GN model, as presented in Fig. 9.

The token α enters GN in place l_1 with an initial characteristic "parameters of GA". The token β enters GN in place l_2 with an initial characteristic "pool of possible parents". Tokens α and β are combined and appear as a token γ in place l_3 with a characteristic "parent 1" and as a token δ in place l_4 with a characteristic "parent 2". The form of the first transition of the GN model is as follows:

$$Z_1 = <\{l_1, l_2, l_{11}\}, \{l_3, l_4\}, r_1, \wedge(l_1, l_2)>$$

$r_1 =$

	l_3	l_4
l_1	*true*	*true*
l_2	*true*	*true*
l_{11}	*true*	*true*

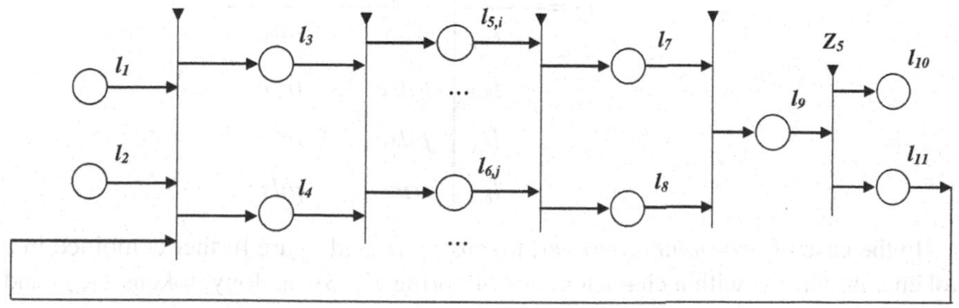

Fig. 9 GN model of crossover techniques

Each of tokens γ and δ is split into a few new tokens, respectively γ_i, $(i = 1 \div 3)$ and δ_j $(j = 1 \div 3)$. The values of i and j depends on the type of the chosen *crossover* operator.

In the case of *one-point crossover* $i = j = 2$, and the tokens will obtain characteristics as follows:

- γ_1 in place $l_{5,1}$ "first string of parent 1";
- γ_2 in place $l_{5,2}$ "second string of parent 1";
- δ_1 in place $l_{6,1}$ "first string of parent 2";
- δ_2 in place $l_{6,2}$ "second string of parent 2".

In the case of *two-point crossover* $i = j = 3$, and the tokens will obtain characteristics as follows:

- γ_1 in place $l_{5,1}$ "first string of parent 1";
- γ_2 in place $l_{5,2}$ "second string of parent 1";
- γ_3 in place $l_{5,3}$ "third string of parent 1";
- δ_1 in place $l_{6,1}$ "first string of parent 2";
- δ_2 in place $l_{6,2}$ "second string of parent 2";
- δ_3 in place $l_{6,3}$ "third string of parent 2".

In both cases the form of the second transition of the GN model is as follows:

$$Z_2 = <\{l_3, l_4\}, \{l_{5,i}, l_{6,j}\}, r_2, \wedge(l_3, l_4)>$$

$r_2 =$

	$l_{5,i}$	$l_{6,j}$
l_3	*true*	*true*
l_4	*true*	*true*

In the case of *one-point crossover*, tokens γ_1 and δ_2 are further combined in a token ε in place l_7 with a characteristic "offspring 1". By analogy, tokens δ_1 and γ_2 are further combined in a token σ in place l_8 with a characteristic "offspring 2". The form of the third transition of the GN model in that case is as follows:

$$Z_3 = <\{l_{5,1}, l_{5,2}, l_{6,1}, l_{6,2}\}, \{l_7, l_8\}, r_3, \wedge(l_{5,1}, l_{5,2}, l_{6,1}, l_{6,2})>$$

$r_3 =$

	l_7	l_8
$l_{5,1}$	*true*	*false*
$l_{5,2}$	*false*	*true*
$l_{6,1}$	*false*	*true*
$l_{6,2}$	*true*	*false*

In the case of *two-point crossover*, tokens γ_1, δ_2 and γ_3 are further combined in a token ε in place l_7 with a characteristic "offspring 1". By analogy, tokens δ_1, γ_2 and

δ_3 are further combined in a token σ in place l_8 with a characteristic "offspring 2". The form of the third transition of the GN model in that case is as follows:

$$Z_3 = \langle \{l_{5,1}, l_{5,2}, l_{5,3}, l_{6,1}, l_{6,2}, l_{6,3}\}, \{l_7, l_8\}, r_3, \wedge(l_{5,1}, l_{5,2}, l_{5,3}, l_{6,1}, l_{6,2}, l_{6,3})\rangle$$

$r_3 =$

	l_7	l_8
$l_{5,1}$	*true*	*false*
$l_{5,2}$	*false*	*true*
$l_{5,3}$	*true*	*false*
$l_{6,1}$	*false*	*true*
$l_{6,2}$	*true*	*false*
$l_{6,3}$	*false*	*true*

Tokens ε and σ are then combined in a token η in place l_9 with a characteristic "new population". The form of the fourth transition of the GN model is as follows:

$Z_4 = \langle \{l_7, l_8\}, \{l_9\},$

	l_9
l_7	*true*
l_8	*true*

$, \wedge(l_7, l_8)\rangle$

After the creation of the new population in place l_9, the token η could pass to place l_{10} with a characteristic "bad individual" or in place l_{10} with a characteristic "good individual". The form of the fifth transition of the GN model is as follows:

$Z_5 = \langle \{l_9\}, \{l_{10}, l_{11}\},$

	l_{10}	l_{11}
l_9	$W_{9,10}$	$W_{9,11}$

$, \wedge(l_9)\rangle,$

where $W_{9,10}$ is "*fit the fitness function*" and $W_{9,11} = \neg\, W_{9,10}$.

As it was explained in [25], the GN model for describing *one-point crossover* could be also used for a description of the technique "*cut and splice*". The GN model to describe the "*cut and splice*" technique is not explicitly presented here because it is equivalent to the GN model which describes *one-point crossover*. The tokens, characteristics, as well as the transitions will be absolutely the same. The main difference between these crossover techniques is the length of the offspring individuals that are obtained, because of the different *crossover* point in both parents. While in the *one-point crossover* the offspring individuals obtained have a length equal to that of their parents, in the case of the "*cut and splice*" technique the offspring individuals obtained have different lengths. However, this fact will not be reflected in the structure of the GN model which describes these techniques.

3.3 GN Model of Mutation Operator

The GN model of the *mutation* operator of the BGA is shown in Fig. 10. The proposed GN model generates a matrix "*Chrom*" with the real representation of the individuals in the current population, mutates the individuals with given *mutation* probability (p_m) and returns the resulting population (*NewChrom*)—the same number of randomly initialized real valued individuals.

The transition Z_1 has the following definition:

$Z_1 = <\{l_1, l_2\}, \{l_3, l_4, l_5\}, r_1, \wedge(l_1, l_2)>$

$r_1 =$

	l_3	l_4	l_5
l_1	W_1	*false*	W_3
l_2	*false*	W_2	*false*

where W_1 = "*estimation of the parameter range*"; W_2 = "*evaluation of uniformly distributed random numbers* α_i"; W_3 = "*generation of a matrix Chrom*".

After the transition Z_1 the tokens take on the following characteristics in position:

- l_1 the preliminary parameters are given: individuals number (*Nvar*) and matrix of the boundaries of each individual—*upper-bound* and *lower-bound*;
- l_2 the initial parameters for *mutation* operator: probability for *mutation* of a variable (p_m); *mutation* type—added (+) or subtracted (–) and *accur*;
- l_3 the value of the *range* is evaluated (according to Eq. (6));
- l_4 the values of the α_i parameters are evaluated (according to Eq. (8));
- l_5 the matrix *Chrom* is formed.

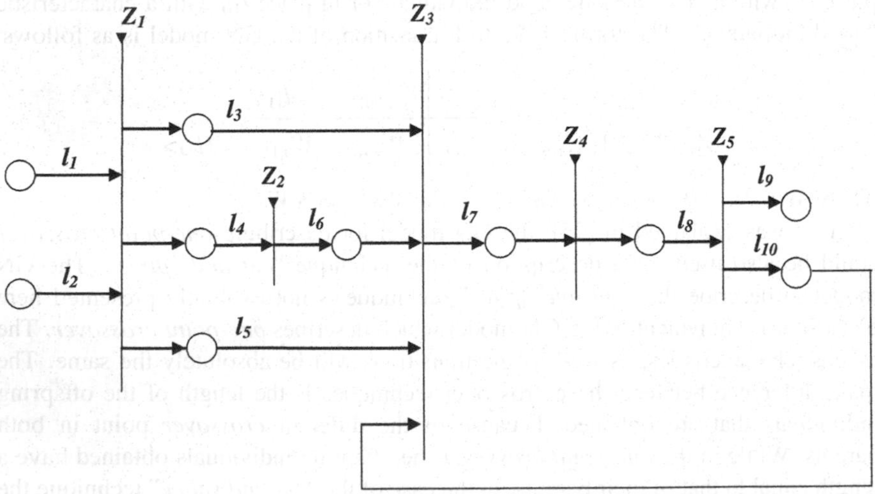

Fig. 10 Generalized net model of mutation operator

The form of the transition Z_2 is:

$$Z_2 = <\{l_4\}, \{l_6\}, r_2, \wedge(l_4)>,$$

$$r_2 = \begin{array}{c|c} & l_6 \\ \hline l_4 & W_4 \end{array}$$

where W_4 = "*δ value evaluation*".

In position l_6 the δ value is obtained (according to Eq. (6)).

The transition Z_3 has the following formal definition:

$$Z_3 = <\{l_3, l_5, l_6, l_{10}\}, \{l_7\}, r_3, \wedge(l_3, l_5, l_6)>,$$

$$r_3 = \begin{array}{c|c} & l_7 \\ \hline l_3 & W_5 \\ l_5 & W_5 \\ l_6 & W_5 \\ l_{10} & W_6 \end{array}$$

where W_5 = "*mutation, based on standard BGA mutation operator*"; W_6 = "*mutation, based on standard BGA mutation operator, if boundaries are not satisfied*".

In position l_7 the matrix of *NewChrom* is obtained. The matrix has the same format as *OldChrom* and contains the chromosomes of the population after *mutation*.

The next transition Z_4 provides control of the variable boundaries validity, compared to lower and upper boundaries. The form of the transition Z_4 is:

$$Z_4 = <\{l_7\}, \{l_8\}, r_4, \wedge(l_7)>,$$

$$r_4 = \begin{array}{c|c} & l_8 \\ \hline l_7 & W_7 \end{array}$$

where W_7 = "*control of variable boundaries*".

The form of the transition Z_5 is:

$$Z_5 = <\{l_8\}, \{l_9, l_{10}\}, r_5, \wedge(l_8)>,$$

$$r_5 = \begin{array}{c|cc} & l_9 & l_{10} \\ \hline l_8 & W_8 & \neg W_8 \end{array}$$

where W_8 = "*end of mutation process*".

In position l_9 the new chromosome is ready for further examination of the GA, if the boundaries of the chromosome are in the appropriate range. Otherwise, if the boundaries of the new chromosome are not in the appropriate range, the result is returned to the transition Z_3 for a new *mutation* operation.

4 Conclusions

Generalized nets developed here have been applied to a description of basic GA operators, namely *selection*, *crossover* and *mutation*. GN models of two *selection* functions – *roulette wheel selection method* and *stochastic universal sampling* have also been developed. It was shown that both GN models are very similar and it might be possible for them to be combined into a "generalized" GN model of the *selection* function. Such a "generalized" GN model was presented here to describe three *crossover* techniques—*one-point crossover*, *two-point crossover* as well as the technique "*cut and splice*". The identical logic for the different techniques of *crossover* operator permits the development of such a "generalized" GN model of *crossover* operator. A GN model has been also constructed to realize the *mutation* operation of a GA. This proposed GN model performs *mutation*, based on the standard Breeder genetic algorithm *mutation* operator. The resulting GN models of the three basic genetic operators, namely *selection*, *crossover* and *mutation*, could be considered as separate modules, but they can also be accumulated into one GN model for modelling a whole GA.

References

1. Abuiziah, I., Shakarneh, N.: A review of genetic algorithm optimization: operations and applications to water pipeline systems. Int. J. Phys. Nucl. Sci. Eng. **7**(12), 341–347 (2013)
2. Atanassov, K.: Generalized Nets. World Scientific, Singapore (1991)
3. Atanassov, K., Aladjov, H.: Generalized Nets in Artificial Intelligence: Generalized nets and Machine Learning, vol. 2. Prof. M. Drinov Academic Publishing House, Sofia (2000)
4. Atanassov, K.: On Generalized Nets Theory. Prof. M. Drinov Academic Publishing House, Sofia (2007)
5. Baker, J.: Reducing bias and inefficiency in the selection algorithm. In: Proceedings of the Second International Conference on Genetic Algorithms and their Application, pp. 14–21. Hillsdale, New Jersey (1987)
6. Bies, R., Muldoon, M., Pollock, B., Manuck, S., Smith, G., Sale, M.: A genetic algorithm-based, hybrid machine learning approach to model selection. J. Pharmacokinet. Pharmacodyn. **33**, 196–221 (2006)
7. Chipperfield A., Fleming, P.J., Pohlheim, H., Fonseca, C.M.: Genetic Algorithm Toolbox for Use with MATLAB. Technical Report No. 512, Department of Automatic Control and Systems Engineering, University of Sheffield (1994)
8. Crisan, C., Mühlenbein, H.: The Breeder Genetic Algorithm for Frequency Assignment. Lecture Notes in Computer Science, vol. 1498, p. 897 (1998)
9. Davis, L.: Handbook of Genetic Algorithms, Van Nostrand Reinhold, New York (1991)
10. Fogel, D.: Evolutionary Computation: Toward a New Philosophy of Machine Intelligence, 3rd edn. IEEE Press, Hoboken (2006)
11. Goldberg, D.E.: Genetic Algorithms in Search, Optimization and Machine Learning. Addison Wesley, Reading (1989)
12. Holland, J.: Adaptation in Natural and Artificial Systems. University of Michigan Press, Ann Arbor (1975)

13. Houck, C., Joines, J., Kay, M.: A genetic algorithm for function optimization: a Matlab implementation. http://citeseerx.ist.psu.edu/viewdoc/download?doi=10.1.1.22.4413&rep=rep1&type=pdf. Accessed 22 Oct 2015
14. http://www.doc.ic.ac.uk/~nd/surprise_96/journal/vol1/hmw/article1.html. Accessed 22 Oct 2015
15. Krawczak, M.: A novel modeling methodology: Generalized nets. In: Lecture Notes in Computer Science, vol. 4029, p. 1160 (2006)
16. Larranaga, P., Karshenas, H., Bielza, C., Santana, R.: A review on evolutionary algorithms in Bayesian network learning and inference tasks, Inf. Sci. **233**, 109–125 (2013)
17. Malhotra, R., Singh, N., Singh, Y.: Genetic algorithms: concepts, design for optimization of process controllers. Comput. Inf. Sci. **4**(2), 39–54 (2011)
18. MathWorks: Genetic Algorithm Toolbox User's Guide for MATLAB
19. Michalewicz, Z.: Genetic Algorithms + Data Structures = Evolution Programs, 2nd edn. Springer, Berlin (1994)
20. Montiel, O., Castillo, O., Melin, P., Sepulveda, R.: Application of a Breeder genetic algorithm for filter optimization. Nat. Comput. Int. J. Arch. **4**(1), 11–37 (2005)
21. Montiel, O., Castillo, O., Sepulveda, R., Melin, P.: Application of a Breeder genetic algorithm for finite impulse filter optimization. Inf. Sci. **161**, 139–158 (2004)
22. Mühlenbein, H., Schlierkamp-Voosen, D.: Predictive model for Breeder genetic algorithm. Evol. Comput. **1**, 25–49 (1993)
23. Pencheva, T., Atanassov, K., Shannon, A.: Generalized net model of selection function choice in genetic algorithms. In: Recent Advances in Fuzzy Sets, Intuitionistic Fuzzy Sets, Generalized Nets and Related Topics, Applications, vol. II, pp. 193–201. Systems Research Institute, Polish Academy of Sciences, Warsaw (2011)
24. Pencheva, T., Atanassov, K.: Generalized Net Model of Simple Genetic Algorithms Modifications. In: Kacprzyk, J., Krawczak, M., Szmidt, E. (eds.) Issues in Intuitionistic Fuzzy Sets and Generalized Nets, vol. 10, pp. 97–106. Wydawnictwo WSISiZ, Warszawa (2013)
25. Pencheva T., Roeva, O., Shannon, A.: Generalized net models of crossover operator of genetic algorithm. In: Proceedings of Ninth International Workshop on Generalized Nets, pp. 64–70. Sofia, 04 July 4 2008
26. Riolo, R., McConaghy, T., Vladislavleva, E. (eds.): Genetic Programming Theory and Practice VIII (Genetic and Evolutionary Computation), 276 p. Springer (2011)
27. Roeva, O. (ed.): Real-World Application of Genetic Algorithms. In Tech, Rijeka (2012)
28. Roeva, O., Atanassov, K., Shannon, A.: Generalized net for evaluation of the genetic algorithm fitness function. In: Proceedings of the Eighth International Workshop on Generalized Nets, pp. 48–55. Sofia, 26 June 2007
29. Roeva, O., Pencheva, T., Shannon, A., Atanassov, K.: Generalized Nets in Artificial Intelligence, Generalized Nets and Genetic Algorithms, vol. 7. Prof. M. Drinov Academic Publishing House, Sofia (2013)
30. Roeva, O., Pencheva, T.: Generalized net model of a multi-population genetic algorithm. In: Kacprzyk, J., Krawczak, M., Szmidt, E. (eds.) Issues in Intuitionistic Fuzzy Sets and Generalized Nets, vol. 8, pp. 91–101. Wydawnictwo WSISiZ, Warszawa (2010)
31. Roeva, O., Pencheva, T., Atanassov, K.: Generalized net of a genetic algorithm with intuitionistic fuzzy selection operator. In: New Developments in Fuzzy Sets, Intuitionistic Fuzzy Sets, Generalized Nets and Related Topics, Foundations, vol. I, pp. 167–178. Systems Research Institute, Polish Academy of Sciences, Warsaw (2012)
32. Tasan, S.O., Tunali, S.: A review of the current applications of genetic algorithms in assembly line balancing. J. Intell. Manuf. **19**(1), 49–69 (2008)

Generalized Nets in Medicine: An Example of Telemedicine for People with Diabetes

Maria Stefanova-Pavlova, Velin Andonov, Violeta Tasseva, Antoaneta Gateva and Elissaveta Stefanova

Abstract In the present paper, an overview of the Generalized Nets (GNs) models in medicine and telecare/telehealth is given. The apparatus of GNs has been used in the modelling of physiological processes, diagnostics of diseases, organisational and administrative processes in hospitals. Recently, in a series of papers, GNs have been used to model telecare/telehealth services. On the basis of these models, a GN model of telemedicine for patients with diabetes is proposed. The sensors included in the model are blood pressure monitor, weight scale, pulse oximeter and blood glucose monitor. Smart filtering of false positive alarm messages is included which reduces the number of events for which the health care person has to take a decision. The GN model can be used to develop a decision support tool for telemedicine for people with diabetes.

M. Stefanova-Pavlova (✉)
Center for Innovation and Technology Transfer—Global,
Jaroslav Veshin Str., Bloc 32/B, 1408 Sofia, Bulgaria
e-mail: stefanova-pavlova@citt-global.net

V. Andonov · V. Tasseva
Institute of Biophysics and Biomedical Engineering, Bulgarian Academy of Sciences,
Acad. G. Bonchev Str., Bl. 105, 1113 Sofia, Bulgaria
e-mail: velin_andonov@yahoo.com

V. Tasseva
e-mail: vtaseva@gmail.com

A. Gateva
Clinic of Endocrinology, University Hospital Alexnadrovska, Medical University—Sofia,
1 Bul. G. Sofijski, Sofia, Bulgaria
e-mail: tony_gateva@yahoo.com

E. Stefanova
Clinic of Endocrinology, Diabetes and Genetics, University Children's Hospital, Medical
University—Sofia, 11 Ivan Geshov Str., 1606 Sofia, Bulgaria
e-mail: elissaveta.stephanova@abv.bg

P. Angelov and S. Sotirov (eds.), *Imprecision and Uncertainty in Information Representation and Processing*, Studies in Fuzziness and Soft Computing 332,
DOI 10.1007/978-3-319-26302-1_20

1 Introduction

Generalized Nets (GNs) are powerful tool for Discrete Event Simulation (DES) and parallel processes flow representation. For the formal definition of GN see the Appendix. The apparatus of GNs is equally well suited for modelling simple systems and for modelling large, complex systems. DES is a method used to model real world systems able to be decomposed into a set of logically separate processes autonomously progressing through time. A major strength of discrete event simulation is its ability to model random events and to predict the effects of the complex interactions between these events. GN-models could be used as a quick method of analyzing and solving complex problems. This reduces the risk and uncertainty associated with important decision making, and increases confidence by supporting the decision with forecasted data. Up to now generalized nets are applied to healthcare delivery systems, general and internal medicine. Many GN models were built which represent various types of organizational and patient workflows, diseases, symptoms and treatments, organs or states of human body. This is possible due to the existence of a proof that every dynamical system and every collection of dynamical systems can be described by a GN (see [7]). GNs have been used as a modelling tool in expert systems and artificial intelligence; computer science; economics, industry and transport; medicine (see [1, 6, 17, 28, 80–82]).

As it is shown in [8, 29, 86, 101, 102, 113] the GN-models in medicine can be used for:

- simulation of real processes with educational aims;
- control of the corresponding hospital processes in real time;
- prognosis of the actual processes in hospital for the purposes of the hospital administration. These models can also help:
- specialists in studying the logic of the processes related to diagnoses;
- medical students and new specialists in acquiring knowledge and diagnostic skills;
- lecturers in medical students examinations with real-time simulations;
- administrative personnel in taking decisions related with planning, management, organization and allocation of the available resources (materials, specialized apparatuses, personnel) and scheduling of the medical specialists.

Up to now, processes in medicine in several directions are modelled with the apparatus of GNs.

2 Modelling of Physiological Processes

Living organisms are featured by a variety of processes flowing in parallel. Some of them, taken separately, are already described by specific mathematical tools, mostly by means of systems theory. These mathematical tools however do not reflect adequately the parallel flow of the processes. In [86], for example, are considered some parallel endocrine processes. These processes are related with production of insulin by the pancreas, introduction of artificial insulin, processing of both types of insulin in the organism, as well as with possible new conditions of production of insulin by the pancreas.

Some other models of physiological processes are presented in [11, 31, 37–39, 41, 52, 71, 72, 80–84, 95–97, 114, 116–118, 123, 124, 127, 134].

3 GN Interpretations of Informational Models of Diseases and Human Body Systems

Modern ideas in every new science with new principles and laws manifestly modify the medical science as well. The attempt to peace together medical informatics and clinical medicine meets some difficulties but as a whole has a lot of benefits.

As it is written in [94] the category "information" in medicine is used for the description of the activities of sensory organs, in genetics, and partly in physiology, pathophysiology and biochemistry. It is found rarely in endocrinology, neurology and psychiatry (where there is a notion of informational disease, illustrating a neurosis due to sensory overloading). In clinical works, the information exchange of the organism is of a very narrow practical significance. Usually, the description of diseases contains morphological changes, dysfunction, etiopathogenesis, metabolism and energy exchange, and relation of symptoms. The fundamental medical sciences seem to support such an approach. The development of GN interpretations of informational model brings medical informatics closer to clinical disciplines and hence raises their effectiveness.

The informational model of diseases proposed by Ivan Dimitrov [52] is based on the following:

1. Cells and organs of the living organism represent a common mechanism, in which they function jointly and in mutual coordination. To achieve this, the elements of the organism communicate, i.e.; exchange of information is carried out among them.
2. In disease, the exchange of information is modified.
3. In the description of diseases, along with metabolism and the exchange of energy, the exchange of information should be included as well.

The development of such an informational model could bring medical informatics closer to clinical disciplines and hence raise their effectiveness. On the other hand, this could enable clinical medicine to improve its models, and perhaps the treatment and prevention of diseases. Undoubtedly, diagnostics can be supported in a decisive way through a profound investigation of information exchange disorders in an organism. The generalized net interpretations of these models are shown in [94]. The basic systems of the human body are:

- Central neurological system;
- Cardiovascular system;
- Respiratory system;
- Gastrointestinal system;
- Endocrine system;
- Hematopoetic system;
- Musculoskeletal system;
- Renal and urological system;
- Reproductive system.

When building a GN model of the human body (see [32]) each of the listed systems can be represented by a transition. These transitions have the simplest form: one input, one output and one input-output place. The last one depicts the interior processes of the respective organ/ system, and contains a token which will have as a current characteristic the status of the corresponding organ/ system.

For completeness of the model input and output places which represent separate organs related to the inputs and outputs of the human body (derma, nose, mouth and tongue, eyes, ears) are added. The tokens which correspond to the exterior factors move through the net and have as characteristics type of the effect and its parameters (power, continuity, volume, etc.).

4 Diagnostics of Diseases

So far GN models are built for diagnostics in nephrology and adult and child neurology. The GN-models in neurology use as a basis some previously made models [53, 138] of the processes in this area.

Decision graphs for diagnostics of isolated (136 in total) neurological diseases are described in [138]. As a whole, the charts in [138] have the form of binary graphs—an initial node (representing the arrival of a patient with a neurological symptom) with two successors representing the alternatives (the patient has/does not have a given symptom). Each arc leads to another node again giving rise to successors corresponding to the presence or absence of symptoms, etc. The graphs make it possible to trace the individual steps of each of these processes. These charts can be used for training of students or professionals. In this sense, the role of the graphs is similar to that of expert systems in subject-oriented areas of medicine [24].

The so made models are used as a basis for construction of GN models, describing the diagnostic processes in adult and child neurology [10, 12–20, 34–36, 44–51, 64–66, 120–122, 125].

At the beginning of the medical diagnostic reasoning or decision making process the practitioner must recognize if a sign or a symptom is significant. As a first step, a detailed patients personal and family history, and complete physical examination are of paramount importance, both in determining their medical significance and in directing evaluation.

Next, the physician begins to sort the data, keeping some pieces of information and ignoring others. The practitioner must first cluster or link some or all of the collected signs and symptoms, and determine any emerging patterns, meaningful groups, and formulate hypotheses. This phase is often referred to as hypothesis formulation (initial, preliminary diagnosis). The formulation of hypotheses or tentative conclusions helps focus further data collection efforts on a manageable group of possibilities.

During the next stage of diagnostic reasoning, the physician focuses on gathering data (laboratory tests, X-ray pictures, and so on) to support or reject the previously generated hypotheses. Once the physician is satisfied that all reasonable explanations for the initial set of signs and symptoms have been thoroughly investigated, each hypothesis must be evaluated in the light of the new evidence that has been collected and a final diagnosis or conclusion reached.

Depending on the course of the final phase of decision making the practitioner determines which explanation has the most supporting data and chooses this hypothesis as the diagnosis. In some cases, however, the clinician can only eliminate hypotheses until only the one with the highest probability remains. A global GN-model for the purpose of diagnosing a definite disease entity is fully described in [17].

Using binary graphs for medical information representation has some advantages. This form of description is very easy for understanding and close to the medical specialists thinking. GNs offer convenience when the specialist's answer is not definitely "yes" or "no". They adequately represent the parallel flow of the processes. In this sense GN-models give possibility for simultaneously examination of several decision paths. This leads to minimization of the times for examinations and decision making. The tokens collect and store in their characteristics all the data that is related to the corresponding examination or patient status. In every moment the whole necessary information is stored in the generalized net and in case of need could easily be obtained. Other models of diseases diagnostic are presented in [21–23, 25–27, 33, 43, 56–62, 64–66, 74–79, 85, 87–90, 92, 93, 98, 103–112, 119, 126, 135, 136].

5 Organisational and Administrative Processes in Hospital Institutions

Modelling such processes with GN offers:

- understanding of the patient flow;
- optimization of the duration of stay of patients in hospital institutions in order to avoid bad-blockage and queuing;
- planning/reassigning of the tasks and activities of medical staff (especially during emergency events epidemics, disasters etc.);
- tracing and monitoring of the treatment of each patient, his/her redirection to another hospital institution;
- detecting bottlenecks in the workflow of the health care structure;
- help in resource planning and allocation, etc.

Medicine is a profession concerned with preserving and improving patients lives. The considerations are thus obvious: the approach has to be patient—centred, and not merely for the convenience of the healthcare providers or administrators. In general, it is difficult to deny the benefits of a more efficient information management system. However, at a less macroscopic level, such generic benefits cannot be assumed for all healthcare delivery systems. Confounding factors such as technical competency of staff, acceptance and adoption by doctors and patients, and intrinsic design—related features can impair rather than facilitate medical care and doctor—patient relationship in some settings.

The automated planning of the resources (necessary equipment and specialists), medical staff working schedules, patients' reception, vastly facilitate the health care units' administration. The store and the processing of the patients' personal data and examination results in data bases and expert systems aids the decision making. The appearance of the electronic health care improves the people informativeness in a low price.

The construction of models of the parallel processes which flow in medicine allows their full and correct understanding. This leads to minimization of the waiting times and the times for decision making. The simulation of the made models with real experimental data allows status evaluations, prognosis for critical moments and situations, planning of the medical personnel, material equipment and specialized apparatuses allocation. Examination of different what-if analyses of real situations give possibility for finding of the healthcare system's bottlenecks.

The GN-models of the organizational and administrative processes in hospital institutions are in several directions:

1. patient flow modelling;
2. information flow modelling;
3. modelling of the resources allocation.

Some of these models represent the processes in a specific health care unit—for example patient flow in a clinic. Thus the GN allows accumulation of data, which concern particular patient—used materials and specialized apparatuses, made examinations, engaged medical stuff in the rehabilitation process and so on, all traced into the time. Using the results of the GNs work it is possible to make conclusions about the dynamics of the process, number of the treated patients, case history, economical assessment of outlays. These results could be used in the process of decision making and further diagnostics.

5.1 Patient Flow Modelling

Planning of hospital resources has always been a matter of great importance. One of the major elements in improving efficiency in the delivery of health care services is optimizing the patient flow and length of stay. Modelling patient flow in health care systems is considered to be vital in understanding the operation of the system and may therefore prove to be useful in improving the functionality of the health care system. Better understanding of system operation is needed to predict and support health care activities in every medical clinic. The understanding and modelling of patient flow can offer information to health care providers about the patients disease progression or recovery status. The effective resource allocation and capacity planning is dependant on patient flow because it is equivalent to the need of health care services. If there is an understanding of patient flows, this knowledge can be used to improve and optimize the activities of health care system. And the resource planning, scheduling and utilization optimization can affect the quality of health care services and patient flow.

In [54] it is proved, with no doubt, the big need of simulation models of health care institutions. The aim of this paper is to asses the benefits of a model that examines the impact of bed blockage, occupancy and emptiness on patient flow in a geriatric inpatient unit. Departments of geriatric medicine provide an acute, rehabilitative and long-stay service for older people with complex medical and social needs. Simulation modelling gives an opportunity for predicting the situation of bed crisis in some months of the year or for epidemics [137]. That is why when there is insufficient amount of beds available to admit ill people in hospitals, queues are formed. This may be fatal for some patients. The movement of patients through hospitals can be seen to occur in streams. Wards such as acute, rehabilitation and long stay are dependent on the dimensions of time and performance. But the availability of hospital beds for admission depends on patients leaving the system. In [137] it is demonstrated using simulation model that the cause of the crisis is a breakdown in the discharge of dependent patients from the medium-stay stream. Clinically, bed blocking occurs when patients are kept waiting in one ward or hospital until free beds are available in a more suitable ward or hospital. For instance, rehabilitation or long—stay patients can be kept waiting in the acute wards until beds become available elsewhere, effectively blocking the availability of the beds for other patients.

A patient flow begins when a patient needs health consultation and goes to his/her personal doctor, who gives him/her diagnose, or when the patient is admitted to a health care system. Similarly, when the patient is discharged from a health care unit this is the patient flow exit. Between these two points there is a set of conditions, activities, services, or locations that the patient may pass. Within these points, the patient requires a variety of health care resources (e.g., beds, examining rooms, physicians, nurses, medical apparatuses and procedures). That is why the patient flow can be depicted as a network. The basic network elements represent the patient statuses in health care system (nodes) and indicate the flow between nodes (arcs). An important characteristic of the patient flow is its random nature [55]. For a given health care service, not all of the elements may be applicable to all patients. And the time that patients spend at each node and in the whole network also contains a degree of randomness. Once patients state is estimated it is possible to assign what resources (for instance medical staff and equipment) will be required. Of course health care clinic has capacity and resource limitations, so queuing for services occurs. Queuing characteristics, such as time in the system and traffic intensity, correspond to the patient flow characteristic.

The study of problems with patient flow distribution has priority importance because it is one of the major elements in improving efficiency in the delivery of health care services. GNs are quite promising for representing patient flow and cases history. Understanding of patient flow is needed to support health care activities in every medical clinic. The effective resource allocation and capacity planning are dependant on the patient flow because it is equivalent to the need of health care services. The patient flow optimization can affect quality of health care services and can have positive effects on patient and clinical staff satisfaction.

Some other GN-models of patient flow are presented in [128–132] .

5.2 *Information Flow Modelling*

Although confidentiality issues have long existed before the arrival of the computer and the Internet, the use of IT that is capable of transmitting large amounts of data in very short time intervals, and of bypassing the conventional physical barriers and safeguards, certainly heightens public anxiety. Based on the medical needs of an individual, several medical professionals (of different specialties) may have been visited by him for medical care. Each of the medical professionals visited keeps information about their patients. Similarly, hospitals keep all the records for patients that require hospitalization for treatment. In addition, the patients fill their prescriptions at different pharmacies. All the bits and pieces of information that are scattered at various places may be necessary for providing effective healthcare to an individual. There are several ways of keeping this information handy and ready for use when needed. It is not practical for every individual to carry this information with him in paper form all the time. Also, legal aspects associated with the medical records require that the medical information should not be altered. Therefore, all mechanisms used for gathering, disseminating, or transporting medical information must adhere to all the legal requirements. Advances in information technology have provided many options for individuals to have their medical history available whenever it is needed.

The information systems and computer networks allow the information of the patient to be accessible for short periods of time even on long distances and they facilitate the putting of diagnose. For that purpose the development of secure and fast connection is required. Personal computers in a specific healthcare unit have to be connected in a local network and connection between healthcare units is also needed. Building a network in a healthcare unit can improve integration of clinical, financial, and administrative data for the various stakeholders, improve patient outcomes, increase administrative efficiency and can reduce the likelihood of medical errors and lower overall costs.

The purpose of the modelling of the information flows is gaining knowledge about the happening inside the health care system. The need for concise and accurate capture or representation of the patient flow and length of stay information assets are important for the delivery of effective and in time healthcare services. Modelling of the processes in the healthcare domain offers the opportunity to detect bottlenecks and to suggest effective changes in case of critical condition. In this way the knowledge and the experience gained from experts could become available to all stakeholders concerned in the quality and effectiveness of healthcare services and contribute towards more effective resource allocation and use.

A GN-model for representation of links and interactions between particular wards in a hospital as well as separate sections in a single ward are described in [99, 100]. The GN-model of the information net is needed, because as the investigations shows it is important that health policy makers and doctors have accurate information about a safety net and the data considering their patients. With such an information system the whole available data for the diseases, symptoms, case history of the patients—their disease progression or recovery status as well as test results, will be easily accessible for the specialists who work in the health care sphere. It is clear that this information play a critically important role in decision making and making primary care available. Information GNs could be a software tool for modelling and simulating real time parallel processes witch runs in a single health care unit as well as in a whole hospital system. The purpose is to develop information technology with which we could easily represent and simulate complex health care systems and apply to medicine and pharmacy.

There exist different ways for the construction of a network architecture in the frames of particular healthcare unit and between the separate medical institutions. Having in mind the area which is to be covered, the optimal trade off between cost and quality has to be derived. This includes decision making for: number and parameters of the servers, their disposal, the connections between them, the communication protocol, the time intervals for updating the information, etc. On that basis are developed two GN-models that represent a process of information exchange (inserting and requesting data) between several healthcare units [99] and within particular medical centre [100]. In [99] there is one global server-repository as in the case of data warehouses, which will save all the data for the patients. Every unit works on its own server-source, which communicates with the central one and sends the data to it.

5.3 Modelling of the Resources Allocation

Health care technology is subject to constant improvement. This is often accompanied by complex interactions between result, efficiency, staff training, equipment maintenance, patient risk and cost of treatment. Implementation of novelties is a complex task requiring the evaluation of the specifics and the benefits and risks related to the most part of medical technology.

Modelling of the processes which flow in the health care system could be used for planning of the resources allocation (buildings, equipment and specialists), as well as for determination of the bottlenecks in the system. Beds, personnel work load and available apparatuses determine the capacity of a given unit. Built model of the system could answer to the question if there is a free position for a new patient or no.

A GN-model of Intensive Care Unit (ICU) workflow is presented in [132]. The goal of the workflow organization in an ICU is to assign a medical team and specialized equipment to each patient. At any moment the department is in a certain state with respect to available beds, patients condition, available staff and equipment. This information is stored in the unit's databases. The medical staff database contains information on the qualifications and work load of the staff. The hierarchy is respected when making a decision in the unit the decision is made by the highest-ranking authority present at the moment. For example, if the head of the unit is absent, the responsibility is transferred to the department head, while if he is in turn not available to the on duty or in charge physician. Despite this strict hierarchy, there are cases when the decision could only be made by the department or unit head. Dedicated databases are used to store the protocols with recommendations for specific actions, as well as past treatments and conditions of the patients.

The description so far reflects the limitation on staff workload only that is, bed or equipment availability were assumed sufficient. However, the amount of beds is actually always limited. It may be necessary to discharge a patient in an unstable status with high risk due to the need to accept another. A decision of this type could only be made by the head of the unit or the department. In most cases, equipment is also limited. This influences the work flow in the department and may lead to difficult decisions. The GN-models describing the process of resource allocation could be used to determine the optimum level of staff and equipment [30, 40, 63, 69, 91, 115, 130, 133].

6 Modelling of Telecare/Telehealth

Telehealth is the remote monitoring of patients' vital signs and symptoms in their own home—proven to enhance the quality of life and clinical outcomes for people with long-term conditions. It also helps people understand and manage their health, enabling them to stay out of hospital and enjoy life with their family and friends.

The evaluation of telehealth/telecare solutions of UK Department of Health's Whole System Demonstrator Program shows the following results:

Commissioning benefits:

- 45 % reduction in mortality rates;
- 20 % reduction in emergency admissions;
- 15 % reduction in A&E visits;
- 14 % reduction in elective admissions;
- 14 % reduction in bed days;
- 8 % reduction in tariff costs.

Clinical benefits:

- Encourages self-management;
- Enables early identification of exacerbations;

- Aids medication compliance;
- Identifies trends over time to aid proactive care planning;
- Helps clinicians make more informed medication management decisions;
- Supports efficient caseload management (Tunstall Healthcare, see [143]).

GNs have been used as a tool for modeling processes in telecare and telemedicine [2–4, 70]. A GN model of telecare is presented in [3]. It can be used as a decision support tool to enhance the work of the specialists in the telehealth center. Smart filtering of alarm messages is proposed in [4]. In the model developed in [70] traces the logical stages of the final part of the process of communication between the sensors connecting mobile adult patients and the staff of the respective hospital unit. The developed model can be used for simulation of the processes of decision making of the appropriate specialists, who must either visit the respective adult patient or transport him/her to the hospital unit. The model permits simulation of different scenarios e.g. the situation, in which many patients simultaneously require medical assistance. Finally, the model in [2] is an example of telemedicine based on body temperature sensors.

7 GN Model of Telemedicine for People with Diabetes

Diabetes mellitus (DM) includes a group of metabolic diseases, characterized by high blood glucose, either because the pancreas does not produce enough insulin (absolute insulin deficiency), or because cells do not respond to the insulin that is secreted (insulin resistance) or both. There are two main types of DM:

Type 1 DM results from the body's failure to produce insulin because of autoimmune destruction of insulin producing cells in the pancreas and requires insulin injections at least 4 times daily or a use of insulin pump. Hypoglycemia and weight gain are the most common adverse effects of insulin therapy.

Type 2 DM results from insulin resistance, a condition in which cells fail to use insulin properly because of metabolic disturbances, most frequently caused by obesity. In the early disease stages, insulin production is normal or increased in absolute terms, but disproportionately low for the degree of insulin sensitivity, which is typically reduced. However the ability of the pancreatic β-cells to release insulin in phase with rising glycemia, are profoundly compromised [142]. In the beginning type 2 diabetes is treated with oral medications which either decrease insulin resistance or increase insulin secretion. Treatment choices depend on many factors, most important of which are body weight and concomitant diseases. Some of the oral antidiabetic drugs can also induce hypoglycemia. At the end stage of the disease when absolute insulin deficiency is developed, patients usually need insulin injections 1–4 times daily. Weight reduction improves glycemic control and other cardiovascular risk factors in patients with type 2 diabetes. Modest weight loss (5–10 %) contributes meaningfully to achieving improved glucose control. Accordingly, establishing a goal of weight reduction, or at least weight maintenance, is recommended

[141]. On the other hand most of the drugs used to treat diabetes, including insulin can lead to significant weight gain. This warrants individualisation of treatment choices, especially in patients with obesity.

Ideally, the principle of diabetes treatment is the achievement of as normal a glycemic profile as possible without unacceptable weight gain or hypoglycemia. The American Diabetes Assotiation Standards of Medical Care in Diabetes recommends lowering HbA 1c to 7.0 % in most patients to reduce the incidence of microvascular disease. This can be achieved with a mean plasma glucose of 8.3–8.9 mmol/l (150–160 mg/dL); ideally, fasting and premeal glucose should be maintained at <7.2 mmol/l (<130 mg/dL) and the postprandial glucose at <10 mmol/l (<180 mg/dL) [139]. Plasma glucose <3.9 mmol/l in patients with diabetes is generally considered hypoglycemia. Patients on insulin therapy and oral agents that can cause hypoglycemia should be instructed in techniques for self-monitoring of blood glucose. Initially, blood glucose levels should be checked at least four times a day in patients taking multiple insulin injections. Generally, these measurements are taken before each meal and at bedtime. In addition, patients should be taught to check their blood glucose level whenever they develop symptoms that could represent a hypoglycemic episode [140].

Acute complications of diabetes.

Hypoglycemia is a life-threatening acute complication of antidiabetic treatment. Clinical hypoglycemia is, by definition, a plasma glucose concentration low enough to cause symptoms or signs, including impairment of brain function. The glycemic thresholds for symptoms and signs of hypoglycemia are dynamic; for example, they shift to lower plasma glucose concentrations in patients with recurrent hypoglycemia and to higher concentrations in those with poorly controlled diabetes. All of the manifestations of hypoglycemia are rapidly relieved by glucose administration. Patients with symptoms of hypoglycemia who are conscious and able to swallow should eat or drink orange juice, glucose tablets, or any sugar-containing beverage or food. In patients that are unconscious the preferred treatment is 50 mL of 50 % glucose solution given rapidly over 3–5 min intravenously. If trained personnel are not available to administer intravenous glucose, the treatment of choice is for a family member or friend to administer 1 mg of glucagon intramuscularly, which usually restores the patient to consciousness within 10–15 min [141]. If a patient develops severe hypoglycemia after use of long-acting antidiabetic medications, that induce insulin secretion, he should be observed in hospital for at least 24 h to prevent recurrent hypoglycemia.

Diabetic ketoacidosis and diabetic hyperosmolar coma are acute complications of diabetes, that require hospital treatment. As opposed to the acute onset of hypoglycemic coma, diabetic ketoacidosis is usually preceded by a day or more of polyuria and polydipsia associated with marked fatigue, nausea, and vomiting. Eventually, mental stupor ensues and can progress to frank coma. High blood glucose (usually >15–18 mmol/l), ketonuria, ketonemia, low arterial blood pH, and low plasma bicarbonate (5–15 mEq/L) are typical laboratory findings in diabetic

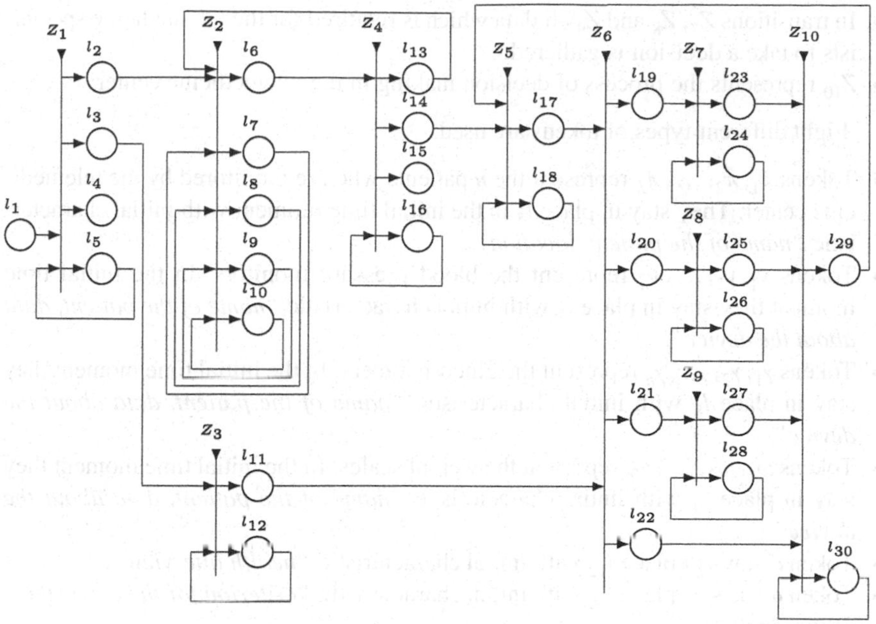

Fig. 1 GN model of telemedicine for people with diabetes

ketoacidosis. The onset of the hyperglycemic, hyperosmolar, nonketotic state may be preceded for days or even weeks by symptoms of weakness, polyuria, and polydipsia. A history of reduced fluid intake is common, whether due to inappropriate absence of thirst, gastrointestinal upset, or, in the case of elderly or bedridden patients, lack of access to water. In diabetic hyperosmolar state there are no ketones in the urine but the blood glucose is very high (>25−30 mmol/l).

The telehealth package for people living with diabetes consists of a blood pressure monitor, weight scale, pulse oximeter and blood glucose monitor. Intelligent health interview is also a necessity (LifeLink Telehealth, see [144]).

Telemedicine has been widely used to bring healthcare to patients living in distant locations. In [42] it is demonstrated how modem technologies can be used as a tool for providing telemedicine for people with diabetes. This approach has been proven to be cost—effective [67, 68]. On the basis of the models presented in [2–4, 70], here we propose a GN model of telemedicine for people with diabetes. The GN consists of ten transitions (see Fig. 1). They have the following meaning:

- Z_1 represents the patients.
- Z_2 represents the collecting of data from the sensors.
- Z_3 represents the process of taking a health interview from the patient.
- In Z_4 the signals from the sensors are checked for correctness.
- Z_5 represents the database with the patients' history and the decisions taken by the specialists in the telemedicine center.
- Z_6 represents the differentiation of the signals depending on the glucose level.

- In transitions Z_7, Z_8 and Z_9 all data which is required for the telemedicine specialists to take a decision is gathered.
- Z_{10} represents the process of decision making in the telemedicine center.

Eight different types of tokens are used.

- Tokens $\pi_1, \pi_2, \ldots, \pi_n$ represent the n patients who are monitored by the telemedicine center. They stay in place l_5 in the initial time moment with initial characteristic *"name of the patient, location"*.
- Tokens $v_1, v_2, \ldots, v_n$ represent the blood pressure monitors. In the initial time moment they stay in place l_7 with initial characteristic *"name of the patient, data about the device"*.
- Tokens $\gamma_1, \gamma_2, \ldots, \gamma_n$ represent the glucose meters. In the initial time moment they stay in place l_8 with initial characteristic *"name of the patient, data about the device"*.
- Tokens $\omega_1, \omega_2, \ldots, \omega_n$ represent the weight scales. In the initial time moment they stay in place l_9 with initial characteristic *"name of the patient, data about the device"*.
- Token ζ stays in place l_{12} with initial characteristic *"health interview"*.
- Token α stays in place l_{16} with initial characteristic *"criterion for the correctness of the signals"*.
- Token β stays in place l_{18} with initial characteristic *"database with data about the patients and the decisions taken by the telemedicine person"*.
- Tokens $\delta_1, \delta_2, \ldots, \delta_k$ represent the telemedicine specialists (telemedicine nurses) who make decisions in the telemedicine center. They stay in place l_{30} in the initial time moment with characteristic *"name, decisions taken, duration of the shift"*.

During the functioning of the net new π-tokens may enter the net through place l_1. These new tokens represent the new patients who are included in the model. Also, some of the π-tokens in place l_5 may leave the net through place l_4 which reflects the fact that the patients corresponding to these tokens are no longer monitored by the system.

The proposed GN is a reduced one, i.e. not all of the components from the definition of GN are present (see [5]). What follows is a description of the transitions of the net.

$$Z_1 = \langle \{l_1, l_5\}, \{l_2, l_3, l_4, l_5\}, r_1, \Box_1 \rangle,$$

where

$$r_1 = \begin{array}{c|cccc} & l_2 & l_3 & l_4 & l_5 \\ \hline l_1 & true & false & false & true \\ l_5 & W_{2,5} & W_{5,3} & W_{5,4} & W_{5,5} \end{array}$$

and

$W_{5,2}$ = *"sensor data about the current patient must be sent to the telemedicine center"*;

$W_{5,3}$ = *"health interview with the current patient has to be conducted"*;

$W_{5,4}$= *"the current patient must leave the system"*;
$W_{5,5} = \neg W_{5,4}$.
Here and below $\neg W_{i,j}$ is the negation of the predicate $W_{i,j}$.
$\square_1 = \vee(l_1, l_5)$.

When the truth value of the predicate $W_{5,2}$ becomes *true* the current π-token in place l_5 splits into two identical tokens—the original that remains in l_5 and a new one π' which enters l_2 without new characteristic. The new π-token in l_1 splits into two identical tokens one of which enters place l_5 with characteristic *"name of the patient, location"*. The other one enters place l_2 with characteristic *"name of the patient, data about the sensors"*. When the truth value of $W_{5,3}$ is true the π-token splits into two identical tokens—the original which stays in l_5 and a new π'' which enters l_3 without new characteristic.

$$Z_2 = \langle \{l_2, l_7, l_8, l_9, l_{10}, l_{13}\}, \{l_6, l_7, l_8, l_9, l_{10}\}, r_2, \square_2 \rangle,$$

where

$$r_2 = \begin{array}{c|ccccc} & l_6 & l_7 & l_8 & l_9 & l_{10} \\ \hline l_2 & false & W_{2,7} & W_{2,8} & W_{2,9} & W_{2,10} \\ l_7 & W_{7,6} & true & false & false & false \\ l_8 & W_{8,6} & false & true & false & false \\ l_9 & W_{9,6} & false & false & true & false \\ l_{10} & W_{10,6} & false & false & false & W_{10,10} \\ l_{13} & false & false & false & false & true \end{array}$$

and

$W_{2,7}$= *"the current π' -token represents a new patient"*;
$W_{2,8} = W_{2,9} = W_{2,7}$;
$W_{2,10} = \neg W_{2,7}$;
$W_{7,6}$= *"the blood pressure of the current patient has been measured"*;
$W_{8,6}$= *"for the current γ-token there is corresponding π'-token in place l_{10}"* & *"the glucose level of the patient corresponding to the current γ-token has been measured"*;
$W_{9,6}$= *"the weight of the patient corresponding to the current ω-token has been measured"*;
$W_{10,6}$= *"the glucose level of the current patient has been measured"*;
$W_{10,10} = \neg W_{10,6}$.
$\square_2 = \vee(l_2, l_7, l_8, l_9, l_{10}, l_{13})$.

If the current π'-token in place l_2 represents a new patient, then it splits into three tokens v_i, γ_i, ω_i which enter places l_7, l_8 and l_9 respectively with characteristic *"name of the patient, data about the respective sensor"*.

When the glucose level the patient has been measured the current γ-token in l_8 splits into two tokens—the original and a new identical one γ' which enters place l_6 where it obtains the characteristic *"glucose level of the patient"*. When the glucose level of the patient corresponding to the π'-token in l_{10} has been measured, this same π'-token enters place l_6 where it merges with the γ-token into a new π'-token.

When the truth value of the predicate $W_{7,6}$ becomes *true* the current ν-token in l_7 splits into two tokens—the original which remains in l_7 and new identical one ν' which enters l_6 with characteristic *"blood pressure of the corresponding patient"*.

When the truth value of the predicate $W_{9,6}$ becomes *true* the current ω-token splits into two tokens—the original which remains in l_9 and new identical one ω' which enters place l_6 with characteristic *"weight of the corresponding patient"*.

$$Z_3 = \langle \{l_3, l_{12}\}, \{l_{11}, l_{12}\}, r_3, \Box_3 \rangle,$$

where

$$r_3 = \begin{array}{c|cc} & l_{11} & l_{12} \\ \hline l_3 & true & false \\ l_{12} & false & true \end{array}$$

$\Box_3 = \wedge(l_3, l_{12})$.

In l_{11} the π''-tokens obtain the characteristic *"answers to the interview questions"*.

$$Z_4 = \langle \{l_6, l_{16}\}, \{l_{13}, l_{14}, l_{15}, l_{16}\}, r_4, \Box_4 \rangle,$$

where

$$r_4 = \begin{array}{c|cccc} & l_{13} & l_{14} & l_{15} & l_{16} \\ \hline l_6 & W_{6,13} & W_{6,14} & W_{6,15} & false \\ l_{16} & false & false & false & true \end{array}$$

and

$W_{6,13}$ = *"the criterion shows that the signal has to be confirmed"*;
$W_{6,14}$ = *"the criterion shows that the signal is correct"*;
$W_{6,15}$ = *"the criterion shows that the signal is incorrect"*.
$\Box_4 = \wedge(l_6, l_{16})$.

In places l_{13}, l_{14} and l_{16} the tokens do not obtain new characteristics. In place l_{15} the tokens obtain the characteristic *"incorrect signal"*.

$$Z_5 = \langle \{l_{14}, l_{18}, l_{29}\}, \{l_{17}, l_{18}\}, r_5, \Box_5 \rangle,$$

where

$$r_5 = \begin{array}{c|cc} & l_{17} & l_{18} \\ \hline l_{14} & W_{14,17} & W_{14,18} \\ l_{18} & false & true \\ l_{29} & false & true \end{array}$$

and

$W_{14,18}$ = "*the current token in* l_{14} *is of type* ω";
$W_{14,17} = \neg W_{14,18}$.
$\square_5 = \vee(\wedge(l_{14}, l_{18}), l_{29})$.

The ω-token in place l_{14} enters l_{18} where it unites with the β-token. The other type of tokens enter l_{17} with characteristic *"data about the current patient"*.

$$Z_6 = \langle \{l_{11}, l_{17}\}, \{l_{19}, l_{20}, l_{21}, l_{22}\}, r_6, \square_6 \rangle,$$

where

$$r_6 = \begin{array}{c|cccc} & l_{19} & l_{20} & l_{21} & l_{22} \\ \hline l_{11} & W_{11,19} & W_{11,20} & W_{11,21} & \mathit{false} \\ l_{17} & W_{17,19} & W_{17,20} & W_{17,21} & W_{17,22} \end{array}$$

and

$W_{11,19}$ = "*there is* π' *token in place* l_{24} *which is corresponding to the current* π'' *in place* l_{11}";
$W_{11,20}$ = "*there is* π' *token in place* l_{26} *which is corresponding to the current* π'' *in place* l_{11}";
$W_{11,21}$ = "*there is* π' *token in place* l_{28} *which is corresponding to the current* π'' *in place* l_{11}";
$W_{17,19}$ = "*the glucose level of the current patient is less than or equal to* 4 mmol/l";
$W_{17,20}$ = "*the glucose level of the current patient is greater than* 4 mmol/l *and less than or equal to* 10 mmol/l";
$W_{17,21}$ = "*the glucose level of the current patient is greater than* 10 mmol/l *and less than or equal to* 18 mmol/l";
$W_{17,22}$ = "*the glucose level of the current patient is greater than or equal to* 18 mmol/l".
$\square_6 = \vee(l_{11}, l_{17})$.

In places l_{19}, l_{20}, l_{21} and l_{22} the tokens obtain characteristic *"time of arrival of the signal"*.

$$Z_7 = \langle \{l_{19}, l_{24}\}, \{l_{23}, l_{24}\}, r_7, \square_7 \rangle,$$

where

$$r_7 = \begin{array}{c|cc} & l_{23} & l_{24} \\ \hline l_{19} & W_{19,23} & W_{19,24} \\ l_{24} & W_{24,23} & W_{24,24} \end{array}$$

and

$W_{19,23}$ = "*the current patient is unconscious*";
$W_{19,24} = \neg W_{19,23}$;
$W_{24,23}$ = "*all required data about the patient has been collected or the maximum time of waiting has been reached*";

$W_{24,24} = \neg W_{24,23}$.
$\square_7 = \vee(l_{19}, l_{24})$.

In place l_{24} the tokens obtain the characteristic *"waiting for blood pressure measurement and/or results from the health interview; duration of the waiting"*. In place l_{23} the tokens do not obtain new characteristic.

$$Z_8 = \langle \{l_{20}, l_{26}\}, \{l_{25}, l_{26}\}, r_8, \square_8 \rangle,$$

where

$$r_8 = \begin{array}{c|cc} & l_{25} & l_{26} \\ \hline l_{20} & W_{20,25} & W_{20,26} \\ l_{26} & W_{26,25} & W_{26,26} \end{array}$$

and

$W_{20,25} =$ *"the current patient is unconscious"*;
$W_{20,26} = \neg W_{20,25}$;
$W_{26,25} =$ *"all required data about the patient has been collected or the maximum time of waiting has been reached"*;
$W_{26,26} = \neg W_{26,25}$.
$\square_8 = \vee(l_{20}, l_{26})$.

In place l_{26} the tokens obtain the characteristic *"waiting for blood pressure measurement and/or results from the health interview; duration of the waiting"*. In place l_{25} the tokens do not obtain new characteristic.

$$Z_9 = \langle \{l_{21}, l_{28}\}, \{l_{27}, l_{28}\}, r_9, \square_9 \rangle,$$

where

$$r_9 = \begin{array}{c|cc} & l_{27} & l_{28} \\ \hline l_{21} & W_{21,27} & W_{21,28} \\ l_{28} & W_{28,27} & W_{28,28} \end{array}$$

and

$W_{21,27} =$ *"the acetone of the current patient is positive"*;
$W_{21,28} = \neg W_{21,27}$;
$W_{28,27} =$ *"all required data about the patient has been collected or the maximum time of waiting has been reached"*;
$W_{28,28} = \neg W_{28,27}$.
$\square_9 = \vee(l_{21}, l_{28})$.

In place l_{28} the tokens obtain the characteristic *"waiting for interview results; duration of the waiting"*. In place l_{27} the tokens do not obtain new characteristic.

$$Z_{10} = \langle \{l_{22}, l_{23}, l_{25}, l_{27}, l_{30}\}, \{l_{29}, l_{30}\}, r_{10}, \square_{10} \rangle,$$

where

$$r_{10} = \begin{array}{c|cc} & l_{29} & l_{30} \\ \hline l_{22} & true & false \\ l_{23} & true & false \\ l_{25} & true & false \\ l_{27} & true & false \\ l_{30} & false & true \end{array}$$

and

$\square_{10} = \wedge(\vee(l_{22}, l_{23}, l_{25}, l_{27}), l_{30})$. In places l_{24}, l_{26} and l_{28} the π''-tokens, the γ'-tokens and the π'-tokens corresponding to one and the same patient merge into a new π'-token. In place l_{29} the tokens obtain the characteristic *"decision taken by the telemedicine person"*. In place l_{30} the δ-tokens obtain the characteristic *"decision taken, duration of the shift"*.

8 Conclusion

The modelling and the simulation of the processes in the health care system as a whole allow better description, control in real time and prognosis. The usage of GNs in medicine holds out an opportunity for a new approach toward modelling and simulation of the information, patient and work-load flows in health care units and health care system. The so made simulation models give a new look over the problems related with restructuring, managing, planning and organization of the health care services. The GN models developed in medicine contribute to:

- early finding of pathological deviations and determining of the reasons;
- start from simpler methods (disease history, thorough examination, simple laboratory tests), available to every physician at each level of the health-care system and if necessary, proceed to newer and more expensive methods;
- showing (using) the most informative methods of study and treatment at a given stage of treatment. This, depending on the level of the physician, facilitates the generation of diagnosis or approximation to the most probable one;
- avoiding redundant studies, assist the decision of using more expensive methods; direct the patient to the corresponding specialist or hospital for precise assessment of his/her status; timely initiation and proper carrying out of the treatment and follow-up;
- determining the bottlenecks for the process of providing health care services and evaluating how changes to clinic design increase or reduce queues, time in system, and number of patients in the clinic (the different what-if scenarios could provide useful information to the hospital administrators for making management decision).

On the basis of the GN model of telemedicine for people with diabetes described in this paper a decision support tool can be developed. The model can be easily extended to include estimations of the costs of the telemedicine center. Computer simulation of the model can be used to determine the optimal number of specialists in the telemedicine center.

Appendix: Short Remark on Generalized Nets

GNs [5, 9] are extensions of Petri Nets [73]. They are defined in a way that is principally different from the ways of defining the other types of Petri nets. We shall first give an example of a GN and make remarks about the notation. A GN is shown in Fig. 2. The *places* are marked with ○. Each part of the net which looks like the one shown on Fig. 3., is called *transition* (more precisely graphic structure of transition). Transition's conditions are denoted by |. GNs, like other nets, contain tokens which are transferred from place to place. Every token enters the net with an initial characteristic. During each transfer, the token receives new characteristics. So, they accumulate their *"history"*. This is the first essential difference with the other types of Petri nets.

Every GN-place has at most one arc entering and at most one arc leaving it. The places with no entering arcs are called *input places* for the net (l_1, l_2 on Fig. 2.) and those with no leaving arcs are called *output places* (l_{14} and l_{15} on Fig. 2.). The *input places* are always at the transition's left, and the *output places* are always at the

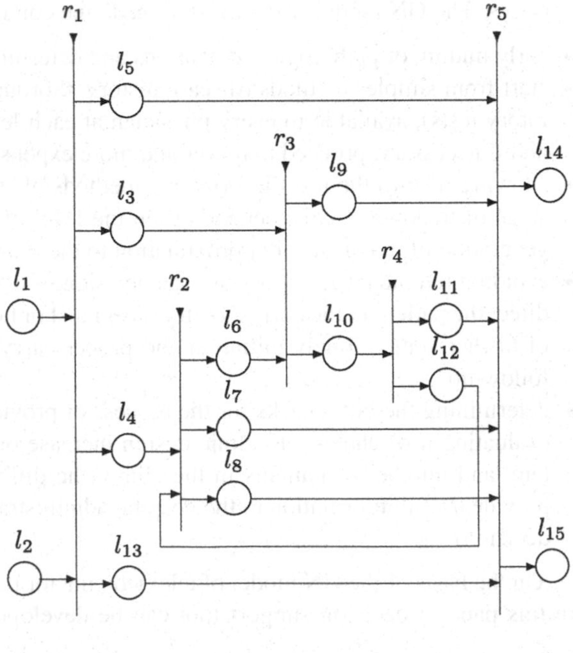

Fig. 2 Generalized net

Fig. 3 Transition

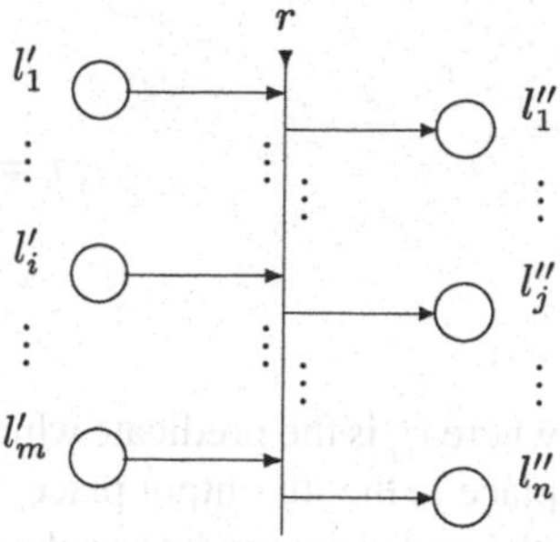

transition's right side. When tokens enter the input place of a transition, it becomes *potentially fireable* and at the moment of their transfer towards the transition's output places, it is being *fired.* The transition becomes active at a given time-moment and remains active up to another predefined moment.

The second basic difference between GNs and the ordinary Petri nets is the "place—transition" relation. Here, transitions are objects of a more complex nature. A transition may contain m input and n output places where $m, n \geq 1$.

The third basic difference is related to the time during which the GN functions. The time *can be* determined from some global time-scale and in this case the net is not invariant about the time-parameters. When we have GN models of some (different, but connected) processes that flow in parallel at time, we can use many time-scales or a single one, accounting the moments of the separate events in the processes. In the present form of the GN-definition, time is discrete. It increases with discrete steps. We can see the status of the GN model in each current time-moment.

Formally, every transition is described by a seven-tuple:

$$Z = \langle L', L'', t_1, t_2, r, M, \Box \rangle,$$

where:

(**a**) L' and L'' are finite, non-empty sets of places (the transition's input and output places, respectively); for the transition in Fig. 3 these are

$$L' = \{l'_1, l'_2, \ldots, l'_m\}$$

and

$$L'' = \{l''_1, l''_2, \ldots, l''_n\};$$

(**b**) t_1 is the current time-moment of the transition's firing;

(**c**) t_2 is the current value of the duration of its active state;

(**d**) r is the transition's *condition* determining which tokens will transfer from the transition's inputs to its outputs. Parameter r has the form of an IM:

$$r = \begin{array}{c|c} & l''_1 \ldots l''_j \ldots l''_n \\ \hline l'_1 & \\ \vdots & r_{i,j} \\ l'_i & (r_{i,j} \ - \text{ predicate}) \\ \vdots & (1 \le i \le m, 1 \le j \le n) \\ l'_m & \end{array} \;;$$

where $r_{i,j}$ is the predicate which expresses the condition for transfer from the ith input place to the jth output place. When $r_{i,j}$ has truth-value "*true*", then a token from the ith input place can be transferred to the jth output place; otherwise, this is impossible;

(e) M is an IM of the capacities of transition's arcs:

$$M = \begin{array}{c|c} & l''_1 \ldots l''_j \ldots l''_n \\ \hline l'_1 & \\ \vdots & m_{i,j} \\ l'_i & (m_{i,j} \ge 0 - \text{ natural number or } \infty) \\ \vdots & (1 \le i \le m, 1 \le j \le n) \\ l'_m & \end{array} \;;$$

(f) $\square$ is called transition type and it is an object having a form similar to a Boolean expression. It may contain as variables the symbols that serve as labels for transition's input places, and it is an expression constructed of variables and the Boolean connectives $\wedge$ and $\vee$ determining the following conditions:

$\wedge(l_{i_1}, l_{i_2}, \ldots, l_{i_u})$ – every place $l_{i_1}, l_{i_2}, \ldots, l_{i_u}$ must contain at least one token,

$\vee(l_{i_1}, l_{i_2}, \ldots, l_{i_u})$ – there must be at least one token in the set of places $l_{i_1}, l_{i_2}, \ldots, l_{i_u}$, where $\{l_{i_1}, l_{i_2}, \ldots, l_{i_u}\} \subset L'$.

When the value of a type (calculated as a Boolean expression) is "*true*", the transition can become active, otherwise it cannot.

The ordered four-tuple

$$E = \langle\langle A, \pi_A, \pi_L, c, f, \theta_1, \theta_2\rangle, \langle K, \pi_K, \theta_K\rangle, \langle T, t^0, t^*\rangle, \langle X, \Phi, b\rangle\rangle$$

is called a *Generalized Net* if:

(a) A is a set of transitions (see above);

(b) π_A is a function giving the priorities of the transitions, i.e., $\pi_A : A \to \mathcal{N}$;

(c) π_L is a function giving the priorities of the places, i.e., $\pi_L : L \to \mathcal{N}$, where

$$L = pr_1 A \cup pr_2 A$$

and obviously, L is the set of all GN-places;

(d) c is a function giving the capacities of the places, i.e., $c : L \to \mathcal{N}$;

(e) f is a function that calculates the truth values of the predicates of the transition's conditions;

(f) θ_1 is a function giving the next time-moment, for which a given transition Z can be activated, i.e., $\theta_1(t) = t'$, where $pr_3 Z = t, t' \in [T, T + t^*]$ and $t \leq t'$; the value of this function is calculated at the moment when the transition terminates its functioning;

(g) θ_2 is a function giving the duration of the active state of a given transition Z, i.e., $\theta_2(t) = t'$, where $pr_4 Z = t \in [T, T + t^*]$ and $t' \geq 0$; the value of this function is calculated at the moment when the transition starts functioning;

(h) K is the set of the GN's tokens. In some cases, it is convenient to consider this set in the form

$$K = \bigcup_{l \in Q^I} K_l,$$

where K_l is the set of tokens which enter the net from place l, and Q^I is the set of all input places of the net;

(i) π_K is a function giving the priorities of the tokens, i.e., $\pi_K : K \to \mathcal{N}$;

(j) θ_K is a function giving the time-moment when a given token can enter the net, i.e., $\theta_K(\alpha) = t$, where $\alpha \in K$ and $t \in [T, T + t^*]$;

(k) T is the time-moment when the GN starts functioning; this moment is determined with respect to a fixed (global) time-scale;

(l) t^0 is an elementary time-step, related to the fixed (global) time-scale;

(m) t^* is the duration of the GN functioning;

(n) X is a function which assigns initial characteristics to every token when it enters input place of the net;

(o) Φ is a characteristic function that assigns new characteristics to every token when it makes a transfer from an input to an output place of a given transition;

(p) b is a function giving the maximum number of characteristics a given token can receive, i.e., $b : K \to N$.

For the algorithms of transition and GN functioning the reader can refer to [9].

References

1. Alexieva, J., Choy, E., Koycheva, E.: Review and bibloigraphy on generalized nets theory and applications. In: Choy, E., Krawczak, M., Shannon, A., Szmidt, E. (eds.) A Survey of Generalized Nets, vol. 10, pp. 207–301. Raffles KvB Monograph, Sydney (2007)
2. Andonov, V., Stephanova, D., Esenturk, M., Angelova, M., Atanassov, K.: Generalized net model of telemedicine based on body temperature sensors. In: Proceedings of the 14th International workshop on Generalized Nets, pp. 78–89. Burgas (2013)
3. Andonov, V., Stefanova-Pavlova, M., Stojanov, T., Angelova, M., Cook, G., Klein, B., Atanassov, K., Vassilev, P.: Generalized net model for telehealth services. In: Proceedings of the 6th IEEE International Conference Intelligent Systems, pp. 221–224. Sofia (2012)

4. Andonov, V., Stojanov, T., Atanassov, K., Kovachev, P.: Generalized net model for telecommunication processes in telecare services. In: Proceedings of the First International Conference on Telecommunications and Remote Sensing, pp. 158–162. Sofia (2012)
5. Atanassov, K.: Generalized Nets. World Scientific, Singapore (1991)
6. Atanassov, K.: Applications of Generalized Nets. World Scientific Publishing Co., Singapore (1993)
7. Atanassov, K.: Generalized Nets and Systems Theory. Academic Publishing House Prof. M. Drinov, Sofia (1997)
8. Atanassov, K.: Generalized Nets in Artificial Intelligence. Generalized Nets and Expert Systems, vol. 1. Academic Publishing House Prof. M. Drinov, Sofia (1998)
9. Atanassov, K.: On Generalized Nets Theory. Prof. M. Drinov Academic Publishing House, Sofia (2007)
10. Atanassov, K., Daskalov, M.: Generalized nets models in neurology (NGN60: patient with acute headache). Social Medicine **2**, 37–38 (1996)
11. Atanassov, K., Shannon, A., Wong, C., Owens, D.: A generalized net for endogenous and exogenous insulin kinetics. In: Proceedings of the International Workshop Bioprocess Engineering'95, Sofia, 2-5 Oct 1995
12. Atanassov, K., Bustince, H., Daskalov, M., Georgiev P., Sorsich J.: Generalized net models in neurology (NGN17: clumsiness and incordination (Ataxia) suspected). In: Proceedings of the International Workshop Bioprocess Engineering'95, Sofia, 2–5 Oct 1995
13. Atanassov, K., Bustince, H., Daskalov, M., Kim, S., Shannon, A., Sorsich, J.: Generalized Net Models in Neurology (Introduction, Second Part). Preprint MRL-MFAIS-2-95, Sofia, 6, 14–19 Oct 1995)
14. Atanassov, K., Bustince, H., Daskalov, M., Kim, S., Shannon, A., Sorsich, J.: Generalized Net Models in Neurology (NGN84: Guillain-Barre Syndrome Suspected). Preprint MRL-MFAIS-2-95, Sofia, 6, 20–25 Oct 1995
15. Atanassov, K., Bustince, H., Daskalov, M., Kim, S., Shannon, A., Sorsich, J.: Generalized Net Models in Neurology (NGN100: Findings of Chorea and Athetoid Movements). Preprint MRL-MFAIS-2-95, Sofia, 6, 26–29 Oct (1995)
16. Atanassov, K., Bustince, H., Daskalov, M., Sorsich, J.: Generalized net models in neurology (introduction), Preprint MRL-MFAIS-1-95, pp. 1–4 (1995)
17. Atanassov, K., Daskalov, M., Georgiev, P., Kim, S., Kim, Y., Nikolov, N., Shannon, A., Sorsich, J.: Generalized Nets in Neurology. Prof. M. Drinov Academic Publishing House, Sofia (1997)
18. Atanassov, K., Shannon, A., Sorsich, J.: Generalized nets in neurology (NGN115: Acute bacterial meningitis in adults). In: Proceedings of the International Symposium Bioprocess Systems'97, pp. VI.1–VI.3. Sofia, 14–16 Oct 1997
19. Atanassov, K., Shannon, A., Sorsich, J.: Generalized nets in neurology (NGN116: subacute and chronic meningitis in adults). In: Proceedings of the International Symposium Bioprocess Systems'97, pp. VI.4–VI.5. Sofia 14-16 Oct 1997
20. Atanassov, K., Shannon, A., Sorsich, J.: Generalized nets in neurology (NGN117: focal intracranial bacterial disease). In: Proceedings of the International Symposium Bioprocess Systems'97, pp. VI.6–VI.9. Sofia, 14–16 Oct (1997)
21. Atanassov, K., Sorsich, J., Bustince, H.: Generalized nets models in internal medicine (GNGIM109: Chronic meningitis and GNGIM110: Aseptic meningitis syndrome). In: Proceedings of the Conference Bioprocess Systems'97, pp. VII.1–VII.3. Sofia, 29 Sept–2 Oct 1997
22. Atanassov, K., Sorsich, J., Bustince, H.: Generalized nets models in internal medicine (GNGIM124: hematuria). In: Proceedings of the Conference Bioprocess Systems'97, pp. VII.4-VII.6. Sofia, 29 Sept–2 Oct 1997
23. Atanassov, K., Sorsich, J., Bustince, H.: Generalized nets models in internal medicine (GNGIM127: metabolic acidosis and GNGIM128: Metabolic akalosis). In: Proceedings of the Conference Bioprocess Systems'97, pp. VII.7-VII. Sofia, 10, 29 Sept—2 Oct 1997

24. Atanassov, K., Weisberg, L., Garsia, K., Daskalov, M., Pajeva, I., Strub, R., Shannon, A., Sorsich, J.: Generalized nets in the biology and medicine. BAS Mag. **1–2**, 44–49 (1998)
25. Atanassov, K., Sorsich, J., Bustince, H.: Generalized nets models in internal medicine (GNGIM198: Caustic ingestion and exposure, and GNGIM201: Hypothermia). In: Proceedings of the International Symposium Bioprocess Systems'99, pp. IV.1–IV.3. Sofia, 18–21 Oct 1999
26. Atanassov, K., Sorsich, J., Bustince, H.: Generalized nets models in internal medicine (GNGIM238: Use and evaluation of serum drug levels). In: Proceedings of the International Symposium Bioprocess Systems'99, pp. IV.4–IV.5, Sofia, 18–21 Oct 1999
27. Atanassov, K., Sorsich, J., Bustince, H.: Generalized nets models in internal medicine (GNGIM242: Inpatient enteral nutrition). In: Proceedings of the International Symposium Bioprocess Systems'99, pp. IV.6–IV.10. Sofia, 18–21 Oct 1999
28. Atanassov, K., Radeva, V., Shannon, A., Sorsich, J.: Generalized Nets Interpretations of Ivan Dimitrov's Informational Theory of Diseases. Prof. M. Drinov Academic Publishing House, Sofia (2001)
29. Atanassov, K., Matveev, M., Tasseva, V.: On the generalized nets and their applications in medicine. In: Proceedings of the Ninth National Conference on Biomedical Physics and Engineering, pp. 250–253. 14–16 Oct 2004
30. Atanassov, K., Matveev, M., Shannon, A., Tasseva, V.: Information model of workflow and resources in general hospital. In: Proceedings of International Conference Advanced Information and Telemedicine Technologies for Health AITTH'2005, Minsk, vol. 1, pp. 42–46. 8–10 Nov 2005
31. Atanassov, K., Chakarov, V., Shannon, A.: Principal generalized net model of the human muscle-skeletal system. In: Proceedings of the Scientific Conference of Free University of Burgas, III, pp. 392–398 (2006)
32. Atanassov, K., Chakarov, V., Shannon, A., Sorsich, J.: Generalized Net Models of the Human Body. Prof. M. Drinov Academic Publishing House, Sofia (2008)
33. Bjankova B., Sorsich, J., Kim, S.-Ki, Atanassov, K.: Application of the generalized net in nephrology (approach to patient with recurrent urinary symptoms). In: Proceedings of the International Workshop Bioprocess Engineering'95, pp. 107–110. Sofia, 2–5 Oct 1995
34. Bustince, H., Kim, S.-Ki, Kim, Y., Nikolov, N.: Generalized nets in neurology (NGN121: Neurosyphilis). In: Proceedings of the International Symposium Bioprocess Systems'97, pp. VI.20–VI.21. Sofia, 14–16 Oct 1997
35. Bustince, H., Kim, S.-Ki, Kim, Y., Nikolov, N.: Generalized nets in neurology (NGN122: Tetanus). In: Proceedings of the International Symposium Bioprocess Systems'97, pp. VI.22–VI.24. Sofia, 14-16 Oct 1997
36. Bustince, H., Kim, S.-Ki, Kim, Y., Nikolov, N.: Generalized nets in neurology (NGN123: Rabies). In: Proceedings of the International Symposium Bioprocess Systems'97, pp. VI.25–VI.27. Sofia, 14-16 Oct 1997
37. Chakarov, V., Atanassov, K., Shannon, A.: Principal generalized net model of an excretory system. In: Proceedings of the Fifth International Workshop on Generalized Nets, Sofia, 10, 13–18 Nov 2004
38. Chakarov, V., Atanassov, K., Tasseva, V.: Application of the generalized nets in medicine (Edemas). First European Conference on Health Care Modelling and Computation Craiova, pp. 79–86. 31 Aug–2 Sept 2005
39. Chakarov, V., Shannon, A., Atanassov, K.: Generalized net model of human hematopoietic system. Int. Electron. J. Bioautomation **2**, 107–114 (2005)
40. Chakarov, V., Atanassov, K., Tasseva, V., Matveev, M., El-Darzi, E., Chountas, P., Petrounias, I.: Generalized net model for some basic clinical administrative decision making. First European Conference on Health Care Modelling and Computation Craiova, pp. 72–78. 31 Aug–2 Sept 2005
41. Chakarov, V., Shannon, A., Sorsich, J., Atanassov, K.: Generalized net model of the endocrine system. Comptes Rendus de l'Academie bulgare des Sci. **61**(6), 705–712 (2008)

42. Chase, H., Pearson, J., Wightman, C., Roberts, M., Oderberg, A., Garg, S.: Modem transmission of glucose values reduces the costs and need for clinic visits. Diabetes Care **26**, 1475–1479 (2003)
43. Choy, Y. H., Shannon, A.: Generalized nets for diabetic nephropathy. In: Proceedings of the Sixth Scientific Session of Mathematical Foundations of Artificial Intelligence Seminar, Sofia, 10, 1–8 June 1998
44. Daskalov, M., Atanassov, K.: Generalized net models in neurology (NGN33: Tinnitus). In: Proceedings of the Seventh National Conference on Biomedical Physics and Engineering, pp. 182–184. Sofia, 17–19 Oct 1996
45. Daskalov, M., Atanassov, K., Bustince, H.: Generalized net models in neurology (coma). Preprint MRL-MFAIS-1-95, pp. 13–16 (1995)
46. Daskalov, M., Bustince, H., Atanassov, K.: Generalized net models in neurology (first headache). Preprint MRL-MFAIS-1-95, pp. 5–8 (1995)
47. Daskalov, M., Bustince, H., Atanassov, K.: Generalized net models in neurology (chronic headache). Preprint MRL-MFAIS-1-95, pp. 9–12 (1995)
48. Daskalov, M., Nikolov, N., Bustince, H.: Generalized net models in neurology (NGN34: dizziness). In: Proceedins of the Seventh National Conference on Biomedical Physics and Engineering, pp. 185–188. Sofia, 17–19 Oct 1996
49. Daskalov, M., Georgiev, P., Kacprzyk, J., Szmidt, E.: Generalized nets in neurology (NGN118: tuberculous meningitis). In: Proceedings of the International Symposium Bioprocess Systems'97, pp. VI.10–VI.12. Sofia, 14–16 Oct 1997
50. Daskalov, M., Georgiev, P., Kacprzyk, J., Szmidt, E.: Generalized nets in neurology (NGN119: cryptococcal infection of the central nervous system). In: Proceedings of the International Symposium Bioprocess Systems'97, pp. VI.13–VI.15. Sofia, 14–16 Oct 1997
51. Daskalov, M., Georgiev, P., Kacprzyk, J., Szmidt, E.: Generalized nets in neurology (NGN120: acute viral encephalitis). In: Proceedings of the International Symposium Bioprocess Systems'97, pp. VI.16–VI.19. Sofia, 14-16 Oct 1997
52. Dimitrov, I.: Informational Theory of Diseases. Albo, Sofia (1993)
53. Dunn, D., Epstein, L.: Decision Making in Child Neurology. B. C. Decker, Toronto (1987)
54. El-Darzi, E., Vasilakis, C., Chaussalet, T., Millard, P.H.: A simulation modelling approach to evaluating length of stay, occupancy, emptiness and bed blocking in a hospital geriatric department. Health Care Manag. Sci. **1**, 143–149 (1998)
55. El-Darzi, E., Vasilakis, C., Chaussalet, T., Millard, P. H.: A simulation model to evaluate the interaction between acute, rehabilitation, long stay care and the community. In: Decision Making: Recent developments and Worldwide applications Zanakis, S.H., Doukidis, G., Zopounidis, C. (eds.), pp. 475–485. Kluwer Academic Publishers (2000)
56. Georgiev, P., Kacprzyk, J., Szmidt, E., Aladjov, H.: Generalized nets models in internal medicine (GNGIM119: hepatitis exposure). In: Proceedings of the Conference Bioprocess Systems'98, pp. VII.20–VII.22. Sofia, 29 Sept–2 Oct 1998
57. Georgiev, P., Kacprzyk, J., Szmidt, E., Aladjov, H.: Generalized nets models in internal medicine (GNGIM126: renal cysts and masses). In: Proceedings of the Conference Bioprocess Systems'98, pp. VII.23–VII.25. Sofia, 29 Sept–2 Oct (1998)
58. Georgiev, P., Kacprzyk, J., Szmidt, E., Aladjov, H.: Generalized nets models in internal medicine (GNGIM131: hypokalemia and GNGIM132: Hyperkalemia). In: Proceedings of the Conference Bioprocess Systems'98, pp. VII.26–VII.28. Sofia, 29 Sept–2 Oct 1998
59. Georgiev, P., Kacprzyk, J., Szmidt, E., Aladjov, H.: Generalized nets models in internal medicine (GNGIM196: acute pulse less extremity and GNGIM197: foreign body ingestion). In: Proceedings of the Conference Bioprocess Systems'99, pp. IV.23–IV.26. Sofia, 18–21 Oct 1999
60. Georgiev, P., Kacprzyk, J., Szmidt, E., Aladjov, H.: Generalized nets models in internal medicine (GNGIM233: non surgical antimicrobial prophylaxis). In: Proceedings of the Conference Bioprocess Systems'99, Sofia, IV.27–IV.29, 18–21 Oct 1999
61. Georgiev, P., Kacprzyk, J., Szmidt, E., Aladjov, H.: Generalized nets models in internal medicine (GNGIM234: antimicrobial prophylaxis in surgical patients). In: Proceedings of the Conference Bioprocess Systems'99, pp. IV.30–IV.32. Sofia, 18–21 Oct (1999)

62. Jordanova, B., Sorsich, J.: Modelling of diagnostic and therapeutic processes in medicine by Generalized nets. In: Atanassov, K. (ed.) Applications of Generalized Nets, pp. 291–297. Singapore, New Jersey, London (1993)
63. Kalinkin, V., Prokopov, D., Iliev, D., Mandjukov, I., Sorsich, J., Atanassov, K., Nikolov, N., Georgiev, P., Zaharieva, S., Aladjov, H.: Modelling of the hospital activities using generalized Nets. In: Proceedings of the International Scientific Conference on Energy and Information Systems and Technologies. vol. III, pp. 752–757. Bitola, 7–8 June 2001 (in Bulgarian)
64. Kim, S.-K., Atanassov, K., Shannon, A.: Generalized nets in neurology: an example of mathematical modelling. Int. J. Math. Educ. Sci. Technol. **31**(2), 173–179 (2000)
65. Koleva, Tz., Sorsich, J., Shannon, A., Kim, S.-K., Kim, Y.: Generalized net models in child neurology CN061: discontinuation of medication in children with epilepsy). In: Proceedings of Second International Workshop on Generalized Nets, pp. 54–55. Sofia, 26 and 27 June 2001
66. Koleva, Tz., Sorsich, J., Shannon, A., Kim, S.-K., Kim, Y.: Generalized net models in child neurology (CN062: the use of anticonvulsant drugs). In: Proceedings of Second International Workshop on Generalized Nets, pp. 56–58, Sofia, 26 and 27 June 2001
67. Klonoff, D.: Diabetes and telemedicine: is the technology sound, effective, cost-effective, and practical? (Editorial). Diabetes Care **26**, 1626–1628 (2003)
68. Levin, K., Madsen, J., Petersen, I., Wanscher, C., Hangaard, J.: Telemedicine diabetes consultations are cost-effective, and effects on essential diabetes treatment parameters are similar to conventional treatment: 7-year results from the svendborg telemedicine diabetes project. J. Diabetes Sci. Technol. **7**(3), 587–595 (2013)
69. Matveev, M., Atanassov, K., Pazvanska, E., Tasseva, V.: Dynamic model of intensive care unit workflow based on generalized nets. Int. Electron. J. Bioautomation **2005**(2), 85–92 (2005)
70. Matveev, M., Andonov, V., Atanassov, K., Milanova, M.: Generalized Net Model for Telecommunication Processes in Telecare Services. In: Proceedings of the 2013 International Conference on Electronics and Communication Systems, pp. 142–145, Rhodes Island (2013)
71. Mengov, G., Hadjitodorov, S., Shannon, A.: Modelling cognitive brain processes with a generalized net. In: Proceedings of Second International Workshop on Generalized Nets, pp. 59–61, Sofia. 26 and 27 June 2001
72. Mengov, G., Pulov, S., Atanassov, K., Georgiev, K., Trifonov, T.: Modelling neural signals with a generalized net. Adv. Stud. Contemp. Math. **7**(2), 155–166 (2003)
73. Murata, T.: Petri nets: properties, analysis and applications. Proc. IEEE **77**(4), 541–580 (1989)
74. Nikolov, N., Shannon, A., Radeva, V.: Generalized nets models in internal medicine (GNGIM114: pulmonary infections in the HIV-infected patient and GNGIM115: Central nervous system infections in the HIV-infected patient). In: Proceedings of the Conference Bioprocess Systems'98, pp. VII.11–VII. Sofia, 13, 29 Sept–2 Oct 1998
75. Nikolov, N., Shannon, A. Radeva, V.: Generalized nets models in internal medicine (GNGIM121: chronic renal failure). In: Proceedings of the Conference Bioprocess Systems'98, pp. VII.14–VII. Sofia, 16, 29 Sept–2 Oct 1998
76. Nikolov, N., Shannon, A., Radeva, V.: Generalized nets models in internal medicine (GNGIM 129: hyponatremia and GNGIM130: Hypernatremia). In: Proceedings of the Conference Bioprocess Systems'98, pp. VII.17–VII.19. Sofia, 29 Sept–2 Oct 1998
77. Nikolov, N., Shannon, A., Kim, S.-K.: Generalized nets models in internal medicine (GNGIM202: drowning and near-drowning and GNGIM205: Abnormal vaginal bleeding). In: Proceedings of the Conference Bioprocess Systems'99, pp. IV.11–IV.14. Sofia, 18–21 Oct 1999
78. Nikolov, N., Shannon, A., Kim, S.-K.: Generalized nets models in internal medicine (GNGIM235: choosing appropriate antimicrobial therapy). In: Proceedings of the Conference Bioprocess Systems'99, pp. IV.15–IV.17. Sofia, 18–21 Oct 1999
79. Nikolov, N.,Shannon, A., Kim, S.-K.: Generalized nets models in internal medicine (GNGIM241: inpatient parenteral nutrition). In: Proceedings of the Conference Bioprocess Systems'99, pp. IV.18–IV.21. Sofia, 18–21 Oct 1999

80. Ribagin, S., Chakarov, V., Atanassov, K.: Generalized net model of the upper limb in relaxed position. New Developments in Fuzzy Sets, Intuitionisitc Fuzzy Sets, Generalized Nets and Related Topics, vol. 2, pp. 201–210. Applications, SRI, Polish Academy of Sciences, Warsaw (2012)
81. Ribagin, S., Chakarov, V., Atanassov, K.: Generaized net model of the upper limb vascular system. In: Proceedings of the 6th IEEE International Conferance Intelligent Systems, pp. 229–232. Sofia, 6–8 Sept 2012
82. Ribagin, S., Andonov, V., Chakarov, V.: Possible applications of generalized nets with characteristics of the places. A medical example. In: Proceedings of the 14th International Workshop on Generalized Nets, pp. 29–30. Burgas, 56–64 Nov 2013
83. Sgurev, V., Gluhchev, G., Atanassov, K.: Generalized net interpretation of information process in the brain. In: Proceedings of the International Conference Automatics and Informatics, pp. A-147–A-150. Sofia, 31 May–2 June 2001
84. Shannon, A., Atanassov, K.: Applications of generalized nets in the modelling of biomedical processes. In: Proceedings of the International Workshop Bioprocess Engineering'95, Sofia, Oct 2–5 1995
85. Shannon, A., Sorsich, J., Atanassov, K.: Application of the generalized net in nephrology (remark on the global generalized net). Preprint MRL-MFAIS-2-95, pp. 20–25. Sofia, 6 Oct 1995
86. Shannon, A., Sorsich, J., Atanassov, K.: Generalized Nets in Medicine. Academic Publishing House Prof. M. Drinov, Sofia (1996)
87. Shannon, A., Sorsich, J., Atanassov, K.: Generalized nets in internal medicine (GNGIM108: acute and subacute meningitis). Preprint MRL-MFAIS-1-98, Sixth Scientific Session of Mathematical Foundations of Artificial Intelligence Seminar, pp. 8–10. Sofia, 10 June 1998
88. Shannon, A., Sorsich, J., Atanassov, K.: Generalized nets in internal medicine (GNGIM111: sexually transmitted diseases). Preprint MRL-MFAIS-1-98, Sixth Scientific Session of Mathematical Foundations of Artificial Intelligence Seminar, pp. 11–15. Sofia, 10 June 1998
89. Shannon, A., Sorsich, J., Atanassov, K.: Generalized nets in internal medicine (GNGIM113: the acutely ill HIV-positive patient). Preprint MRL-MFAIS-1-98, Sixth Scientific Session of Mathematical Foundations of Artificial Intelligence Seminar, pp. 16–18. Sofia, 10 June 1998
90. Shannon, A., Sorsich, J., Atanassov, K., Nikolov, N., Georgiev, P.: Generalized Nets in General and Internal Medicine, vol. 1. Prof. M. Drinov Academic Publishing House, Sofia (1998)
91. Shannon, A., Iliev, D., Szmidt, E., Aladjov, H., Bustince, H., Kacprzyk, J., Sorsich, J., Atanassov, K., Nikolov, N., Georgiev, P., Radeva, V.: Generalized net model of health unit activities (an intuitionistic fuzzy approach). In: Kacprzyk, J., Atanassov, K. (Eds.) Proceedings of the Second International Conference on Intuitionistic Fuzzy Sets, Vol. 2; Notes on Intuitionistic Fuzzy Sets, vol. 4, No. 3, pp. 79–84 (1998)
92. Shannon, A., Sorsich, J., Atanassov, K., Nikolov, N., Georgiev, P.: Generalized Nets in General and Internal Medicine, vol. 2. Prof. M. Drinov Academic Publishing House, Sofia (1999)
93. Shannon, A., Sorsich, J., Atanassov, K., Nikolov, N., Georgiev, P.: Generalized Nets in General and Internal Medicine, vol. 3. Prof. M. Drinov Academic Publishing House, Sofia (2000)
94. Shannon, A., Atanassov, K., Sorsich, J., Radeva, V.: Generalized Net Interpretations of Ivan Dimitrovs Informational Theory of Diseases. Academic Publishing House Prof. M. Drinov, Sofia (2001)
95. Shannon, A., Atanassov, K., Chakarov, V.: Global generalized net model of a human body: an intuitionistic fuzzy approach. In: Kacprzyk, J., Atanassov, K. (eds.) Proceedings of the Eight International Conference on Intuitionistic Fuzzy Sets, vol. 2, pp. 75–81. Sofia, 20–21 June 2004
96. Shannon, A., Atanassov, K., Chakarov, V.: Principal generalized net model of the human gastrointestinal tract. In: Proceedings of the Ninth National Conference on Biomedical Physics and Engineering, pp. 278–283. 14-16 Oct 2004
97. Shannon, A., Atanassov, K., Chakarov, V.: Generalized net model of the gastrointestinal system of the human body. Adv. Stud. Contemp. Math. **10**(2), 101–110 (2005)

98. Shannon, A., El-Darzi, E., Atanassov, K., Chountas, P., Chakarov, V., Tasseva, V.: Principal generalized net model of diagnostic and therapeutic processes in medicine. In: Proceedings of the Seventh Int. Workshop on Generalized Nets, pp. 30–38. Sofia, 14-15 July 2006
99. Shannon, A., El-Darzi, E., Peneva, D., Atanassov, K., Matveev, M., Chountas, P., Vassilev, P., Tasseva, V.: The generalized net modelling of information healthcare system. In: Proceedings of the International Conference Automatics and Informatics'06, pp. 119–122. Sofia (2006)
100. Shannon, A., Peneva, D., El-Darzi, E., Atanassov, K., Chountas, P., Tasseva, V.: On the generalized net modelling of healthcare local area network using intuitionistic fuzzy estimations. In: Proceedings of the Tenth International Conferenc on IFSs, NIFS vol. 12, No. 3, pp. 60–68. Sofia, 28–29 Oct 2006
101. Sorsich, J.: An example of application of generalized nets in medicine. In: Proceedings of II International Symposium Automation and Scientific Instrumentation, pp. 387–389. Varna (1983)
102. Sorsich, J.: Application of the generalized nets in medicine (Elevated blood used nitrogen). First Science Session of the Mathematics Found. AI Seminar, Sofia, Oct. 10: Preprint IM-MFAIS-7-89, pp. 57–59, Sofia (1989)
103. Sorsich, J., Atanassov, K.: Application of Generalized nets in medicine (Diagnostics of arterial hypertension of renal origin). In: Proceedings of III International School Automation and Scientific Instrumentation, pp. 233–236. Varna, Oct 1984
104. Sorsich, J., Atanassov, K.: Application of Generalized Nets in Medicine (Renal Colic). Lecture Notes in Medical Informatics, vol. 24, pp. 352–355 (1984)
105. Sorsich, J., Atanassov, K.: Application of Generalized nets in medicine (Haematuria). In: Proceedings of III International Symposium Automation and Scientific Instrumentation, pp. 163–166, Varna, Oct 1985
106. Sorsich, J., Atanassov, K.: Application of Generalized nets in medicine (Renovascular hypertension). In: Proceedings of III International Symposium Automation and Scientific Instrumentation, pp. 167–169. Varna, Oct 1985
107. Sorsich, J., Atanassov, K.: Generalized Nets and their application in Medicine (permanent proteinurea). In: Proceedings of National Scientific Session Automation of the Biotechnical Processes, pp. 138–141. Sofia, (in Bulgarian) Oct 1985
108. Sorsich, J., Atanassov, K.: Application of Generalized nets in medicine (Dysuria). In: Proceedings of III Symposium International Ingenieur Biomedical, pp. 639–642. Madrid, Oct 1987
109. Sorsich, J., Atanassov, K.: Application of Generalized nets in medicine (Acute Attack of Gouty Arthritis). In: Proceedings of III Symposium International Ingenieur Biomedical, pp. 643–645. Madrid, Oct 1987
110. Sorsich, J., Atanassov, K.: Applications of the generalized nets in nephrology. In: Atanassov, K. (ed.) Applications of Generalized Nets, pp. 220–290. World Scientific Publishing Co., Singapore (1993)
111. Sorsich, J., Atanassov, K.: Application of the generalized net in medicine (polyuria). In: Proceedings of the Seventh National Conference on Biomedical Physics and Engineering, pp. 192–194. Sofia 17–19 Oct 1996
112. Sorsich, J., Atanassov, K.: Application of generalized nets in medicine (renal mass lesions). Model. Meas. Control C **55**(1–2), 39–49 (1997)
113. Sorsich, J., Bustince, H.: Application of generalized nets in medicine (Acute back pain). Preprint MRL-MFAIS-1-95, pp. 17–20. Sofia (1995)
114. Sorsich, J., Shannon, A., Atanassov, K.: Application of the generalized net in medicine (modelling of the management of blood cholesterol in the adult general population). In: Proceedings of the Seventh National Conference on Biomedical Physics and Engineering, pp. 189–191. Sofia, 17–19 Oct 1996
115. Sorsich, J., Atanassov, K., Nikolov, N., Georgiev, P.: Modelling of the hospital activities using generalized nets. Autom. Inf. **III**(3), 23–26 (1999)
116. Sorsich, J., Shannon, A., Atanassov, K.: A global generalized net model of the human body. In: Proceedings of the Conference Bioprocess Systems'2000, pp. IV.1-IV.4. Sofia, 11–13 Sept 2000

117. Sorsich, J., Shannon, A., Atanassov, K.: Generalized net model of a renal and urological system. In: Proceedings of the Conference Bioprocess Systems'2000, pp. IV.5–IV.8, Sofia, 11-13 Sept 2000
118. Sorsich, J., Shannon, A., Atanassov, K.: Generalized net model of the cardiovascular system (An intuitionistic fuzzy approach). In: Kacprzyk, J., Atanassov, K. (eds.) Proceedings of the Fourth International Conference on Intuitionistic Fuzzy Sets, Vol. 2; Notes on Intuitionistic Fuzzy Sets, vol. 6 , No. 4, pp. 59–63 (2000)
119. Sorsich, J., Shannon, A., Atanassov, K., Kim, S.-K.: Generalized net of rheumatism diagnosis (a general model). In: Proceedings of the First International Workshop on Generalized Nets, pp. 6–8. Sofia, 9 July 2000
120. Sorsich, J., Atanassov, K., Nikolov, N., Koleva, Tz.: Generalized net models in child neurology (CN053: Febrile seizures and CN054: Infantile spasms). In: Proceedings of the Conference Bioprocess systems'2001, pp. III.19–III.20 Sofia, 1–3 Oct 2001
121. Sorsich, J., Atanassov, K., Nikolov, N., Koleva, Tz.: Generalized net models in child neurology (CN055: Minor motor seizures and CN056: absence seizures). In: Proceedings of the Conference Bioprocess systems'2001, pp. III.21–III.22. Sofia, 1–3 Oct 2001
122. Sorsich, J., Atanassov, K., Nikolov, N., Koleva, Tz.: Generalized net models in child neurology (A global generalized net model of epilepsy diseases). In: Proceedings of the Conference Bioprocess systems'2001, pp. III.25–III.26. Sofia, 1–3 Oct 2001
123. Sorsich, J., Shannon, A., Atanassov, K.: Generalized net model of reproductive system. Part 1. In: Proceedings of the Third International Workshop on GNs, pp. 25–31. Sofia, 1 Oct 2002
124. Sorsich, J., Atanassov, K., Shannon, A.: A generalized net model of endocrine system. Part 1: thyroid gland. In: Proceedings of the Conference Bioprocess systems'2002, pp. III.27-III.28. Sofia, 28-29 Oct 2002
125. Sorsich, J., Shannon, A., Atanassov, K.: Generalized Nets in Child Neurology. Academic Publishing House Prof. M. Drinov, Sofia (2002)
126. Sorsich, J., Chakarov, V., Shannon, A., Atanassov, K.: Generalized Net Model in use for medical diagnosing. In: Proceedings of the Fourth International Workshop on Generalized Nets, pp. 8–11. Sofia, 23 Sept 2003
127. Sorsich, J., Atanassov, K., Shannon, A., Szmidt, E., Kacprzyk, J.: Generalized net model of the interconnection effect of the neurological, endocrine and cardiovascular systems on metabolism in human body. In: Atanassov, K., Kacprzyk, J., Krawczak, M. (eds.) Issues in Intuitionistic Fuzzy Sets and Generalized Nets, vol. 2, pp. 47–52. Wydawnictwo WSISiZ, Warszawa (2004)
128. Tasseva, V.: Generalized net model of intensive care unit. In: Proceedings of 2nd International IEEE Conference Intelligent Systems, pp. 58–63. Varna, Bulgaria, 22–24 June 2004
129. Tasseva, V., Peneva, D., Dimova, K., El-Darzi, E., Chountas, P.: Generalised Nets Model for National health Insurance in Bulgaria. Part 1, Fifth International Workshop on GNs, pp. 32–40. Sofia, 10 Nov 2004
130. Tasseva, V., Peneva, D., Dimova, K., El-Darzi, E., Chountas, P.: Generalised Nets Model for National health Insurance in Bulgaria. Part 2. Sixth International Workshop on GNs, pp. 20–28. Sofia, 17 Dec 2005
131. Tasseva, V., Atanassov, K., Matveev, M., El-Darzi, E., Chountas, P., Gorunescu. F.: Modelling the flow of patient through intensive care unit using generalized nets. First European Conference on Health Care Modelling and Computation Craiova, pp. 290–299. 31 Aug–2 Sept 2005
132. Tasseva, V., Peneva, D., El-Darzi, E., Chountas, P.: Generalised nets model for national health Insurance in Bulgaria. Part 3. First International Workshop on Intuitionistic Fuzzy Sets, Generalized Nets and Knowledge Engineering, pp. 88–96. London, 6–7 Sep 2006
133. Tetev, M., Sorsich, J., Atanassov, K.: Generalized net model of hospital activities. Adv. Model. Anal. AMSE Press **17**(1), 55–64 (1993)
134. Todorova, L., Sorsich, J.: Generalized net model of the mechanical ventilation process. In: Proceedings of Second International Workshop on Generalized Nets, pp. 34–39. Sofia, 26 and 27 June 2001

135. Todorova, L., Bentes, I., Barroso, J., Temelkov, A.: A generalized net model of the treatment of mechanically ventilated patients with nosocomial pneumonia. In: Proceedings of the 10th ISPE International Conference on Concurrent Engineering Advanced Design, Production and Management Systems, pp. 1057–1063. Madeira, 26–30 July 2003
136. Todorova, L., Temelkov, A., Antonov, A.: GN model of transition to spontaneous breathing after long term mechanical ventilation. First European Conference on Health Care Modelling and Computation Craiova, pp. 300–308. 31 Aug–2 Sept 2005
137. Vasilakis, C., El-Darzi, E.: A simulation study of the winter bed crisis. Health Care Manag. Sci. J. **4**(1), 31–36 (2001)
138. Weisberg, L., Strub, R., Garcia, C.: Decision Making in Adult Neurology. B. C. Decker, Toronto (1993)
139. Position Statement of the American Diabetes Association (ADA) and the European Association for the Study of Diabetes (EASD). Management of Hyperglycemiain Type 2 Diabetes: a Patient-Centered Approach. DIABETES CARE, vol. 35, June 2012
140. ACE Comprehensive Diabetes Management, Endocr Pract. 19 (Suppl 2) (2013)
141. Gardner, D.G., Shoback, D. (eds.) Greenspans Basic and Clinical Endocrinology (8th edn.). The McGraw-Hill Medical, New York (2007)
142. Melmed, S. et al. (eds.) Williams Textbook of Endocrinology (12th edn.). Saunders/Elsevier, Philladelphia (2011)
143. http://www.tunstallhealthcare.com.au/Uploads/Documents/141861elehealth_web2025_6_13.pdf
144. http://www.lifelinkresponse.com.au/telehealth/

Part IV
Hybrid Approaches

Differential Evolution with Fuzzy Logic for Dynamic Adaptation of Parameters in Mathematical Function Optimization

Oscar Castillo, Patricia Ochoa and José Soria

Abstract The proposal described in this paper uses the Differential Evolution (DE) algorithm as an optimization method in which we want to dynamically adapt its parameters using fuzzy logic control systems, with the goal that the fuzzy system gives the optimal parameter of the DE algorithm to find better results, depending on the type of problems the DE is applied.

1 Introduction

The use of fuzzy logic in evolutionary computing is becoming a common approach to improve the performance of the algorithms. Currently the parameters involved in the algorithms are determined by trial and error. In this aspect we propose the application of fuzzy logic which is responsible in performing the dynamic adjustment of mutation and crossover parameters in the Differential Evolution (DE) algorithm. This has the goal of providing better performance to the Differential Evolution algorithm.

Fuzzy logic or multi-valued logic is based on fuzzy set theory proposed by Zadeh in 1965 which helps us in modeling knowledge, through the use of if-then fuzzy rules. The fuzzy set theory provides a systematic calculus to deal with linguistic information, and that improves the numerical computation by using linguistic labels stipulated by membership functions [12]. Differential Evolution (DE) is one of the latest evolutionary algorithms that have been proposed. It was created in 1994 by Price and Storn in, attempts to resolve the problem of Chebychev polynomial. The following year these two authors proposed the DE for optimization of nonlinear and non- differentiable functions on continuous spaces.

The DE algorithm is a stochastic method of direct search, which has proven effective, efficient and robust in a wide variety of applications such as learning of a

O. Castillo (✉) · P. Ochoa · J. Soria
Tijuana Institute of Technology, Tijuana, Mexico
e-mail: ocastillo@tectijuana.mx

P. Angelov and S. Sotirov (eds.), *Imprecision and Uncertainty in Information Representation and Processing*, Studies in Fuzziness and Soft Computing 332,
DOI 10.1007/978-3-319-26302-1_21

neural network, a filter design of IIR, aerodynamically optimized. The DE has a number of important features which make it attractive for solving global optimization problems, among them are the following: it has the ability to handle non-differentiable, nonlinear and multimodal objective functions, usually converges to the optimal uses with few control parameters, etc.

The DE belongs to the class of evolutionary algorithms that is based on populations. It uses two evolutionary mechanisms for the generation of descendants: mutation and crossover; finally a replacement mechanism, which is applied between the vector father and son vector determining who survive into the next generation. There exist works where they currently use fuzzy logic to optimize the performance of the algorithms, to name a few articles such as:

Optimization of Membership Functions for Type-1 and Type 2 Fuzzy Controllers of an Autonomous Mobile Robot Using PSO [1], Optimization of a Fuzzy Tracking Controller for an Autonomous Mobile Robot under Perturbed Torques by Means of a Chemical Optimization Paradigm [2], Design of Fuzzy Control Systems with Different PSO Variants [4], A Method to Solve the Traveling Salesman Problem Using Ant Colony Optimization Variants with Ant Set Partitioning [6], Evolutionary Optimization of the Fuzzy Integrator in a Navigation System for a Mobile Robot [7], Optimal design of fuzzy classification systems using PSO with dynamic parameter adaptation through fuzzy logic [8], Dynamic Fuzzy Logic Parameter Tuning for ACO and Its Application in TSP Problems [10], Bio-inspired Optimization Methods on Graphic Processing Unit for Minimization of Complex Mathematical Functions [15].

Similarly as there are papers on Differential Evolution (DE) applications that uses this algorithm to solve real problems. To mention a few:

A fuzzy logic control using a differential evolution algorithm aimed at modelling the financial market dynamics [5], Design of optimized cascade fuzzy controller based on differential evolution: Simulation studies and practical insights [11], Eliciting transparent fuzzy model using differential evolution [3], Assessment of human operator functional state using a novel differential evolution optimization based adaptive fuzzy model [14].

This paper is organized as follows: Sect. 2 shows the concept of the differential evolution algorithm as applied to the technique for parameter optimization. Section 3 describes the proposed methods. Section 4 shows the simulation results. Section 5 offers the conclusions.

2 Differential Evolution

Differential Evolution (DE) is an optimization method belonging to the category of evolutionary computation applied in solving complex optimization problems.

DE is composed of 4 steps:

- Initialization.
- Mutation.
- Crossover.
- Selection.

This is a non-deterministic technique based on the evolution of a vector population (individuals) of real values representing the solutions in the search space. The generation of new individuals is carried out by differential crossover and mutation operators [13].

The operation of the algorithm is explained below:

2.1 Population Structure

The differential evolution algorithm maintains a pair of vector populations, both of which contain Np D-dimensional vectors of real-valued parameters [8].

$$P_{x,g} = (\mathbf{x}_{i,g}), i = 0, 1, \ldots, Np, g = 0, 1, \ldots, g_{max} \quad (1)$$

$$\mathbf{x}_{i,g} = (x_{j,i,g}), j = 0, 1, \ldots, D - 1 \quad (2)$$

where:
P_x = current population.
g_{max} = maximum number of iterations.
i = index population.
j = parameters within the vector.

Once the vectors are initialized, three individuals are selected randomly to produce an intermediate population, $P_{v,g}$, of Np mutant vectors, $\boldsymbol{v}_{i,g}$.

$$P_{v,g} = (\boldsymbol{v}_{i,g}), i = 0, 1, \ldots, Np - 1, g = 0, 1, \ldots, g_{max} \quad (3)$$

$$\boldsymbol{v}_{i,g} = (v_{j,I,g}), j = 0, 1, \ldots, D - 1 \quad (4)$$

Each vector in the current population are recombined with a mutant vector to produce a trial population, P_u, the NP, mutant vector $\mathbf{u}_{i,g}$:

$$P_{v,g} = (\boldsymbol{u}_{i,g}), i = 0, 1, \ldots, Np - 1, g = 0, 1, \ldots, g_{max} \quad (5)$$

$$\boldsymbol{u}_{i,g} = (u_{j,I,g}), j = 0, 1, \ldots, D - 1 \quad (6)$$

2.2 Initialization

Before initializing the population, the upper and lower limits for each parameter must be specified. These 2D values can be collected by two initialized vectors, D-dimensional, b_L y b_U, to which subscripts L and U indicate the lower and upper limits respectively. Once the initialization limits have been specified number generator randomly assigns each parameter in every vector a value within the set range. For example, the initial value (g = 0) of the j-th vector parameter is ith:

$$x_{j,i,0} = rand_j(0,1) \cdot \left(b_{j,U} - b_{j,L}\right) + b_{j,L} \tag{7}$$

2.3 Mutation

In particular, the differential mutation adds a random sample equation showing how to combine three different vectors chosen randomly to create a mutant vector.

$$\boldsymbol{v}_{i,g} = \mathbf{x}_{r0,g} + F \cdot \left(\mathbf{x}_{r1,g} - \mathbf{x}_{r2,g}\right) \tag{8}$$

The scale factor, F ∈ (0,1) is a positive real number that controls the rate at which the population evolves. While there is no upper limit on F, the values are rarely greater than 1.0.

2.4 Crossover

To complement the differential mutation search strategy, DE also uses uniform crossover. Sometimes known as discrete recombination (dual). In particular, DE crosses each vector with a mutant vector:

$$U_{i,g} = (u_{j,i,g}) = \begin{Bmatrix} v_{j,i,g} & if\left(rand_j(0,1) \leq Cr \, or \, j = j_{rand}\right) \\ x_{j,i,g} & otherwise. \end{Bmatrix} \tag{9}$$

2.5 Selection

If the test vector, $\boldsymbol{U}_{i,g}$ has a value of the objective function equal to or less than its target vector, $\mathbf{X}_{i,g}$. It replaces the target vector in the next generation; otherwise, the target retains its place in population for at least another generation [2].

$$X_{i,g+1} = \begin{cases} U_{i,g} & \text{if } f(U_{i,g}) \leq f(X_{i,g}) \\ X_{i,g} & \text{otherwise.} \end{cases} \tag{10}$$

The process of mutation, recombination and selection are repeated until the optimum is found, or terminating pre criteria specified is satisfied. DE is a simple, but powerful search engine that simulates natural evolution combined with a mechanism to generate multiple search directions based on the distribution of solutions in the current population. Each vector *i* in the population at generation *G*, **x***i*,*G*, called at this moment of reproduction as the target vector will be able to generate one offspring, called trial vector (**u***i*,*G*). This trial vector is generated as follows: First of all, a search direction is defined by calculating the difference between a pair of vectors *r*1 and *r*2, called "*differential vectors*", both of them chosen at random from the population. This difference vector is also scaled by using a user defined parameter called "$F \geq 0$". This scaled difference vector is then added to a third vector *r*3, called "*base vector*". As a result, a new vector is obtained, known as the mutation vector. After that, this mutation vector is recombined with the target vector (also called parent vector) by using discrete recombination (usually binomial crossover) controlled by a crossover parameter $0 \leq CR \leq 1$ whose value determines how similar the trial vector will be with respect to the target vector. There are several DE variants. However, the most known and used is DE/rand/1/bin, where the base vector is chosen at random, there is only a pair of differential vectors and a binomial crossover is used. The detailed pseudocode of this variant is presented in Fig. 1 [9].

Fig. 1 "DE/rand/1/bin" pseudocode rand [0, 1) is a function that returns a real number between 0 and 1. Randint (min, max) is a function that returns an integer number between min and max. *NP*, *MAX GEN*, *CR* and *F* are user-defined parameters *n* is the dimensionality of the problem [9]

Begin
  G=0
  Create a random initial population $\mathbf{x}_{i,G}\ \forall i, i = 1, \ldots, NP$
  Evaluate $f(\mathbf{x}_{i,G})\ \forall i, i = 1, \ldots, NP$
  For G=1 to MAX_GEN Do
    For i=1 to NP Do
      Select randomly $r_1 \neq r_2 \neq r_3$:
      $j_{rand} = \text{randint}(1, D)$
      For j=1 to n **Do**
        If $(rand_j[0,1) < CR$ **or** $j = j_{rand})$ **Then**
          $u_{i,j,G+1} = x_{r_3,j,G} + F(x_{r_1,j,G} - x_{r_2,j,G})$
        Else
          $u_{i,j,G+1} = x_{i,j,G}$
        End If
      End For
      If $(f(\mathbf{u}_{i,G+1}) \leq f(\mathbf{x}_{i,G}))$ **Then**
        $\mathbf{x}_{i,G+1} = \mathbf{u}_{i,G+1}$
      Else
        $\mathbf{x}_{i,G+1} = \mathbf{x}_{i,G}$
      End If
    End For
    $G = G + 1$
  End For
End

2.6 *Illustrative Example of the Classic DE Algorithm*

A simple numerical example adopted is presented to illustrate the classic DE algorithm. Let us consider the following objective function for optimization:

$$\text{Minimize } f(x) = x_1 + x_2 + x_3$$

The initial population is chosen randomly between the bounds of decision variables, in this case *x1, x2* and *x3* ϵ [*0, 1*]. The population along with its respective objective function values is shown in Table 1. The first member of the population, "Individual 1", is set as the target vector. In order to generate the mutated vector, three individuals ("Individual 2", "Individual 4" and "Individual 6") from the population size are selected randomly (ignoring "Individual 1", since it is set as the target vector). The weighted difference between "Individual 2" and "Individual 4" is added to the third randomly chosen vector "Individual 6" to generate the mutated vector. The weighting factor *F* is chosen as 0.80 and the weighted difference vector is obtained in Table 2 and the mutated vector in Table 3 [16].

The mutated vector does a crossover with the target vector to generate the trial vector, as shown in Table 4. This is carried out by (1) generating random numbers equal to the dimension of the problem (2) for each of the dimensions: if random number > *CR*; copy the value from the target vector, else copy the value from the mutated vector into the trial vector. In this example, the crossover constant *CR* is chosen as 0.50.

Table 1 An illustrative example [16]

	Population size NP = 6 (user define), D = 3					
	Individual 1	Individual 2	Individual 3	Individual 4	Individual 5	Individual 6
x_1	0.68	0.92	0.22	0.12	0.40	0.94
x_2	0.89	0.92	0.14	0.09	0.81	0.63
x_3	0.04	0.33	0.40	0.05	0.83	0.13
f(x)	1.61	2.17	0.76	0.26	2.04	1.70

Table 2 Calculation of the weighted difference vector for the illustrative example [16]

	Individual 2	Individual 4	Difference vector		Weighted difference vector
x_1	0.92	0.12	=0.80		=0.64
x_2	0.92	0.09	=0.83	***xF*** (F = 0.80)	=0.66
x_3	0.33	0.05	=0.28		=0.22

Table 3 Calculation of the mutated vector for the illustrative example [16]

	Weighted difference vector		Individual 6	Muted vector
x_1	0.64		0.94	=1.58
x_2	0.66	+	0.63	=1.29
x_3	0.22		0.13	=0.35

Table 4 Generation of the trial vector for the illustrative example [16]

	Target vector		Mutated vector	Trial vector
x_1	0.68		**1.58**	=1.58
x_2	**0.89**	Crossover	1.29	=0.89
x_3	**0.04**	(CR = 0.50)	0.35	=0.04
f(x)	1.61		3.22	2.51

Table 5 New populations for the next generation in the illustrative example [16]

	New population for the next generation					
	Individual 1	Individual 2	Individual 3	Individual 4	Individual 5	Individual 6
x_1	0.68					
x_2	0.89					
x_3	0.04					
f(x)	1.61					

The objective function of the trial vector is compared with that of the target vector and the vector with the lowest value of the two (minimization problem) becomes "Individual 1" for the next generation. To evolve "Individual 2" for the next generation, the second member of the population is set as target vector (see Table 5) and the above process is repeated. This process is repeated *NP* times until the new population set array is filled, which completes one generation. Once the termination criterion is met, the algorithm ends.

3 Proposed Method

The Differential Evolution (DE) Algorithm is a powerful search technique used for solving optimization problems. In this paper a new algorithm called Fuzzy Differential Evolution (FDE) with dynamic adjustment of parameters for the optimization of controllers is proposed. The main objective is that the fuzzy system will provides us with the optimal parameters for the best performance of the DE algorithm. In addition the parameters that the fuzzy system optimizes are the crossover and mutation, as shown in Fig. 2.

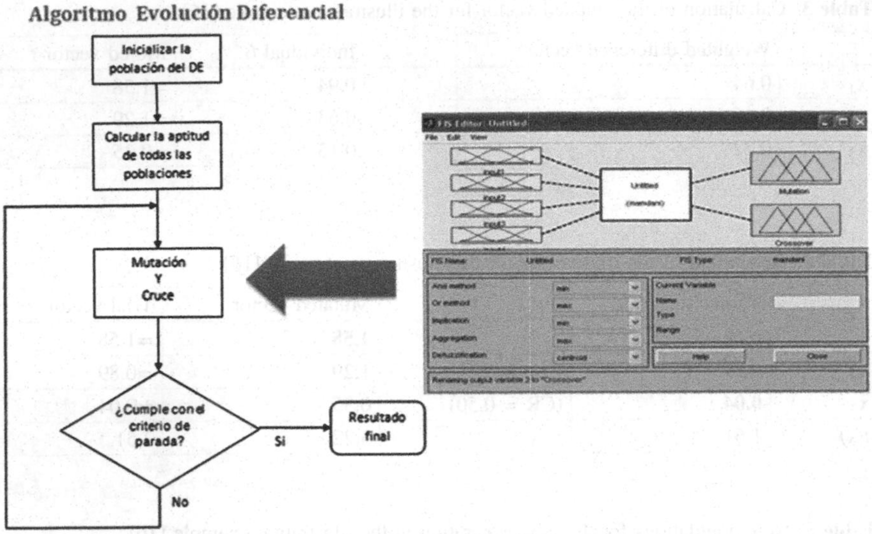

Fig. 2 The proposed is an algorithm of differential evolution (DE) by integrating a fuzzy system to dynamically adapt parameters

4 Simulation Results

This paper presents the current results obtained from the experiments using the scale factor F from the Differential Evolution (DE) algorithm. This helps us visualize how the algorithm performs modifications to this variant which helps create mutations. This helps us visualize how the algorithm performs modifications to this variant which helps create mutations. The Rosenbrock mathematical function was used to carry out these experiments. Rosenbrock is a classic optimization problem, also known as banana function or the second function of De Jong. The global optimum lies inside a long, narrow, parabolic shaped flat valley. To find the valley is trivial, however convergence to the global optimum is difficult and hence this problem has been frequently used to test the performance of optimization algorithms. Function has the following definition (Fig. 3).

$$f(x) = \sum_{i=1}^{n-1} \left[100\left(x_{i+1} - x_i^2\right)^2 + (1 - x_i)^2\right]. \quad (11)$$

Test area is usually restricted to hypercube $-2.048 \leq x_i \leq 2.048$, $i = 1,\ldots, n$.Its global minimum equal $f(x) = 0$ *is obtainable for* $x_i = 0$, $i = 1,\ldots,n$.

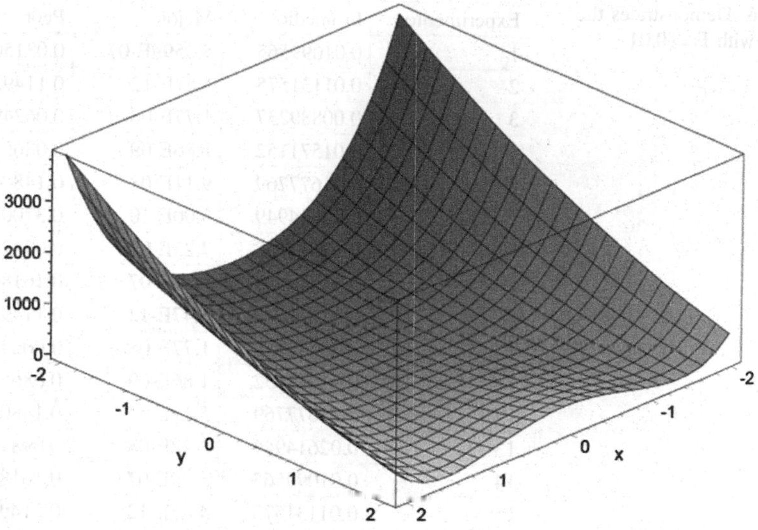

Fig. 3 Rosenbrock's in 2D, $f(x,y) = 100(y - x^2)^2 + (1 - x)^2$

30 experiments were done each one 30 times, using the following parameters:

Parameters:
D = 50
NP = 250
F = 0.1
CR = 0.1
GEN = 6000
L = −500
H = 500

where:
D = Vector Dimension
NP = Size of population
F = scale factor
CR = crossover
GEN = Maximum number of generations
L = Lower Limits
H = Upper Limits

Table 6 shows the simulation results when F = 0.01. Table 7 shows the results when F = 0.02. The experiments with F = 0.03 and 0.04 are shown on Tables 8 and 9, respectively.

Table 6 Demonstrates the results with F = 0.01

Experimento	Promedio	Mejor	Peor
1	0.01099565	5.3599E-07	0.03150259
2	0.01131575	4.47E-12	0.1149224
3	0.00889237	1.77E-08	0.06249856
4	0.01571152	1.86E-09	0.08654202
5	0.01677769	9.11E-08	0.1480454
6	0.02614949	5.00E-10	0.37007527
7	0.01057668	7.23E-07	0.09474443
8	0.01099565	5.36E-07	0.16184713
9	0.01131575	4.47E-12	0.1149224
10	0.00889237	1.77E-08	0.06249856
11	0.01571152	1.86E-09	0.08654202
12	0.01677769	5.14E-07	0.1480454
13	0.02614949	3.48E-08	0.09871141
14	0.01099565	5.79E-07	0.16184713
15	0.01131575	4.47E-12	0.1149224
16	0.00889237	4.47E-06	0.06249856
17	0.01571152	1.86E-09	0.08654202
18	0.01677769	9.11E-08	0.06250744
19	0.02614949	5.00E-10	0.37007527
20	0.01099565	5.79E-07	0.16184713
21	0.01131575	4.47E-12	0.1149224
22	0.00889237	1.77E-08	0.06249856
23	0.01571152	1.43E-11	0.08654202
24	0.01677769	9.11E-08	0.1480454
25	0.02614949	5.00E-10	0.11850916
26	0.01057668	7.23E-07	0.09474443
27	0.01931193	4.17E-08	0.10814125
28	0.01402754	3.80E-09	0.03090101
29	0.01099565	5.79E-07	0.16184713
30	0.01131575	4.47E-12	0.1149224

Table 7 Demonstrates the results with F = 0.02

Experimento	Promedio	Mejor	Peor
1	0.01099565	5.3599E-07	0.03150259
2	0.01131575	4.47E-12	0.1149224
3	0.00889237	1.77E-08	0.06249856
4	0.01571152	1.86E-09	0.08654202
5	0.01677769	9.11E-08	0.1480454
6	0.02614949	5.00E-10	0.37007527
7	0.01057668	7.23E-07	0.09474443
8	0.01099565	5.36E-07	0.16184713
9	0.01131575	4.47E-12	0.1149224
10	0.00889237	1.77E-08	0.06249856
11	0.01571152	1.86E-09	0.08654202
12	0.01677769	5.14E-07	0.1480454
13	0.02614949	3.48E-08	0.09871141
14	0.01099565	5.79E-07	0.16184713
15	0.01131575	4.47E-12	0.1149224
16	0.00889237	4.47E-06	0.06249856
17	0.01571152	1.86E-09	0.08654202
18	0.01677769	9.11E-08	0.06250744
19	0.02614949	5.00E-10	0.37007527
20	0.01099565	5.79E-07	0.16184713
21	0.01131575	4.47E-12	0.1149224
22	0.00889237	1.77E-08	0.06249856
23	0.01571152	1.43E-11	0.08654202
24	0.01677769	9.11E-08	0.1480454
25	0.02614949	5.00E-10	0.11850916
26	0.01057668	7.23E-07	0.09474443
27	0.01931193	4.17E-08	0.10814125
28	0.01402754	3.80E-09	0.03090101
29	0.01099565	5.79E-07	0.16184713
30	0.01131575	4.47E-12	0.1149224

Table 8 Demonstrates the results with F = 0.03

Experimento	Promedio	Mejor	Peor
1	6.66E-04	1.73E-17	0.01785224
2	1.39E-02	5.39E-18	0.01226117
3	1.06E-02	3.04E-14	0.00209507
4	2.62E-03	5.02E-13	0.05327532
5	6.66E-04	1.73E-17	0.01785224
6	1.15E-03	1.59E-17	0.01705839
7	1.34E-03	5.49E-17	0.01487419
8	1.39E-02	1.61E-20	0.38046604
9	9.61E-03	6.21E-15	0.17429945
10	6.36E-03	8.17E-17	0.06436387
11	1.06E-02	3.06E-15	0.31396925
12	9.78E-04	9.51E-15	0.02470255
13	6.66E-04	1.73E-17	0.01785224
14	2.62E-03	4.85E-16	0.05327532
15	1.39E-02	6.56E-17	0.38046604
16	1.06E-02	7.07E-13	0.31396925
17	2.62E-03	4.85E-16	0.05327532
18	6.66E-04	1.73E-17	0.01785224
19	1.15E-03	1.59E-17	0.01705839
20	1.39E-02	5.39E-18	0.01226117
21	1.06E-02	3.06E-15	0.31396925
22	2.62E-03	4.85E-16	0.05327532
23	1.15E-03	1.59E-17	0.01705839
24	1.34E-03	5.49E-17	0.01487419
25	1.39E-02	6.56E-17	0.38046604
26	9.61E-03	6.21E-15	0.17429945
27	6.36E-03	8.17E-17	0.06436387
28	1.06E-02	3.06E-15	0.31396925
29	9.78E-04	9.37E-15	0.02470255
30	3.73E-04	4.85E-17	0.00940052

Table 9 Demonstrates the results with F = 0.04

Experimento	Promedio	Mejor	Peor
1	0.01179166	4.34E-13	0.12649579
2	0.00689591	3.79E-11	0.07022387
3	0.00884897	8.91E-16	0.02545798
4	0.00786076	3.11E-12	0.08015746
5	0.01261989	4.97E-11	0.27651624
6	0.01652249	2.22E-10	0.25001424
7	0.01179166	4.34E-13	0.12649579
8	0.00884897	8.91E-16	0.17181307
9	0.00786076	3.11E-12	0.08015746
10	0.01261989	4.97E-11	0.27651624
11	0.01652249	3.49E-11	0.25001424
12	0.02675393	5.16E-13	0.17059178
13	0.00347209	1.08E-13	0.02806518
14	0.04041566	1.80E-15	0.74575319
15	0.00468022	9.61E-13	0.05179923
16	0.01471826	1.73E-12	0.21308918
17	0.01179166	4.34E-13	0.06084437
18	0.00884897	8.91E-16	0.02545798
19	0.00786076	3.11E-12	0.08015746
20	0.01652249	3.49E-11	0.25001424
21	0.02675393	5.16E-13	0.3559589
22	0.00347209	1.08E-13	0.02806518
23	0.04041566	1.80E-15	0.36500605
24	0.00468022	9.61E-13	0.05179923
25	0.01471826	3.93E-09	0.21308918
26	0.00247429	7.88E-13	0.01730106
27	0.02497659	2.84E-10	0.0582611
28	0.00639566	3.73E-13	0.04982843
29	0.0372501	6.99E-11	0.61885365
30	0.00866303	7.95E-12	0.11939131

5 Conclusions

To conclude this paper, the preliminary results will help us better understand with more clarity the Differential Evolution (DE) algorithm. In the same way we can discover the way of exploring and then exploitating the algorithm. The results will be observed to see if the majority are good although in very few cases the errors are very high. We can tentatively conclude when Fuzzy logic is applied we can get better results from Differential Evolution (DE) avoiding high errors, like those obtained when trial and error is used.

References

1. Aguas-Marmolejo, S.J., Castillo, O.: Optimization of membership functions for Type-1 and Type 2 fuzzy controllers of an autonomous mobile robot using PSO. In: Recent Advances on Hybrid Intelligent Systems, pp. 97–104 (2013)
2. Astudillo, L., Melin, P., Castillo, O.: Optimization of a fuzzy tracking controller for an autonomous mobile robot under perturbed torques by means of a chemical optimization paradigm. In: Recent Advances on Hybrid Intelligent Systems, pp. 3–20 (2013)
3. Eftekhari, M., Katebi, S.D., Karimi, M., Jahanmir, A.H.: Eliciting transparent fuzzy model using differential evolution. Appl. Soft Comput. School of Engineering, Shiraz University, Shiraz, Iran, **8**, 466–476 (2008)
4. Fierro, R., Castillo, O.: Design of fuzzy control systems with different PSO variants. In: Recent Advances on Hybrid Intelligent Systems, pp. 81–88 (2013)
5. Hachicha, N., Jarboui, B., Siarry, P.: A fuzzy logic control using a differential evolution algorithm aimed at modelling the financial market dynamics. Institut Supérieur de Commerce et de Comptabilité de Bizerte, Zarzouna 7021, Bizerte, Tunisia, Information Sciences **181**, 79–91 (2011)
6. Lizárraga, E., Castillo, O., Soria, J.: A method to solve the traveling salesman problem using ant colony optimization variants with ant set partitioning. In: Recent Advances on Hybrid Intelligent Systems, pp. 237–2461 (2013)
7. Melendez, A., Castillo, O.: Evolutionary optimization of the fuzzy integrator in a navigation system for a mobile robot. In: Recent Advances on Hybrid Intelligent Systems, pp. 21–31 (2013)
8. Melin, P., Olivas, F., Castillo, O., Valdez, F., Soria, J., García, J.: Optimal design of fuzzy classification systems using PSO with dynamic parameter adaptation through fuzzy logic. Expert Syst. Appl. **40**(8), 3196–3206 (2013)
9. Mezura-Montes, E. Palomeque-Ortiz, A.: Self-adaptive and deterministic parameter control in differential evolution for constrained optimization. Efren Mezura-Montes, Laboratorio Nacional de Inform´atica Avanzada (LANIA A.C.), R´ebsamen 80, Centro, Xalapa, Veracruz, 91000, MEXICO (2009)
10. Neyoy, H., Castillo, O., José Soria: Dynamic Fuzzy Logic Parameter Tuning for ACO and Its Application in TSP Problems. Recent Advances on Hybrid Intelligent Systems, pp. 259–271 (2013)
11. Oh, S.-K., Kim, W.-D., Pedrycz, W.: Design of optimized cascade fuzzy controller based on differential evolution: Simulation studies and practical insights, Department of Electrical Engineering, The University of Suwon. Eng. Appl. Artif. Intell. **25**, 520–532 (2012)
12. Olivas F., Castillo O.: Particle Swarm Optimization with Dynamic Parameter Adaptation Using Fuzzy Logic for Benchmark Mathematical Functions. Recent Advances on Hybrid Intelligent Systems 2013: 247–258
13. Price, Storn R., Lampinen, J. A., Differential Evolution, Kenneth V., Springer, Berlin (2005)
14. Raofen, W., Zhang, J., Zhang, Y., Wang, X.: Assessment of human operator functional state using a novel differential evolution optimization based adaptive fuzzy model, Lab for Brain-Computer Interfaces and Control, East China University of Science and Technology, Shanghai 200237, PR China, Biomedical Signal Processing and Control 7 (2012) 490– 498
15. Valdez, F., Melin, P., Castillo, O.: Bio-inspired optimization methods on graphic processing unit for minimization of complex mathematical functions. In: Recent Advances on Hybrid Intelligent Systems, pp. 313–322 (2013)
16. Vucetic, D.: Fuzzy Differential Evolution Algorithm. The University of Western Ontario, London, Ontario, Canada (2012)

Ensemble Neural Network with Type-1 and Type-2 Fuzzy Integration for Time Series Prediction and Its Optimization with PSO

Patricia Melin, Martha Pulido and Oscar Castillo

Abstract This paper describes the design of ensemble neural networks using Particle Swarm Optimization (PSO) for time series prediction with Type-1 and Type-2 Fuzzy Integration. The time series that is being considered in this work is the Mackey-Glass benchmark time series. Simulation results show that the ensemble approach produces good prediction of the Mackey-Glass time series.

Keywords Ensemble neural networks · Particle swarm · Optimization · Time series prediction

1 Introduction

Time Series is defined as a set of measurements of some phenomenon or experiment recorded sequentially in time. The first step in analyzing a time series is to plot it, this allows: to identify the trends, seasonal components and irregular variations. A classic model for a time series can be expressed as a sum or product of three components: trend, seasonality and random error term.

Time series predictions are very important because based on them we can analyze past events to know the possible behavior of futures events and thus we can take preventive or corrective decisions to help avoid unwanted circumstances.

The contribution of this paper is the proposed approach for ensemble neural network optimization using particle swarm optimization. The proposed models are also used as a basis for statistical tests [1–4, 9, 10, 12, 14, 15, 19–22].

The rest of the paper is organized as follows: Sect. 2 describes the concepts of optimization, Sect. 3 describes the concepts of particle swarm optimization, Sect. 4

P. Melin (✉) · M. Pulido · O. Castillo
Tijuana Institute of Technology, Tijuana, Mexico
e-mail: pmelin@tectijuana.mx

P. Angelov and S. Sotirov (eds.), *Imprecision and Uncertainty in Information Representation and Processing*, Studies in Fuzziness and Soft Computing 332,
DOI 10.1007/978-3-319-26302-1_22

describes the concepts of Fuzzy Systems as Methods of integration, Sect. 5 describes the problem and the proposed method of solution, Sect. 6 describes the simulation results of the proposed method, and Sect. 7 shows the conclusions.

2 Optimization

Regarding optimization, we have the following situation in mind: there exists a search space *V*, and a function:

$$g\colon V \rightarrow \mathbb{R}$$

and the problem is to find

$$\underset{v \in V}{\arg\min}\, \mathbf{g}.$$

Here, ***V** is vector of decision variables,* and ***g** is the objective function.* In this case we have assumed that the problem is one of minimization, but everything we say can of course be applied *mutatis mutandis* to a maximization problem. Although specified here in an abstract way, this is nonetheless a problem with a huge number of real-world applications.

In many cases the search space is discrete, so that we have the class of *combinatorial optimization problems* (COPs). When the domain of the **g** function is continuous, a different approach may well be required, although even here we note that in practice, optimization problems are usually solved using a computer, so that in the final analysis the solutions are represented by strings of binary digits (bits) [32].

There are several optimization techniques that can be applied to neural networks, some of these are: evolutionary algorithms [18], ant colony optimization [5] and Particle swarm [7].

3 Particle Swarm Optimization

The Particle Swarm Optimization algorithm maintains a swarm of particles, where each particle represents a potential solution. In analogy with evolutionary computation paradigms, a swarm is a population, while a particle is similar to an individual. In simple terms, the particles are "flown" through a multidimensional search space where the position of each particle is adjusted according to its own experience and that of their neighbors. Let x_i(t) denote the position of particle *i* in the search space at time step t unless otherwise selected, t denotes discrete time steps. The position of the particle is changed by adding a velocity, $v_i(t)$ to the current position i.e.

$$x_i(t+1) = x_i(t) + v_i(t+1) \tag{1}$$

with $x_i(0) \sim U(X_{min}, X_{max})$.
It is the velocity vector the one that drives of the optimization process, and reflects both the experimental knowledge of the particles and the information exchanged in the vicinity of particles. The experimental knowledge of a particle which is generally known as the cognitive component, which is proportional to the distance of the particle from its own best position (hereinafter, the personal best position particles) that are from the first step. Socially exchanged information is known as the social component of the velocity equation.

For the gbest PSO, the particle velocity is calculated as:

$$v_{ij}(t+1) = v_{ij}(t) + c_1 r_1 \left[y_{ij}(t) - x_{ij}(t)\right], + c_2 r_2(t)\left[\hat{y}_j(t) - x_{ij}(t)\right] \tag{2}$$

where $v_{ij}(t)$ is the velocity of the particle i in dimension j at time step t, c_1 y c_2 are positive acceleration constants used to scale the contribution of cognitive and social skills, respectively, y $r_{1j}(t)$, y $r_{2j}(t) \sim U(0, 1)$ are random values in the range [0, 1].

The best personal position in the next time step $t + 1$ is calculated as:

$$y_i(t+1) = \begin{cases} y_i(t) & \text{if } f(x_i(x_i(t+1)) \geq f\,y_i(t)) \\ x_i(t+1) & \text{if } f(x_i(x_i(t+1)) > f\,y_i(t)) \end{cases} \tag{3}$$

where $f: \mathbb{R}^{nx} \rightarrow \mathbb{R}$ is the fitness function, as with EAs, measuring fitness with the function will help find the optimal solution, for example the objective function quantifies the performance, or the quality of a particle (or solution).

The overall best position, $\hat{y}(t)$ at time step t, is defined as:

$$\hat{y}(t) \quad \epsilon\{y_0(t), \ldots, y_{ns}(t)\} f(y(t)) = \min\{f(y_0(t)), \ldots f(y_{ns}(t)),\} \tag{4}$$

where n_S is the total number of particles in the swarm. Importantly, the above equation defining and establishing $\hat{y}$ the best position is uncovered by either of the particles so far as this is usually calculated from the best position best personal [5, 6, 10].

The overall best position may be selected from the actual swarm particles, in which case:

$$\hat{y}(t) = \min\{f(x_o(t)), \ldots f(x_{ns}(t)),\} \tag{5}$$

4 Fuzzy Systems as Methods of Integration

Fuzzy logic was proposed for the first time in the mid-sixties at the University of California Berkeley by the brilliant engineer Lofty A. Zadeh., who proposed what it's called the principle of incompatibility: "As the complexity of system increases, our ability to be precise instructions and build on their behavior decreases to the threshold beyond which the accuracy and meaning are mutually exclusive characteristics." Then introduced the concept of a fuzzy set, under which lies the idea that the elements on which to build human thinking are not numbers but linguistic labels. Fuzzy logic can represent the common knowledge as a form of language that is mostly qualitative and not necessarily a quantity in a mathematical language [29].

Type-1 Fuzzy system theory was first introduced by Zadeh [13] in 1965, and has been applied in many areas such as control, data mining, time series prediction, etc.

The basic structure of a fuzzy inference system consists of three conceptual components: a rule base, which contains a selection of fuzzy rules, a database (or dictionary) which defines the membership functions used in the rules, and reasoning mechanism, which performs the inference procedure (usually fuzzy reasoning) [14].

Type-2 Fuzzy systems were proposed to overcome the limitations of a type-1 FLS, the concept of type-1 fuzzy sets was extended into type-2 fuzzy sets by Zadeh in 1975. These were designed to mathematically represent the vagueness and uncertainty of linguistic problems; thereby obtaining formal tools to work with intrinsic imprecision in different type of problems; it is considered a generalization of the classic set theory. Type-2 fuzzy sets are used for modeling uncertainty and imprecision in a better way [15–17].

5 Problem Statement and Proposed Method

The objective of this work is to develop a model that is based on integrating the responses of an ensemble neural network using type-1 and type-2 fuzzy systems and their optimization. Figure 1 represents the general architecture of the proposed method, where historical data, analyzing data, creation of the ensemble neural network and integrate responses of the ensemble neural network with type-2 fuzzy system integration and finally obtaining the outputs as shown. The information can be historical data, these can be images, time series, etc., in this case we show the application to time series prediction of the Dow Jones where we obtain good results with this series.

Figure 2 shows a type-2 fuzzy system consisting of 5 inputs depending on the number of modules of the neural network ensemble and one output. Each input and output linguistic variable of the fuzzy system uses 2 Gaussian membership functions. The performance of the type-2 fuzzy integrators is analyzed under different levels of uncertainty to find out the best design of the membership functions and consist of 32 rules. For the type-2 fuzzy integrator using 2 membership functions,

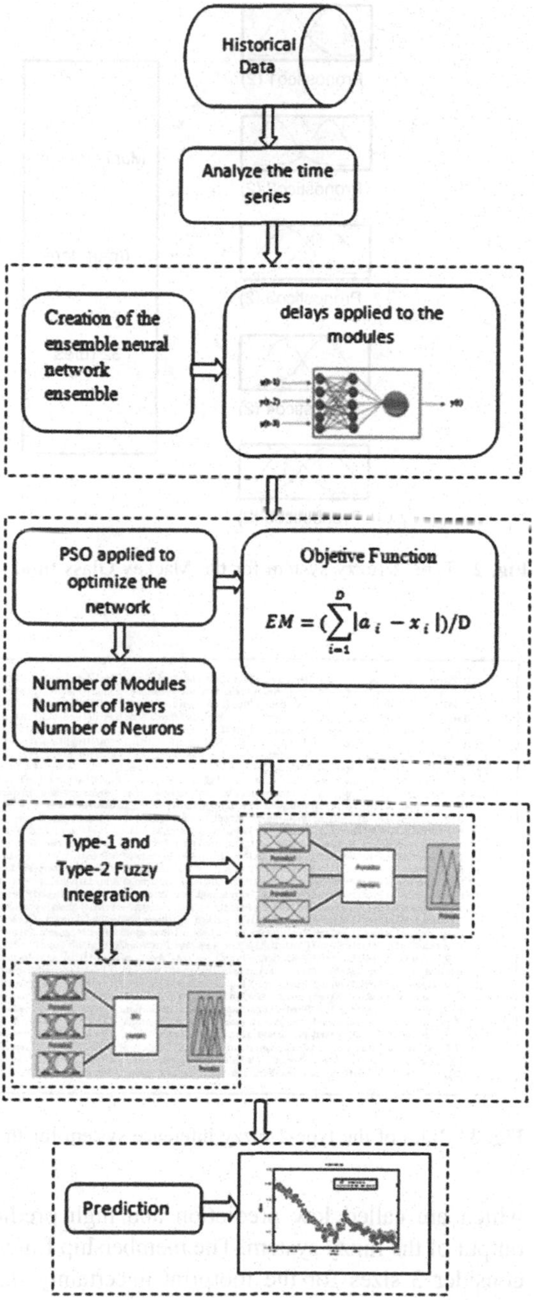

Fig. 1 General architecture of the proposed method

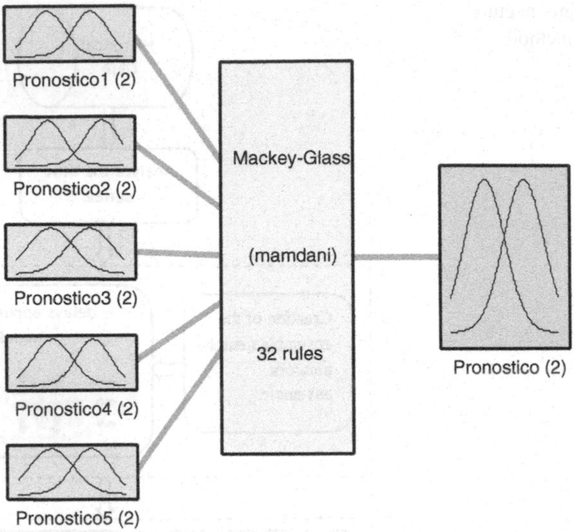

Fig. 2 Type-2 fuzzy system for the Mackey Glass time series

1. If (Prrediction1 is Pred1Low) and (Prediction2 is Pron2Low) and (Prediction3 is Pred3Low) and (Prediction4 is Pred4Low) and (Prediction5is Pred5Low) then (Prediction is Low)
2. If (Prrediction1 is Pred1High) and (Prediction2 is Pron2High) and (Prediction3 is Pred3High)and (Prediction4 is Pred4Low) and (Prediction5is Pred5High) then (Prediction is High)
3. If (Prrediction1 is Pred1Low) and (Prediction2 is Pron2Low) and (Prediction3 is Pred3Low) and (Prediction4 is Pred4Low) and (Prediction5is Pred5High) then (Prediction is Low)
4. If (Prrediction1 is Pred1High) and (Prediction2 is Pron2High) and (Prediction3 is Pred3High)and (Prediction4 is Pred4Low) and (Prediction5is Pred5Low) then (Prediction is High)
5. If (Prrediction1 is Pred1Low) and (Prediction2 is Pron2Low) and (Prediction3 is Pred3Low) and (Prediction4 is Pred4High) and (Prediction5is Pred5High) then (Prediction is Low)
6. If (Prrediction1 is Pred1High) and (Prediction2 is Pron2High) and (Prediction3 is Pred3High)and (Prediction4 is Pred4Low) and (Prediction5is Pred5Low) then (Prediction is High)
7. If (Prrediction1 is Pred1Low) and (Prediction2 is Pron2Low) and (Prediction3 is Pred3High) and (Prediction4 is Pred4High) and (Prediction5is Pred5High) then (Prediction is High)
8. If (Prrediction1 is Pred1High) and (Prediction2 is Pron2High) and (Prediction3 is Pred3Low)and (Prediction4 is Pred4Low) and (Prediction5is Pred5Low) then (Prediction is Low)
9. If (Prrediction1 is Pred1Low) and (Prediction2 is Pron2Highand (Prediction3 is Pred3High) and (Prediction4 is Pred4High) and (Prediction5is Pred5High) then (Prediction is High)
10. If (Prrediction1 is Pred1High) and (Prediction2 is Pron2Low) and (Prediction3 is Pred3Low)and (Prediction4 is Pred4Low) and (Prediction5is Pred5Low) then (Prediction is Low)
11. If (Prrediction1 is PredLow) and (Prediction2 is Pron2High) and (Prediction3 is Pred3Low)and (Prediction4 is Pred4High) and (Prediction5is Pred5Low) then (Prediction is Low)
12. If (Prrediction1 is Pred1High) and (Prediction2 is Pron2Low) and (Prediction3 is Pred3H¡gh)and (Prediction4 is Pred4Low) and (Prediction5is Pred5High) then (Prediction is High)
13. If (Prrediction1 is PredLow) and (Prediction2 is Pron2High) and (Prediction3 is Pred3Low)and (Prediction4 is PredLow) and (Prediction5is Pred5Low) then (Prediction is Low)
14. If (Prrediction1 is Pred1High) and (Prediction2 is Pron2Low) and (Prediction3 is Pred3H¡gh)and (Prediction4 is Pred4Low) and (Prediction5is Pred5High) then (Prediction is High)
15. If (Prrediction1 is PredLow) and (Prediction2 is Pron2High) and (Prediction3 is Pred3Low)and (Prediction4 is PredLow) and (Prediction5is Pred5Low) then (Prediction is Low)
16. If (Prrediction1 is Pred1High) and (Prediction2 is Pron2Low) and (Prediction3 is Pred3H¡gh)and (Prediction4 is Pred4Low) and (Prediction5is Pred5High) then (Prediction is High)
17. If (Prrediction1 is PredLow) and (Prediction2 is Pron2Low) and (Prediction3 is Pred3High)and (Prediction4 is PredLow) and (Prediction5is Pred5Low) then (Prediction is Low)
18. If (Prrediction1 is Pred1High) and (Prediction2 is Pron2High) and (Prediction3 is Pred3Lowh)and (Prediction4 is Pred4Low) and (Prediction5is Pred5High) then (Prediction is High)
19. If (Prrediction1 is PredLow) and (Prediction2 is Pron2Low) and (Prediction3 is Pred3High)and (Prediction4 is Pred4High) and (Prediction5is Pred5Low) then (Prediction is Low)
20. If (Prrediction1 is Pred1High) and (Prediction2 is Pron2High) and (Prediction3 is Pred3Lowh)and (Prediction4 is Pred4Low) and (Prediction5is Pred5High) then (Prediction is High)
21. If (Prrediction1 is Pred1Low) and (Prediction2 is Pron2High) and (Prediction3 is Pred3High)and (Prediction4 is Pred4Low) and (Prediction5is Pred5Low) then (Prediction is Low)
22. If (Prrediction1 is Pred1High) and (Prediction2 is Pron2Low) and (Prediction3 is Pred3Low)and (Prediction4 is Pred4High) and (Prediction5is Pred5High) then (Prediction is High)
23. If (Prrediction1 is Pred1Low) and (Prediction2 is Pron2Low) and (Prediction3 is Pred3High)and (Prediction4 is Pred4High) and (Prediction5is Pred5Low) then (Prediction is Low)
24. If (Prrediction1 is Pred1High) and (Prediction2 is Pron2High) and (Prediction3 is Pred3Low)and (Prediction4 is Pred4Low) and (Prediction5is Pred5High) then (Prediction is High)
25. If (Prrediction1 is Pred1Low) and (Prediction2 is Pron2High) and (Prediction3 is Pred3High)and (Prediction4 is Pred4Low) and (Prediction5is Pred5High) then (Prediction is High)
26. If (Prrediction1 is Pred1Low) and (Prediction2 is Pron2High) and (Prediction3 is Pred3Low)and (Prediction4 is Pred4High) and (Prediction5is Pred5Low) then (Prediction is Low)
27. If (Prrediction1 is Pred1Low) and (Prediction2 is Pron2High) and (Prediction3 is Pred3Low)and (Prediction4 is Pred4High) and (Prediction5is Pred5Low) then (Prediction is Low)
28. If (Prrediction1 is Pred1Low) and (Prediction2 is Pron2High) and (Prediction3 is Pred3Low)and (Prediction4 is Pred4High) and (Prediction5is Pred5Low) then (Prediction is Low)
29. If (Prrediction1 is Pred1Low) and (Prediction2 is Pron2High) and (Prediction3 is Pred3High)and (Prediction4 is Pred4High) and (Prediction5is Pred5Low) then (Prediction is High)
30. If (Prrediction1 is Pred1High) and (Prediction2 is Pron2Low) and (Prediction3 is Pred3Low)and (Prediction4 is Pred4Low) and (Prediction5is Pred5High) then (Prediction is Low)
31. If (Prrediction1 is Pred1Low) and (Prediction2 is Pron2High) and (Prediction3 is Pred3Low)and (Prediction4 is Pred4High) and (Prediction5is Pred5High) then (Prediction is High)
302 If (Prrediction1 is Pred1High) and (Prediction2 is Pron2Low) and (Prediction3 is Pred3High)and (Prediction4 is Pred4Low) and (Prediction5is Pred5Low) then (Prediction is Low)

Fig. 3 Rules of the type-2 fuzzy inference system for the Dow Jones time series

which are called low prediction and high prediction for each of the inputs and output of the fuzzy system. The membership functions are of Gaussian type, and we consider 3 sizes for the footprint uncertainty 0.3, 0.4 and 0.5 to obtain a better prediction of our time series.

In this Fig. 3 shows the possible rules of a type-2 fuzzy system.

Number of Modules	Number of Layers 1	Neurons 1	Neurons ... n

Fig. 4 Particle structure to optimize the ensemble neural network

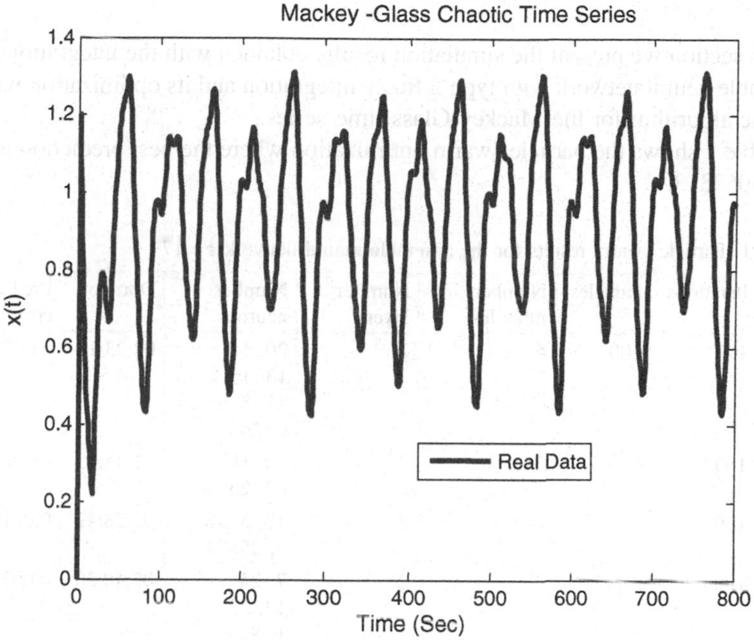

Fig. 5 Mackey Glass time series

Figure 4 represents the Particle Structure to optimize the ensemble neural network, where the parameters that are optimized are the number de modules, number of layers, number of neurons.

Data of the Mackey-Glass time series was generated using Eq. (6). We are using 800 points of the time series. We use 70 % of the data for the ensemble neural network trainings and 30 % to test the network.

The Mackey-Glass Equation is defined as follows:

$$\dot{x}(t) = \frac{0.2x(t-\tau)}{1+x^{10}(t-\tau)} - 0.1x(t) \qquad (6)$$

where it is assumed x(0) = 1.2, $\tau = 17$, $\tau = 34$, and 68 x(t) = 0 for t < 0. Figure 5 shows a plot of the time series for these parameter values.

This time series is chaotic, and there is no clearly defined period. The series does not converge or diverge, and the trajectory is extremely sensitive to the initial conditions. The time series is measured in number of points, and we apply the fourth order Runge-Kutta method to find the numerical solution of the equation [10, 11].

6 Simulation Results

In this section we present the simulation results obtained with the integration of the ensemble neural network with type-2 fuzzy integration and its optimization with the genetic algorithm for the Mackey-Glass time series.

Table 1 shows the particle swarm optimization where the best prediction error is of 0.0063313.

Table 1 Particle swarm results for the ensemble neural network $\tau = 17$

No.	Iterations	Panicles	Number modules	Number layers	Number neurons	Duration	Prediction error
1	100	100	4	2	20, 14 13, 16 17, 8 6, 26	02:23:18	0.0076048
2	100	100	2	2	12, 16 12, 26	01:45:45	0.0063313
3	100	100	2	3	17, 5, 18 6, 25, 24	01:28:42	0.0018838
4	100	100	4	2	7, 24 14, 22 1, 8 15, 23	02:40:20	0.0073005
5	100	100	4	2	14, 9 11, 26 27, 16 11, 13	02:11:34	0.0081418
6	100	100	4	2	16, 16 9, 19 6, 6 9, 12	01:34:05	0.0087983
7	100	100	2	3	11, 23, 26 15, 15, 5	02:09:17	0.0076315
8	100	100	2	2	14, 10 14, 21	01:23:28	0.0061291
9	100	100	3	2	9, 5 23, 20 22, 13	02:17:06	0.0053679
10	100	100	3	3	23, 14, 16 19, 10, 23 22, 12, 11	02:20:04	0.0061983

Table 2 Results of Type-1 fuzzy integration for $\tau = 17$

Experiment	Prediction error with fuzzy integration Type-1
Experiment 1	0.1879
Experiment 2	0.1789
Experiment 3	0.2221
Experiment 4	0.1888
Experiment 5	0.1521
Experiment 6	0.2561
Experiment 7	0.1785
Experiment 8	0.1942
Experiment 9	0.2536
Experiment 10	0.1965

Table 3 Results of Type-2 fuzzy integration for $\tau = 17$

Experiment	Prediction error 0.3 Uncertainty	Prediction error 0.4 Uncertainty	Prediction error 0.5 Uncertainty
Experiment 1	0.2385	0.2385	0.3952
Experiment 2	0.2489	0.2231	0.3909
Experiment 3	0.2482	0.2226	0.3642
Experiment 4	0.2214	**0.1658**	0.3856
Experiment 5	0.2658	0.2234	0.3857
Experiment 6	0.2756	0.2592	**0.3134**
Experiment 7	**0.1785**	0.2352	0.3358
Experiment 8	0.1825	0.2546	0.4561
Experiment 9	0.2018	0.2373	0.3394
Experiment 10	0.2076	0.2003	0.3687

Fuzzy integration is performed initially by implementing a type-1 fuzzy system in which the best result was in the experiment of row number 5 of Table 2 with an error of: 0.1521.

Fuzzy integration is performed by implementing a type-1 fuzzy system in which the results were as follows: for the best evolution with a degree of uncertainty of 0.3 a forecast error of 0.1785 was obtained, and with a degree of uncertainty of 0.4 a forecast error of 0.1658 and with a degree of uncertainty of 0.5 a forecast error of 0.3134 was obtained, as shown in Table 3.

Table 4 shows the particle swarm optimization where the best prediction error is of 0.0019726.

Fuzzy integration is performed by implementing a type-1 fuzzy system in which the best result was in the experiment of row number 2 of Table 5 with an error of: 0.4586.

Table 4 Particle swarm results for the ensemble neural network for $\tau = 34$

No.	Iterations	Particles	Number modules	Number layers	Number neurons	Duration	Prediction error
1	100	100	4	3	12, 23, 12 9, 19, 7	02:45:14	0.0019726
2	100	100	4	3	19, 11, 11 16, 11, 14 15, 24, 19 22, 13, 27	01:28:06	0.0063623
3	100	100	3	2	4, 9 9, 20 10, 11 23, 20	02:03:06	0.0046644
4	100	100	4	2	14, 18 12, 19 20, 17 10, 6	03:22:13	0.0072153
5	100	100	3	2	7, 6 10, 15 12, 16	01:39:13	0.0075658
6	100	100	3	3	14, 20, 18 15, 21, 12 19, 17, 26	03:08:02	0.0047515
7	100	100	2	2	4, 24 9, 26	02:00:10	0.003601
8	100	100	2	2	24, 17 14, 23	02:27:21	0.0065506
9	100	100	3	3	7, 11, 8 23, 21, 21 17, 8, 11	02:03:12	0.0037758
10	100	100	2	3	20, 28, 15 15, 12, 24	02:04:18	0.0066375

Table 5 Results of Type-1 fuzzy integration for $\tau = 34$

Experiment	Prediction error with fuzzy integration Type-1
Experiment 1	0.9587
Experiment 2	**0.4586**
Experiment 3	0.5871
Experiment 4	1.2569
Experiment 5	0.9517
Experiment 6	1.556
Experiment 7	1.0987
Experiment 8	1.9671
Experiment 9	1.698
Experiment 10	1.4626

Table 6 Results of Type-2 fuzzy integration for $\tau = 34$

Evolution	Prediction error 0.3 Uncertainty	Prediction error 0.4 Uncertainty	Prediction error 0.5 Uncertainty
Evolution 1	**0.6036**	0.8545	0.4570
Evolution 2	1.5862	1.0021	1.3533
Evolution 3	0.8002	0.6943	**0.3893**
Evolution 4	1.4032	0.9617	0.9665
Evolution 5	0.8658	0.8299	0.6358
Evolution 6	1.3986	0.1052	1.2354
Evolution 7	1.465	1.3566	0.6646
Evolution 8	1.7453	0.8966	0.8241
Evolution 9	0.9866	**0.6524**	0.6661
Evolution 10	1.4552	0.9956	0.7557

Table 7 Particle swarm results for the ensemble neural network for $\tau = 68$

No.	Iterations	Particles	Number of modules	Number of layers	Number of neurons	Duration	Prediet ion error
1	**100**	**100**	**2**	**3**	**17, 5, 18** **6, 25, 19**	**02:05:53**	**0.0019348**
2	100	100	2	2	7, 8 6, 20	04:1936	0.0041123
3	100	100	2	3	21, 11, 16 5, 10, 10	02:23:02	0.0042367
4	100	100	4	3	15, 7, 4 11, 22, 5 24, 19, 22 4, 14, 11	02:37:06	0.0050847
5	100	100	3	2	22, 23 2, 21 10, 2	01:5	0.0037132
6	100	100	4	3	10, 13, 22 24, 8, 17 13, 16, 20 7, 24, 17	02:10:27	0.0057235
7	100	100	2	2	8, 20 15, 23		0.0033082
8	100	100	3	2	28, 6 2, 16 18, 10	01:40:18	0.0057402
9	100	100	3	2	22, 17 10, 10 21, 12	02:45:31	0.0047309
10	100	100	2	3	22, 11, 18 27, 7, 14	01:35:13	0.0044649

Table 8 Results of Type-1 fuzzy integration for $\tau = 68$

Experiment	Prediction error with fuzzy integration Type-1
Experiment 1	0.8753
Experiment 2	0.3625
Experiment 3	0.6687
Experiment 4	**0.3254**
Experiment 5	0.5489
Experiment 6	1.3183
Experiment 7	1.8972
Experiment 8	1.6977
Experiment 9	1.5879
Experiment 10	0.9652

Fuzzy integration is performed by implementing a type-2 fuzzy system in which the results were as follows: for the best evolution with a degree of uncertainty of 0.3 a forecast error of 0.6036 was obtained, and with a degree of uncertainty of 0.4 a forecast error of 0.6524 and with a degree of uncertainty of 0.5 a forecast error of 0.3893 was obtained, as shown in Table 6.

Table 7 shows the particle swarm optimization where the prediction error is of 0.0019348.

Fuzzy integration is performed by implementing a type-1 fuzzy system in which the best result was in the experiment of row number 4 of Table 8 with an error of: 0.32546.

Fuzzy integration is also performed by implementing a type-2 fuzzy system in which the results were as follows: for the best evolution with a degree of uncertainty of 0.3 a forecast error of 0.6825 was obtained, and with a degree of uncertainty of 0.4 a forecast error of 0.7652 and with a degree of uncertainty of 0.5 a forecast error of 0.6581 was obtained, as shown in Table 9.

Table 9 Results of Type-2 fuzzy integration for $\tau = 68$

Evolution	Prediction error 0.3 Uncertainty	Prediction error 0.4 Uncertainty	Prediction error 0.5 Uncertainty
Evolution 1	0.7895	0.9631	0.7365
Evolution 2	0.9875	1.2365	1.564
Evolution 3	0.9874	0.7965	**0.6581**
Evolution 4	1.5325	0.9874	0.9723
Evolution 5	0.7763	0.9723	0.9858
Evolution 6	0.8694	0.9235	1.3697
Evolution 7	**0.6825**	1.4263	0.6646
Evolution 8	1.336	0.8963	0.8288
Evolution 9	0.9852	**0.7652**	0.7234
Evolution 10	1.365	1.4224	1.5984

7 Conclusions

Using the technique of PSO particle we can reach the conclusion that this algorithm is good for reducing the execution time compared to other techniques such as genetic algorithms, and also architectures for ensemble neural network are small and they applied to the time series, as in this case the time series of Mackey-Glass. Also the outputs results obtained integrating the results of the neural network with type-1 and type-2 fuzzy systems and integrated type-2 the best results with type 2 are very good.

Acknowledgments We would like to express our gratitude to the CONACYT, Tijuana Institute of Technology for the facilities and resources granted for the development of this research.

References

1. Castillo, O., Melin, P.: Type-2 Fuzzy Logic: Theory and Applications. Neural Networks, pp. 30–43. Springer, New York (2008)
2. Castillo, O., Melin, P.: Hybrid intelligent systems for time series prediction using neural networks, fuzzy logic, and fractal theory. IEEE Trans. Neural Netw. **13**(6), 1395–1408 (2002)
3. Castillo, O., Melin, P.: Simulation and forecasting complex economic time series using neural networks and fuzzy logic. In: Proceedings of the International Neural Networks Conference, vol. 3, pp. 1805–1810 (2001)
4. Castillo, O., Melin, P.: Simulation and forecasting complex financial time series using neural networks and fuzzy logic. In: Proceedings the IEEE the International Conference on Systems, Man and Cybernetics, vol. 4, pp. 2664–2669 (2001)
5. Eberhart, R.C., Kennedy, J.: A new optimizer particle swarm theory. In: Proceedings of the sixth Symposium on Micromachine and Human Science, pp. 39–43 (1995)
6. Eberhart, R.C.: Fundamentals of Computational Swarm Intelligence, pp. 93–129. Wiley, New York (2005)
7. Jang, J.S.R, Sun, C.T., Mizutani, E.: Neuro-Fuzzy and Sof Computing, Prentice Hall, Englewood Cliffs (1996)
8. Kennedy, J., Eberhart, R.C.: Particle swarm optimization. In: Proceedings Intelligent Symposium, pp. 80–87. April 2003
9. Krogh, A., Vedelsby, J.: Neural network ensembles, cross validation, and active learning. in: Tesauro, G., Touretzky, D., Leen, T. (eds.) Advances in Neural Information Processing Systems, vol. 7, pp. 231–238, 1001. MIT Press, Cambridge, Denver (1995)
10. Mackey, M.C.: Adventures in Poland: having fun and doing research with Andrzej Lasota. Mat. Stosow **8**, 5–32 (2007)
11. Mackey, M.C., Glass, L.: Oascillation and chaos in physiological control systems. Science **197**, 287–289 (1997)
12. Maguire, L.P., Roche, B., McGinnity, T.M., McDaid, L.J.: Predicting a chaotic time series using a fuzzy neural network. Adv. Soft Comput. **12**(1–4), 125–136 (1998)
13. Multaba, I.M., Hussain, M.A.: Application of Neural Networks and Other Learning. Technologies in Process Engineering, Imperial Collage Press, London (2001)
14. Plummer, E.A.: Time series forecasting with feed-forward neural networks: guidelines and limitations. University of Wyoming, July 2000

15. Pulido, M., Mancilla, A., Melin, P.: An ensemble neural network architecture with fuzzy response integration for complex time series prediction. Evolutionary Design of Intelligent Systems in Modeling, Simulation and Control, vol. 257/2009, pp. 85–110. Springer, Berlin (2009)
16. Sharkey, A.: One combining Artificial of Neural Nets. Department of Computer Science, University of Sheffield, Sheffield (1996)
17. Sharkey, A.: Combining Artificial Neural Nets: Ensemble and Modular Multi-Net Systems. Springer, London (1999)
18. Sollich, P., Krogh, A.: Learning with ensembles: how over-fitting can be useful. In: Touretzky, D.S., Mozer, M.C., Hasselmo, M.E. (eds.) Advances in Neural Information Processing Systems, vol. 8, pp. 190–196. MIT Press, Denver, Cambridge (1996)
19. Yadav, R.N., Kalra, P.K., John, J.: Time series prediction with single multiplicative neuron model. Soft Comput. Time Ser. Predict. Appl. Soft Comput. **7**(4), 1157–1163 (2007)
20. Yao, X., Liu, Y.: Making use of population information in evolutionary artificial neural networks. IEEE Trans. Syst. Man Cybern. Part B Cybern. **28**(3), 417–425 (1998)
21. Zhao, L., Yang, Y.: PSO-based single multiplicative neuron model for time series prediction. Expert Syst. Appl. **36**(2 Part 2), 2805–2812 (2009)
22. Zhou, Z.-H., Jiang, Y., Yang, Y.-B., Chen, S.-F.: Lung cancer cell identification based on artificial neural network ensembles. Artif. Intell. Med. **24**(1), 25–36 (2002)

A Generalized Net Model for Evaluation Process Using InterCriteria Analysis Method in the University

Evdokia Sotirova, Veselina Bureva and Sotir Sotirov

Abstract In the paper is constructed a generalized net model which describes the process of evaluation of objects using a set of criteria. For calculating evaluations is used InterCriteria analysis method that detects possible correlations between pairs of criteria. The objects can be lecturers, students, Ph.D. candidates, problems solved by students, disciplines, and so on. The model can be used for monitoring and analysis of the process of assessment.

Keywords Generalized net · Intercriteria analysis · Intuitionistic fuzziness · Index matrices · University

1 Introduction

The present paper is a continuation of previous investigations into the modelling of an information flow in a university. In a series of research, the authors study some of the most important processes of functioning of universities [1–13]. For modelling are used Generalized Nets (GNs, see [14, 15]). For assessment the processes is used theory of intuitionistic fuzzy sets (IFSs, [16, 17]). In [3] is constructed generalized net model of e-learning evaluation with intuitionistic fuzzy estimations. In papers [6, 7] is modelled the process of lecturers' evaluation of student work with intuitionistic fuzzy estimations. In [8] is described the process of ordering of university subjects. In [9, 10] theories of generalized nets and intuitionistic fuzzy sets are applied for evaluation of lecturers and student's course. In [11] is proposed

E. Sotirova (✉) · V. Bureva · S. Sotirov
"Prof. Asen Zlatarov" University, 8010 Burgas, Bulgaria
e-mail: esotirova@btu.bg

V. Bureva
e-mail: vbureva@btu.bg

S. Sotirov
e-mail: ssotirov@btu.bg

P. Angelov and S. Sotirov (eds.), *Imprecision and Uncertainty in Information Representation and Processing*, Studies in Fuzziness and Soft Computing 332,
DOI 10.1007/978-3-319-26302-1_23

a method for evaluation of the rating of the university subjects using intuitionistic fuzzy estimations. The main focus in [12] is to analyse the processes in the preparation of Ph.D. candidates. In [13] is investigated how to apply some data mining techniques for clustering and classification the assessment of the different publications and articles. The aim of [18] is to use the techniques of self-organizing map in the process of e-learning to assess the students' knowledge on relevant topics in intuition-istic fuzzy form.

In this paper we construct a generalized net model for evaluation process of objects in typical university using a set of criteria. The objects can be lecturers, students, Ph.D. candidates, problems solved by students, disciplines, and so on. For calculating evaluations is used a new approach named InterCriteria Analysis (ICA) [2]. It is based on theory of intuitionistic fuzzy sets [16, 17] and index matrices [19, 20].

Via InterCriteria Analysis method we can reduce a number of criteria through calculating a correlations for each pair of criteria in the form of intuitionistic fuzzy pairs of values [1, 21]. The intuitionistic fuzzy pairs of values are the intuitionistic fuzzy evaluation in the interval [0, 1] of the relations that can be established between any two criteria C_w and C_t.

2 Realization

Let us have a number of E_i experts, $i = 1, \ldots, p$; a number of C_j criteria, $j = 1, \ldots, q$; and a number of O_k objects, $k = 1, \ldots, s$. So we use the following sets: a set of criteria $C_j = \{C_1, \ldots, C_q\}$, a set of objects $O_k = \{O_1, \ldots, O_s\}$, and a set of experts $E_i = \{E_1, \ldots, E_p\}$.

The experts evaluate the objects using the criteria. As a result we obtain an index matrix that contains two sets of indexes, one for rows and another for columns [19, 20]:

$$M = \begin{array}{c|ccccc} & O_1 & \ldots & O_k & \ldots & O_s \\ \hline C_1 & a_{C_1,O_1} & \ldots & a_{C_1,O_k} & \ldots & a_{C_1,O_s} \\ \ldots & \ldots & \ldots & \ldots & \ldots & \ldots \\ C_j & a_{C_j,O_1} & \ldots & a_{C_j,O_k} & \ldots & a_{C_j,O_s} \\ \ldots & \ldots & \ldots & \ldots & \ldots & \ldots \\ C_q & a_{C_q,O_1} & \ldots & a_{C_q,O_k} & \ldots & a_{C_q,O_s} \end{array}. \quad (1)$$

The next step is applying the InterCriteria Analysis method for calculating evaluations. Now we obtain a new index matrix M^* with intuitionistic fuzzy pairs $\langle \mu_{C_w,C_t}, \nu_{C_w,C_t} \rangle$ that represents an intuitionistic fuzzy evaluation of the relations between every pair of criteria C_w and C_t:

$$M^* = \begin{array}{c|ccc} & C_1 & \dots & C_q \\ \hline C_1 & \langle \mu_{C_1,C_1}, \nu_{C_1,C_1} \rangle & \dots & \langle \mu_{C_1,C_q}, \nu_{C_1,C_q} \rangle \\ \dots & \dots & \dots & \dots \\ C_q & \langle \mu_{C_q,C_1}, \nu_{C_q,C_1} \rangle & \dots & \langle \mu_{C_q,C_q}, \nu_{C_q,C_q} \rangle \end{array} \tag{2}$$

If we work with many criteria, we can use some Data Mining technique for analysis the obtained estimations, for example classification or clusterization.

The GN-model (Fig. 1) contains 7 transitions and 24 places, collected in five groups and related to the five types of the tokens that will enter respective types of places:

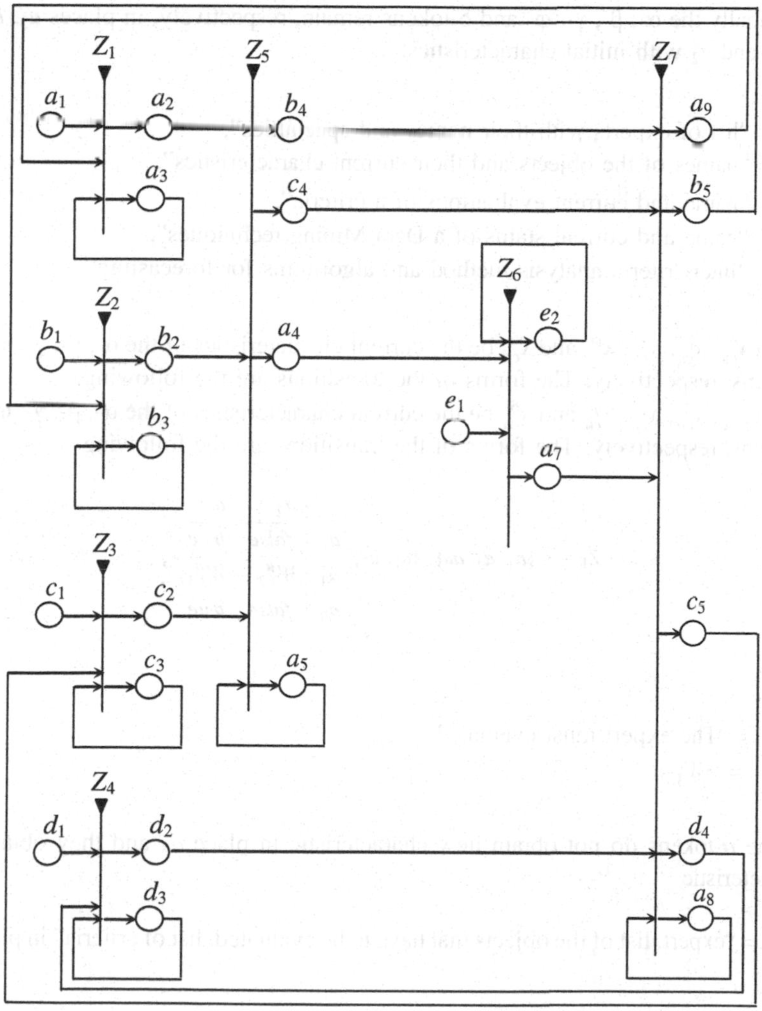

Fig. 1 GN Model for evaluation process using InterCriteria Analysis method

- α-tokens and *a*-places represent the experts and their activities,
- β-tokens and *b*-places represent the objects and connected with them tasks, functions, activities,
- γ-tokens and *c*-places represent the criteria and their correlations,
- φ-tokens and *d*-places represent the Data Mining techniques for analysis of the intuitionistic fuzzy pairs of values representing correlations for each pair of criteria,
- δ-tokens and *e*-places represent the InterCriteria analysis method and algorithms for forecasting.
- For brevity, we shall use the notation α-, β-, γ-, φ- and δ-tokens instead of α_i-, β_j-, γ_k-, φ_l- and δ_m- tokens, where *i*, *j*, *k*, *l*, *m* are numerations of the respective tokens.
- Initially the α-, β-, γ-, φ- and δ-tokens remain, respectively, in places a_3, b_3, c_3 d_3 and e_2 with initial characteristics:

x_0^{α} = "list of experts with their names and specialties",
x_0^{β} = "names of the objects and their current characteristics",
x_0^{γ} = "name and current evaluations of a criteria",
x_0^{φ} = "name and current status of a Data Mining techniques",
x_0^{δ} = "InterCriteria analysis method and algorithms for forecasting".

Let x_{cu}^{α}, x_{cu}^{β}, x_{cu}^{γ}, x_{cu}^{φ} and x_{cu}^{δ} be the current characteristics of the α-, β-, γ-, φ- and δ-tokens, respectively. The forms of the transitions are the following.

Let x_{cu}^{α}, x_{cu}^{β}, x_{cu}^{γ}, x_{cu}^{φ} and x_{cu}^{δ} be the current characteristics of the α-, β-, γ-, φ- and δ-tokens, respectively. The forms of the transitions are the following.

$$Z_1 = \langle\{a_1, a_3, a_9\}, \{a_2, a_3\}, \begin{array}{c|cc} & a_2 & a_3 \\ \hline a_1 & false & true \\ a_3 & W_{3,2}^{a} & W_{3,3}^{a} \\ a_9 & false & true \end{array}\rangle,$$

where:

$W_{3,2}^{a}$ ="The expert must evaluate"
$W_{3,3}^{a} = \neg W_{3,2}^{a}$

The *α*-tokens do not obtain new characteristic in place a_3 and they obtain the characteristic

x_{cu}^{α} = "expert, list of the objects that have to be evaluated, list of criteria" in place a_2

$$Z_2 = <\{ b_1, b_3, b_5\}, \{b_2, b_3\}, \begin{array}{c|cc} & b_2 & b_3 \\ \hline b_1 & false & true \\ b_3 & W^b_{3,2} & W^b_{3,3} \\ b_5 & false & true \end{array}>,$$

where:

$W^b_{3,2}$ = "The objects are included in x^{α}_{cu}",
$W^b_{3,3}$ $= \neg W^b_{3,2}$

The β-tokens do not have new characteristic in place b_3 and they obtain the characteristic

x^{β}_{cu} = "current objects that have to be evaluated" in place b_2

$$Z_3 = <\{c_1, c_3, c_5\}, \{c_2, c_3\}, \begin{array}{c|cc} & c_2 & c_3 \\ \hline c_1 & false & true \\ c_3 & W^c_{3,2} & W^c_{3,3} \\ c_5 & false & true \end{array}>,$$

where:

$W^c_{3,2}$ = "The criteria are included in x^{α}_{cu}",
$W^c_{3,3}$ $= \neg W^c_{3,2}$

The γ-tokens do not have new characteristic in place c_3 and they obtain the characteristic

x^{γ}_{cu} = "current evaluating criteria for the evaluated objects" in place c_2

$$Z_4 = <\{ d_1, d_3, d_4\}, \{d_2, d_3\}, \begin{array}{c|cc} & d_2 & d_3 \\ \hline d_1 & false & true \\ d_3 & W^d_{3,2} & W^d_{3,3} \\ d_4 & false & true \end{array}>,$$

where:

$W^d_{3,2}$ = "The intuitionistic fuzzy pairs for correlations of criteria must be evaluated by Data mining technique",
$W^d_{3,3}$ $= \neg W^d_{3,2}$

The φ-tokens do not obtain new characteristic in place d_3 and they obtain the characteristic

x^{φ}_{cu} = "Data mining technique" in place d_2

$$Z_5 = <\{a_2, a_5, b_2, c_2\}, \{a_4, a_5, b_4, c_4\}, \begin{array}{c|cccc} & a_4 & a_5 & b_4 & c_4 \\ \hline a_2 & false & true & false & false \\ a_5 & W^a_{5,4} & W^a_{5,5} & false & false \\ b_2 & false & false & true & false \\ c_2 & false & false & false & true \end{array}>,$$

where:

$W^a_{5,4}$ = "All objects are evaluated via criteria by the current expert",
$W^a_{5,5} = \neg W^a_{5,4}$

The β- and γ-tokens do not have new characteristic in places b_4 and c_4, respectively, while α-tokens obtain characteristic
"The index matrix M with evaluation of the O_k objects using a set of criteria C_j" in place a_5.

$$Z_6 = <\{a_4, e_1, e_2\}, \{a_7, e_2\}, \begin{array}{c|cc} & a_7 & e_2 \\ \hline a_4 & W^a_{4,7} & false \\ e_1 & false & true \\ e_2 & false & true \end{array}>,$$

where:

$W^a_{4,7}$ = "The matrix with intuitionistic fuzzy pairs is obtained"

The δ-tokens do not obtain new characteristic in places e_2. The α-tokens that enter place a_7 obtain characteristics
"The index matrix M^* with degrees of correspondence between the criteria C_1, …, C_q".

$$Z_7 = <\{a_7, a_8, b_4, c_4, d_2\}, \{a_8, a_9, b_5, c_5, d_4\}, \begin{array}{c|ccccc} & a_8 & a_9 & b_5 & c_5 & d_4 \\ \hline a_7 & W^a_{7,8} & W^a_{7,9} & false & false & false \\ a_8 & false & W^a_{8,9} & false & false & false \\ b_4 & false & false & true & false & false \\ c_4 & false & false & false & true & false \\ d_2 & false & false & false & false & true \end{array}>.$$

$W^a_{7,8}$ = "There is a token in place d_2 and x^{φ}_{cu} is included in x^{a}_{cu}",
$W^a_{7,9} = \neg W^a_{7,8}$,

$W^a_{8,9}$ = "The intuitionistic fuzzy pairs of values are evaluated with Data mining technique"

The α-, β-, γ- and φ-tokens do not obtain new characteristic in places a_8, b_5, c_5 and d_4.

The α-tokens that enter place a_9 obtain characteristics

"Results from analysis of the intuitionistic fuzzy pairs"

3 Implementation of the InterCriteria Analysis Method

As an input data of the InterCriteria Analysis method we use the students' marks for 81 students.

Realization 1. Correlations between different disciplines.

In this section are discussed students' marks obtained in 9 disciplines (Table 1). Here we use 81 × 9 table and a software application that implements the InterCriteria Analysis algorithm and returns the results in the form of two index matrices in Tables 2 and 3, containing the membership and the non-membership parts of the intuitionistic fuzzy correlations detected between each pair of criteria (36 pairs). In this way we construct matrix M^* that gives the relations among the criteria. From practical considerations, it has been more flexible to work with two index matrices M_μ and M_ν, rather than with the index matrix M^* of intuitionistic fuzzy pairs.

From the Tables 1 and 2 we obtain the following correlations, based on the membership coefficient μ in descending order (Table 4).

The analysis can help for logical ordering of the study subjects, accounting on the needs of the students' university training. The correlations can be used for proper constructing of an academic curriculum with connected couples of disciplines, and providing opportunity for easier assimilation of the obtained material and continuity from one semester to another. Through properly analysis of the relations among disciplines can be obtained sustainability of the knowledge.

Table 1 Evaluated disciplines

I.	Mathematics 1
II.	Mathematics 2
III.	Mathematics 3
IV.	Discrete structures
V.	Informatics 1
VI.	Informatics 2
VII.	Programming and using of computers
VIII.	Synthesis and analysis of algorithms
IX.	Programming languages

Table 2 Membership part of the intuitionistic fuzzy pairs, giving the InterCriteria correlations

μ	I	II	III	IV	V	VI	VII	VIII	IX
I	1.000	0.795	0.754	0.569	0.467	0.467	0.344	0.610	0.426
II	0.795	1.000	0.836	0.487	0.446	0.467	0.405	0.508	0.426
III	0.754	0.836	1.000	0.549	0.467	0.508	0.344	0.569	0.467
IV	0.569	0.87	0.549	1.000	0.569	0.549	0.528	0.856	0.651
V	0.467	0.446	0.467	0.569	1.000	0.836	0.590	0.569	0.774
VI	0.467	0.467	0.508	0.549	0.836	1.000	0.487	0.569	0.672
VII	0.344	0.405	0.344	0.528	0.590	0.487	1.000	0.508	0.774
VIII	0.610	0.508	0.569	0.856	0.569	0.569	0.508	1.000	0.672
IX	0.426	0.426	0.467	0.651	0.774	0.672	0.774	0.672	1.000

Table 3 Non-membership part of the intuitionistic fuzzy pairs, giving the InterCriteria correlations

ν	I	II	III	IV	V	VI	VII	VIII	IX
I	0.000	0.000	0.000	0.082	0.246	0.164	0.205	0.103	0.185
II	0.000	0.000	0.000	0.123	0.185	0.123	0.144	0.123	0.144
III	0.000	0.000	0.000	0.062	0.123	0.082	0.164	0.062	0.103
IV	0.082	0.123	0.062	0.000	0.041	0.062	0.082	0.000	0.021
V	0.246	0.185	0.123	0.041	0.000	0.000	0.041	0.062	0.000
VI	0.164	0.123	0.082	0.062	0.000	0.000	0.103	0.062	0.062
VII	0.205	0.144	0.164	0.082	0.041	0.103	0.000	0.123	0.000
VIII	0.103	0.123	0.062	0.000	0.062	0.062	0.123	0.000	0.062
IX	0.185	0.144	0.103	0.021	0.000	0.062	0.000	0.062	0.000

The precise distribution of the order of the university subjects is very important for the quality of the training process and costs a lot of time and resources.

Realization 2. Correlations for level of significance of the different disciplines in average grade.

In this section are discussed students' marks obtained in the same 9 disciplines from Table 1 and the average grade (row X in Table 5). We work with data for 81 students, obtained against 10 criteria. We use 81 × 10 table and a software application that implements the InterCriteria Analysis algorithm and returns the results in the form of two index matrices in Tables 5 and 6, containing the membership and the non-membership parts of the intuitionistic fuzzy correlations detected between each pair of criteria (45 pairs).

From the Tables 5 and 6 we obtain the following correlations between average grade and disciplines, based on the membership coefficient μ in descending order (Table 7).

The correlations between disciplines and average grade based on membership coefficient μ can be called sensitization of the training process. The students have to choose one or more subjects during their study. Some of them prefer to choose a

Table 4 Correlation between disciplines based on membership part of the intuitionistic fuzzy pairs

IV—Discrete structures	VIII—Synthesis and analysis of algorithms
V—Informatics 1	VI—Informatics 2
IV—Discrete structures	III—Mathematics 3
I—Mathematics 1	II—Mathematics 2
I—Programming and using of computers	II—Programming languages

Table 5 Membership part of the intuitionistic fuzzy pairs, giving the InterCriteria correlations

μ	I	II	III	IV	V	VI	VII	VIII	IX	X
I	1.000	0.795	0.754	0.569	0.467	0.467	0.344	0.610	0.426	0.713
II	0.795	1.000	0.836	0.487	0.446	0.467	0.405	0.508	0.426	0.549
III	0.754	0.836	1.000	0.549	0.467	0.508	0.344	0.569	0.467	0.549
IV	0.569	0.487	0.549	1.000	0.569	0.549	0.528	0.856	0.651	0.774
V	0.467	0.446	0.467	0.569	1.000	0.836	0.590	0.569	0.774	0.631
VI	0.467	0.467	0.508	0.549	0.836	1.000	0.487	0.569	0.672	0.569
VII	0.344	0.405	0.344	0.528	0.590	0.487	1.000	0.508	0.774	0.528
VIII	0.610	0.508	0.569	0.856	0.569	0.569	0.508	1.000	0.672	0.754
IX	0.426	0.426	0.467	0.651	0.774	0.672	0.774	0.672	1.000	0.631
X	0.713	0.549	0.549	0.774	0.631	0.569	0.528	0.754	0.631	1.000

Table 6 Non-membership part of the intuitionistic fuzzy pairs, giving the InterCriteria correlations

ν	I	II	III	IV	V	VI	VII	VIII	IX	X
I	0.000	0.000	0.000	0.082	0.246	0.164	0.205	0.103	0.185	0.123
II	0.000	0.000	0.000	0.123	0.185	0.123	0.144	0.123	0.144	0.082
III	0.000	0.000	0.000	0.062	0.123	0.082	0.164	0.062	0.103	0.041
IV	0.082	0.123	0.062	0.000	0.041	0.062	0.082	0.000	0.021	0.000
V	0.246	0.185	0.123	0.041	0.000	0.000	0.041	0.062	0.000	0.082
VI	0.164	0.123	0.082	0.062	0.000	0.000	0.103	0.062	0.062	0.103
VII	0.205	0.144	0.164	0.082	0.041	0.103	0.000	0.123	0.000	0.062
VIII	0.103	0.123	0.062	0.000	0.062	0.062	0.123	0.000	0.062	0.000
IX	0.185	0.144	0.103	0.021	0.000	0.062	0.000	0.062	0.000	0.021
X	0.123	0.082	0.041	0.000	0.082	0.103	0.062	0.000	0.021	0.000

few specific subjects, the other choose subjects that answer their needs. Usually students can choose subjects from list with available subjects for the current semester. When disciplines with a high coefficient of sensitization are chosen and successfully completed, students can raise their average grade.

Table 7 Correlation between average grade and disciplines based on membership part of the intuitionistic fuzzy pairs

IV—Discrete structures
VIII—Synthesis and analysis of algorithms
I—Mathematics 1
V—Informatics 1
II—Programming languages
VI—Informatics 2
II—Mathematics 2
III—Mathematics 3
I—Programming and using of computers

4 Conclusions

The constructed generalized net model of the process of evaluation represents an InterCriteria analysis method that detects possible correlations between pairs of criteria. It can be applied for monitoring and analysis of the process of assessment in a university. The evaluated objects can be lecturers, students, PhD candidates, problems solved by students, disciplines, and so on. Two implementations based on membership part of the intuitionistic fuzzy pairs are given. In the first correlations between different disciplines are found for proper constructing of an academic curriculum. In the second correlations for level of significance of the different disciplines in average grade are analyzed. The two realizations can be helpful for achieving of the sustainability and sensitization of the learning process.

Acknowledgements The authors are thankful for the support provided by the Bulgarian National Science Fund under Grant Ref. No. DFNI-I-02-5 *"InterCriteria Analysis: A New Approach"*.

References

1. Atanassov, K., Mavrov, D., Atanassova, V.: InterCriteria decision making. A new approach for multicriteria decision making, based on index matrices and intuitionistic fuzzy sets. Issues IFS GN **11**, 1–7 (2014)
2. Atanassova, V., Mavrov, D., Doukovska, L., Atanassov, K.: Discussion on the threshold values in the InterCriteria decision making approach. Notes Intuit. Fuzzy Sets **20**(2), 94–99 (2014)
3. Melo-Pinto, P., Kim, T., Atanassov, K., Sotirova, E., Shannon, A., Krawczak, M.: Generalized net model of e-learning evaluation with intuitionistic fuzzy estimations. Issues in the Representation and Processing of Uncertain and Imprecise Information, pp. 241–249. Warszawa (2005)
4. Shannon, A., Langova-Orozova, D., Sotirova, E., Petrounias, I., Atanassov, K., Krawczak, M., Melo-Pinto, P., Kim, T.: Generalized Net Modelling of University Processes. KVB Visual Concepts Pty Ltd, Monograph No. 7, Sydney (2005)

5. Shannon, A., Atanassov, K., Sotirova, E., Langova-Orozova, D., Krawczak, M., Melo-Pinto, P., Petrounias, I., Kim, T.: Generalized Nets and Information Flow Within a University, Warsaw School of Information Technology, Warsaw (2007)
6. Shannon, A., Sotirova, E., Petrounias, I., Atanassov, K., Krawczak, M., Melo-Pinto, P., Kim, T.: Intuitionistic fuzzy estimations of lecturers' evaluation of student work. In: First International Workshop on Intuitionistic Fuzzy Sets, Generalized Nets and Knowledge Engineering, pp. 44–47. University of Westminster, London, 6–7 Sept 2006
7. Shannon, A., Sotirova, E., Petrounias, I., Atanassov, K., Krawczak, M., Melo-Pinto, P., Kim, T.: Generalized net model of lecturers' evaluation of student work with intuitionistic fuzzy estimations. In: Second International Workshop on Intuitionistic Fuzzy Sets. Notes on IFS, vol. 12, No. 4, pp. 22–28. Banska Bystrica, Slovakia, 03 Dec 2006
8. Shannon, A., Sotirova, E., Petrounias, I., Atanassov, K., Krawczak, M., Melo-Pinto, P., Sotirov, S., Kim, T.: Generalized net model of the process of ordering of university subjects. In: Seventh Int. Workshop on GNs, Sofia, pp. 25–29. 14–15 July 2006
9. Shannon, A., Dimitrakiev, D., Sotirova, E., Krawczak, M.: Kim, T.: Towards a model of the digital university: generalized net model of a lecturer's evaluation with intuitionistic fuzzy estimations. Cybernetics and Information Technologies, vol. 9, No. 2, pp. 69–78. Bulgarian Academy of Sciences, Sofia (2009)
10. Shannon, A., Sotirova, E., Hristova, M., Kim, T.: Generalized Net Model of a student's course evaluation with intuitionistic fuzzy estimations in a digital university. Proc. Jangjeon Math. Soc. **13**(1), 31–38 (2010)
11. Shannon, A., Riečan, B., Sotirova, E., Inovska, G., Atanassov, K., Krawczak, M., Melo-Pinto, P., Kim, T.: A generalized net model of university subjects rating with intuitionistic fuzzy estimations. Notes Intuit. Fuzzy Sets **18**(3), 61–67 (2012)
12. Shannon, A., Riečan, B., Sotirova, E., Krawczak, M., Atanassov, K., Melo-Pinto, P., Kim, T.: Modelling the process of Ph.D. preparation using generalized nets. In: 14th International Workshop on Generalized Nets, pp. 34–38. Burgas, 29–30 Nov 2013
13. Sotirova, E., Sotirov, S., Vardeva, I., Riečan, B., Publication's assessment with intuitionistic fuzzy estimations. Notes Intuit. Fuzzy Sets **18**(4), 26–31 (2012)
14. Atanassov, K.: Generalized Nets. World Scientific, Singapore (1991)
15. Atanassov, K.: On Generalized Nets Theory. Prof. M. Drinov Academic Publishing House, Sofia (2007)
16. Atanassov, K.: Intuitionistic Fuzzy Sets. Springer, Berlin (1999)
17. Atanassov, K.: On Intuitionistic Fuzzy Sets Theory. Springer, Berlin (2012)
18. Sotirova, E.: Classification of the students' intuitionistic fuzzy estimations by a 3-dimensional self organizing map. Notes Intuit. Fuzzy Sets **17**(4), 64–68 (2011)
19. Atanassov, K.: Generalized index matrices. Comptes rendus de l'Academie Bulgare des Sciences **40**(11), 15–18 (1987)
20. Atanassov, K.: On index matrices. Part 1: standard cases. Adv. Stud. Contemp. Math. **20**(2), 291–302 (2010)
21. Atanassov, K., Szmidt, E., Kacprzyk, J.: On intuitionistic fuzzy pairs, Notes Intuit. Fuzzy Sets **19**(3), 1–13 (2013)

ItUitionistic Fuzzy Estimation of the Generalized Nets Model of Spatial-Temporal Group Scheduling Problems

Sotir Sotirov, Evdokia Sotirova, Matthias Werner, Stanislav Simeonov, Wolfram Hardt and Neli Simeonova

Abstract A cyber-physical system (CPS) is a system of collaborating computational elements controlling physical entities. Nowadays cyber-physical systems can be found in areas as diverse as robotics, automotive, chemical processes, civil infrastructure, energy, healthcare, manufacturing, transportation, entertainment, and consumer appliances. In this paper, we address the problem of spatial-temporal group scheduling using Generalized nets (GN). We use GN in order to model the spatial, temporal, ordered and concurrent character of our mobile, distributed system. Our model is based on a discrete topology in which devices can change their location by moving from cell to cell. Using the GN, we model movement in a heterogeneous terrain as well as task execution or access to other resources of the devices. Intuitionistic Fuzzy Logic (IFL) are defined as extensions of ordinary fuzzy sets. All results which are valid for fuzzy sets can be transformed here too. Also, all research, for which the apparatus of fuzzy sets can be used, can be used to describe the details of IFL. In this paper we use it to obtain the Intuitionistic Fuzzy Estimation (IFE) for obtaining the degree of effectiveness, the degree of ineffectiveness of the robot and uncertainty during the robot movement.

S. Sotirov (✉) · E. Sotirova · S. Simeonov · N. Simeonova
Intelligent Systems Laboratory, "Prof. Assen Zlatarov" University, Bourgas, Bulgaria
e-mail: ssotirov@btu.bg

E. Sotirova
e-mail: esotirova@btu.bg

S. Simeonov
e-mail: stanislav_simeonov@btu.bg

N. Simeonova
e-mail: neli_simeonova@btu.bg

M. Werner
Operating Systems Group, Chemnitz University of Technology, Chemnitz, Germany
e-mail: matthias.werner@informatik.tu-chemnitz.de

W. Hardt
Chemnitz University of Technology, Chemnitz, Germany
e-mail: hardt@cs.tu-chemnitz.de

P. Angelov and S. Sotirov (eds.), *Imprecision and Uncertainty in Information Representation and Processing*, Studies in Fuzziness and Soft Computing 332,
DOI 10.1007/978-3-319-26302-1_24

Keywords Generalized net · Mobile distributed systems · Spatial-temporal group scheduling · Intuitionistic fuzzy logic

1 Introduction

Cyber-Physical Systems (CPS) [1] are integrations of computation, networking, and physical processes. Embedded computers and networks monitor and control the physical processes, usually with feedback loops where physical processes affect computations and vice versa. The potential of such systems is greater than what has been realized. There are considerable challenges, particularly because the physical components of such systems introduce safety and reliability requirements qualitatively different from those in general-purpose computing. Moreover, the standard abstractions used in computing do not fit the physical parts of the system well.

Some researchers [2] are focused on the challenges of modeling cyber-physical systems that arise from the intrinsic heterogeneity, concurrency, and sensitivity to timing of such systems. It uses a portion of an aircraft vehicle management systems (VMS), specifically the fuel management subsystem, to illustrate the challenges, and then discusses technologies that at least partially address the challenges. Specific technologies described include hybrid system modeling and simulation, concurrent and hetero-geneous models of computation, the use of domain-specific ontologies to enhance modularity, and the joint modeling of functionality and implementation architectures.

Applications of CPS arguably have high potential [1]. They include high confidence medical devices and systems, assisted living, traffic control and safety, advanced automotive systems, process control, energy conservation, environmental control, avionics, instrumentation, critical infrastructure control (electric power, water resources, and communications systems for example), distributed robotics (telepresence, telemedicine), defense systems, manufacturing, and smart structures. Networked autonomous vehicles could dramatically enhance the effectiveness and could offer substantially more effective disaster recovery techniques. In communications, cognitive radio could benefit enormously from distributed consensus about available band-width and from distributed control technologies. Distributed real-time games that integrate sensors and actuators could change the (relatively passive) nature of on-line social interactions.

By focusing on the physical world it becomes obvious that non-computational processes (physical actions) are strongly distributed and concurrent. Thus, designing and programming those systems have to cope with those issues. Since thinking in distributed and concurrent terms is complexity-introducing and often error-prone [3], we have studied this problem and proposed a suitable programming model [4–6] that both abstracts from distribution and concurrency by allowing the programmer to develop sequential object-oriented program code. Besides the imperative code fragments, declarative annotations can be integrated into the source code for defining spatial-temporal constraints that are glued to imperative code fragments and restrict its execution. All this requires a coordination of resources in space and time.

The aim in this paper is to address the problem of spatial-temporal group scheduling by using Generalized nets [7, 8] and Intuitionistic Fuzzy Logic and based on the [9]. Our concept is to map space to time and describe physical locations based on durations needed to change locations. The computation of a schedule is based on those timed transitions.

Intuitionistic Fuzzy Logic (IFL) [10] are defined as extensions of ordinary fuzzy sets. All results which are valid for fuzzy sets can be transformed here too. Also, all research, for which the apparatus of fuzzy sets can be used, can be used to describe the details of IFL. In this paper we use it to obtain the Intuitionistic Fuzzy Estimation (IFE) for obtaining the degree of effectiveness, the degree of ineffectiveness of the robot and uncertainty during the robot movement.

2 Assumptions

In our understanding, a task is associated with duration and, e.g., a deadline at which the task has to be completed depending on hard or soft deadlines. Thus, tasks may have temporal constraints. In this paper, we extend the classical view by a new dimension: space.

A task t is described by a set of properties $\{d, p, p', r, T'\}$, with d indicating the duration of the task and p and p' the beginning and ending location of the task, respectively. A task may also bound to a fixed location—in that case p and p' are identical. A location is a physical position on a 2D surface. In addition, we address the problem of performing tasks jointly, i.e., a given amount of robots $r \in R$, with $|R|$ denote the total number of robots, is required to perform a task that have to be coordinated in space and time. For simplification, we assume tasks are non-interruptable. Finally, the execution of t depends on the result of the set of predecessor tasks T' that need to be executed prior to t.

The 2D surface in which the robots operate is discretized and mapped to a specific topology. Each cell c_i in the topology indicate a space in which an arbitrary amount of robots can be placed: $c_i \in \{x \in N \mid 0 \leq x \leq |R|\}$ and $c_i = |R|$. We support different topologies as shown in Fig. 1 with respect to the geometry of the surface, the discretization (cell shape) and the multi-plicity of movements. A robot can change its location by moving in discrete steps to a neighboring cell along the indicated arrows. On the left hand side of the figure a cell is represented by a square and exhibits four possible movements of a robot. The middle topology doubles the degree of freedom by allowing diagonal movements. Finally, the topology on the right shows a discretization that is based on hexagons which allows for six different types of movements. During each time step a robot has different options:

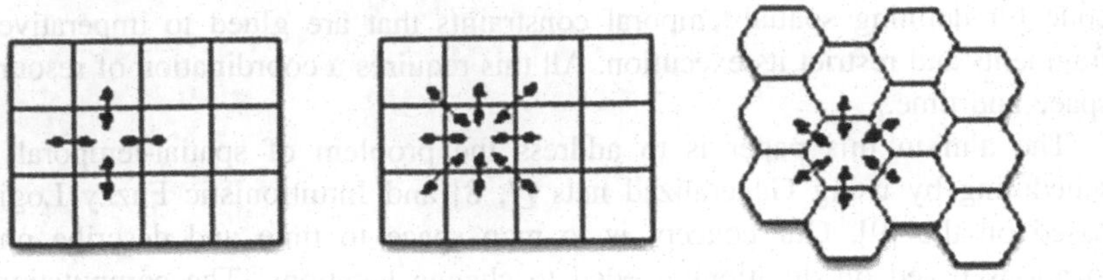

Fig. 1 Examples for discrete topologies

- Stay in the current cell (idle)
- Move along the arrows towards a neighboring cell
- Execute a task (if the task involves movement, the robot moves towards the tasks' ending location while executing it at the same time).

The movement model is based on a binary state: The robot does not move (idle) or simply moves (speed is not incorporated in the model). If a robot decides to move to a neighboring cell, the cell transition is associated with a given amount of time required for reaching the other cell (this again represents the worst case time needed for moving the robot). The topology does not have a homogeneous terrain, thus, times between cell transitions may vary. With this, we are able to model accessible and non-accessible obstacles. Driving uphill takes more time to process the transition than driving downhill. On the other hand, a solid formation, e.g. rocks, are not accessible and, thus, the robots have to take the longer way in terms of geographic distance. Altogether, this approach allows us to model the important properties of robots moving in a terrain without the need to deal with the physics of the actual movement actions—these are represented by the time needed for transitions between the cells.

Now, the overall goal is to find a schedule with minimal makespan such that all tasks t_i are executed according to their requirements of beginning and ending location and the number of robots which includes physical positioning of robots.

3 Intuitionistic Fuzzy Logic

Intuitionistic Fuzzy Logic [10] (IFL) are defined as extensions of ordinary fuzzy sets. All results which are valid for fuzzy sets can be transformed here too. Also, all research, for which the apparatus of fuzzy sets can be used, can be used to describe the details of IFL.

On the other hand, there have been defined over IFL not only operations similar to those of ordinary fuzzy sets, but also operators that cannot be defined in the case of ordinary fuzzy sets.

Fig. 2 Geometrical interpretation of one IFS

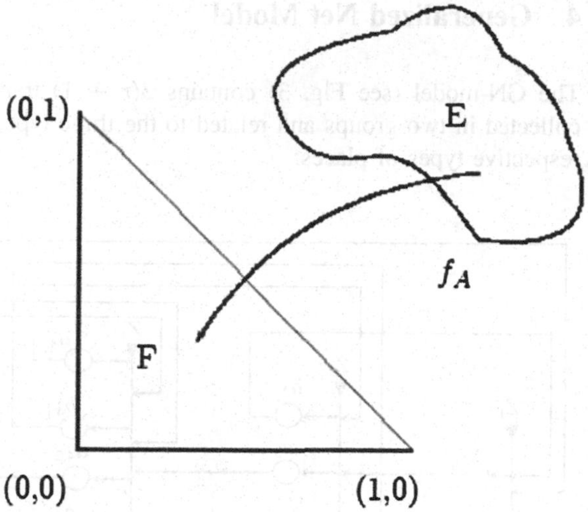

Let a set E be fixed. An IFS A in E is an object of the following form:

$$A = \{ < x, \mu_A(x), \nu_A(x) > | x \in E \},$$

where functions $\mu_A : E$ [0, 1] and $\nu_A : E$ [0, 1] define the degree of membership and the degree of non-membership of the element $x \in E$, respectively, and for every $x \in E$:

$$0 \leq \mu_A(x) + \nu_A(x) \leq 1,$$

For every $x \in E$, let

$$\pi_A(x) = 1 - \mu_A(x) - \nu_A(x).$$

Therefore, the function π determines the degree of uncertainty.

Obviously, for every ordinary fuzzy set $\pi_A(x) = 0$ for each $x \in E$, these sets have the form:

$$\{ < x, \mu_A(x), 1 - \mu_A(x) > | x \in E \}.$$

Let a universe E be given. One of the geometrical interpretations of the IFL uses figure F on Fig. 2.

4 Generalized Net Model

The GN-model (see Fig. 3) contains $3(r + 1)$ transitions and $9(r + 1)$ places, collected in two groups and related to the three types of the tokens that will enter respective types of places:

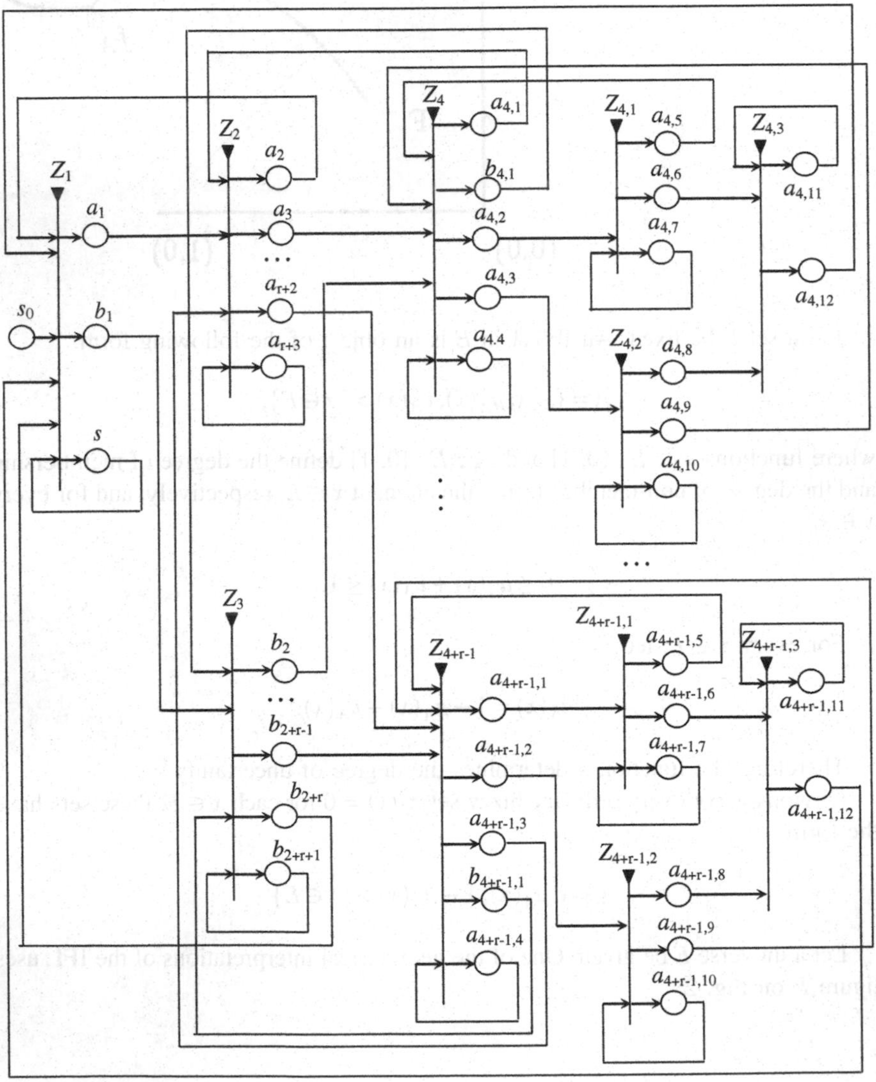

Fig. 3 Generalized net model

- α-tokens and a-places represent the robots and its positions;
- β-tokens and b-places represent the tasks of the robots;
- γ-tokens and S-places represent the functions of the intelligent system.

For brevity, we shall use the notation α- and β-tokens instead of α_i- and β_j- tokens, where i and j are numerations of the respective tokens.

In the beginning α-, β- and γ- tokens stay, respectively, in places a_{r+3}, b_{2+r+1} and s with initial characteristics:

x_0^{α} = "robot r_i, position of the robot r_i (initial position p and end positions p')" in place a_{r+3},
x_0^{β} = "task t for the robot r_i; duration d of the task", in place b_{2+r+1},
for $i \in [1, \ldots, r]$,
x_0^{γ} = "intelligent system", in place s.

Via place s_0 enter γ-token with initial characteristics:

x_0^{γ} = "robot r_i, position of the robot r_i (initial position p and end positions p'), task t for the robot r_i; duration d of the task, predecessor tasks T'".

Let $x_{cu}^{\alpha}, x_{cu}^{\beta}$ and x_{cu}^{γ} be current α-, β-, and γ-tokens' characteristics, respectively. Generalized net is presented by a set of transitions:

$$A = \{Z_1, Z_2, Z_4, Z_{4,1}, Z_{4,2}, Z_{4,3}, \ldots, Z_{4+r-1}, Z_{4+r-1,1}, Z_{4+r-1,2}, Z_{4,3}, Z_{4+r-1,3}\},$$

where transitions describe the following processes:

- Z_1—Task of the intelligent system for control of the robots;
- Z_2—Activity of the robots;
- Z_3—Task performed of the robots;
- Z_4—Choice of action for the robot 1;

…

- Z_{4+r-1}—Choice of action for the robot r;
- $Z_{4,1}$—Movement of the robot 1;
- $Z_{4,2}$—Execution of a task of the robot 1;
- $Z_{4,3}$—Calculating the intuitionistic fuzzy estimation of the execution of a task of the robot 1;

…

- $Z_{4+r-1,1}$—Movement of the robot r;
- $Z_{4+r-1,2}$—Execution of a task of the robot r.
- $Z_{4+r-1,3}$—calculating the intuitionistic fuzzy estimation of the execution of a task of the robot r.

The forms of the transitions are the following.

$$Z_1 = \langle \{a_2, a_{4,12}, a_{4+r-1,12}, b_{2+r}, s_0, s\}, \{a_1, b_1, s\}, R_1, \vee(a_2, a_{4,12}, a_{4+r-1,12}, b_{2+r}, s_0, s)\rangle$$

where:

$$R_1 = \begin{array}{c|ccc} & a_1 & b_1 & s \\ \hline a_2 & False & False & True \\ a_{4,12} & False & False & True \\ a_{4+r-1,12} & False & False & True \\ b_{2+r} & False & False & True \\ s_0 & False & False & True \\ s & W_{s,a1} & W_{s,b1} & True \end{array}$$

and

- $W_{s,a1}$ = "The tasks are ordered in the proper way",
- $W_{s,b1} = W_{s,a1}$.

The α-token obtains characteristic "robot r_i, initial position p and the end positions p' of the robot r_i" in place a_1. The β-token obtains characteristic "task t for the robot r_i; duration d of the task, predecessor tasks T'" in place b_1.

The β-token obtains characteristic "task t for the robot r_i; duration d of the task, predecessor tasks T'" in place b_1.

Initially when still no information for robot tasks has been obtained, all estimations are given initial values of <0, 0>. When $k \geq 0$, the $(k + 1)$-st estimation for the effectiveness of the system is calculated on the basis of the previous estimations according to the recurrence relation

$$<\mu_{k+1}, \nu_{k+1}> = <\frac{\mu_k k + m}{k+1}, \frac{\nu_k k + l}{k+1}>,$$

where:

- $m = \frac{\sum_{i=1}^{r} m_i}{r}$,
- $n = \frac{\sum_{i=1}^{r} n_i}{r}$
- $<\mu_k, \nu_k>$ is the previous estimation, and $<m, l>$ is the estimation of the latest measurement, for $m, l \in [0, 1]$ and $m + l \leq 1$.

$$Z_2 = \langle \{a_{4,1}, a_1, a_{4+r-1,3}, a_{r+3}\}, \{a_2, a_3, \ldots, a_{r+2}, a_{r+3}\}, R_2, \vee(a_{4,1}, a_1, a_{4+r-1,3}, a_{r+3})\rangle$$

where:

	a_2	a_3	…	a_{r+2}	a_{r+3}
$a_{4,1}$	*False*	*False*	…	*False*	*True*
$R_2 = \; a_1$	*False*	*False*	…	*False*	*True*
$a_{4+r-1,3}$	*False*	*False*	…	*False*	*True*
a_{r+3}	$W_{r+3,a2}$	$W_{r+3,a3}$	…	$W_{r+3,r+2}$	*False*

and

- $W_{r+3,a2}$ = "There is a feedback from robot r_i",
- $W_{r+3,a3}$ = "There is a task for robot r_1", …
- $W_{r+3,r+2}$ = "There is a task for robot r_r".

The α-token obtains characteristic "robot r_i, end positions p' of the robot r_i" in place a_2. The α-token obtains characteristic "robot r_1, end positions p' of the robot r_1" in place a_3. The α-token obtains characteristic "robot r_n, end positions p' of the robot r_r" in place a_{r+2}. The α token obtains characteristic "robot r_i, end positions p' of the robot r_i" in place a_{r+3}.

$$Z_3 = \langle \{b_{4,2}, b_1, b_{4+r-1,1}, b_{2+r+1}\}, \{b_2, \ldots, b_{2+r-1}, b_{2+r}, b_{2+r+1}\}, R_3, \vee (b_{4,2}, b_1, b_{4+r-1,1}, b_{2+r+1})\rangle$$

where:

	b_2	…	b_{2+r-1}	b_{2+r}	b_{2+r+1}
$b_{4,2}$	*False*	…	*False*	*False*	*True*
$R_3 = \; b_1$	*False*	…	*False*	*False*	*True*
$b_{4+r-1,1}$	*False*	…	*False*	*False*	*True*
b_{2+r+1}	$W_{2+r+1,b2}$	…	$W_{2+r+1,2+r-1}$	$W_{2+r+1,2+r}$	*False*

and

- $W_{2+r+1,b2}$ = "There is a task for robot r_1 for time d", …
- $W_{2+r+1,2+r-1}$ = "There is a task for robot r_r for time d".
- $W_{2+r+1,2+r}$ = "There is a feedback from robot r_i",

The β-token obtains characteristic "task for robot r_1, time d" in place b_2. The β-token obtains characteristic "task for robot r_r, time d" in place b_{2+r-1}. The β-token obtains characteristic "task for robot r_i, time d" in place b_{2+r}.

$$Z_4 = \langle \{a_{4,5}, a_{4,7}, a_3, b_2, a_{4,4}\}, \{a_{4,1}, b_{4,1}, a_{4,2}, a_{4,3}, a_{4,4}\}, R_4, \vee (a_{4,5}, a_{4,7}, \vee (a_3, b_2), a_{4,4})\rangle$$

where:

$$R_4 = \begin{array}{c|ccccc} & a_{4,1} & b_{4,1} & a_{4,2} & a_{4,3} & a_{4,4} \\ \hline a_{4,5} & False & False & False & False & True \\ a_{4,7} & False & False & False & False & True \\ a_3 & False & False & False & False & True \\ b_2 & False & False & False & False & True \\ a_{4,4} & W_{4,4,a4,1} & W_{4,4,b4,1} & W_{4,4,a4,2} & W_{4,4,a4,3} & False \end{array}$$

and

- $W_{4,4,a4,1}$ = “There is a feedback from robot r_1”,
- $W_{4,4,b4,1}$ = “The current task for robot r_1 is executed”.
- $W_{4,4,a4,2}$ ="There is a command for movement for robot 1"
- $W_{4,4,a4,3}$ ="There is a task for robot 1"

The α-token obtains characteristic “robot r_1, end positions p'''” in place $a_{4,1}$. The β-token obtains characteristic “robot r_1, executed task” in place $b_{4,1}$. The α-token obtains characteristic “robot r_1, end positions p' of the robot r_1” in place $a_{4,2}$. The α-token obtains characteristic “task for the robot r_1” in place $a_{4,3}$.

$$Z_{4,1} = \langle \{a_{4,2}, a_{4,7}\}, \{a_{4,5}, a_{4,6}, a_{4,7}\}, R_{4,1}, \vee(a_{4,2}, a_{4,6}, a_{4,7}) \rangle$$

where:

$$R_{4,1} = \begin{array}{c|ccc} & a_{4,5} & a_{4,6} & a_{4,7} \\ \hline a_{4,2} & False & False & True \\ a_{4,7} & W_{4,6,4,5} & W_{4,6,4,6} & False \end{array} >,$$

and $W_{4,7,4,5} = W_{4,7,4,6}$ “The robot 1 performed the movement to positions p'''”.

The α-tokens obtain characteristic “the result of the performans of the movement to positions p'''” in places $a_{4,5}$ and $a_{4,6}$.

$$Z_{4,2} = \langle \{a_{4,3}, a_{4,10}\}, \{a_{4,8}, a_{4,9}, a_{4,10}\}, R_{4,2}, \vee(a_{4,8}, a_{4,9}, a_{4,10}) \rangle$$

where:

$$R_{4,2} = \begin{array}{c|ccc} & a_{4,8} & a_{4,9} & a_{4,10} \\ \hline a_{4,3} & False & False & True \\ a_{4,10} & W_{4,10,4,8} & W_{4,10,4,9} & False \end{array}$$

and $W_{4,10,4,8} = W_{4,10,4,9}$ “The robot 1 performed the task”.

The α-tokens obtain characteristic "the result of the execution of the task" in places $a_{4,8}$ and $a_{4,9}$.

$$Z_{4,3} = \langle \{a_{4,6}, a_{4,8}\}, \{a_{4,11}, a_{4,12}\}, R_{4,3}, \vee(a_{4,11}, a_{4,12}) \rangle$$

where:

$R_{4,3} =$

	$a_{4,11}$	$a_{4,12}$
$a_{4,6}$	*False*	*True*
$a_{4,8}$	*False*	*True*
$a_{4,11}$	*True*	$W_{4,8,4,12}$

and $W_{4,8,4,12}$ = "The ituitionistic fuzzy estimation was calculated",

The α-tokens obtain characteristic: "*intuitionistic fuzzy estimation* $\langle m_1, l_1 \rangle$".

The estimations $\langle m_1, l_1 \rangle \in [0, 1] \times [0, 1]$ reflects the degree of effectiveness (m_1) and the degree of ineffectiveness of the first robot (l_1) for a time t.

$$m_1 = \frac{p_1}{n_1}, l_1 = \frac{s_1}{n_1},$$

where:

- p_1 is the number of successfully movements and tasks performans of the robot 1,
- s_1 is the number of unsuccessfully movements and tasks performans of the robot 1,
- n_1 is the total number of movements and tasks performans of the robot 1.

The degree of uncertainty $\pi_1 = 1 - m_1 - n_1$ reflects the cases when the robot have not completed a movement or task, which is the number of α-tokens in both places $a_{4,7}$ and $a_{4,10}$.

$$Z_{4+r-1} = \langle a_{4+r-1,5}; a_{4+r-1,7}, a_{r+2}, b_{2+r-1}, a_{4+r-1,4}\},$$
$$\{a_{4+r-1,1}, a_{4+r-1,2}, a_{4+r-1,3}, b_{4+r-1,1}, a_{4+r-1,4}\},$$
$$R_{4+r-1}, \vee(a_{4+r-1,5}, a_{4+r-1,7}, \vee(a_{r+2}, b_{2+r-1}), a_{4+r-1,4})\rangle$$

where:

$R_{4+r-1} =$

	$a_{4+r-1,1}$	$a_{4+r-1,2}$	$a_{4+r-1,3}$	$b_{4+r-1,1}$	$a_{4+r-1,4}$
$a_{4+r-1,5}$	*False*	*False*	*False*	*False*	*True*
$a_{4+r-1,7}$	*False*	*False*	*False*	*False*	*True*
a_{r+2}	*False*	*False*	*False*	*False*	*True*
b_{2+r-1}	*False*	*False*	*False*	*False*	*True*
$a_{4+r-1,4}$	$W_{14,11}$	$W_{14,12}$	$W_{14,13}$	$W_{14,b11}$	*False*

and

- $W_{14,11}$ = "There is a feedback from robot r_r",
- $W_{14,11}$ = "The current task for robot r_r is executed".
- $W_{14,13}$ = "There is a task for movement for robot r"
- $W_{14,b11}$ = "There is a task for robot r"

The α-token obtains characteristic "robot r_r, end positions p''' in place $a_{4+r-1,1}$. The β-token obtains characteristic "robot r_r, executed task" in place $b_{4+r-1,1}$. The α-token obtains characteristic "robot r_r, end positions p' of the robot r_1" in place $a_{4+r-1,3}$. The α-token obtains characteristic "task for the robot r_r" in place $a_{4+r-1,2}$.

$$Z_{4+r-1,1} = \langle \{a_{4+r-1,1}, a_{4+r-1,6}\}, \{a_{4+r-1,5}, a_{4+r-1,6}, a_{4+r-1,7}\}, R_{4+r-1,1}, \vee(a_{4+r-1,5}, a_{4+r-1,6}, a_{4+r-1,7})\rangle$$

where:

$$R_{4+r-1,1} = \begin{array}{c|ccc} & a_{4+r-1,5} & a_{4+r-1,6} & a_{4+r-1,7} \\ \hline a_{4+r-1,1} & False & True & True \\ a_{4+r-1,7} & W_{17,15} & W_{17,16} & False \end{array}$$

and $W_{17,15} = W_{17,16}$ = "The robot n performed the movement to positions p'''".

The α-tokens obtain characteristic "performed movement to positions p''' in places $a_{4+r-1,5}$ and $a_{4+r-1,6}$.

$$Z_{4+r-1,2} = \langle \{a_{4+r-1,2}, a_{4+r-1,10}\}, \{a_{4+r-1,8}, a_{4+r-1,9}, a_{4+r-1,10}\}, R_{4+r-1,2}, \vee(a_{4+r-1,8}, a_{4+r-1,9}, a_{4+r-1,10})\rangle$$

where:

$$R_{4+r-1,2} = \begin{array}{c|ccc} & a_{4+r-1,8} & a_{4+r-1,9} & a_{4+r-1,10} \\ \hline a_{4+r-1,2} & False & True & True \\ a_{4+r-1,10} & W_{110,18} & W_{110,19} & False \end{array}$$

and $W_{110,18} = W_{110,19}$ = "The robot r performed the task". The α-tokens obtain charac-teristic "performed task" in places $a_{4+r-1,8}$ and $a_{4+r-1,9}$.

$$Z_{4+r-1,3} = \langle \{a_{4+r-1,6}, a_{4+r-1,8}\}, \{a_{4+r-1,11}, a_{4+r-1,12}\}, R_{4+r-1,3}, \vee(a_{4+r-1,11}, a_{4+r-1,12})\rangle$$

where:

$$R_{4+r\text{-}1,3} = \begin{array}{c|cc} & a_{4+r\text{-}1,11} & a_{4+r\text{-}1,12} \\ \hline a_{4+r\text{-}1,6} & False & True \\ a_{4+r\text{-}1,8} & False & True \\ a_{4+r\text{-}1,11} & True & W_{4,8,4,12} \end{array}$$

and $W_{4,8,4,12}$ = "The ituitionistic fuzzy estimation was calculated".

The α-tokens obtain characteristic: "*intuitionistic fuzzy estimation* $\langle m_r, l_r\rangle$".

The estimations $\langle m_r, l_r\rangle \in [0, 1] \times [0, 1]$ reflects the degree of effectiveness (m_r) and the degree of ineffectiveness of the robot with number n (l_r) for a time t.

$$m_r = \frac{p_r}{n_r}, lr = \frac{s_r}{n_r},$$

where:

- p_r is the number of successfully movements and tasks performans of the robot n,
- s_r is the number of unsuccessfully movements and tasks performans of the robot n,
- n_r is the total number of movements and tasks performans of the robot n.

The degree of uncertainty $\pi_r = 1 - m_r - n_r$ reflects the cases when the robot have not completed a movement or task, which is the number of α-tokens in both places $a_{4+r-1,7}$ and $a_{4+r-1,10}$.

5 Conclusion

The proposed GN introduces model the spatial, temporal, ordered and concurrent character of mobile, distributed system. The model is based on a discrete topology in which devices can change their location by moving from cell to cell. With GN, we model movement in a heterogeneous terrain as well as task execution or access to other resources of the devices.

On the other side a cyber-physical system (CPS) is a system of collaborating computational elements controlling physical entities. Nowadays cyber-physical systems can be found in areas as diverse as robotics, auto-motive, chemical processes, civil infrastructure, energy, healthcare, manufacturing, transportation, entertainment, and consumer appliances. In this paper, we address the problem of spatial-temporal group scheduling using Generalized nets (GN). The IFE is used to estimate the degree of effectiveness, the degree of ineffectiveness of the robot and uncertainty during the robot movement.

Acknowledgments The authors are grateful for the support provided by the project NTS Germania 11, funded by the DAAD.

References

1. Lee, E.A.: CPS foundations. In: Proceedings of the 47th Design Automation Conference (DAC), pp. 737–742. ACM (2010)
2. Derler, P., Lee, E.A., Sangiovanni-Vincentelli, A:. Modeling cyber-physical systems. Proc. IEEE **100**(1), 13–28 (2012) (special issue on CPS)
3. Lee, E.A.: The problem with threads. Computer **39**, 33–42 (2006)
4. Graff, D., Richling, J., Stupp, T.M., Werner, M.: Context-aware annotations for distributed mobile applications. In: Wolfgang Karl, D.S. (ed.) ARCS'11 Workshop Proceedings: Second Workshop on Context-Systems Design, Evaluation and Optimisation (CoSDEO 2011), pp. 357–366. VDE, Feb 2011
5. Graff, D., Richling, J., Stupp, T.M., Werner, M.: Distributed active objects—a systemic approach to distributed mobile applications. In: Sterrit, R. (ed.) 8th IEEE International Conference and Workshops on Engineering of Autonomic and Autonomous Systems, pp. 10–19. IEEE Computer Society, Apr 2011
6. Graff, D., Werner, M., Parzyjegla, H., Richling, J., Mühl, G.: An object-oriented and context-aware approach for distributed mobile applications. In: Workshop on Context-Systems Design, Evaluation and Optimisation (CoSDEO 2010) at ARCS 2010—Architecture of Computing Systems, pp. 191–200 (2010)
7. Atanassov, K.: Generalized Nets. World Scientific, Singapore (1991)
8. Atanassov, K.: On Generalized Nets Theory. "Prof. M. Drinov" Academic Publishing House, Sofia (2007)
9. Sotirov S., Werner, M., Simeonov, S., Hardt, W., Sotirova, E., Simeonova, N.: Using generalized nets to model spatial-temporal group scheduling problems. Issues in IFS and GNs, vol. 11, pp. 42–54 (2014)
10. Atanassov, K.T.: On Intuitionistic Fuzzy Sets Theory. Springer, Berlin (2012)